A Textbook of
HYDROLOGY

AF519032

A Textbook of
HYDROLOGY

(In M.K.S. and S.I. Units)

[For Degree, U.P.S.C., (Engg. Services) and A.M.I.E. Indian Students]

By

Dr. P. Jaya Rami Reddy

Principal

G. Pulla Reddy Engg. College

Kurnool

(A.P.)

(An Imprint of Laxmi Publications Pvt. Ltd.)

An ISO 9001:2015 Company

BENGALURU • CHENNAI • GUWAHATI • HYDERABAD • JALANDHAR

KOCHI • KOLKATA • LUCKNOW • MUMBAI • RANCHI

NEW DELHI

A TEXTBOOK OF HYDROLOGY

Copyright © by Laxmi Publications (P) Ltd.
All rights reserved including those of translation into other languages. In accordance with the Copyright (Amendment) Act, 2012, no part of this publication may be reproduced, stored in a retrieval system, or transmitted in any form or by any means, electronic, mechanical, photocopying, recording or otherwise. Any such act or scanning, uploading, and or electronic sharing of any part of this book without the permission of the publisher constitutes unlawful piracy and theft of the copyright holder's intellectual property. If you would like to use material from the book (other than for review purposes), prior written permission must be obtained from the publishers.

Printed and bound in India
Typeset at : Goswami Associates, Delhi
Third Edition : 2011, *Reprint* : 2012, 2013, 2015, *Edition* : 2019
ISBN : 978-93-80856-04-9

Limits of Liability/Disclaimer of Warranty: The publisher and the author make no representation or warranties with respect to the accuracy or completeness of the contents of this work and specifically disclaim all warranties. The advice, strategies, and activities contained herein may not be suitable for every situation. In performing activities adult supervision must be sought. Likewise, common sense and care are essential to the conduct of any and all activities, whether described in this book or otherwise. Neither the publisher nor the author shall be liable or assumes any responsibility for any injuries or damages arising here from. The fact that an organization or Website if referred to in this work as a citation and/or a potential source of further information does not mean that the author or the publisher endorses the information the organization or Website may provide or recommendations it may make. Further, readers must be aware that the Internet Websites listed in this work may have changed or disappeared between when this work was written and when it is read.

All trademarks, logos or any other mark such as Vibgyor, USP, Amanda, Golden Bells, Firewall Media, Mercury, Trinity, Laxmi appearing in this work are trademarks and intellectual property owned by or licensed to Laxmi Publications, its subsidiaries or affiliates. Notwithstanding this disclaimer, all other names and marks mentioned in this work are the trade names, trademarks or service marks of their respective owners.

Branches

Bengaluru	080-26 75 69 30	
Chennai	044-24 34 47 26	
Guwahati	0361-254 36 69	
Hyderabad	040-27 55 53 83,	27 55 53 93
Jalandhar	0181-222 12 72	
Kochi	0484-237 70 04,	405 13 03
Kolkata	033-22 27 43 84	
Lucknow	0522-220 99 16	
Ranchi	0651-224 24 64	

PUBLISHED IN INDIA BY

UNIVERSITY SCIENCE PRESS
(An Imprint of Laxmi Publications Pvt. Ltd.)

An ISO 9001:2015 Company
113, GOLDEN HOUSE, GURUDWARA ROAD, DARYAGANJ,
NEW DELHI - 110002, INDIA
Telephone : 91-11-4353 2500, 4353 2501
www.laxmipublications.com
info@laxmipublications.com

C—00342/019/06

CONTENTS

6. STREAMFLOW MEASUREMENT 154–200

7. EVAPORATION AND EVAPOTRANSPIRATION 201–240

8. INFILTRATION 241–264

9. GROUNDWATER 265–310

Preface to the Third Edition

In this third edition some new topics have been added making the book more comprehensive. Additional solved examples have been included to help the reader understand the application of the principles to the field situations. The end of the chapter exercises have also been increased.

The author shall be grateful to the suggestions that would improve the standard and enhance the usefulness of the book.

–Author

Preface to the First Edition

Hydrology is no longer treated as an empirical science. Many institutions are offering this as a core course in the under graduate Civil Engineering Programme. Though there is no dearth of books on this subject, I felt the need for a simple one written in clearly understandable terms that can be easily digested by the students.

Based on my experience of teaching this subject for more than a decade for both under graduate and post graduate engineering students, I endeavoured, in this book, to present the subject matter of hydrology in a lucid and comprehensive manner. The various processes of hydrologic cycle are treated systematically and in a proper sequence. The physics of each process is described first. The new terminology is precisely defined. The techniques of measuring all the hydrologic variables are described in detail. The methods of analysis and interpretation of hydrologic data are presented in an elaborate manner and are supported by well formulated worked out examples. Some topics like depth area duration analysis, moving boat technique, Penman's evaporation theory, and the theory of instantaneous unit hydrograph which are not usually found in all the textbooks have been presented in this book. Numerous Problems (along with answers in many cases) are given under exercises at the end of each chapter to help the readers to test their comprehension of the principles. As the coverage of the topics is fairly exhaustive, this book can be used by both under graduate and graduate students.

I am indebted to my teacher Dr. R. Raghavendran (presently visiting Professor, Centre for water resources, Anna University, Madras) for introducing me to the areas of Hydrology and Water Resources. I am grateful to all the authors whose works I have consulted.

I wish to thank Dr. I. Gopal Reddy, our Principal, for readily providing the support for the preparation of the manuscript.

Finally I thank my wife Premeela and our children Bugga Rami Reddy and Raajitha for their tolerance and understanding during the last three years while the manuscript was under preparation.

Any suggestions which would improve the standard of the book and enhance its usefulness will be thankfully received.

–Author

TO
my wife
PRAMEELA
and our children
BUGGA RAMI REDDY
and
RAAJITHA

Chapter 1 INTRODUCTION

1.1 DEFINITION AND SCOPE

The word *hydrology* is derived from the Greek words, *hydro*, which means water, and *logos*, which means science. Thus in this broad sense hydrology is the science which is concerned with all waters on Earth, its occurrence, distribution and circulation, its physical and chemical properties, its effects on the environment and on life of all forms. However, in customary usage, the word hydrology is rather used in a restricted sense in many respects. For example, it may not cover all studies of ocean waters and it is not concerned with medical uses of water. There are various elaborate definitions of hydrology. But from the view point of practical applications, Wisler and Brater have aptly defined hydrology as *the science that deals with processes governing the depletion and replenishment of water resources of the land areas of the earth.* It is concerned with the transportation of water through the air, over the ground surface and through the strata of the earth. It is science that treats of the various phases of the hydrologic cycle. Hydrology finds many applications in *Hydraulics, Agricultural water resources* and other branches of engineering. Therefore, the names like *Engineering Hydrology* and *Applied Hydrology* are also often used.

The discussion of the ensuing chapters would indicate that hydrology is a highly interdisciplinary science. For its investigations it draws many principles from other branches of science like *Physics, Chemistry, Biology, Geology, Fluid Mechanics, Mathematics, Probability* and *Statistics* and *Operations Research.* As the water is transported into air, hydrology interacts with sciences like *Hydrometeorology, Meteorology* and *Climatology.* When dealing with surface water, the supporting sciences required are *Potamology* (science of surface streams), *Limnology* (science of lakes), *Cryology* (science of snow and ice), *Glaciology* and *Oceanology.* Similarly when dealing with ground water, hydrology encompasses the domains of *Agronomy, Hydrogeology* (emphasising the geologic aspects), *Geohydrology* (emphasising the hydrologic aspects) and *Geomorphology.* As water affects plants as well as animals, hydrology extends itself into *Plant Ecology* (ecology is a branch of biology dealing with living organisms, habits, modes of life and relations to their surroundings), *Silviculture* (the growing and tending of trees as a branch of forestry), *Biohydrology* (emphasising the hydrologic aspects) and *Hydrobiology* (emphasising biological aspects). Since hydrology is a science that underlies the development and control of water resources it has its important influence on *Agriculture, Forestry, Geography, Watershed Management, Political Science, Economics* and *Sociology.* It has practical applications in *Structural Design, Water Supply, Wastewater Disposal and Treatment, Irrigation, Drainage, Hydropower, Flood Control, Navigation, Erosion and Sediment Control, Salinity Control, Pollution Abatement, Recreational uses of water, Fish and Wild Life Preservation, Insect Control and Coastal Works.* Some of these practical applications will be discussed later.

1.2 HYDROLOGIC CYCLE

Water exists on planet earth in its three forms *viz.* gaseous, liquid and solid forms and is circulated mainly by *Solar* and *Planetary forces.* The *Sun* provides the energy for the evaporation of sea water and earth's gravitational field and ***coriolis*** force contributes to the circulation of water. The

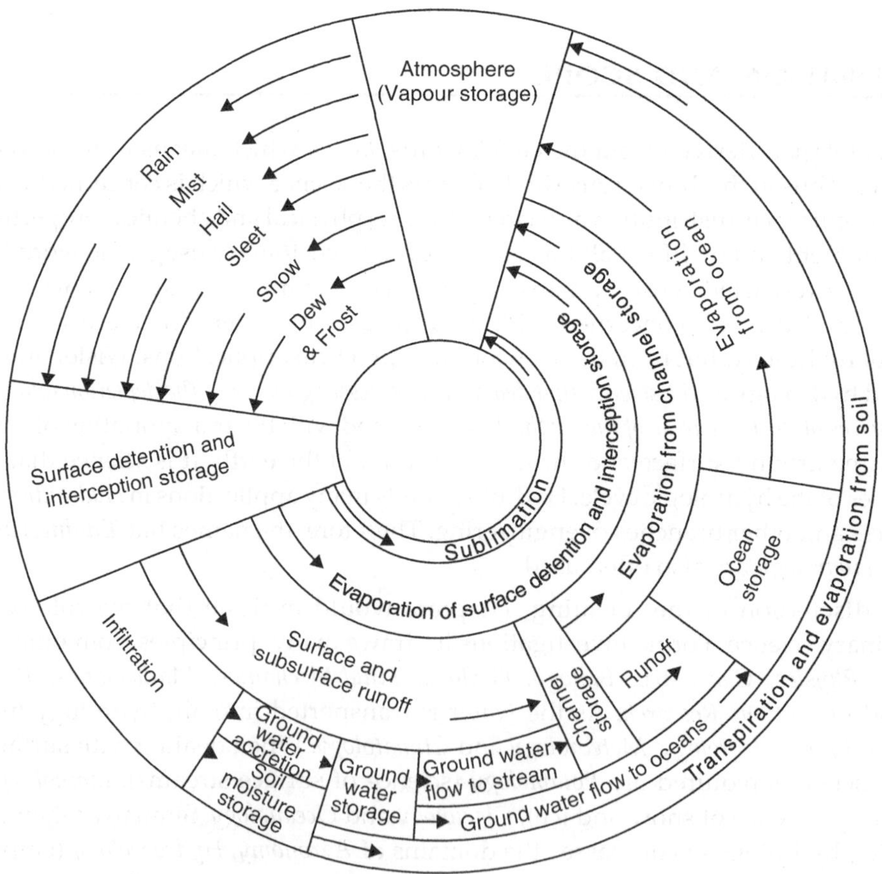

FIGURE 1.1 Qualitative representaiton of hydrologic cycle.

group of numerous arcs which represent the different paths through which the water in nature circulates and is transformed is what is known as *hydrologic cycle.* These arcs penetrate into the three parts of the total earth system, namely, *Atmosphere, Hydrosphere,* and *Lithosphere.* The atmosphere is the gaseous envelope above the hydrosphere. Hydrosphere is the bodies of water that cover the surface of the earth. Lithosphere is the solid rock below the hydrosphere. The activities of water extend through these three parts of the earth system from an average depth of about 1 km in the lithosphere to a height of about 15 km in the atmosphere.

The hydrologic cycle can be represented in many different ways in pictorial or diagrammatic forms. It depends on the view point it represents and is simplified in general as it is difficult to represent all aspects of the cycle. Fig.1.1 is a qualitative representation of the hydrologic cycle where in the various processes involved in the circulation of water are depicted in the form of a circle. Fig. 1.2 is a descriptive representation of the hydrologic cycle where many of the details are skipped. Fig. 1.3 is an engineering representation of the hydrologic cycle where the evaporation is shown by wavy arrows. Though these figures are self- explanatory, a qualitative description of the hydrologic cycle is given below.

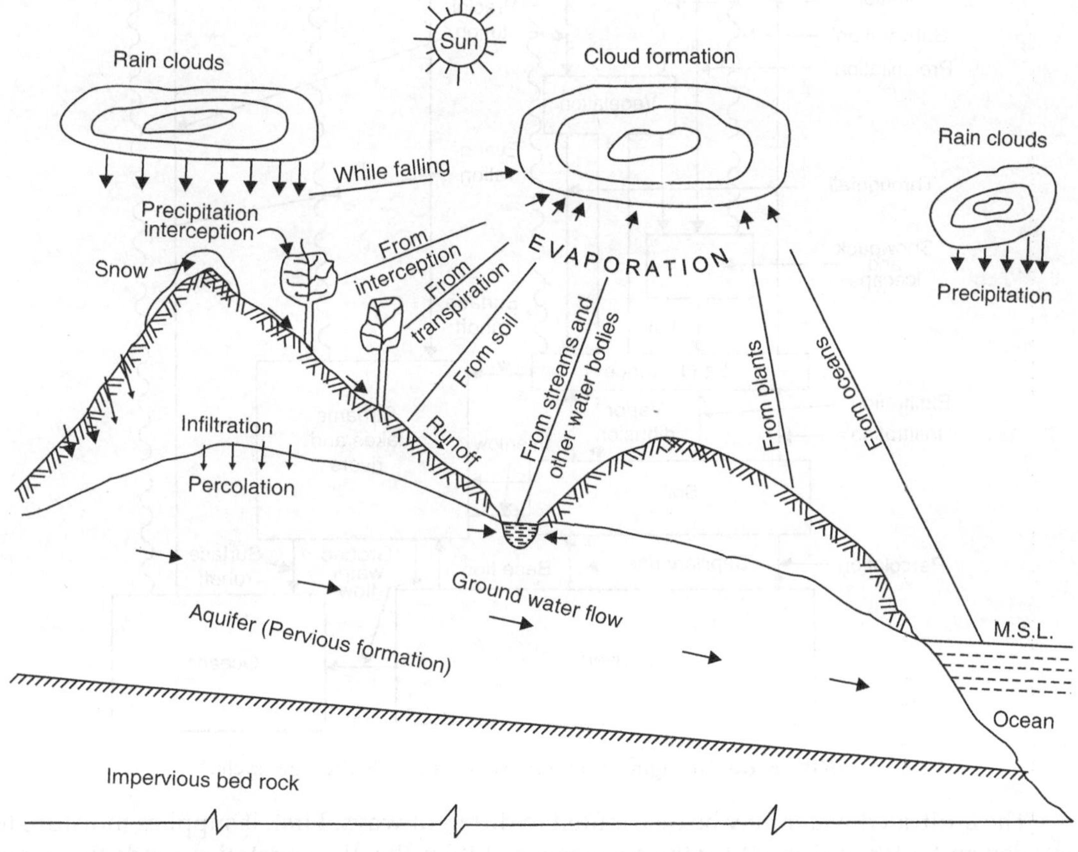

FIGURE 1.2 Descriptive representation of hydrologic cycle.

The hydrologic cycle has no beginning or end as the water in nature is continuously kept in cyclic motion. However, for the purpose of description the cycle may be visualised to commence with the precipitation from the atmosphere. Precipitation may take place in liquid form as *rain* and also in soild form as *hail, snow, dew, frost* etc. While precipitation is taking place, a part of it may evaporate and reach back the atmosphere. Some more precipitation is intercepted by the trees and vegetation and the rest of it only would reach the ground. The intercepted precipitation eventually evaporates into the atmosphere. The precipitation reaching the ground surface is called the *throughfall*. Considerable portion of the throughfall gets infiltrated into the ground and that in excess of *infiltration* would be detained temporarily on the ground before it becomes *overland flow* and subsequently *surface runoff.* The precipitation falling directly over the streams is called the *channel precipitation* and it readily becomes runoff without any delay.

The precipitation falling on the water bodies like ponds and lakes may be disposed of either as surface runoff to streams if the water bodies overflow or as evaporation or as infiltration. The evaporation would also be taking place from stream surfaces.

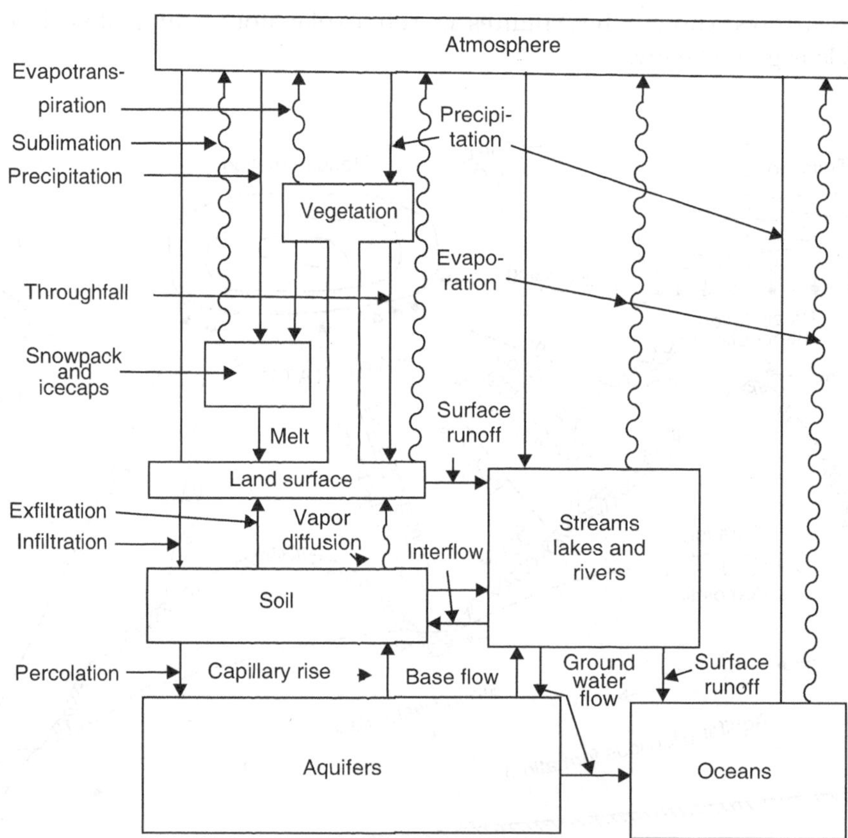

FIGURE 1.3 An engineering representation of hydrologic cycle.

The infiltrated water may be distributed in different ways. First, it supplies moisture to the vegetation and after utilising it for the sustenance of their life, the vegetation sends this moisture back into the atmosphere through the leaves by a process known as *transpiration*. Secondly, the infiltrated water may percolate deep and become groundwater supply to surface streams known as the *groundwater runoff* , or it may become groundwater supply to oceans. The groundwater runoff is sometimes referred to as the *baseflow* or the *interflow*. The total streamflow which is the sum of the surface runoff and the groundwater runoff ultimately joins the oceans wherefrom it again evaporates into the atmosphere thus completing the hydrologic cycle. The entire cycle repeats when the atmospheric moisture precipitates on to the ground after cloud formation. Thus the hydrologic cycle consists of various complicated processes such as precipitation, interception, evaporation and transpiration, infiltration, percolation, storage and runoff. These processes are the subjects of detailed discussion in the ensuing chapters.

1.3 THE GLOBAL WATER BUDGET

Water is the most important natural resource and is vital for all life on earth. The well-being and the development of our society is dependant on the availability of water. This most precious resource is sometimes scarce somteimes abundant and is always very unevenly distributed, both in space and time. For instance, Europe and Asia together have only 27% of world's fresh water though they accommodate about 76% of the world's population. Estimates of total amount of water available on earth vary. Table 1.1 indicates the approximate distribution of world's water resources among separate elements of the hydrologic cycle.

Table 1.1. Distribution of World's Water Resources

Location	*Water volume in km³*	*Percentage of total water*
Fresh water lakes	125000	0.009
Saline lakes	104200	0.008
Stream channels	1200	0.0001
Ground water (< 0.8 km deep)	4168200	0.31
Ground water (> 0.8 km deep)	4168200	0.31
Soil moisture etc.,	66700	0.005
Ice caps and glaciers	29177300	2.15
Atmosphere	12900	0.001
Oceans	1321310000	97.2
Approx totals	13.6×10^8	100

The oceans which cover 71% of the surface area of earth and which have an average depth of 3.8 km hold as large as 97% of the earth's water, while 2% is frozen in ice caps. The deep groundwater accounts for 0.31%. This 99.31% of water is of no use to man. The remaining 0.69% which is of the order of 4.374×10^6 km^3 represents the fresh water resources with which the man has to deal. Surprisingly, at any given instant of time rivers and lakes hold only 3% of fresh water (or 0.0093% of total water) and atmosphere holds only 0.3% of fresh water (or 0.00093% of total water). The global annual precipitation and evaporation are estimated to be equal and each being of the order of 100 cm. Therefore the global average annual precipitaiton volume (based on surface area of 510×10^6 km^2 of the earth whose mean radius is 6370 km) works out to 510000 km^3. The average moisture available in the atmosphere at any time is only 12900 km^3. This implies that the entire atmospheric moisture must be replaced approximately 40 times every year. Interpreted in another way, the average residence period of the moisture in the atmosphere is only slightly more than 9 days.

1.4 INDIA'S WATER BUDGET

The basic source of water for India is the rainfall over the most part of the country and snowfall in the Northern region. While the rainfall is recorded by more than 3000 rain gauges set up by Indian Meteorological Department and the State Governments, the snowfall record is scanty. The rainfall varies from place to place and from year to year. The average annual rainfall for the country is about 119.4 cm and with the country's geographical area of 3.28×10^6 km^2 it is equivalent to 3916 km^3 and together with snowfall which is not yet fully assessed the annual water resources may be approximated to 4000 km^3. Out of this, 700 km^3 is lost to atmosphere, 2150 km^3 is soaked into the ground and the balance 1150 km^3 becomes the direct surface runoff to the streams. According to Irrigation Commission of India (1972), the total annual surface waterflow in the country is about 1800 km^3. This includes 200 km^3 of runoff brought in by the streams and rivers from the catchments lying outside the country, 450 km^3 of groundwater runoff carried by the streams during non-rainy periods and 1150 km^3 of direct surface runoff from precipitation (out of which 100 km^3 is the contribution by the snowfall). Out of the total surface water of 1800 km^3, about 150 km^3 is stored in various reservoirs and tanks, another 150 km^3 is utilised through diversion works and direct pumping and the remaining 1500 km^3 goes to the sea and some adjoining countries. Even after full irrigation development in the country, the use of water through diversions and direct pumping is expected to increase to only 450 km^3 and the balance 1050 km^3 would continue to flow to the sea and outside the country . Out of 2150 km^3 of infiltrated water only 500 km^3 percolates deeply and becomes groundwater while the remaining 1650 km^3 is retained as soil moisture and goes back to the atmosphere as evaporation and transpiration. During floods the water surface in the streams is higher than the adjoining water-table along most of its course and therefore some river water seeps into the adjoining area and adds to the groundwater. Such contributions to groundwater during floods have been estimated to be of the order of 50 km^3. Similarly the additions from the irrigation systems to the groundwater are estimated to be 120 km^3. Thus the total groundwater available comes to 670 km^3. This, on full development of water resources is likely to increase to 850 km^3.

1.5 PRACTICAL APPLICATIONS OF HYDROLOGY

Some of the practical applications of hydrology which were mentioned earlier are briefly discussed below.

Design of Hydraulic Structures

The design of any structure related to water such as spillway, dam, culvert, highway bridge, rail bridge etc. may be considered to consist of three parts, *viz., hydrologic design, hydraulic design* and *structural design.* Hydrologic design deals with the estimation of the quantities of water to be handled at the site of the structure ; specifically their time distribution, time of occurrence and frequency of occurrence. Hydraulic design provides the best suitable shape and section of the structure to cope with the waters estimated and the structural design ensures the stability and safety of the chosen section against the water pressure and other pressures. So, hydrology plays an important role in the design of any structure. For example, if we consider the design of rail or highway bridges or culverts, their cost adds up to considerable amount in the total cost of the transportation system. It is not possible to design all of them to withstand the severemost and rare floods. Hence, the problem is to find not only the magnitude of the flood discharge but also its probability of occurrence so that the structure may be designed for such a discharge (depending on its relative importance) at which the additional cost of protection against higher flood discharges may be balanced against the possible cost of replacing the structure. Hydrology provides basic data for such studies. Coming to the design of a spillway, the hydrologist must evaluate not only the magnitude and probability of occurrence of the design flood but also the effect of the storage reservoir behind the spillway on the distribution of the flood. This point is to be emphasised because it is very often overlooked and the structures are over-designed. For instance, if the magnitude of the design flood is 20000 m^3/s, say, it may be sufficent to provide a spillway of capacity 18000 m^3/s because of the *attenuating* effect of the storage reservoir on the peak discharge. This problem comes under the purview of *flood routing*. Similarly the hydrology of urban areas play an important role in the efficient and economic design of drainage structures such as storm-sewers.

Municipal and Industrial Water Supply

The availability of water is often the most important factor in locating the major industries and it has considerable effect on the growth of municipalities. Hydrologist would answer such questions as whether the flow in the nearby stream is sufficient to meet the needs of the municipal city or the industry, whether the groundwater in that region can be tapped economically to meet the needs without over-exploitation or whether the streamflow and the groundwater can be utilised conjunctively.

Irrigation

Provision of adequate storage facilities at the irrigation and other multipurpose projects either to irrigate the ayacut under the project or to meet the demands of other purposes is the essential part of the design. In arriving at the storage capacity of the reservoir the evaporation, seepage and other losses must be properly accounted for. This requires the information on the hydrological variables such as runoff, infiltration and evaporation. Till recent times the practice has been to design the project in isolation with the available data at the proposed site. It has been proved beneficial to use the *systems approach* (or the integrated approach) in designing a system as a whole in its entirety. A large river basin is taken as a system and all the projects in the system are together optimally designed and operated to achieve the stipulated objectives. Hydrology of the basin provides the required information regarding the streamflows at various projects sites, their time distribution and frequency distribution which is essential in fixing the storage capacities and finalising the operation policies. Hydrologist may also be called upon to help the planners and engineers in considering the alternatives such as *artificial recharge* in the case of lift irrigation schemes.

Hydropower

Hydrologic studies are essential for the planning of any water power development. To determine the feasibility of the *run-of-river* plant operating with pondage, a reliable prediction is needed of the absolute minimum daily flow that may be expected in the stream and of the percentages of time that various other flows may be expected to exist. In other words, a reliable *flow duration* curve is required. The absolute minimum flow decides the prime capacity (firm power) of the plant while additional flow data is useful in estimating the amount of power that will have to be obtained from other sources for any increase in the installed capacity. For storage plants, however, low seasonal flows rather than low daily flows are important. After commissioning, the storage plant is continuously faced with problems in the economics of operation. Whether the reservoir can be emptied boldly permitting the shut down of the other thermal plants in the grid resulting in fuel saving, or should it be drawn on sparingly so that there will be ample water for future dry periods are the questions which call for reliable hydrologic predications of the stream behaviour.

Flood Control

Reservoirs, levees, channel improvements and channel diversions, or a combination of them are the most commonly used flood control structures. Flood control problems are complicated because any type of flood control project modifies the natural regime of the stream and thus in the process of protecting one area it may increase the flood damage in another. Hence the technique of flood routing is essential to the intelligent and economic planning of flood control projects. Flood control measures will really become very effective when they are associated with flood forecasts. The forecasts which are the joint work of the hydrologists and the meteorologists could be profitably utilised in planning the advanced evacuation of the threatened areas, in organising the stand-by crews for the emergency works on rail roads and highways and in putting emergency controls into effect at municipal water works and other public utilities. When adequate hydrometeorological information is available, *flood plain zoning* may also be considered as one of the flood control measures.

Navigation

Hydrologic problems in navigation projects require answers to such questions as how much water will be required for lock gates and to maintain minimum draft, wherefrom this water can be obtained and preserved and what will be the effect of the navigation structures on the stages in the river especially during floods etc. If the rivers are carrying silt they increase the complexity of the problems.

Erosion and Sediment Control

Excessive erosion in the catchment feeds sediment into the runoff. This leads to many undesirable effects. The reservoirs may lose their capacity at a faster rate reducing their economic life span drastically. Tons and tons of fertile top soil will be lost every year resulting in reduced crop yields. The problem of erosion control is mainly linked with the phenomena of overland flow and infiltration. Hydrology of the catchment alongwith the knowledge of the existing watershed management practices will help in finding out the effective erosion control measures suitable for the given soil conditions. These measures include fixing of judicious crop pattern and cropping procedures, formation of contour bunds, afforestation etc. Effective erosion control measures would not only decrease the sediment load in streams but also reduce the peak flood discharges because of the increased infiltration opportunities in the catchment.

Pollution Abatement

Indiscriminate disposal of sewage from cities and industries into the nearby streams, which has been the general tendency, results in health hazards to the public and the destruction of the fish and the other wildlife. However, complete prevention of river pollution is not economically feasible. As the stream is a natural water purification system, it is generally considered premissible to allow disposal of certain amount of sewage into streams. The purification in the stream is a result of bacterial action and aeration. After a thorough study of streamflow and its variability a hydrologist can help the sanitary engineer regarding the level to which the sewage has to be treated before it is disposed of into the stream and also the quantity of sewage that can be allowed into the river without causing any pollution to the downstream regions. To keep the sewage concentration within the allowable limits sometimes augmentation of streamflow by way of storage dam in the upstream may prove to be more economical than additional sewage treatment plants.

1.6 HISTORICAL DEVELOPMENT

There is evidence of utilisation of water resources in the history like that of drawing of groundwater from wells and the early irrigation systems which date back to prehistoric times. But it is not known who perceived the concept of hydrologic cycle first as we know it today. The appearance of streamflow in the rivers even long after the cessation of rainfall was rather a puzzle to the prehistoric men. The speculation of the early philosophers, including Plato (427-347 B.C.) and Aristotle (384-322 B.C.) led to the erroneous concepts such as the subterranean condensation and the underground sea which in their opinion were responsible for the streamflow even during the dry periods. It is Vitruvius (who lived at the time of Christ) who postulated that groundwater is derived only from the precipitation, the concept as we accept today. Leonordo da Vinci (1452-1519) and Bernard Pallissy (1509-1589) were among the first to advocate the infiltration theory to explain the occurrence of water in springs and the dry weather flow in rivers.

It is only in the 17th century that the quantitative measurement of hydrologic phenomena began. Pierre Perrault (1608-1680) showed that the estimated streamflow of Siene River was approximately only one-sixth of the measured rainfall volume on the basin. Mariotte (1620-1684) made detailed velocity measurements using floats and made computations of discharge and established the findings of perrault on the basis of more exact runoff data. Later English Astronomer Halley (1656-1742) conducted experiments on evaporation of salt water and established that evaporation from the sea is adequate to supply all the water that the rivers discharge into the sea. These works once for all dispelled the erroneous concepts of subterranean condensation and subteranean flow regarding groundwater. Thus according to Meinzer (who is the author of one of the early text books on hydrology) Perrault, Mariotte and Halley are to be considered the founders of hydrology while Vitruvius, Vinci and Pallissy were discoverers and pioneers in advocating correct hydrologic principles.

The development of velocity tube by pitot and the derivation of the flow formula by Chezy were the significant contributions to hydrology in the 18th century. Weir formula by Francis, Venturimeter by Herschel, formula by Ganguillet and Kutter to determine the Chezy's C are among the notable developments regarding the surface water in the 19th century while the developments in the groundwater were due to Darcy, Dupuit, Forchheimer, Thiem etc.

Modernisation of the hydrological methods gained momentum in the 20th century with the unit hydrograph theory proposed by Sherman and the use of statistical methods in the hydrologic studies initiated by Hazen. In the later part of the 20th century there has been a spurt of activity in the hydrologic study and research. Use of systems approaches, development of mathematical models, digital simulation and the application of probablistic and stochastic theories to hydrologic phenomena are some of the significant developments in recent years.

Ven Te Chow has classified the historical development of hydrology into eight periods as given below:

1. Period of speculation (Ancient times to 1400 AD).
2. Period of observation (1400 AD to 1600 AD).
3. Period of measurement (1600 AD to 1700 AD).
4. Period of experimentation (1700 AD to 1800 AD).
5. Period of modernisation (1800 AD to 1900 AD).
6. Period of empericism (1900 AD to 1930 AD).
7. Period of rationalisation (1930 AD to 1950 AD).
8. Period of theorisation (1950 AD to 1965 AD).

To this classification can be added the period of mathematical modelling and computerisation (1965 Ad–till date).

In order to focus the attention of developing countries on the importance of hydrology the United Nations Education, Scientific and Cultural Organisation (UNESCO) organised the *International Hydrologic Decade* during 1965 to 1975 and in the year 1975 the same was converted into *International Hydrologic Programme* (*IHP*). Hydrology is now introduced as a full course at the undergraduate level in many of the Universities in India.

REVIEW EXERCISES

Descriptive Questions

1. Give few other elaborate definitions of hydrology.
2. Hydrology is a highly inter-disciplinary science. Justify.
3. Describe the hydrologic cycle with a neat sketch.
4. List out various practical applications of hydrology.
5. Discuss the practical applications of hydrology in fish and wildlife preservation, recreational use of water and coastal works.
6. Collect the information regarding the geographical area and the average annual rainfall of your District or State. Estimate the average rainwater volume received every year.
7. What are the State Government agencies that are undertaking the hydrologic data collection in your state ? What are the Central Government agencies with similar responsibilities?

8. Visit a major water resources project in your state and find out what specific hydrological investigations were involved in its design.
9. Is your state affected by floods or droughts? Explain how hydrology is useful in proposing the measures to mitigate these natural disasters.

Objective Type Questions

1. The science which deals with occurrence, distribution and circulation of water is called
   (*a*) Hydrography (*b*) Hydrometry
   (*c*) Hydrology (*d*) Hydraulics.
2. The science which deals with the surface streams is known as
   (*a*) Limnology (*b*) Cryology
   (*c*) Geomorphology (*d*) Potamology.
3. The science which deals with the lakes is known as
   (*a*) Limnology (*b*) Cryology
   (*c*) Geomorphology (*d*) Potamology.
4. The science which deals with snow and ice is known as
   (*a*) Cryology (*b*) Glaciology
   (*c*) Limnology (*d*) Potamology.
5. The concept of hydrologic was correctly postulated for the first time by
   (*a*) Leonardo da Vinci (*b*) Vitruvius
   (*c*) Plato (*d*) Aristotle.
6. Out of the total water present on the globe the saline water of oceans accounts for
   (*a*) 95% (*b*) 96%
   (*c*) 97% (*d*) 97.2%.
7. IHP stands for
   (*a*) Intensive Hydraulic Press (*b*) International Hydrologic Publications
   (*c*) International Hydrologic Periodicals (*d*) International Hydrologic Programme.

ANSWERS

Objective Type Questions

1. (*c*) 2. (*d*) 3. (*a*) 4. (*a*) 5. (*b*)
6. (*c*) 7. (*d*).

Chapter 2 HYDROMETEOROLOGY

2.1 INTRODUCTION

In the hydrologic cycle water is transported in the air, over the land and below the surface of the earth. Meteorology deals with the atmospheric portion of the hydrologic cycle, that is, with the transportation of water in the air. For this reason meteorology is often defined as the science of the atmospheric phenomena. The word atmosphere is derived from two Greek words *atmos* which means vapour or breath and *spheria* which means a sphere or a ball and it is now taken to refer to gaseous envelope of any heavenly body especially that of the earth. Similarly the word meteorology is derived from two Greek words *meteors* which means lofty and *logos* which means discourse or science. Hydrometeorology is defined as the study of the atmospheric processes which affect the water resources of the earth and are of interest to the hydrologic engineers. A broader definition of hydrometeorology was given by the World Meteorological Organisation (WMO fourth Congress-1963) in stating that hydrometeorology is concerned with the study of the atmospheric and land phases of the hydrologic cycle with the emphasis on the inter-relationships involved. The purpose of this chapter is to given an introductory outline of hydrometeorology. Determination of the *probable maximum precipitation* used in the spillway design, forecast of precipitation for reservoir operation and determination of probable maximum winds over water surfaces for evaluating the resulting waves in connection with the design of dams and levees are some of the hydrological problems in which meteorology plays an important role.

2.2 CONSTITUENTS OF THE ATMOSPHERE

That invisible and odourless something which we breathe and which sustains life and fire is what we call air. Though its composition is exceedingly complex, a simplified analysis is given here in connection with the hydrologic cycle. The permanent constituents of the atmosphere, by volume, are Nitrogen 78%, Oxygen 21% and the rest inert gases. The variable constituents may be in solid, liquid and gaseous forms. Water is the most important of these and can comprise as much as 4% of the atmosphere and exists in all the three forms. When water exists in the gaseous form as pure vapour, it is invisible and is referred to as *humidity.* And when it is in the form of minute drops formed upon condensation, it is called *fog* or clouds depending on the altitude. Ozone and carbon dioxide exist as gases in widely varying amounts. Solid components, besides water as snow and ice, are particles of dust, smoke, salts, and micro-organisms. They reflect, scatter and dissociate radiation and in particular light to produce various colours on the sky and sunset. They also play a role in the condensation of water vapour, in the formation of clouds and also in the precipitation process. The atmosphere has no clearcut upper limit but gradually decreases in density until it becomes indistinguishable at about 600 km above the earth's surface. However, 90% of the atmosphere is below 20 km and two-thirds of it below Mount Everest, *i.e.*, below 9 km.

Ozone. Most of the oxygen present in the atmosphere is diatomic (O_2). A small portion consists of triatomic oxygen (O_3), called ozone. The ozone is mainly found at large altitudes (between 20 and 70 km above the earth's surface). Though present only in minute quantities ozone is important because it prevents harmful ultra-violet radiation from reaching down to levels where biological processes are prominent. Some traces of ozone may sometimes be formed in the lower atmosphere as a result of electric charges. But these are negligible when compared to the amounts found in the upper layers.

Carbon Dioxide. The amount of carbon dioxide (CO_2) present in the air is a variable quantity. It is continuously consumed by the vegetable world, and it is produced by the animal world, by the burning of fuels, by volcanic actions, and by various processes of decay in the soil. Although these processes are not always balanced, the oceans absorb carbon dioxide so readily that only a small portion of it remains in the air. The carbon dioxide in the air is important for the heat budget of the atmosphere because together with water vapour it causes part of the long wave radiation from the earth to be absorbed by the air. It has been observed that the amount of carbon dioxide in the atmosphere has increased by 10% since the beginning of the century and this is attributed to increased burning of fuels by the industries.

Water Vapour. Form the view point of hydrologists, the most important constituent of the atmosphere is the water vapour. The amount of water vapour present in the air is indirectly expressed through what is known as the *vapour pressure.* The partial pressure exerted by the water vapour is called the vapour pressure. Consider certain amount of moist air in a closed container. It will be exerting a total pressure p on the container. Now, if all the water vapour is removed from the container without affecting the temperature, the pressure exerted by the dry air p' will be less than the original pressure exerted by the moist air, and the quantity $(p - p')$ is obviously the pressure contribution due to vapour and is therefore the vapour pressure. The more the vapour present in the air, the higher is the vapour pressure. Vapour pressure is usually denoted by e, and is expressed either in millibars or in mm of mercury. One millibar is one-thousandth of the standard barometric pressure which is taken to be 10^5 N/m^2. Therefore we have that 1 mb = 100 N/m^2 and 1 mm of mercury = 1.33 mb. Mean sea level standard atmospheric pressure at 15°C is 1013 mb.

For any given temperature there is a limit for water vapour that can be held by air. If the air is holding water vapour which is equal to the maximum that it can hold at a given temperature then the air is said to be saturated. The partial pressure exerted by the water vapour under saturated conditions is known as the *saturation vapour pressure* and is denoted by e_s. The saturation vapour pressure increases with increase in temperature. The variation of e_s with temperature is shown in Fig. 2.1 and the same is tabulated in Table 2.1. It can also be approximated by the simple equation given below.

$$e_s = 6.11 \exp.\left(\frac{17.27\,T}{237.3 + T}\right) \qquad ...(2.1)$$

where e_s is in millibars and T is in °C.

Table 2.1. Variation of Saturation Vapour Pressure e_s with Temperature T

T in °C	e_s in mb	T in °C	e_s in mb	T in °C	e_s in mb	T in °C	e_s in mb	T in °C	e_s in mb
0	6.1078	10	12.272	20	23.373	30	42.430	40	73.777
1	6.5662	11	13.119	21	24.861	31	44.927	41	77.802
2	7.0457	12	14.017	22	26.430	32	47.551	42	82.015
3	7.5753	13	14.969	23	28.086	33	50.307	43	86.423
4	8.1294	14	15.977	24	29.831	34	53.200	44	91.034
5	8.7192	15	17.044	25	31.671	35	56.236	45	95.855
6	9.3465	16	18.173	26	33.608	36	59.422	46	100.89
7	10.013	17	19.367	27	35.649	37	62.762	47	106.16
8	10.722	18	20.630	28	37.796	38	66.264	48	111.66
9	11.474	19	21.964	29	40.055	39	69.934	49	117.40

Another measure of water vapour is the *relative humidity* which is defined as the ratio of the actual vapour pressure to the saturation vapour pressure corresponding to the same temperature. Relative humidity is denoted by h and is usually expressed as percentage. Thus

$$h = \frac{100e}{e_s} \qquad ...(2.2)$$

Let the moist air exist at pressure p, temperature T and density ρ_a. Since it has to satisfy the equation of state, we have

$$p = \rho_a R_a T \qquad ...(2.3)$$

or

$$\rho_a = \frac{p}{R_a T} \qquad ...(2.4)$$

where T is temperature in K, R_a is the gas constant for the moist air in Nm/kg K and p is the absolute pressure in N/m^2.

If the vapour pressure of the moist air is e, then the pressure exerted by the dry air would be $(p - e)$. Then from Eq. (2.4) we have

$$\rho_v = \frac{e}{R_v T} \qquad ...(2.5)$$

and

$$\rho_d = \frac{(p-e)}{R_d . T} \qquad ...(2.6)$$

where ρ_v and ρ_d are the densities of water vapour and dry air and R_v and R_d are the gas constants of water vapour and dry air. Also $R_v = R_d/0.622 = 1.6077\ R_d$, where 0.622 is the ratio of molecular weight of water vapour to the average molecular weight of dry air. Now adding Eqns. (2.5) and (2.6), we get

$$\rho_v + \rho_d = \rho_a = \frac{p}{R_d T}\left(1 - 0.378\frac{e}{p}\right) \qquad ...(2.7)$$

At a given pressure p and temperature T, the density of dry air is $\rho_d = (p/R_d\,T)$. Eq. (2.7) thus shows that at a given pressure and temperature, moist air is lighter than dry air.

Again using Eq. (2.3) we can write

$$e = \rho_v R_v T \qquad ...(2.8)$$

$$(p - e) = \rho_d R_d T \qquad ...(2.9)$$

Now adding Eqns. (2.8) and (2.9), one can write

$$p = R_d\, T\, [\rho_d + \rho_v/0.622] \qquad ...(2.10)$$

Consider an atmosphere with vapour pressure e and at temperature T, which can be denoted by the point $P\,(T, e)$ in Fig. 2.1. Since $e < e_s$, this point lies below the curve. If the vapour pressure of this atmosphere is increased by adding water vapour to it, keeping the temperature constant of course, then the point in Fig. (2.1) moves vertically upwards till it meets the curve at P'_s with coordinates (T, e_s). At this condition the air is saturated and any further addition of water vapour results in condensation. The quantity $(e_s - e)$ is known as the *saturation deficit.* On the other hand, if the atmosphere is cooled without changing its pressure and vapour pressure the point P moves to the left horizontally till it meets the curve at P_s'' with coordinates (T_d, e). Here again the air is saturated and further cooling results in condensation. The temperature T_d is called the *dew point temperature.* It is the temperature at which the saturation vapour pressure is equal to the existing vapour pressure. Suppose the evaporation is allowed freely into the air without controlling the temperature. Then as more and more water vapour is added to the air it reduces the temperature of the air and consequently the saturation vapour pressure of the air also decreases. In this case the point P moves neither vertically nor horizontally but follows rather a diagonal path till it meets the curve at P_s''' with coordinates (T_w, e_w). This temperature T_w is called the wet bulb temperature and it is the temperature to which the original air can be cooled by evaporating water into it and is found by the wet bulb thermometer. This principle is utilised in measuring the relative humidity of the air which is explained in Article (2.11).

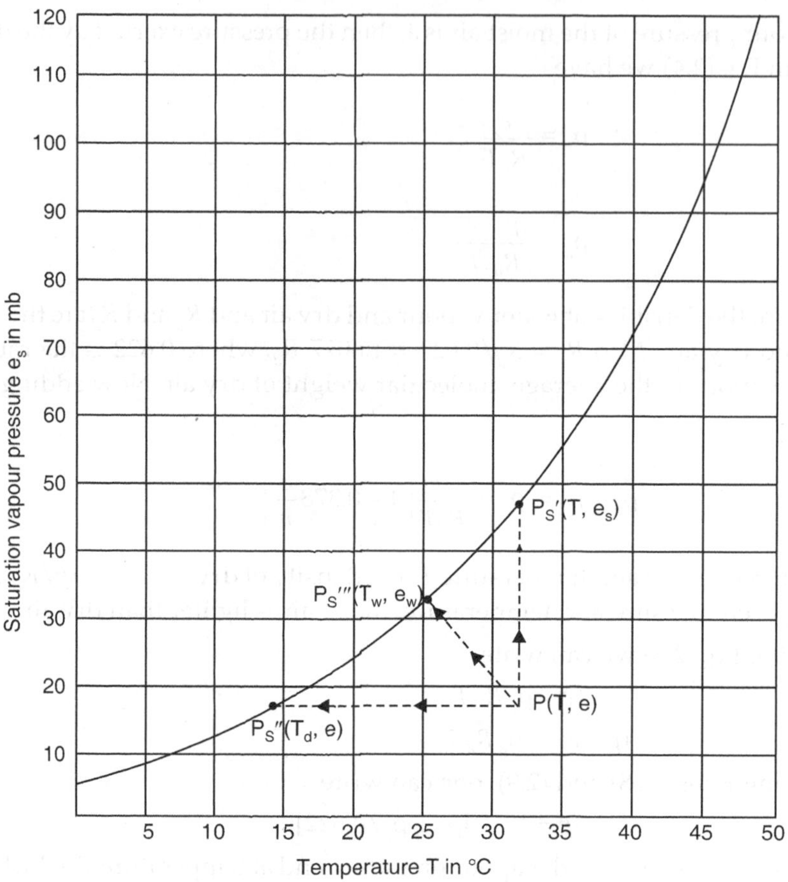

FIGURE 2.1 Variation of saturation vapour pressure with temperature.

The specific humidity of moist air q_h is defined as the ratio of density of water vapour ρ_v to the density of moist air ρ_a. It represents the mass of water vapour present in unit mass of moist air and expressed in kg of water vapour/kg of moist air.

$$q_h = \frac{\rho_v}{\rho_a} \qquad ...(2.11)$$

The relationship between the gas constants of moist air and dry air is given by

$$R_a = R_d\,(1 + 0.608\,q_h) \qquad ...(2.12)$$

Thus we note that the gas constant of moist air increases with specific humidity. But even for a large specific humidity, for example $q_h = 0.03$ kg of water vapour/kg of moist air, the difference between the gas constants of moist and dry air is hardly 2 per cent. Hence we can always take $R_a \approx R_d$. Consider the ratio $\frac{\rho_a}{\rho_v}$ which can be written as $\frac{\rho_d + \rho_v}{\rho_v} = 1 + \frac{\rho_d}{\rho_v}$. But $\frac{\rho_d}{\rho_v}$ is given by $\frac{(p-e)}{R_d . T}\frac{R_v . T}{e}$. With $R_v = 1.6077\,R_d$, we get $\frac{\rho_d}{\rho_v} = 1.6077\left(\frac{p}{e} - 1\right)$.

$$\therefore \quad \frac{\rho_a}{\rho_v} = 1 + \frac{\rho_d}{\rho_v} = 1 + 1.6077\left(\frac{p}{e} - 1\right) = 1.6077\frac{p}{e} - 0.6077.$$

Generally $\frac{p}{e}$ will be very large compared to 0.6077 and hence we can write

$$\frac{\rho_a}{\rho_v} = 1.6077\frac{p}{e};$$

$$q_h = \frac{\rho_v}{\rho_a} = \frac{1}{1.6077}\cdot\frac{e}{p} = 0.622\, e/p$$

$$q_h = \frac{0.622\, e}{p} \quad \text{...(2.13)}$$

The total amount of water vapour present in a column of air is known as precipitable water. It may be expressed in kg or as equivalent depth in mm. There is no natural process capable of precipitating the entire moisture content in the air.

Consider an elemental air column of area A and height dz. The mass of air in this element is given by $\rho_a A\, dz$ and the mass of water in this element is given by $q_h\, \rho_a A\, dz$. The total mass of precipitable water in the column between elevations z_1 and z_2 is then given by

$$m_p = \int_{z_1}^{z_2} q_h\, \rho_a\, A\, dz \quad \text{...(2.14)}$$

Eq. (2.14) is usually numerically integrated using an interval of Δz with an incremental mass of precipitable water in each interval Δz given as

$$\Delta m_p = \bar{q}_h\, \bar{\rho}_a\, A\, .\, \Delta z \quad \text{...(2.15)}$$

where $\bar{q}_h$ and $\bar{\rho}_a$ are the average specific humidity and average density of the air over the interval dz. The mass increments are summed up over the column to give the total precipitable water. Monograms are also available to yield direct solution to Eq. (2.14).

The hydrostatic pressure law and the variation of air temperature with altitude are given by

$$\frac{dp}{dz} = -\rho_a\, g \quad \text{...(2.16)}$$

and

$$\frac{dT}{dz} = -\propto \quad \text{...(2.17)}$$

where $\propto$ is the lapse rate which is taken to be 0.0065°C/m. The negative sign in Eqns. (2.16) and (2.17) indicate that both pressure and temperature of air decrease with elevation. If T_1 and T_2 are the temperatures at elevations z_1 and z_2, from Eq. (2.17) we can write

$$T_2 = T_1 - \propto (z_2 - z_1) \quad \text{...(2.18)}$$

Substituting ρ_a from Eq. (2.4) in Eq. (2.16), we have

$$\frac{dp}{dz} = -\frac{p}{R_a T}\cdot g \quad \text{or} \quad \frac{dp}{p} = \frac{-g}{R_a T}.dz \qquad ...(2.19)$$

From Eq. (2.17) $dz = \frac{-dT}{\propto}$ on substitution of which Eq. (2.19) becomes

$$\frac{dp}{p} = \left(\frac{g}{\propto R_a}\right)\frac{dT}{T} \qquad ...(2.20)$$

Integrating Eq. (2.20) on both sides, we get

$$\ln p = \left(\frac{g}{\propto R_a}\right)\ln T$$

$$\therefore \qquad \frac{p_2}{p_1} = \left(\frac{T_2}{T_1}\right)^{(g/\propto R_a)} \qquad ...(2.21)$$

From the known pressure and temperature p_1 and T_1 at elevation z_1, Eqns. (2.18) and (2.21) can be used to obtain p_2 and T_2 at z_2. These values immediately provide densities of air at z_1 and z_2 from Eq. (2.4) and then the average density $\overline{\rho}_a$. If we know the vapour pressure at z_1 and z_2, they can be used to compute the specific humidities at z_1 and z_2 and also $\overline{q}_h$. Now Eq. (2.15) can be applied to compute the precipitable water present between elevations z_1 and z_2. Example 2.5 illustrates this procedure.

Example 2.1. *Determine the density in kilograms per cubic metre, of (a) dry air at 30°C and a pressure of 900 mb and (b) moist air with relative humidity of 70 per cent at the same temperature and pressure.*

Solution. (*a*) $p = 900 \text{ mb} = 900 \times 100 = 9 \times 10^4 \text{ N/m}^2$

$R = 287$ N-m/kg- K

$T = 30°\text{C} = 30 + 273 = 303$ K

$$\rho = \frac{p}{RT} = \frac{9\times 10^4}{287\times 303} = 1.035 \text{ kg/m}^3$$

$\therefore$ Density of dry air = 1.035 kg/m^3.

(*b*) Relative humidity $h = 70\%$

$$h = \frac{100e}{e_s}; \quad \therefore \quad e = \frac{h}{100}e_s = 0.7\ e_s.$$

From Table (2.1) for a temperature of 30°C, $e_s = 42.43$ mb

$\therefore$ $e = 0.7 \times 42.43 = 29.701$ mb

The density of moist air is given by

$$\rho = \frac{p}{RT}\left(1 - 0.378\frac{e}{p}\right)$$

$$= 1.035\left(1 - 0.378 \times \frac{29.701}{900}\right)$$

$$= 1.0221\ \text{kg/m}^3.$$

Example 2.2. *Determine (i) the saturation vapour pressure (ii) the actual vapour pressure (iii) saturation deficit and (iv) the dew point temperature of the atmosphere corresponding to a temperature of 30° and a relative humidity of 80%.*

Solution. (*i*) From Table 2.1, the saturation vapour pressure at 30°C is obtained as 42.43 mb

$\therefore$ $e_s = 42.43$ mb

(*ii*) $h = 80\%$, $\therefore$ $e = 0.8\, e_s = 33.944$ mb

$\therefore$ Actual vapour pressure = 33.944 mb

(*iii*) Saturation deficit = $(e_s - e)$

$$= (42.43 - 33.944) = 8.486 \text{ mb}$$

(*iv*) From Table 2.1, the temperature at which

$e_s = e = 33.944$ is obtained as 26.165°C

$\therefore$ The dew point temperature, $T_d = 26.165$°C.

Example 2.3. *If the atmospheric pressure, temperature and vapour pressure of the atmosphere are 1005 mb, 28°C and 19.873 mm of mercury, determine the relative humidity, dew point temperature and the density of the atmosphere.*

Solution. From Table 2.1, for 28°C $e_s = 37.796$ mb

$e = 19.873$ mm of mercury $= 19.873 \times 1.33 = 26.43$ mb

$$\therefore \text{Relative humidity } h = \frac{100e}{e_s} = \frac{100 \times 26.43}{37.796} = 69.93\%$$

From Table 2.1, the temperature corresponding to a saturation vapour pressure of 26.43 mb is 22°C. Therefore the dew point temperature $T_d = 22$°C.

$$\rho = \frac{p}{RT}\left(1 - 0.378\frac{e}{p}\right).$$

Here $p = 1005 \text{ mb} = 1.005 \times 10^5\ \text{N/m}^2$

$T = 28°\text{C} = 28 + 273 = 301$ K

$R = 287$ N-m/kg- K

$$\therefore \quad \rho = \frac{1.005 \times 10^5}{287 \times 301}\left(1 - \frac{0.378 \times 26.43}{1005}\right)$$

$$= 1.152\ \text{kg/m}^3.$$

Example 2.4. *What is the relative humidity if the air temperature and the dew point are 32°C and 18°C respectively ?*

Solution. $T = 32°C$ and from Table 2.1, $e_s = 47.551$ mb

$T_d = 18°C$ and from Table 2.1, $e = 20.63$ mb.

Relative humidity
$$h = \frac{100e}{e_s}$$
$$= \frac{100 \times 20.63}{47.551} = 43.385\%.$$

Example 2.5. *The surface air pressure and air temperature at a place are 1010 mb and 32°C respectively. Compute the precipitable water in a saturated air column of height 1000 m over a ground surface of 1 m² area. Take* $\propto = 0.0065°C/m$.

Solution. At the ground surface :

$z_1 = 0$; $p_1 = 1010$ mb $= 1.01 \times 10^5$ N/m² $T_1 = 32°C = 273 + 32 = 305$ K

The air is given to be saturated. Therefore $e_1 = e_s$ at 32°C

From Eq. (2.1)
$$e_1 = 6.11 \exp\left(\frac{17.27 \times 32}{237.3 + 32}\right)$$
$$= 6.11\, e^{2.0521} = 47.56 \text{ mb}$$
$$q_h = 0.622\frac{e_1}{p_1} = \frac{0.622 \times 47.56}{1010} = 0.02929 \text{ kg/kg.}$$
$$\rho_a = \frac{p_1}{R_a T_1} \text{ ; Taking } R_a = R_d = 287 \text{ Nm/kg K,}$$
$$\rho_a = \frac{1.01 \times 10^5}{287 \times 305} = 1.1538 \text{ kg/m}^3.$$

At an elevation of 1000 m :
$$z_2 = 1000 \text{ m} ; T_2 = 32 - 0.0065(1000 - 0)$$
$$= 25.5°C = 25.5 + 273 = 298.5 \text{ K}$$
$$\left(\frac{g}{\propto R_a}\right) = \frac{9.81}{0.0065 \times 287} = 5.2586$$

From Eq. (2.21)
$$\frac{p_2}{p_1} = \left(\frac{T_2}{T_1}\right)^{(g/\propto R_a)} = \left(\frac{298.5}{305}\right)^{5.2586} = 0.8929$$
$$p_2 = p_1 \times 0.8929 = 1.01 \times 10^5 \times 0.8929$$
$$= 0.90183 \times 10^5 \text{ N/m}^2 = 901.83 \text{ mb}$$

Density of air at elevation of 1000 m

$$\rho_a = \frac{p_2}{R_a T_2}$$

$$\therefore \quad \rho_a = \frac{0.90183 \times 10^5}{287 \times 298.5} = 1.0527 \text{ kg/m}^3.$$

Since the column of air is given to be saturated $e_2 = e_s$ at 25.5°C

From Eq. (2.1)

$$e_2 = 6.11 \exp\left(\frac{17.27 \times 25.5}{237.3 + 25.5}\right)$$

$$= 6.11\, e^{1.6757} = 32.6429 \text{ mb}$$

$$q_h = 0.622\frac{e_2}{p_2} = \frac{0.622 \times 32.6429}{901.83} = 0.02251$$

$$\bar{q}_h = \frac{0.02929 + 0.02251}{2} = 0.0259$$

$$\bar{\rho}_a = \frac{1.1538 + 1.0527}{2} = 1.10325 \text{ kg/m}^3$$

$$\Delta m_p = \bar{q}_h\, \bar{\rho}_a A.\, \Delta z$$

$$= 0.0259 \times 1.10325 \times 1 \times 1000 = 28.574 \text{ kg}$$

Expressed in equivalent depths, Δm_p = 28.574 mm.

Example 2.6. *What is the relative humidity and specific humidity of air if its pressure, temperature and vapour pressure are 986 mb, 35°C and 47.5 mb respectively.*

Solution. Relative humidity $h = \dfrac{100e}{e_s}$; e = 47.5 mb

From Eq. (2.1) e_s corresponding to 35°C $= 6.11 \exp.\left(\dfrac{17.27 \times 35}{237.3 + 35}\right)$

$$= 6.11\, e^{2.2198} = 56.2455 \text{ mb}$$

Relative humidity $h = \dfrac{100 \times 47.5}{56.2455} = 84.45\%$

Specific humidity $q_h = 0.622\dfrac{e}{p}$

$$= \frac{0.622 \times 47.5}{986} = 0.03 \text{ kg/kg}.$$

2.3 VERTICAL STRUCTURE OF THE ATMOSPHERE

The average state of the atmosphere has been well explored with the help of meteorological instruments and satellite observations. Fig. 2.2 illustrates the features of the atmosphere wherein one may visualise some concentric shells or spheres with distinct temperature characteristics. These layers are given different names.

The lowest layer which contains about 75 per cent of the mass and almost all the moisture and dust of the atmosphere is called the *troposphere*. All the phenomena which are called weather are confined to the *troposphere*.

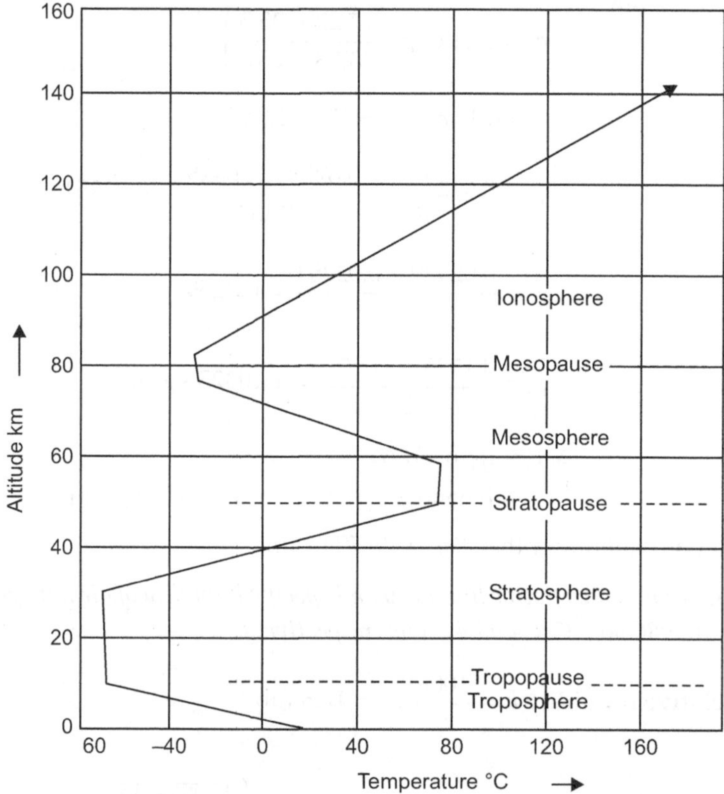

FIGURE 2.2 Idealised vertical structure of the atmosphere.

The top of the troposphere is called the *tropopause*. The height of the tropopause varies from about 8 km over the poles to about 16 km near the equator. In the troposphere the temperature decreases with the elevation at an average rate of 6.5 °C/km. This is known as the *lapse rate*.

Above the tropopause is the *stratosphere*. The overturnings resulting from daily and annual heating of the earth's surface are generally confined to the troposphere and therefore all these disturbances are below the tropopause. Thus the stratosphere is free of these influences and contains very little moisture and dust except when major volcanic eruptions bring them to high altitudes. Stratosphere contains a major portion of the ozone that was discussed earlier. In the lower layers of the stratosphere the temperature is more or less constant with elevation and the stratification is very stable. The rise of temperature in the stratosphere is due to selective absorption of ultra-violet radiation by the ozone. *Stratopause* located at 30 to 50 km above the earth's surface is the upper limit of the stratosphere.

Above the stratosphere is a warm layer which is variously known as the *mesosphere* or the *chemosphere*. About 80 km from the earth's surface is the *mesopause* which is the upperlimit of the mesosphere. The fourth major layer above the mesopause is called the *ionosphere*. The pressures in the ionosphere are very low with about 0.01 mb at 90 km. At such low pressures ionisation is lively resulting in a layer with high concentration of free electrons. The ionosphere is divided into a number of layers with different electrical characteristics and varying intensities. The ionosphere merges gradually into the outermost shell called the *exosphere*. Here the mean free path is very large and the atmosphere has lost the property of continuum.

Troposphere is the zone in which the weather phenomena and atmospheric turbulence are most marked because of their effect on mass transport (water vapour, dust, pollutants), momentum transport (winds) and energy transport. Hydrologists are mainly interested in the troposphere and the processes that take place in it. However, although the weather is considered as resulting from the interchange of the heat energy within the troposphere, processes above this layer giving ozone variations and ionisation etc., may influence the weather conditions in the troposphere to some extent.

2.4 SOLAR RADIATION

The source of energy for all the atmospheric activities is the solar radiation. The radiation is usually measured in cal/cm^2/ minute. The unit cal/cm^2 is known as *langley** abbreviated as ly. The sun with a temperature of 6000 K radiates energy at a rate of 107050 ly/min. However, only a very small portion of this is intercepted by the earth and its atmosphere. The average solar radiation received on a plane unit area normal to the incident radiation at the outer limit of the atmosphere is called the *solar constant*, whose value is generally accepted to be 1.97 ly/min or equivalently 1374 Watt/m^2. The solar radiation received by a unit area on a horizontal surface which is of primary concern to the hydrologists and meteorologists is called *insolation*. The insolation at the outer limit of the earth's atmosphere at any given time depends on the declination of the sun, the latitude of the place, and the position of the sun at that time. The daily amounts of insolation received at the top of the atmosphere are tabulated in terms of the latitude of the place and the time of the year as given in Table 7.1.

A major portion of the insolation is scattered and absorbed in the atmosphere or reflected from the clouds and the earth's surface. Air molecules scatters the radiation with shortest wavelengths most effectively and the blue colour of the sky on clear days is believed to be because of this scattering. The ratio of the amount of radiation reflected to the amount of radiation received

---

*ly/min is the non-SI unit of radiation. In SI units radiation is measured in Watt/m^2. The conversion is 1 ly/min = 697.8 Watt/m^2.

by any surface is called its *albedo*. Clouds reflect the solar radiation back to the space. The radiation reflected by the clouds depends on type of cloud, thickness and water content of the cloud, and also on the elevation of the cloud. After scattering and absorption by the atmosphere and the reflection by the clouds, approximately about half of the insolation reaches the earth's surface. A part of this is again reflected to the atmosphere and the rest is absorbed by the earth. The amount of radiation reflected to the atmosphere by the earth is governed by its albedo which varies widely depending on the nature of the surface and the solar altitude. The average albedo of the earth's surface has been estimated to be about 14% while that for the planet as a whole including the atmosphere and clouds is between 35 and 43 per cent.

The earth with a mean temperature of 287 K radiates into the atmosphere at a rate of 0.56 ly/min. This is essentially in the form of longwave radiation and is approximately 2.5 times the shortwave solar radiation absorbed by it. Though the longwave radiation emitted by the earth is more than shortwave solar radiation absorbed by it, the atmosphere reflects 85% of the longwave radiation back to the earth thus maintaining the heat balance on the earth. This is known as the *green house effect*. Otherwise the temperature of the earth would be – 40°C. A portion of the longwave radiation emitted by the earth is absorbed by the carbon dioxide and water vapour present in the atmosphere.

The solar radiation will be nearly normal to the earth's surface near the equator. Therefore the insolation is high at low latitudes whereas it will be low at high latitudes. These differences in insolation are primarily responsible for the general circulation of the earth's atmosphere.

2.5 THE GENERAL CIRCULATION

If the earth were to be considered as a homogeneous and non-rotating globe receiving uniform solar radiation all over, the convective activity would occur within the atmosphere as shown in Fig. 2.3. The air gets heated at the surface and becomes lighter. The lighter air moves upward to the height of the tropopause before it cools down and returns to the surface.

However, if the earth were to be non-rotating globe receiving non-uniform solar energy with maximum intake of solar radiation near the equator and the minimum near the poles, convection might still be expected. But this time the movement of lighter and warm air at higher altitudes would be from equator to poles while the cooler air on the surface moves from poles to equator. This is shown in Fig. 2.4.

As the earth rotates from west to east making one revolution per day, it changes the direction of the wind movement in the following manner. The radius of earth at equator is 6370 km while at 60° latitude it is half this value. A ring of air near the equator has the same speed as that of the earth which is 463 m/s and at 60° latitude this speed is 231.5 m/s. The angular momentum of the air near the equator about the polar axis is (463 MR) kg-m/s, where M is the mass of the air and R is the radius of the earth. If the ring of the air at the equator is forced to move towards the poles, its radius will be reduced and for its angular momentum about the polar axis to remain the same, its speed will increase from west to east.

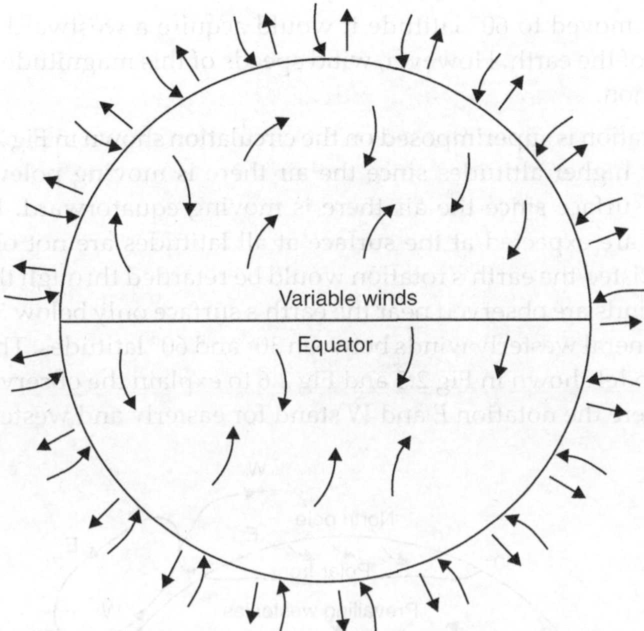

FIGURE 2.3 Circulation over non-rotating uniformly heated earth.

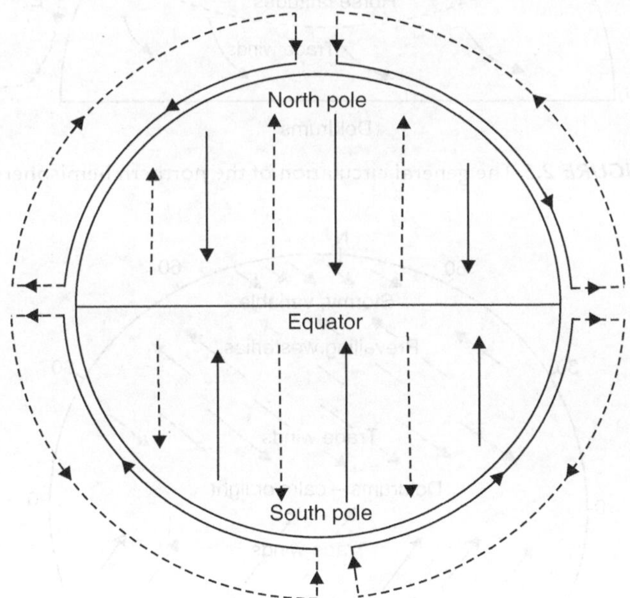

FIGURE 2.4 Circulation over non-rotating and non-uniformly heated earth.

In other words it becomes *westerly*. This is known as the *coriolis* effect which increases with latitude. The force that would be required to produce these changes in the velocity is known as the *coriolis force*. Theoretically if the air at the equator moves to 60° latitude it would acquire an eastward velocity equal to (463 MR/ 0.5 MR) = 926 m/s. Since the speed of earth at 60° latitude is only 231.5 m/s, the velocity of air relative to earth at this latitude is (926 – 231.5) = 694.5 m/s. Similarly, if the air at the poles is moved to 60° latitude it would acquire a westward velocity of 231.5 m/s relative to the surface of the earth. However, wind speeds of this magnitude are never observed in nature because of friction.

If the earth's rotation is superimposed on the circulation shown in Fig 2.4, we should observe the westerly winds at higher altitudes since the air there is moving poleward and the easterly winds on the earth's surface since the air there is moving equatorward. However, the general easterly winds which are expected at the surface at all latitudes are not observed. Obviously if such easterly winds existed the earth's rotation would be retarded through the friction. In practice, the general easterly winds are observed near the earth's surface only below 30° latitude and above 60° latitude and the general westerly winds between 30° and 60° latitudes. This leads to the general *triple cell circulation* model shown in Fig 2.5 and Fig 2.6 to explain the observed general circulation of the atmosphere where the notation E and W stand for easterly and westerly winds.

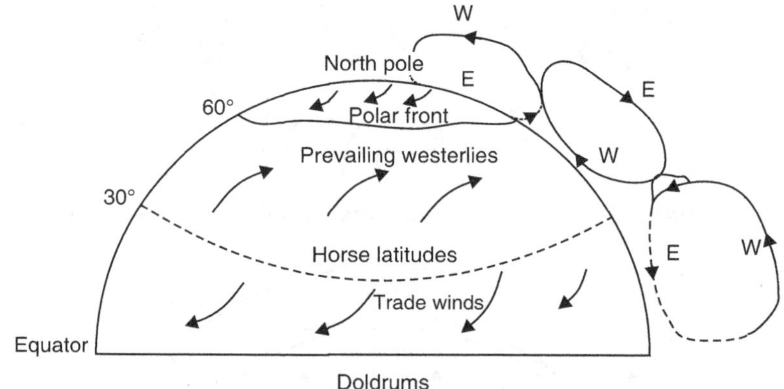

FIGURE 2.5 The general circulation of the northern hemisphere.

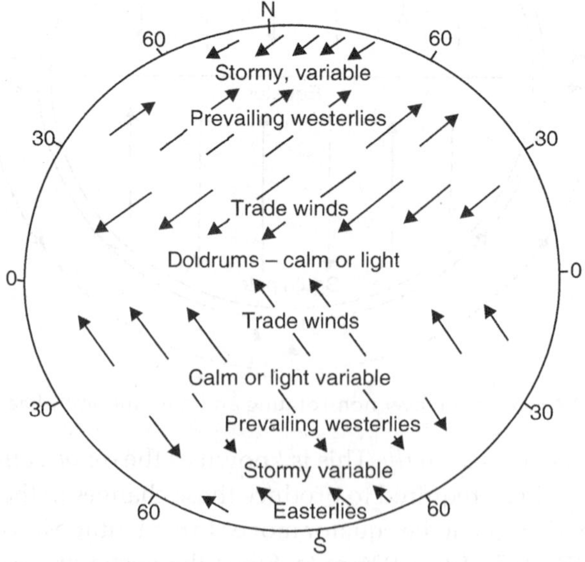

FIGURE 2.6 The triple cell circulation at the earth's surface.

Air rising at the equator and moving towards the north acquires an eastward component because of earth's rotation. By the time this air reaches about 30° latitude, known as the *horse latitudes*, it will have lost sufficient heat so that it tends to subside. The subsiding air divides into two branches— one branch moving towards the south and the other towards the north. The branch moving towards the south attains the westward velocity because of earth's rotation and is called the *North-East Trade Wind*. The branch moving towards the north attains the eastward velocity and is called the westerly winds. The region between the equator and the horse latitude is called the equatorial cell, that between 30° latitude and 60° latitude is called the middle cell and the one between 60° latitude and the pole is called the polar cell. In the polar cell, because of the loss of heat in the lower layers the air subsides. The subsiding air acquires the westward velocity due to earth's rotation as it moves towards the south. While moving southward this air gets heated and at about 60° latitude it rises and returns to the pole as the southwesterly wind aloft where it again subsides. That branch of air which moves northward from the horse latitude in the form of westerly winds meets the westward moving air of the polar cell at the southern extreme of the polar cell at 60° latitude. While the air coming from the pole is generally cooler and denser, the air of the middle cell is warmer and lighter. The region where these two airs with distinct properties meet is called the *polar front*. The westerly wind of the middle cell being lighter is forced upward over the colder air from the polar cell and continues northward as the westerly wind aloft till it reaches the north pole. To maintain the balance in the circulation, air is released from the polar cell by outbreaks through the polar front which is carried southward through the temperate zone of the middle cell and finally into the equatorial cell. Otherwise the accumulation of the air in the polar cell and the loss of the air in the equatorial cell would result. The tripple cell circulation implies the belts of low pressures near the surface at the equator and at about 60° latitude where the warm air rises. Similarly high pressures would be expected at about 30° latitude and at the poles where the air subsides.

The reasons for the triple cell circulation are only partly known. The triple cell circulation model is essentially based on the homogeneous surface for the earth while actually it is not so. Even then, the average pattern of the observed circulation over the earth indicates many of the factors mentioned in the triple cell circulation model. For example:

(*a*) The north east trade winds in the northern hemisphere and the south east trade winds in the southern hemisphere.

(*b*) The prevailing westerlies in the middle latitudes.

(*c*) The *doldrums*, that is the area of low pressure near the equator with strong vertical motion, instability, high temperatures, convection and heavy precipitation. (In the areas of doldrums ships cannot move because there is no wind).

(*d*) The high pressure belts around 30° latitude where the air is usually descending, dry and hot and the climate is arid. (Most of the world's deserts are in this region).

(*e*) The wind aloft westerly everywhere except at the tropics.

(*f*) The polar front region of low pressure, strong vertical motions, frequent precipitation with contrasting air masses and temperatures, and

(*g*) The polar high pressure areas with cold dry air and polar deserts.

In spite of these factors the actual circulation is more complex. The circulation in the Southern Hemisphere, where the land area is only 20% of the earth's surface, is nearer to the ideal triple cell circulation described above than in the Northern Hemisphere which has the land area of about 40%. Isolated cells of low and high pressure zones, rather than the belts, are observed. The presence of the land and vegetation distorts the picture significantly. For example, the Himalayas push the mid-latitude high to the south. There may be local winds because of topography, physiography, land, sea etc. During the day the lower layers of the air over the land are warmer than those over the ocean and a pressure gradient is set up which causes a *sea breeze*, that is, a wind from the ocean to the land. At night the situation is reversed to produce a *land breeze*. Seasonal temperature contrasts between land and water produce large scale land and sea breezes called *monsoons*. In winter, high pressure over relatively cold land surfaces produces the off shore winds, while in summer the reverse takes place. In a typical monsoon climate, the winters are clear and dry and the summers are cloudy and wet. A good example of a monsoon climate is that of India.

Man-made effects like irrigation, agriculture, urbanisation and industrialisation may also affect the circulation locally.

2.6 TRANSITORY SYSTEMS

The features of the general circulation described earlier are rather the mean characteristics averaged over a long time interval like an year. They may be distorted or displaced by the transitory or migratory systems at any given time. The study of *synoptic charts* will reveal the transitory systems which are called the *cyclones* and *anticyclones*. The synoptic charts are prepared from the observations taken on pressure and wind concurrently at different locations of the region. These transitory systems will obviously disappear when presenting the average circulation. The air *masses* and the *air fronts* which sometimes lead to the development of the cyclones are first described.

Air Mass. A vast and deep body of the air in which the temperature and humidity characteristics are relatively homogeneous at any given elevation is called an air mass. Air mass is formed through the heat exchange when the air stagnates for a considerable time interval over a region where the earth's surface is homogeneous. Good source regions of air masses are the plains and vast deserts of the land areas of the earth and oceans.

An air mass is classified as *continental air mass* if it originates from the land areas and as *maritime air mass* if it originates from the seas. The continental air mass is usually dry and fluctuates widely in temperature from relatively cold in winter to hot in summer. The maritime air mass is nearly always moist with more uniform temperature. Further classification of air masses is based on the location of the source region, namely, equatorial (E), tropical (T), polar (P) or arctic (A). Thus E_m denotes the equatorial maritime air mass and P_c denotes the polar continental air mass etc.

The air mass eventually moves away from its source area because of the pressure gradients that exist. The weather conditions likely to occur can be deduced from the properties and the passage of the air mass from its source region into the other areas. The air mass gets modified as it moves and the extent of the modification depends on various factors. If the air mass is colder than the surface over which it moves, it is known as the *cold air mass*. It is heated from below and therefore a cold air mass is often unstable. If the air mass is warmer than the surface over which it moves it is known as the *warm air mass*. It is cooled from below and becomes more stable. Similarly, if the air mass is cooled from above it becomes unstable and when it is warmed above it becomes more stable. Turbulence and convection may occur because of the increased instability and may lead to the cloud formation and precipitation resulting in the removal of water from air mass. If the air mass passes over steep hills, the precipitation process may be initiated and water will be removed from the air mass because of the lifting and the subsequent cooling of the air mass. When the air mass passes over a moist surface, it may gather water vapour by way of evaporation to the extent of the existing vapour pressure gradient. Likewise, passage of the air mass over a dry surface may lead to the condensation of water vapour.

Air Front. The surface of contact between two adjacent air masses or between an air mass and the surrounding atmosphere is called a *frontal surface* or simply air front. The intersection of the frontal surface with the earth's surface is called a *surface front*. When two air masses come into contact, the colder air forms a wedge under the warm air and the frontal surface has a slope of 1 in 100 in normal cases. Usually the warm air ascends over the cold wedge and as a result of adiabatic cooling the clouds are often present in the ascending air. At the earth's surface, the front may sometimes be so sharp that it appears as a discontinuity in the temperature or a sudden shift of wind and other times it may be wide. In the upper air, the front almost always appears as a broad zone of transition. The most prominent frontal zone in the Northern Hemisphere is the polar front which is the zone separating polar air and tropical air.

The air fronts are classified according to their motion. When the cold air is moving into warm air forcing the warm air up over the surface front, then it is called a *cold surface front* or simply a *cold front*. When the warm air is moving up over a static or retreating cold air mass, then it is called a *warm front*. Sometimes along the same surface front there may be a cold front and warm front. On certain occasions a cold front may overtake a warm front in such a manner that the warm air between them is displaced upward. Then it is called an *occluded front*.

Cyclone. A cyclone is a more or less circular area of low atmospheric pressure in which winds blow spirally inward in counter clockwise direction in the Northern Hemisphere and in clockwise direction in the Southern Hemisphere. The pressure gradient exists towards the centre of the cyclone. There are two types of cyclones. They are *tropical cyclones* and *extratropical cyclones*.

The source areas of the tropical cyclones are the tropical seas of the low latitudes. The formation of the tropical cyclones is believed to be due to the proper combination of (*i*) intense sunshine, (*ii*) high humidity and temperature, (*iii*) low surface friction, (*iv*) deflection force of the earth's rotation and (*v*) favourable lapse rate. The tropical cyclones are relatively smaller in area with their diameter ranging from 80 km to 500 km. They develop into violent storms. The centre of the cyclonic storm is called the *eye of the storm*. When they are moving over the seas they gather the moisture and the principal energy of the tropical cyclones is derived from the latent heat of condensation carried by the moist air feeding them. The winds in these storms may develop speeds up to 300 km/h. The energy of the tropical cyclones is rapidly dissipated when they move over the large land areas because of the increased friction and lack of the moisture supply and therefore they die soon. The tropical cyclones are called *hurricanes* in North America and *typhoons* in Japan and simply cyclones in India.

Extratropical cyclones are mainly frontal in nature. They tend to develop wherever air masses of contrasting properties converge. As seen from the description of the general circulation, such areas exist in the middle latitudes (polar front) where the polar and tropical air masses meet. The principal energy source of these extratropical cyclones lies in the temperature and density contrasts between the two conflicting air masses. The pressure gradients developed from these contrasts are generally sufficient to produce wind speeds of 30 to 80 km/h and as the two masses become mixed, the storm dies. Therefore the extratropical cyclones are less violent than the tropical cyclones. But they are several km thick in vertical direction and their diameter may be as large as 2400 km. They are associated with the extensive clouds and, large and sustained precipitation. The life history of a typical extratropical cyclone of frontal nature in the northern hemisphere as seen from the top is shown in Fig. 2.7. For reasons not yet completely understood, a wave is generated on the boundary between the two air masses. Under conditions of dynamic stability this wave moves along the front with little change in the form. It the wave is unstable, it goes through the successive stages developing into a cyclone. As shown in the figure the cold front moves faster and overtakes the warm front resulting in an occluded front.

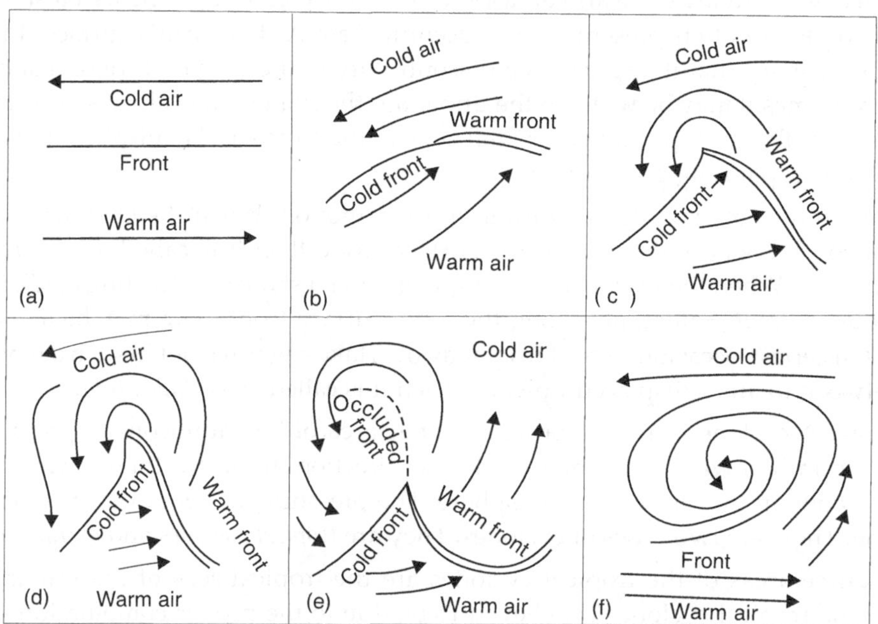

FIGURE 2.7 Wave theory of extratropical cyclone in the Northern Hemisphere.

There is an evidence that some extratropical cyclones are not generated from the frontal disturbances but they grow downward from some upper air disturbances towards the earth's surface as fully developed vortices.

Anticyclone. These are the high pressure areas generally elliptical in shape covering very large areas. The wind blows spirally outward in clockwise direction in the Northern Hemisphere and in counter clockwise direction in the Southern Hemisphere. The pressure gradient exists in outward direction away from the centre. In general, the anticyclones are accompanied by fair weather, though clouds may be present at low levels, particularly at sea and along coasts. The winds are generally moderate except towards the border of well developed cyclones.

Anticyclones may be classified as either warm or cold. The cold anticyclones develop because of the cooling of surface air from beneath by conduction and radiation and are usually only a few km deep. The cold anticyclones usually originate in high latitudes. They generally move equatorward and from west to east in the mid-latitudes, often trailing a cyclone. The warm anticyclones originate in the subtropics and move very slowly or they may not move at all. They are deep disturbances, but their exact origin is not clear.

Thunder storm. Because of the convective instability resulting from unequal heating at the surface or unequal cooling at the top, small scale disturbances of about 250 km^2 area are often created in the moist air in the form of vertical currents. Torrential rain, hail, gusty surface winds, thunder and lightning may all occur in such disturbances. These are called the *thunder storms* or the *convectional showers.*

Tornadoe. Under some conditions the frontal activity may develop into very strong localised convection currents. Such circulatory systems are called the *tornadoes* and normally have diameters of less than 1.5 km. Their peak wind speeds may attain several hundred km/h and their translational velocity may reach 100 km/h. Though they are most destructive their life span is very small often measured in minutes.

2.7 FORMATION OF PRECIPITATION

There are four basic conditions which are to be satisfied for the precipitation to occur. They are:

(*i*) Accumulation of moisture of sufficient intensity to account for the observed rates of precipitation.

(*ii*) Cooling of air to the dew point temperature to produce saturation condition.

(*iii*) Condensation.

(*iv*) Growth of small water droplets to precipitable size.

It is common for the first three conditions to occur but not the fourth, in which case the clouds will gradually dissipate without producing any precipitation.

Accumulation of Moisture. It is evaporation which feeds moisture to the atmosphere. The simple principle of conservation of mass requires that a balance must be maintained between the evaporation and precipitation. In fact the observed evaporation rates support this. So it is by evaporation that the moisture accumulates in the atmosphere.

Cooling. Cooling is of three types, namely, *cyclonic cooling, orographic cooling,* and *convective cooling*.

The cyclonic cooling may be divided into two types as frontal cooling and non-frontal cooling. In the frontal type of cyclonic cooling, the circulation *forces* the air up over a frontal surface. The lifting and subsequent cooling of the air is gradual thus producing moderate rainfall rates often of quite long durations. The non-frontal type of cyclonic cooling results from the convergence and subsequent lifting of air which accompanies a low pressure area. Frontal cyclonic precipitation of the extratropical origin produce rains (or snows) of moderate intensity but of fairly long durations. This type of storm may last from 24 to 72 h and deliver a total of 50 to 150 mm of precipitation in that time. Non-frontal cyclonic precipitation of the tropical origin may deliver as much as 375 mm of rain in a period of 12 to 24 h.

Orographic cooling occurs when the air is lifted up a slope to low pressures corresponding to higher elevations. Clouds first form and then precipitate if the dew point is reached. Of particular interest in the case of orographic cooling is what is known as the *rain shadow* or deficiency in rainfall which may occur on the lee side of the mountain because of removal of the moisture by high mountain. Such a rain shadow region is found to the east of western ghats in India. In general, the windward side of the mountain is more cloudy and rainy with smaller temperature variations. On the leeward side, however, the climate is drier, sunnier, and more variable in temperature.

When vertical instability of moist air is produced by surface heating, a convection current results. Convective precipitation is of very short duration, rarely exceeding 1 h and covering limited areas of 20 to 50 km^2. But the intensities are so high that the total precipitation may amount to 75 to 100 mm in a duration of 0.5 to 1 h.

Condensation. Under certain controlled conditions in the laboratory, lowering of temperature of the air does not produce condensation. Investigations have shown that the air, from which all the foreign particles have been removed, can be cooled until the relative humidity is as much as 1000 per cent before the droplets of the liquid water form. However, condensation takes place even when the air is not saturated provided the air contains small particles of substance that have an affinity to water. These hygroscopic particles are called the *condensation nuclei*. The size of these nuclei is of the order of 0.001 to 10 microns. The condensation is hastened if the clouds possess condensation nuclei in large quantities. Otherwise, though sufficient moisture is present in the clouds, they may not precipitate.

Growth of Droplets. The saturation vapour pressure over a curved surface is more than that over a horizontal surface. As the condensation nuclei attracts water, the size of the droplet increases making it less curved, and further condensation therefore requires less super saturation. On the other hand, the water attracted by the nuclei goes into solution and the attraction for water decreases. These two effects of curvature and hygroscopity which counteract with each other in the initial stages of the growth of the droplet become negligible after the droplet has attained a radius of 1 micron. Computations have shown that under the normal conditions, it takes a few seconds for the nucleus to grow to a droplet of 10 microns, a few minutes to reach 100 microns, 3 h to grow to 1000 microns (1 mm) and about 24 h to become a small drop (of diameter 3 mm). It follows then that pure condensation is not capable of producing rain drops though it may well produce oversize cloud droplets. As the droplets gain mass they begin to have a motion relative to their host cloud. This relative motion has the effect of transporting additional water vapour into the neighbourhood of the droplet. This is known as the *ventilation effect*. When the water droplet moves relative to the cloud it may collide with the other droplets and grow further in size. This phenomenon is called the *coalescence*. As a drop grows to a diameter of 7 mm the fall velocity increases to about 10 m/s. At such high speeds the drop flattens and breaks up into drops of the size of small rain or *drizzle*. Each of these will fall faster than the cloud droplets, with the result the coalescence processes are repeated. It has been shown theoretically that droplets of radius less than 18 microns cannot grow by coalescence, droplets of radius less than 60 microns of nearly uniform size are relatively stable against coalescence and droplets with radius greater than 80 microns of nearly uniform size will have very high capturing efficiencies. This clearly shows that a few drops of relatively large size are most important in the formation of precipitation. This has been the basis for the *cloud seeding* or the *artificial rain making*. Dry ice and silver iodide have been the common materials used as seeds to accelerate the coalescence process. There is a varying degree of success reported about the artificial rain making, but the technique is yet to be accepted for wider use.

2.8 TYPES OF PRECIPITATION

Precipitation is often classified according to the factors responsible for lifting and subsequent cooling. Broadly speaking there are three types of precipitation. They are:

(*i*) Cyclonic Precipitation

(*ii*) Convective Precipitation

(*iii*) Orographic Precipitation.

The mechanisms of cooling have already been discussed in the previous section. However, in nature the effects of these various types of cooling are often inter-related and it may not always be possible to identify the observed precipitation as of any one particular type.

2.9 FORMS OF PRECIPITATION

The number of different forms of precipitation is very large. Only common types are described here.

Drizzle. It is fine sprinkle of very small and rather uniform water drops with diameters between 0.1 and 0.5 mm. The drops are so small that they seem to float in air. To qualify as drizzle (also called *mist* sometimes), the drops must not only be small but they must also be very numerous. The intensity of drizzle rarely exceeds 1 mm/h.

Rain. Rain is the precipitation of liquid water in which the drops are generally larger than 0.5 mm in size.

Glaze. The ice coating formed when rain or drizzle freezes as it comes in contact with cold objects at the ground is called the glaze.

Sleet. When rain drops are frozen while falling through a layer of subfreezing air (below 0°C) near the earth's surface or refreezing of largely melted ice crystals occurs, transparent globular grains of ice known as sleet or *ice pellets* are formed. The pellets are generally between 1 mm and 4 mm in diameter.

Snow. Precipitation in the form of ice crystals is called snow. When the ice crystals fuse together, it is called *snowflake*.

Hail. Precipitation in the form of balls or irregular lumps of ice over 5 mm in diameter is called hail. Hail stones are generally composed of alternating ice and opaque snow like layers as a result of repeated ascents and descents within the cloud during their formation. Hail occurs almost exclusively in violent or prolonged thunderstorms.

Dew. Dew forms directly by condensation on the ground mainly during the night when the surface has been cooled by the outgoing radiation.

2.10 CLIMATE AND WEATHER SEASONS IN INDIA

The state of the atmosphere with respect to the temperature, humidity, wind, cloudiness etc. at any given time is generally referred to as the weather. The climate of a region is the aggregate of the weather. As mentioned in the earlier discussions the earth receives non-uniform solar radiation over its surface. This results in broadly three different climatic zones, namely the *torrid zone*, the *temperate zone* and the *frigid zone*. The region on either side of the equator between the tropics, that is, between 23½°N and 23½°S latitudes is known as the torrid zone. The regions between 23½° and 66½° latitudes are the temperate zones while the remaining areas between 66½°N latitude and North pole and 66½ °S latitude and South pole are called the frigid zones. The surface of the torrid zone is more normal to the incident solar radiation and hence it receives a larger solar energy, which makes it very hot. In the temperate zones the radiation is more slanting and therefore they receive less solar energy. The temperature is generally less and it decreases further towards higher latitudes. The frigid regions are extremely cold. Though half of it lies in the torrid zone and the other half in the temperate zone, the climate in India is generally that of a torrid zone except in few places with high altitudes here and there.

In India, the year may be divided into four weather seasons as given below:

(*i*) Cold weather season : December to February

(*ii*) Hot weather season : March to May

(*iii*) Southwest monsoon season : June to September

(*iv*) Retreating southwest monsoon season : October to November

Cold Weather Season. This is the season during which fine weather prevails over the entire country, though the southern parts along the east coast may be occasionally subjected to cyclones. The north will be in the grip of the cold wave. The day and night temperatures touch their lowest. The Himalayas and Jammu and Kashmir will be experiencing moderate to heavy rainfall and snowfall. The northeast monsoon is fully established over the country. The adjoining water bodies of the Bay of Bengal and the Arabian sea are generally free but for the occasional cyclonic disturbances.

Hot Weather Season. During this season the temperature is on the rise reaching the highest value in the second half of May at most of the places. The highest temperature of about 38°C will be recorded in the Deccan Plateau in March. In April, the highest temperature in the country will be observed in Gujarat and Madhya Pradesh with a value of about 43°C. In May, northwest Rajasthan records the highest temperature with mercury touching 48°C. The area of lowest pressure also lies in this region witnessing the thunderstorms and dust-storms. Local thunderstorms may also occur over the central parts of the country and over the Deccan Plateau. There is very little rainfall over the country except that produced by thunderstorms here and there locally. The depressions and the cyclonic storms start developing both in the Central Bay of Bengal and the southwest Arabian sea. The Bay of Bengal storms move northward first and then northeast and reach the Arakan (along the west coast of Burma) and Chittogong (Bangladesh) coasts by the end of this season. The Arabian sea storms move to the northeast arriving the west coast of Maharashtra and Kerala or they may move northward sometimes arriving at the south Gujarat coast.

Southwest Monsoon Season. The southwest monsoon consists of the two currents which originate in the south Bay of Bengal and the Arabian sea during the hot weather season.

The Bay currents move north or northeast. They are deflected by Arakan hills and the hills of Assam and then by the eastern range of Himalayas. After deflection, these currents which are originally southwesterly will appear as southeasterly or even easterly. Owing to forced ascent over the hills they give very heavy rainfall, for example Chirrapunji in Assam. Monsoon depressions forming in the north or Central Bay and intensifying into cyclonic storms sometimes move to the west-northwest to merge with the seasonal low pressure regions formed in Rajasthan which are called the *western disturbances*. These western disturbances provide the much needed rainfall to the wheat growing region of the country and the snowfall caused by them accumulate to feed the Himalayan rivers in the spring and summer that follow.

The Arabian sea currents as they reach the west coast are forced to ascend the western ghats producing very heavy rainfall in the areas to the west of western ghats and creating a rainshadow region to the east of western ghats. After crossing the ghats they advance over the Deccan Plateau and then Madhya Pradesh producing less rainfall there and then pass into the Bay of Bengal to meet the Bay currents. The other part of the Arabian sea currents crosses the coast of Saurashtra and Kutch and passes over the plains of west Rajasthan, giving little rain in these areas, until it reaches the Aravali hills. It is in this southwest monsoon season that a major portion of the country receives more than 75% of the annual rainfall. However, the southwest monsoon period is not the period of continuous rainfall. There may be breaks of about a week or more with no rainfall activity. The weather is generally cloudy with frequent spells of rainfall and July will be the rainiest month.

Retreating Southwest Monsoon Season. In the post-monsoon period, as the southwest monsoon retreats, storms due to tropical cyclonic disturbances develop in the south or the Central Bay. They are observed to move in any of the following paths:

(*i*) Though they are northeasterly to start with, they may recurve and strike the Arakan coast like the southwest monsoon storms developing in the south Bay have been observed to do.

(*ii*) Moving southwest they may strike the north Tamil Nadu and south Andhra Coast and weaken. (In fact Tamil Nadu and south Coastal Andhra Pradesh regions receive major part of their annual rainfall during this period in contrast to the rest of the country which receives major share of its annual rainfall in the southwest monsoon period.) Then moving through Andhra Pradesh and Orissa they may emerge into the north Bay, intensify there and move to the northeast.

(*iii*) They may strike the Tamil Nadu coast, weaken there and then emerge into the southeast Arabian sea and intensify there. From the southeast Arabian sea they may either move westward or may recurve through the central Arabian sea and strike the south Gujarat coast and then move over to Rajasthan. During this season, the number of cyclones formed in the Bay of Bengal is approximately double the corresponding number in the Arabian sea. They produce intense rainfall along the coasts and cause heavy damage to life and property.

Thus we see that the rainfall over the entire country is mainly confined to southwest monsoon season except the Tamil Nadu and the southern Andhra Pradesh coasts which receive rainfall in the retreating monsoon season due to northeasterly currents. The movements and directions of the monsoon currents also explain the highly non-uniform rainfall distribution over the country ranging from very heavy rainfalls over the west coast and Assam hills and less rainfall over the Deccan Plateau and Central India to almost scanty rainfall over Rajasthan and the Thar deserts.

2.11 METEOROLOGICAL OBSERVATIONS

Periodical measurements on meteorological parameters such as atmospheric pressure, wind velocity and wind direction, temperature, humidity, radiation, sunshine, precipitation and evaporation are essential for the understanding of the atmospheric phenomena. If the observations on some meteorological parameters such as atmospheric pressure, wind velocity and wind direction taken at a concurrent time at different places of the region are plotted at their respective locations on the map, then we can develop what are known as the *synoptic charts*. Such charts would indicate the instantaneous weather conditions. The study of synoptic charts would also help in predicting the activities like cyclones and their movements in advance. The *radar* and *satellite* observations are playing a key role in this regard. Though it is generally difficult to make short term weather predictions, observations over a period of time would render long term predictions on a statistical basis.

Every country institutes an organisation to collect the meteorological data systematically and continuously. In India this responsibility lies with the India Meteorological Department, abbreviated as IMD, with its headquarters at Pune. The WMO, which is under the fold of the United Nations Organisation UNO, engages itself in co-ordinating such agencies, in formulating standards for meteorological data collection and also in providing guidelines for the continued research in the field of meteorology.

The place where all the meteorological instruments are located for collection of the data is called the *meteorological observatory* or the *weather station*. There are specific standards formulated by the IMD for the location, operation and maintenance of these instruments which are to be followed scrupulously. These details can be obtained from the IMD. Generally the observations are taken twice a day, that is, at 08.30 h and at 17.30 h. A brief description of the measurement techniques of the meteorological parameters now follows.

Atmospheric Pressure. The pressure exerted by the atmosphere is a very important characteristic of the atmosphere. This is usually measured by *Fortins Barometer* consisting of an inverted U-tube filled with mercury kept in a cistern, or an *aneroid barometer* which consists of a diaphragm deflected by the atmospheric pressure. The instrument which records the variation of the atmospheric pressure continuously with time is called a *barograph*. The atmospheric pressure is always expressed in millibars. A curve joining points of equal atmospheric pressure is called an *isobar*. With the help of the isobars drawn on the synoptic charts, one can easily locate a region of high pressure or a region of low pressure which in the meteorological terminology are respectively called a *High* and a *Low*.

Example 2.7. *At a place the atmospheric pressure is recorded as 29.8 inches of mercury. Express the same in mb and in N/mm^2.*

Solution. p = 29.8 inches of mercury

= 29.8 × 25.4 mm of mercury

= 756.92 mm of mercury.

Since 1 mm of mercury is equal to 1.33 mb,

p = 756.92 × 1.33 mb

= 1006.7 mb or 100670 N/m^2

= 0.1007 N/mm^2.

Wind Velocity. Air in motion is called the wind. The horizontal component of the air movement parallel to the earth's surface is generally referred to as wind while the vertical components are referred to as the air currents.

Wind velocity is measured by the *cup anemometer*. It consists of 3 or 4 hemispherical cups fixed in a horizontal plane to a sleeve which can freely rotate about a vertical axis. The rotation is transmitted to a gauge, mounted under the cups, which counts the number of revolutions. The counter is so designed that the increase in the reading of the counter divided by the intervening time interval in hours gives the wind velocity directly in km/h. Thus if the counter reading increases by 2.5 in a duration of 3 minutes in an observation, the instantaneous wind velocity at that time of observation is 50 km/h. Similarly the difference in readings of the counter taken at 8.30 h on any day and at 8.30 h of the previous day divided by 24 yields the average wind velocity for the previous day. The *wind direction* is the direction from which it is blowing and it is measured by an instrument called the *wind vane*. As mentioned earlier the data on wind velocity and wind direction are very useful in preparing the synoptic charts. The pattern of atmospheric pressure and wind direction depicted by a synoptic chart and as observed in a cyclone and an anticyclone in the Northern Hemisphere are shown in Fig. 2.8.

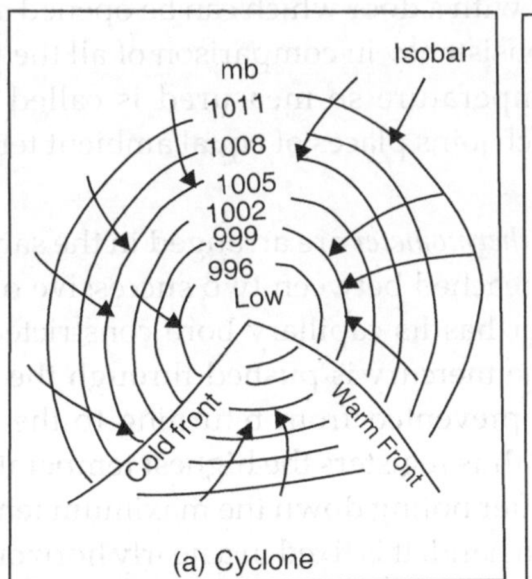

(a) Cyclone

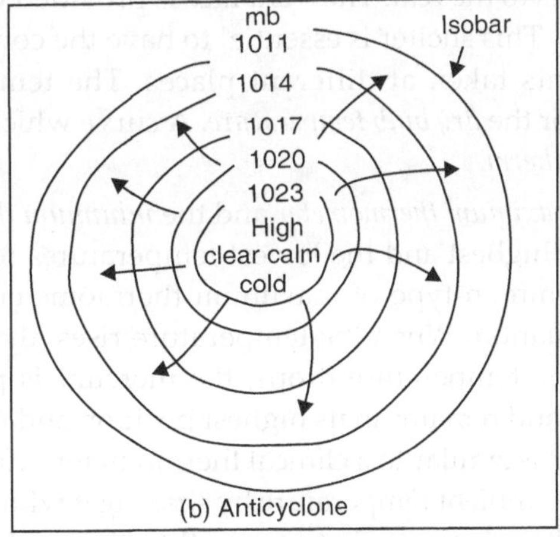

(b) Anticyclone

FIGURE 2.8 Cyclone and anticyclone in northern hemisphere as revealed by synoptic charts.

Data on wind velocity is also required in the evaporation studies. Wind speed varies with the height above the ground. No standard level has been specified for the anemometer. However, the wind speed at any height can be approximately obtained from the known wind speed at the known height of observation. For this purpose, the velocity distribution is assumed to follow either the logarithmic profile or the power-law profile. For most meteorological purposes the power-law profile is usually expressed as

$$\frac{v}{v_0} = \left(\frac{z}{z_0}\right)^{0.15} \qquad \text{...(2.22)}$$

where v_0 is the observed velocity at height z_0 and v is the velocity at any other height z.

Example 2.8. *An anemometer 10 m above the ground recorded a wind velocity of 5.0 m/s. What is the estimated wind velocity at 2 m above the surface ?*

Solution. $z_0 = 10$ m and $v_0 = 5$ m/s

$z = 2$ m $v = ?$

$$\frac{v}{5} = \left(\frac{2}{10}\right)^{0.15} = 0.7855$$

∴ $v = 0.7855 \times 5 = 3.93$ m/s.

Temperature. The temperature is usually measured with mercury-in-glass type of thermometer in degree Celsius (°C). The thermometer is placed in a standard *instrument shelter*, called the *Stevenson Screen*, whose purpose is to see that relatively unobstructed air circulation is there around thermometers and at the same time the thermometers are protected from the direct rays of the sun and from precipitation. The shelter consists of louvered walls made of wood on all the four sides which provide good ventilation and a plain wooden surface at the top and bottom. It is kept about 1.2 m above the ground level. While the bottom surface is horizontal, the top surface slopes to the rear. The front face is provided with a door which can be opened at the time of observation. This shelter is essential to have the consistency in comparison of all the temperature measurements taken at different places. The temperature so measured is called the *ambient temperature* or the *dry bulb temperature*. A curve which joins places of equal ambient temperature is called an *isotherm*.

The *maximum thermometer* and the *minimum thermometer* are arranged in the same shelter to measure the highest and the lowest temperatures reached between two successive observations. The most common type of maximum thermometer has its capillary bore constricted below the lowest graduation. When the temperature rises, the mercury is pushed through the constriction and when the temperature drops, the mercury is prevented from returning to the bulb by the constriction and remains in its highest position and thus registers the highest temperature since its last setting. It is similar to a clinical thermometer. After noting down the maximum temperature, it is reset to the ambient temperature by shaking it with hand. It is fixed in a nearly horizontal position with the bulb end slightly lower than the other end.

The minimum thermometer, usually alcohol-in-glass type, has a small index inside the alcohol column. When the temperature drops, the surface tension force of the meniscus drives the index into its lowest position. When the temperature rises, the alcohol flows fast the index while the index remains at its lowest position to register the lowest temperature reached since its last setting. Once the minimum temperature is recorded, the thermometer is reset to ambient temperature bringing the index back to the meniscus by giving the required tilt. This thermometer should be installed in a perfect horizontal position so that the index does not move by its own weight.

A continuous record of temperature with time can be obtained by an automatic recording instrument called the *thermograph*. The thermograph measures the temperature using the principle that a bimetallic strip changes its shape under the influence of the change in temperature. However, even at the observatories equipped with thermograph, the maximum, minimum and dry bulb thermometers are necessary to supplement and check the record obtained from the thermograph.

The *mean daily temperature* is computed as the arithmetic average of the maximum and minimum temperatures recorded on that day. However, this value may differ slightly from the true daily average obtained from the thermograph chart. The *daily range* in temperature is the difference between the maximum and minimum temperature recorded on a particular day.

The *mean monthly temperature* is computed as the arithmetic average of the mean daily temperatures of all the days in the month. Similarly the *mean annual temperature* for any year is computed as the arithmetic average of the mean monthly temperatures of all the twelve months in that year.

The *normal daily temperature* is the average of the daily mean temperatures for a given date computed for a specific 30 years period. The *normal monthly temperature* and the *normal annual temperature* have similar definitions. For example, the normal monthly temperature for the month of May at any place is the arithmetic average of the mean monthly temperatures of May in the last 30 years at that place. The normal temperatures are revised every decade by deleting the oldest 10 years data and adding the most recent 10 years data.

Humidity. Relative humidity, also known as simply humidity, is usually measured by the *psychrometer*. It consists of two glass thermometers one of them called a *wet bulb* thermometer and the other the *dry bulb* thermometer. The wet bulb thermometer has its mercury bulb wrapped in a wick. The other end of the wick is submerged in a container of distilled water which ensures a continuous moisture supply to the wet bulb thermometer through the effect of surface tension. The dry bulb temperature records the ambient air temperature denoted by T. The thermometers are ventilated by whirling or by the use of a fan. Because of the cooling effect of evaporation the wet bulb thermometer reads a temperature T_w, which is always lower than the dry bulb temperature. The difference between the two, that is $(T - T_w)$, is known as the *wet bulb depression*. From the measured values of the atmospheric pressure p, the temperature T, and the wet bulb depression the relative humidity can be obtained from the psychrometric tables. A psychrometric table corresponding to an atmospheric pressure of 990 mb is shown in Table 2.2.

Table 2.2. Variation of Relative Humidity in per cent with Temperature and Wet-bulb Depression in °C

Atmospheric Pressure = 990 mb

Air temp	*Wet-bulb depression* °C															
°C	0	1	2	3	4	5	6	7	8	9	10	11	12	13	14	15
0	100	81	63	46	29	13										
2	100	84	68	52	37	22	7									
4	100	85	71	57	43	29	16									
6	100	86	73	60	48	35	24	11								
8	100	87	75	63	51	40	29	19	8							
10	100	88	77	66	55	44	34	24	15	6						
12	100	89	78	68	58	48	39	29	21	12	4					
14	100	90	79	70	60	51	42	34	26	18	10	3				
16	100	90	81	71	63	54	46	38	30	23	15	8				
18	100	91	82	73	65	57	49	41	34	27	20	14	7			
20	100	91	83	74	66	59	51	44	37	31	24	18	12	6		
22	100	92	83	76	68	61	54	47	40	34	28	22	17	11	6	
24	100	92	84	77	69	62	56	49	43	37	31	26	20	15	10	5
26	100	92	85	78	71	64	58	51	46	40	34	29	24	19	14	10
28	100	93	85	78	72	65	59	53	48	42	37	32	27	22	18	13
30	100	93	86	79	73	67	61	55	50	44	39	35	30	25	21	17
32	100	93	86	80	74	68	62	57	51	46	41	37	32	28	24	20
34	100	93	87	81	75	69	63	58	53	48	43	39	35	30	26	23
36	100	94	87	81	75	70	64	59	54	50	45	41	37	33	29	25
38	100	94	88	82	76	71	66	61	56	51	47	43	39	35	31	27
40	100	94	88	82	77	72	67	62	57	53	48	44	40	36	33	29

Alternatively, the relative humidity can also be computed from the following equations.

$$e = e'_s - 0.0006606\, p\,(T - T_w)\left(1 + \frac{T_w}{872.78}\right) \quad \text{...(2.23)}$$

$$h = \frac{100e}{e_s} \quad \text{(From Eq. 2.2)}$$

where h is the relative humidity in per cent

e is the actual vapour pressure in mb

p is the atmospheric pressure in mb

e'_s is the saturation vapour pressure in mb corresponding to T_w

e_s is the saturation vapour pressure in mb corresponding to T

T is the dry bulb temperature in °C

T_w is the wet bulb temperature in °C.

The continuous recording of humidity with time is done by an automatic recording instrument called the *hygrograph,* which works on the principle that the human or animal hair reacts to the changes in the air humidity by expanding or contracting. Sometimes the hygrograph is also equipped with thermograph to record the variation of both humidity and temperature on the same graph. Then it is called a *thermohygrograph*. The thermograph and the hygrograph or the thermohygrograph are placed in a shelter similar to Stevenson screen of appropriate size. The dry bulb and wet bulb thermometers are generally installed alongwith the maximum and minimum thermometers in the same shelter.

Example 2.9. *Assuming dry bulb and wet bulb temperatures of 27°C and 17°C respectively, and the atmospheric pressure as 995 mb; determine (a) actual vapour pressure, (b) saturation vapour pressure, (c) relative humidity and (d) dew point temperature.*

Solution. We have T = 27°C and T_w = 17°C

From Table 2.1, the saturation vapour pressure e_s = 35.649 mb and e'_s which is the saturation pressure corresponding to 17°C = 19.367 mb.

$$p = 995 \text{ mb}$$

From Eq. (2.23), the actual vapour pressure e is given by

$$e = 19.367 - 0.0006606 \times 995 \times (27 - 17) \times \left(1 + \frac{17}{872.78}\right)$$

$$= 19.367 - 6.701 = 12.666 \text{ mb.}$$

$$\text{Relative humidity } h = \frac{100e}{e_s} = \frac{100 \times 12.666}{35.649} = 35.53 \text{ per cent.}$$

Again from Table 2.1, temperature corresponding to e_s of 12.666 is read as 10.47°C, which is the dew point temperature.

Example 2.10. *If the air temperature and the wet bulb depression are 30°C and 8.5°C. What is the wet bulb temperature ?*

Solution. Wet bulb depression = $(T - T_w)$ = 8.5°C

Air temperature T = 30°C

$$T_w = T - 8.5 = 30 - 8.5$$

$$= 22.5°\text{C}.$$

Radiation. The shortwave radiation, both direct and diffused, reaching the earth's surface is measured by an instrument called the *pyranometer*. It was formerly called as *pyrheliometer*. But now the term pyranometer is preferred by WMO. This device normally consists of a flat circular plate mounted horizontally within a lime-glass tube. The plate is divided into three portions: (*i*) a central white spot, (*ii*) a middle black ring, and (*iii*) an outer white ring. The black ring absorbs the radiation while the white ring reflects it. Therefore a temperature difference is set up between white and black rings which is measured by means of a mechanism called the thermopile. And a calibrated thermopile provides a measure of shortwave radiation. Reflected shortwave radiation from a surface is measured in the same way by using a downward facing pyranometer. The purpose of lime-glass is to filter out the longwave radiation.

The total incoming or outgoing radiation is measured in the same way but without the filtering effects of the lime glass. The devices is then called a *radiometer*.

Sunshine. The duration of bright sunshine in a day is recorded by an instrument called a *sunshine recorder*. It consists of a hemispherical glass dome underneath which a chart wrapped over a clock driven drum is placed. As the sun's rays are always directed towards the centre of the dome the intense heat developed there will make a black mark by burning the chart when there is bright sunshine. When the sky is cloudy such mark will be absent. The length of the chart which has the burnt mark would indicate the duration of sunshine. The data on sunshine hours is required in the estimation of potential evaporation.

Precipitation. Precipitation is the moisture in any form that reaches the earth's surface from the atmosphere. It is the most important meteorological parameter of interest to the hydrologist. The techniques of measuring precipitation and analysis of precipitation data are discussed in detail in chapter 5.

Evaporation. It is the process by which water from land and water bodies escapes into atmosphere. Evaporation is one of the most complicated hydrologic processes and deserves more elaborate treatment. Hence the physics of the evaporation process and the measurement and estimation of evaporation are dealt with in detail in chapter 7.

REVIEW EXERCISES

Descriptive Questions

1. Define hydrometeorology. Mention some of the hydrological problems in which meteorology plays an important role.
2. Write short notes on the permanent and variable constituents of the atmosphere.
3. Describe the vertical structure of the atmosphere with respect to temperature, with the help of a neat sketch.
4. Explain what do you understood by *green house effect*. Differentiate between the solar constant and insolation.
5. Describe the triple cell general circulation model with the help of neat diagrams.
6. Mention some of the observed circulation patterns over the earth which agree with the triple cell circulation model.
7. Define air mass, air front and the occluded front. Discuss the characteristics of cold air mass and warm air mass.
8. Discuss briefly the classification of the air masses.
9. Write short notes on cyclones and anticyclones.

10. List out the essential requisites for the formation of precipitation and discuss them in detail.
11. What do you understand by *cloud seeding*? How does it help in the artificial rain making?
12. Describe various types and forms of precipitation.
13. Explain how a *rain shadow* regions is formed. Give an example of such a region in India.
14. Explain the weather seasons of India with particular reference to southwest monsoon season.
15. Visit the nearest meteorological observatory in your area and study the working of various meteorological instruments.
16. Distinguish between mean daily temperature and normal daily temperature, mean monthly temperature and normal monthly temperature, and mean annual temperature and normal annual temperature.
17. Collect the data on temperature of your town and compute the normal daily temperature for May 15th, normal monthly temperature for January, and the normal annual temperature.
18. Define isobar and isotherm.
19. Define dew point temperature, wet bulb temperature, wet bulb depression, ambient temperature and daily range.
20. Why is the instrument shelter required for installing thermometers?
21. What is a psychrometer? How is relative humidity obtained from psychrometric tables?
22. Name the instruments, which are used to measure wind direction, atmospheric pressure, shortwave radiation, total radiation and sun shine.

Analytical Questions

1. The atmospheric pressure at a place is recorded as 1.013×10^5 N/m^2. Express the same in mb and mm of mercury.
2. Compute the saturation vapour pressure of air at 20°C and 30°C from Eq. (2.1) and compare them with those from Table 2.1.
3. Show that the density of moist air is always less than that of dry air.
4. Determine the density of air at 35°C, with a relative humidity of 85% when the atmospheric pressure is 980 mb. What is the density of dry air at the same temperature and pressure ?
5. Determine (*i*) the saturation vapour pressure, (*ii*) the actual vapour pressure, (*iii*) saturation deficit, and (*iv*) the dew point temperature of the atmosphere corresponding to a temperature of 28°C and relative humidity of 75%.
6. If the atmospheric pressure, temperature and actual vapour pressure of the atmosphere are 995 mb, 37°C and 1.186 inches of mercury determine the relative humidity, dew point temperature and the density of the atmosphere.
7. What is the relative humidity corresponding to air and dew point temperatures of (*i*) 33°C and 27°C and (*ii*) 29°C and 15°C.
8. Express the radiation energy of 1.25 kW/m^2 in ly/minute.
9. What would be the velocity acquired by a parcel of air at rest relative to the earth's surface at the equator, if it were displaced to 30°N latitude?

10. Express an atmospheric pressure of 0.098 N/mm^2 in mb and mm of mercury.
11. A wind speed of 5.6 m/s is recorded by an anemometer which is at a height of 7.5 m above the ground. An empirical formula for estimating the potential evaporation requires the wind speed at a height of 2 m above the ground. Find this speed.
12. The maximum and minimum temperatures recorded in a day are 32°C and 24.4°C respectively. What is the mean daily temperature and daily range ?
13. A psychrometer recorded the dry bulb and wet bulb temperatures of 41°C and 27°C respectively when the atmospheric pressure was 1001 mb. Determine (*a*) actual vapour pressure, (*b*) saturation vapour pressure, (*c*) relative humidity and (*d*) dew point temperature.
14. What is the wet bulb temperature corresponding to air temperature and wet bulb depression of 38°C and 11°C ?
15. What is the air temperature corresponding to wet bulb temperature and wet bulb depression of 28°C and 7°C ?

Objective Type Questions

1. Hydrometeorology is the science which deals with
   (*a*) water in the atmosphere (*b*) water below the surface of the earth
   (*c*) water in the surface streams (*d*) water in oceans.
2. Relative humidity of the air is defined as the ratio of
   (*a*) actual vapour pressure to the saturation vapour pressure at 0°C
   (*b*) weight of water in unit volume of air to the weight of air in the same volume
   (*c*) actual vapour pressure to the saturation vapour pressure at the air temperature
   (*d*) actual vapour pressure to the atmospheric pressure.
3. A pressure of one mb is equal to
   (*a*) 100 N/m^2 (*b*) 10 N/m^2
   (*c*) 1 N/m^2 (*d*) 0.1 N/m^2.
4. When expressed in mb, the average atmospheric pressure is equal to
   (*a*) 1.013 (*b*) 10.13
   (*c*) 101.3 (*d*) 1013.
5. When the air is saturated with water vapour
   (*a*) the air temperature and dew point temperature are same
   (*b*) the saturation deficit is zero
   (*c*) the dry bulb and wet bulb temperatures are same
   (*d*) all the above.
6. The temperature of the atmosphere remains fairly constant with altitude in
   (*a*) Stratosphere (*b*) Troposphere
   (*c*) Ionosphere (*d*) Mesosphere.

7. Langley is the unit which measures
(*a*) permeability (*b*) radiation
(*c*) infiltration (*d*) thickness of cloud.

8. Isobar is a line which joins the points of equal
(*a*) rainfall depth (*b*) temperature
(*c*) humidity (*d*) atmospheric pressure.

9. Isotherm is a line which joins the points of equal
(*a*) rainfall depth (*b*) temperature
(*c*) humidity (*d*) atmospheric pressure.

10. The ratio of the radiation reflected back by a surface to the radiation received by it is called its
(*a*) radiation coefficient (*b*) absorption coefficient
(*c*) albedo (*d*) Bowen's ratio.

11. IMD is the abbreviation for
(*a*) India Meteorological Department (*b*) Indian Mineral Deposits
(*c*) International Monetary Debt (*d*) Indian Mining Department.

12. In the cyclones of the Northern Hemisphere, wind blows
(*a*) clockwise inward (*b*) anticlockwise inward
(*c*) clockwise outward (*d*) anticlockwise outward.

13. In the anticyclones of the Northern Hemisphere, wind blows
(*a*) clockwise inward (*b*) anticlockwise inward
(*c*) clockwise outward (*d*) anticlockwise outward.

14. The instrument which measures the wind velocity is known as
(*a*) current meter (*b*) atmometer
(*c*) aerometer (*d*) anemometer.

15. Which of the following is different from the rest
(*a*) rain (*b*) drizzle
(*c*) hail (*d*) fog.

16. Which is the odd one in the following
(*a*) snow (*b*) sleet
(*c*) rain (*d*) hail.

17. Rainshadow region in India is found to the
(*a*) west of western ghats (*b*) east of western ghats
(*c*) west of eastern ghats (*d*) south of Himalayas.

18. The instrument which measures the variation of the atmospheric humidity with time is known as
(*a*) Barograph (*b*) Thermograph
(*c*) Hygrograph (*d*) Thermohygrograph.

19. The instrument which measures the variation of the atmospheric temperature with time is known as

(*a*) Barograph (*b*) Thermograph

(*c*) Hygrograph (*d*) Thermohygrograph.

20. The instrument which records the variation of both humidity and temperature with time is called

(*a*) Barograph (*b*) Thermograph

(*c*) Hygrograph (*d*) Thermohygrograph.

21. Pyranometer is the instrument which measures the

(*a*) shortwave radiation (*b*) sun shine hours

(*c*) humidity (*d*) albedo.

22. The convective precipitation is caused when

(*a*) vertical instability of moist air is produced by surface heating

(*b*) a disturbance on the air front develops into a cyclone

(*c*) the colder air mass forms a wedge and lifts the warm air mass

(*d*) the orographic cooling takes place over a mountainous slope.

ANSWERS

Analytical Questions

1. 1013 mb ; 761.6 mm of mercury
2. At 20°C ; from Eq. (2.1) = 23.39 mb ; from Table 2.1 = 23.373 mb
   At 30°C ; from Eq. (2.1) = 42.444 mb ; from Table 2.1 = 42.43 mb
4. 1.107 kg/m^3 ; 1.088 kg/m^3
5. 37.796 mb ; 28.347 mb ; 9.449 mb, 23.15°C
6. 63.84%, 29°C, 1.1015 kg/m^3
7. (*i*) 70.86%, (*ii*) 42.55%
8. 1.79 ly/min
9. 133.66 m/s
10. 980 mb, 736.84 mm
11. 4.592 m/s
12. 28.2°C, 7.6°C
13. 26.105 mb, 77.802 mb, 33.55%, 21.793°C
14. 27°C
15. 35°C

Objective Type Questions

1. (*a*)	2. (*c*)	3. (*a*)	4. (*d*)	5. (*d*)
6. (*a*)	7. (*b*)	8. (*d*)	9. (*b*)	10. (*c*)
11. (*a*)	12. (*b*)	13. (*c*)	14. (*d*)	15. (*d*)
16. (*c*)	17. (*b*)	18. (*c*)	19. (*b*)	20. (*d*)
21. (*a*)	22. (*a*).			

Chapter 3

PROBABILITY AND STATISTICS

3.1 INTRODUCTION

Strictly speaking, in any physical process, there should be a definite relation between the *cause* and the *effect*. Once this relation is known, the outcome of the process can be precisely predicted given the information regarding the factors affecting the process. For example, from the relationship between the head H causing the flow through an orifice and the velocity through the orifice V, given by the equation $V = \sqrt{2gH}$, we can predict the velocity for any given head. Similarly the declination of the sun, the time of sunset and sunrise on any day can be uniquely predicted in advance. Such processes are known as the *deterministic processes.* But in some cases there is an element of uncertainty or unpredictability regarding the outcome. Under seemingly identical conditions , the outcome of the process is different during different observations. For example, though the average weather conditions are same from year to year, the yearly rainfall at a given place is not the same every year and it cannot be predicted uniquely. The reasons for our inability to predict the outcome could be that we may not know all the causative factors, we may not have enough data about the initial conditions of the process, the causative factors are so many in number and so complicated that calculations of their combined effect is not feasible, there may be some indeterminancy in the problem, or some of the causative factors themselves are unpredictable. In all such cases, the process is treated as a *random process,* or *probabilistic process,* or *random phenomenon* wherein we would not try to find out the deterministic relationship which relates the causative factors and the corresponding effects, but instead we treat the process as a *chance phenomenon,* the outcome of which is governed by some *probability law.* Once this probability law is identified, the probability theory can be used to derive the probabilities of the complex events and statistical averages.

Most of the hydrologic processes such as precipitation, runoff, evaporation etc., are rather random processes. Probability and statistics are therefore essentially required to analyse and interpret the observed data of such processes. A brief discussion of these topics is presented here to make the reader familiar with them. For a comprehensive treatment, the interested readers may consult the literature on these topics which is available in abundance.*

3.2 PROBABILITY AND RANDOM VARIABLE

If an experiment is conducted N times, or if the outcome of a process is observed N times, and if a particular attribute A occurs n times, then the limit of n/N as N becomes large is defined as the probability of the event A. The probability of any event is thus a number which always lies between 0 and 1. This is what is known as the frequency interpretation of probability. The frequency interpretation is the one that is most commonly used by the modern statisticians and the engineers.

It may not always be possible or convenient to work out the probabilities only in terms of events because events describe the outcome only qualitatively. A variable may therefore be defined to quantify the outcome of the process. Such a variable associated with a random phenomenon is called a *random variable* since its value cannot be predicted uniquely. In engineering applications, the probabilities defined for random variables are more meaningful than the probabilities for the events.

A random variable is classified into two ways : *discrete random variable* and *continuous random variable*. A discrete random variable is one that can take on only a countable number of values. The examples of discrete random variables in hydrology are the number of storms in a year, the number of rainy days in a given month, the number of floods (exceeding a specified magnitude) in a year etc. On the other hand, a continuous random variable may take on any value in a given interval which is either bounded or unbounded. Continuous random variables result when we are dealing with the measured rather than the counted data. The discharge in a river, the outflow from a lake, the depth to groundwater table, the annual rainfall at a given place etc., are some of the continuous random variables in hydrology. Further discussion here is limited to only the continuous random variables.

A random variable is usually denoted by an upper case letter such as X, while any value it takes is denoted by the corresponding lower case letter x. The behaviour of the random variable is completely described by its probability law which in turn may be characterised by either the *probability density function*, abbreviated as p.d.f. or the *cumulative distribution function*, abbreviated as C.D.F.

Probability Density Function (p.d.f.). The probability that a continuous random variable takes any specific value in its range or interval cannot be assigned. Because in any given interval it can take infinite number of values. Therefore, the probabilities are assigned only for intervals through the p.d.f. Let $f_X(x)$ be the p.d.f. of a continuous random variable X. Then

$$P\left[x_1 < X \le x_2\right] = \int_{x_1}^{x_2} f_X(x) \,.\, dx \qquad ...(3.1)$$

*Interested readers may consult the book *Stochastic Hydrology* by the author published by Laxmi Publications, New Delhi.

The suffix X inthe p.d.f. is only to indicate that it pertains to the random variable X, and hence it will be dropped hereafter. $P[\cdot]$ denotes the probability for the event specified in the parenthesis. It is obvious that the p.d.f. has to satisfy the following two conditions :

(*i*) Since negative probabilities cannot be assigned to any event,

$$f(x) \geq 0 \qquad ...(3.2)$$

(*ii*) Since it is certain that the random variable takes some value or other in its entire range,

$$\int_{-\infty}^{\infty} f(x)\,.dx = 1 \qquad ...(3.3)$$

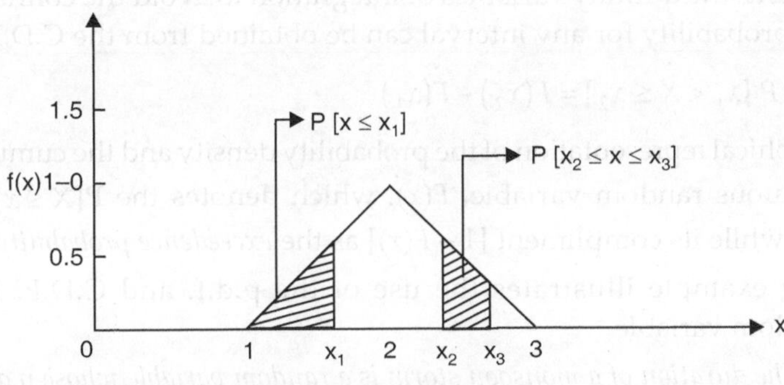

(a) Probability density function

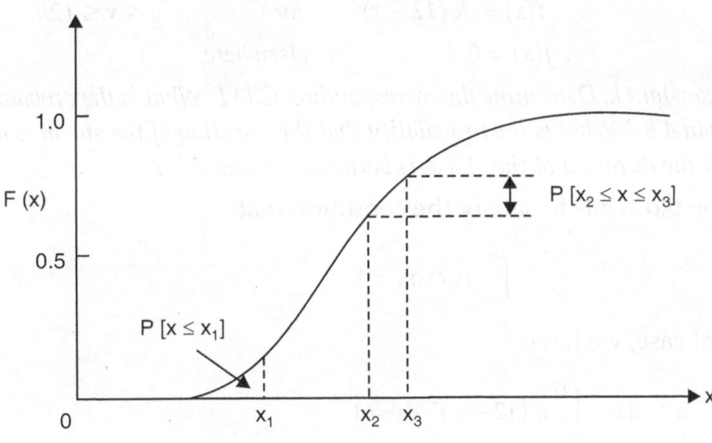

(b) Cumulative distribution function

FIGURE 3.1 Graphical representation of probability density and cumulative distribution functions.

Cumulative Distribution Function (C.D.F.). The cumulative distribution function, also known as the *probability distribution function* sometimes, is an alternative form by which the probability law of the random variable can be specified. If $F_x(x)$ is the C.D.F. of a random variable X, then we have

$$F_x(x) = P[X \leq x] \qquad ...(3.4)$$

Where again the suffix X is to denote that the C.D.F. pertains to the random variable X, and hence it will be dropped hereafter. Without loss of generality the right hand side of Eq. (3.4) may be written as $P[-\infty < X \le x]$ and thus we have that

$$F(x) = P[-\infty < X \le x] \quad \text{...(3.5)}$$

But, from the definition of the p.d.f., we also have that

$$P[-\infty < X \le x] = \int_{-\infty}^{x} f(u)\,.\,du \quad \text{...(3.6)}$$

where u has been used as the dummy variables of integration to avoid the confusion with the limit of integration x. The probability for any interval can be obtained from the C.D.F. as given below.

$$P[x_1 < X \le x_2] = F(x_2) - F(x_1) \quad \text{...(3.7)}$$

Fig. 3.1 is a graphical representation of the probability density and the cumulative distribution functions of a continuous random variable. $F(x)$, which denotes the $P[X \le x]$ is known as the *cumulative probability* while its compliment $[1 - F(x)]$ as the *exceedence probability.*

The following example illustrates the use of the p.d.f. and C.D.F. in evaluating the probabilities of a random variable.

Example 3.1. *The duration of a monsoon storm is a random variable whose p.d.f. is given by*

$$f(x) = kx^2 \quad \textit{for} \quad 0 \le x \le 6h$$

$$f(x) = k(12 - x)^2 \quad \textit{for} \quad 6 \le x \le 12h$$

$$f(x) = 0 \quad \textit{elsewhere.}$$

Evaluate the constant k. Determine the corresponding C.D.F. What is the probability that the duration of the storm is less than 4 h ? What is the probability that the duration of the storm is more than 8 h ? What is the probability that the duration of the storm is between 5 h and 7 h ?

Solution. The p.d.f. has to satisfy the condition that

$$\int_{-\infty}^{\infty} f(x)\,dx = 1$$

In the present case, we have

$$\int_0^6 k\,.\,x^2\,.\,dx + \int_6^{12} k\,(12 - x)^2\,dx = 1$$

$$k\left[\frac{x^3}{3}\right]_0^6 - k\left[\frac{(12-x)^3}{3}\right]_6^{12} = 1$$

$$72k + 72k = 1$$

or

$$k = \frac{1}{144}.$$

The C.D.F. is given by

$$F(x) = \int_{-\infty}^{x} f(u)\,.\,du$$

In the present case, we have

$$F(x) = \int_0^x \frac{1}{144}u^2\,.\,du = \frac{x^3}{432} \quad \text{for } 0 \le x \le 6$$

and $$F(x) = \int_0^6 \frac{1}{144}\,.\,u^2\,.\,du + \int_6^x \frac{1}{144}(12-u)^2\,du$$

$$= 0.5 + \frac{1}{144}\left[72 - \frac{(12-x)^3}{3}\right] \quad \textit{for } 6 \le x \le 12$$

(i) $P\,[0 \le X \le 4]$ $$= \int_0^4 f(x)\,dx$$

$$= \frac{1}{144}\int_0^4 x^2\,.\,dx = 0.148$$

(ii) $P\,[X \ge 8]$ $$= \int_8^{12} f(x)\,.\,dx$$

$$= \frac{1}{144}\int_8^{12} (12-x)^2\,.\,dx = 0.148$$

(iii) $P\,[5 \le X \le 7]$ $$= \int_5^7 f(x)\,.\,dx$$

$$= \int_5^6 \frac{1}{144}\,.\,x^2\,.\,dx + \int_6^7 \frac{1}{144}(12-x)^2\,dx$$

$$= 0.21065 + 0.21065 = 0.4213$$

From the above results it can be interpreted that if we observe 100 monsoon storms, about 15 of them will last for less than 4 *h*, another 15 of them will have durations of more than 8 *h*, and about 42 of them will have duration between 5 *h* and 7 *h*.

The probabilities of the above events can also be worked out from the C.D.F. as follows.

(i) $P\,[0 \le X \le 4] = F(4) = \dfrac{4^3}{432} = 0.148$

(ii) $P\left[X \ge 8\right] = 1 - P\left[0 \le X \le 8\right] = 1 - F(8)$

$$= 1 - \left[0.5 + \frac{1}{144}\left\{72 - \frac{(12-8)^3}{3}\right\}\right]$$

$$= 1 - [0.5 + 0.3518] = 0.148$$

(*iii*) $P[5 \le X \le 7] = F(7) - F(5)$

$$= \left[0.5 + \frac{1}{144}\left\{72 - \frac{(12-7)^3}{3}\right\}\right] - \frac{5^3}{432}$$

$$= [0.5 + 0.21065] - 0.28935 = 0.4213.$$

In the above discussion we have attempted to show that the p.d.f. associated with any random variable offers the complete description of the probabilistic behaviour of the random variable. The next task is to identify the p.d.f. which can satisfactorily describe the random variable and to estimate the parameters which appear in the p.d.f. The observed data on the random variable forms the basis to achieve this task.

Moments and Expectations. Strictly speaking, to predict the behaviour of a random variable one must have the complete knowledge of its p.d.f. But certain descriptors like mean, variance and skewness summarising only the dominant features of the distribution are often very useful and provide considerable information on the behaviour of the random variable. These descriptors are defined in terms of the moments of the distribution. One of the most important single descriptive values is the point around which the distribution is centered and it is taken as a measure of the central tendency. The best known measure of the central tendency is the *expected value,* which is more frequently called as the *arithmetic mean,* or the *average value,* or simply the *mean.* The other commonly used measures of central tendency are the *median* and the *mode.*

The expected value or the mean is nothing but the first moment of the distribution about the origin. It is denoted by either $E[X]$ or μ, and is defined by

$$E(X) = \mu = \int_{-\infty}^{\infty} x \, . \, f(x) \, . \, dx \qquad \text{...(3.8)}$$

The mode of the distribution is defined as that value at which the p.d.f. has the maximum ordinate. Some distributions may have more than one mode. The median of the distribution is defined as a point z, such that

$$\int_{-\infty}^{z} f(x) \, . \, dx = 0.5 \qquad \text{...(3.9)}$$

Therefore, from this definition, the median is that value such that half the area under p.d.f. is to its left and the remaining half to its right.

If the moment is taken about the mean of the distribution instead of the origin, then it is called the *central moment.* The kth central moment is denoted by μ_k and is defined by

$$\mu_k = \int_{-\infty}^{\infty} (x - \mu)^k \, . \, f(x) \, . \, dx \qquad \text{...(3.10)}$$

The second central moment, known as the *variance* of the distribution and denoted by Var. [X], is a measure of the dispersion. The positive square root of the variance is called the *standard deviation* and is denoted by σ. Thus, we have

$$\text{Var}\,.\,[X] = \sigma^2 = \int_{-\infty}^{\infty} (x-\mu)^2 \,.\, f(x) \,.\, dx \qquad \text{...(3.11)}$$

The ratio of the standard deviation to the mean is defined as the *coefficient of variation.*

The ratio of the third central moment to the cube of the standard deviation is known as the *skewness coefficient* which is denoted by γ_1 and is taken to be the measure of the asymmerty of the distribution. Therefore,

$$\gamma_1 = \frac{\mu_3}{\sigma^3} \qquad \text{...(3.12)}$$

A single peaked distribution with $\gamma_1 < 0$ is said to be skewed to right and has a long left hand tail. If $\gamma_1 > 0$, the distribution is skewed to the left with a long right hand tail. For a symmetric distribution $\gamma_1 = 0$, and also mean, mode and median are one and the same. Pearson's skewness coefficient is one of the other measures of asymmetry which is frequently used. It is given by

$$\gamma_1 = \frac{\mu - m}{\sigma} \qquad \text{...(3.13)}$$

where μ, m and σ respectively are the mean, mode and standard deviation. If mode can be accurately determined, this is a convenient definition. The skewness of a distribution is shown graphically for the three cases of negative, zero and positive skewness in Fig. 3.2

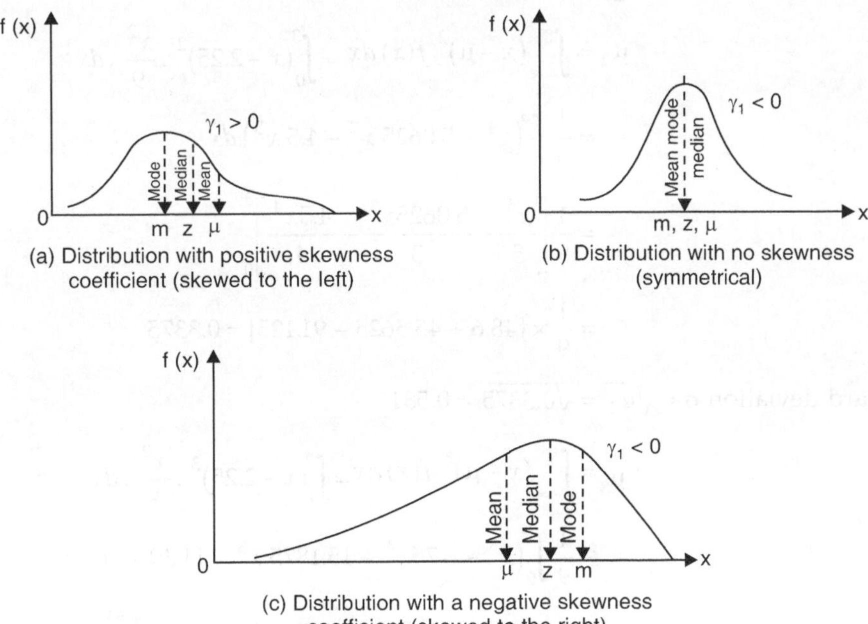

FIGURE 3.2 Definition sketch of the skewness of a distribution.

Example 3.2. *The p.d.f. of a random variable is given by* $f(x) = \frac{x^2}{9}$ *for* $0 \le x \le 3$. *Find the mean, mode, median, standard deviation and skewness coefficient of the distribution.*

Solution. Mean $\mu = \int_{-\infty}^{\infty} x \,.\, f(x)\,.\, dx$

$$= \int_0^3 x \,.\, \frac{x^2}{9} \,.\, dx = \frac{1}{9}\left[\frac{x^4}{4}\right]_0^3 = 2.25$$

In the range of the random variable, the p.d.f. has the maximum ordinate at $x = 3$. Therefore, mode $m = 3$.

The median is obtained from

$$\int_{-\infty}^{z} f(x)\,.\, dx = 0.5$$

$$\int_0^z \frac{x^2}{9}\,.\, dx = 0.5\,;\, \frac{1}{9}\left[\frac{x^3}{3}\right]_0^z = 0.5$$

$$\frac{z^3}{27} = 0.5\,;\ \text{or } z = 2.381$$

$$\mu_2 = \int_{-\infty}^{\infty} (x-\mu)^2 f(x)\, dx = \int_0^3 (x - 2.25)^2 \,.\, \frac{x^2}{9} \,.\, dx$$

$$= \frac{1}{9}\int_0^3 \left(x^4 + 5.0625\, x^2 - 4.5\, x^3\right) dx$$

$$= \frac{1}{9}\left[\frac{x^5}{5} + \frac{5.0625x^3}{3} - \frac{4.5x^4}{4}\right]_0^3$$

$$= \frac{1}{9} \times [48.6 + 45.5625 - 91.125] = 0.3375$$

Standard deviation $\sigma = \sqrt{\mu_2} = \sqrt{0.3375} = 0.581$

$$\mu_3 = \int_{-\infty}^{\infty} (x-\mu)^3 f(x)\, dx = \int_0^3 (x - 2.25)^3 \,.\, \frac{x^2}{9} \,.\, dx$$

$$= \frac{1}{9}\int_0^3 \left(x^5 - 6.75x^4 + 15.1875\, x^3 - 11.39\, x^2\right) .\, dx$$

$$= \frac{1}{9}\left[\frac{x^6}{6} - \frac{6.75\, x^5}{5} + \frac{15.1875\, x^4}{4} - \frac{11.39\, x^3}{3}\right]_0^3$$

$$= \frac{1}{9}[121.5 - 328.05 + 307.547 - 102.51] = \frac{-1.513}{9}$$

$$= -0.168$$

Skewness coefficient $\gamma_1 = \dfrac{\mu_3}{\sigma_3}$

$$= \frac{-0.168}{(0.581)^3} = -0.857.$$

3.3 DISTRIBUTION FUNCTIONS

There are many probability distribution functions which have been found to be useful in describing the hydrologic variables. The most important of them, which are commonly used, are discussed here.

Normal Distribution. A random variable is said to follow the normal (also called the *Gaussian*) distribution when its p.d.f. is given by

$$f(x) = \frac{1}{\sigma\sqrt{2\pi}} e^{\frac{-(x-\mu)^2}{2\sigma^2}} \qquad \text{...(3.14)}$$

where μ and σ are called the parameters of the distribution. By taking the relevant moments, it can be shown that μ and σ are nothing but the mean and standard deviation of the distribution respectively. The distribution curve is bell-shaped and it is symmetrical about the vertical through its mean.

If a random variable X is a result of the addition of *n* other independent random variables, then according to the *Central limit theorem* the variable X tends to follow the normal distribution as *n* tends to be large. Hydrologic random variables related to a large time interval such as annual runoff, annual precipitation, annual evaporation etc., are approximately normally distributed. If daily events such as daily runoff or daily precipitation are considered to be random variables, the corresponding annual events is a result of summation of 365 random variables. Though the daily events are not truly independent, since 365 is fairly a large number, invoking the central limit theorem, annual hydrologic variables are likely to follow the normal distribution.

If a new variable Z is defined as $Z = (X - \mu)/\sigma$, then it can be shown that Z also follows normal distribution with its mean and variance equal to 0 and 1 respectively. The p.d.f. of the variable Z is given by

$$f(z) = \frac{1}{\sqrt{2\pi}} e^{\frac{-z^2}{2}} \qquad \text{...(3.15)}$$

Eq. (3.15) is called the standardised normal distribution function. The corresponding C.D.F. is given by

$$F(z) = \frac{1}{\sqrt{2\pi}} \int_{-\infty}^{z} e^{\frac{-u^2}{2}} \,.\, du \qquad \text{...(3.16)}$$

It is rather difficult to evaluated Eq. (3.16) analytically. But it can be evaluated by employing numerical integration and tabulated as given in Table 3.1. Since the distribution is symmetrical about the mean (which is 0), it will be sufficient if it is tabulated for one half. It is usually tabulated for $z > 0$. Let us suppose we want $F(-z)$. This can be obtained as follows. Because of symmetry the area of the curve between $-\infty$ to $-z$ is same as the area of the curve between $+z$ to $+\infty$. Again, the area between $+z$ to $+\infty$ may be written as [1 – (area between $-\infty$ to $+z$)] since the total area under the curve is 1. But the area of the curve between $-\infty$ to $+z$ is nothing but $F(z)$. Therefore,

$$F(-z) = 1 - F(z) \qquad ...(3.17)$$

Eq. (3.16) can be conveniently utilised to determine the probabilities of any normal variable with any mean and any standard deviation. Consider the inequality

$$x_1 < X \le x_2 \qquad ...(a)$$

Subtracting μ from each term in the inequality and then dividing by σ throughout

$$\frac{x_1 - \mu}{\sigma} < \frac{X - \mu}{\sigma} \le \frac{x_2 - \mu}{\sigma} \qquad ...(b)$$

The inequalities (*a*) and (*b*) are one and the same. Therefore,

$$P[x_1 < X \le x_2] = P\left[\frac{x_1 - \mu}{\sigma} < \frac{X - \mu}{\sigma} \le \frac{x_2 - \mu}{\sigma}\right] \qquad ...(3.18)$$

and the right hand side of the above equation can be readily evaluated as $\left[F(z_2) - F(z_1)\right]$ referring to the C.D.F. table of the standardised normal distribution, where $z_1 = \frac{x_1 - \mu}{\sigma}$ and $z_2 = \frac{x_2 - \mu}{\sigma}$.

Table 3.1. The Cumulative Distribution Function of the Standardised Normal Distribution

$$\mathbf{F(z) = \frac{1}{\sqrt{2\pi}} \int_{-\infty}^{z} e^{\frac{-u^2}{2}} . du; \text{ and } F(-z) = 1 - f(z)}$$

z	*F(z)*	*z*	*F(z)*	*z*	*F(z)*
0.00	0.500	1.10	0.864	2.05	0.980
0.05	0.520	1.15	0.875	2.10	0.982
0.10	0.540	1.20	0.885	2.15	0.984
0.15	0.560	1.25	0.894	2.20	0.986
0.20	0.579	**1.282**	**0.900**	2.25	0.988
0.25	0.599	1.30	0.903	2.30	0.989
0.30	0.618	1.35	0.911	**2.326**	**0.990**
0.35	0.637	1.40	0.919	2.35	0.991
0.40	0.655	1.45	0.926	2.40	0.992
0.45	0.674	1.50	0.933	2.45	0.993

0.50	0.691	1.55	0.939	2.50	0.994
0.55	0.709	1.60	0.945	2.55	0.995
0.60	0.726	**1.645**	**0.950**	**2.576**	**0.995**
0.65	0.742	1.65	0.951	2.60	0.995
0.70	0.758	1.70	0.955	2.65	0.996
0.75	0.773	1.75	0.960	2.70	0.997
0.80	0.788	1.80	0.964	2.75	0.997
0.85	0.802	1.85	0.969	2.80	0.997
0.90	0.816	1.90	0.971	2.85	0.998
0.95	0.829	1.95	0.974	2.90	0.998
1.00	0.841	**1.960**	**0.975**	2.95	0.998
1.05	0.853	2.00	0.977	3.00	0.999

Example 3.3. *The annual runoff of a stream is assumed to follow normal distribution with a mean of 2390 ha-m and a standard deviation of 567 ha-m. Find the probability that the annual runoff in any year is greater than 3000 ha-m. What is the probability that the annual runoff in any year will be between 1800 and 2800 ha-m ?*

Solution. If X denotes the annual runoff, then $\left(\frac{X - 2390}{567}\right)$ is the standardised normal variable.

$$(i)\ P[X \geq 3000] = P\left[Z \geq \frac{3000 - 2390}{567}\right]$$

$$= P\left[Z \geq 1.076\right] = 1 - F(1.076)$$

$$= 1 - 0.859 = 0.141$$

Therefore the probability that annual runoff exceeds 3000 ha-m in any year is 0.141. In other words, on the average in 14 years out of 100 years the annual runoff in the stream will be more than 3000 ha-m.

$$(ii)\ P[1800 < X \leq 2800] = P\left[\frac{1800 - 2390}{567} < Z < \frac{2800 - 2390}{567}\right]$$

$$= P[-1.0406 < Z \leq 0.7231]$$

$$= F(0.7231) - F(-1.0406)$$

$$= F(0.7231) - [1 - F(1.0406)]$$

$$= 0.76493 - (1 - 0.85074)$$

$$= 0.76493 - 0.14926 = 0.616$$

The probability that the annual runoff in the stream in any year will be anything between1800 and 2800 ha-m is 0.616. In other words, on the average roughly in 62 years out of 100 years the annual runoff in the stream will be between 1800 and 2800 ha-m.

Example 3.4. *It is estimated that the annual rainfall at a place exceeds 720 mm with a probability of 0.4, and it exceeds 780 mm with a probability of 0.325. Assuming the annual rainfall to be a normal random variable, determine the probability that it exceeds 900 mm in any year.*

Solution. Let μ and σ be the mean and standard deviation of the annual rainfall.

Let $$z_1 = \frac{720 - \mu}{\sigma} \text{ and } z_2 = \frac{780 - \mu}{\sigma}.$$

Then from the information given in the problem

$$P\,[Z \geq z_1] = 0.4 \text{ and } P\,[Z \geq z_2] = 0.325$$

$$P\,[Z \geq z_1] = 1 - F(z_1) = 0.4$$

$F(z_1) = 0.6$ and from Table 3.1, $z_1 = 0.25$

$$P\,[Z \geq z_2] = 1 - F(z_2) = 0.325$$

$F(z_2) = 0.675$ and from Table 3.1 $z_2 = 0.45$

$$z_1 = \frac{720 - \mu}{\sigma} = 0.25 \text{ or } 720 - \mu = 0.25\,\sigma$$

$$z_2 = \frac{780 - \mu}{\sigma} = 0.45 \quad \text{or } 780 - \mu = 0.45\sigma$$

Solving for μ and σ, we obtain σ = 300 and μ = 645

Let $$z = \frac{900 - \mu}{\sigma} = \frac{900 - 645}{300} = 0.85$$

$$P\,[Z \geq 0.85] = 1 - F\,(0.85) = 1 - 0.802$$
$$= 0.198$$

Therefore, the probability that the annual rainfall is more than 900 mm in any year is 0.198.

Extreme Value (Gumbel's) Distribution. Let $X_1, X_2, \ldots\ldots, X_n$ be n independent random variables and let each of these variables have a distribution whose right tail is unbounded and of exponential type. Let X be another random variable which is the largest of these n random variables. Then it is shown that in the limiting case as n becomes large, the variable X follows the *extreme value type-III* distribution whose p.d.f. is given by

$$f(x) = \alpha e^{-\alpha(x-\beta) - e^{-\alpha(x-\beta)}} \quad \ldots(3.19)$$

where α and β are called the parameters of the distribution. The C.D.F. is given by

$$F(x) = e^{-e^{-\alpha(x-\beta)}} = e^{-e^{-y}} \quad \ldots(3.20)$$

where $y = \alpha(x - \beta)$ is called the *reduced variate*. The mean and standard deviation of the variable μ and σ are related to its parameters through the following equations.

$$\alpha = \frac{1.28255}{\sigma} \quad ...(3.21)$$

$$\beta = \mu - 0.45005\sigma \quad ...(3.22)$$

It may be noted that β is the mode of the distribution.

This distribution was extensively used by Gumbel to describe the annual peak discharges in the rivers. Hence it is also known as the Gumbel's distribution. The use of this distribution to describe the annual peak discharge is justified on the following assumptions.

(*i*) The daily discharge follows an exponential type of distribution.

(*ii*) The original number of elements (365 daily discharges) form which the peak value is taken is sufficiently large.

(*iii*) The daily discharges are independent.

Although the last assumption is strictly not valid, experience has shown that extreme value distribution is a good fit for the annual peak discharges and therefore it is extensively used all over the world.

Example 3.5. *The annual peak discharge of a river follows the Gumbel's extreme value distribution with a mean of 10000 m³/s and a standard deviation of 3000 m³/s. What is the probability that the annual peak discharge is more than 15000 m³/s ? What is the magnitude of the peak discharge with an exceedence probability of 0.1 ?*

Solution. $\mu = 10000 \text{ m}^3/\text{s} \;;\; \sigma = 3000 \text{ m}^3/\text{s}$

$$\alpha = \frac{1.28255}{\sigma} = \frac{1.28255}{3000} = \frac{1}{2339}$$

$$\beta = \mu - 0.45005\,\sigma = 10000 - 0.45005 \times 3000 = 8650.$$

(*i*) When $x = 15000$, $y = \alpha(x - \beta) = \frac{1}{2339}(15000 - 8650) = 2.7148$

$$F(x) = P[X \le x] = e^{-e^{-y}} = e^{-e^{-2.7148}} = 0.9359$$

$$P[X > 15000] = 1 - F(x) = 1 - 0.9359 = 0.0641.$$

Therefore the probability that the annual peak discharge in any year will be more than 15000 m³/s is 0.0641.

(*ii*) The exceedence probability = 0.1

$$\therefore \quad 1 - F(x) = 0.1$$

$$F(x) = 0.9 = e^{-e^{-y}}$$

Taking logarithm to the base *e* twice,

$$y = -\ln[-\ln(0.9)] = 2.25037$$

$$y = \frac{1}{2339}(x - 8650)$$

$$x = 2339\,y + 8650$$
$$= 2339 \times 2.25037 + 8650 = 13913.6 \text{ m}^3/\text{s}.$$

Therefore, the magnitude of the annual peak discharge which is exceeded in any year with a probability of 0.1 is 13914 m^3/s.

3.4 SELECTION OF DISTRIBUTION FUNCTION AND ESTIMATION OF PARAMETERS

So far we have worked out the probabilities of a random variable assuming that its p.d.f. is known to us. In practice, however, we will not be knowing the from of the p.d.f. of the random variable under consideration nor the values of its parameters. The task of identifying the proper distribution function to describe the random variable and the estimation of the parameters of the distribution identified must be accomplished with the help of the data of the random variable observed in the past. Whenever possible the selection of the distribution should be based on an understanding of the underlying physical phenomenon. For example, for describing the annual hydrologic events one can try the normal distribution whereas to describe the annual peak discharges perhaps Gumbel's distribution would be the best choice. Similarly other distributions such as *Log-normal distribution, Gamma distribution* etc., can be justified under certain conditions. Only as a last resort the selection should be done by the trial and error. In a trial and error procedure there are three steps : (*i*) Selection of the distribution, (*ii*) Estimation of the parameter and (*iii*) Testing the goodness of fit of the distribution by the means of statistical tests such as the *chi- square* test or the *Smirnov - Kolmogorov* test. Whichever distribution passes the test best is taken to be the correct distribution to describe the random variable. There are number of methods like graphical method, least square method, method of moments and the maximum likelihood method which are employed to estimate the parameters. The method of moments, which is the most commonly used, simply makes use of the relationships between the parameters and the moments of the distribution.

A collection of n independent observations on a random variable X, which are denoted as $(x_1, x_2, \ldots\ldots\ldots, x_n)$ may be called a *sample.* Any value computed as a function of the sample data may be called a *statistic,* more specifically a *sample statistic.* The mean, standard deviation and the other moments of a sample are defined in a manner analogous to the moments of the distribution. In contrast, the moments of the distribution are known as the *population moments.* The sample mean, $\overline{x}$ is given by

$$\overline{x} = \frac{x_1 + x_2 + \ldots.. + x_n}{n} = \frac{1}{n}\sum_{i=1}^{n} x_i \qquad \ldots(3.23)$$

The sample mean $\overline{x}$, is taken to be an estimate of the population mean μ of the distribution. The sample standard deviation, denoted by s_x, is given by

$$s_x = \sqrt{\frac{1}{(n-1)}\sum_{i=1}^{n}(x_i - \overline{x})^2} \qquad \ldots(3.24)$$

and is taken to be an estimate of the population standard deviation σ of the distribution. The term $(n - 1)$ instead of n in the denominator of Eq. (3.24) is to make the estimate unbiased. The skewness coefficient of the sample, denoted by g_1, is given by

$$g_1 = \frac{n}{(n+1)(n-2)\,s_x^3} \sum_{i=1}^{n} (x_i - \overline{x})^3 \qquad \text{...(3.25)}$$

and is taken to be an estimate of the population skewness coefficient γ_1, of the distribution. The *range* is obtained as the difference between the maximum and minimum values of the sample, while the sample coefficient of variation is obtained as $\left(\frac{s_x}{\overline{x}}\right)$.

After the sample moments are computed from the observed data, the parameters of the chosen distribution are estimated using the relations that exist between the parameters and the moments. For example, in the case of Gumbel's extreme value distribution, these relations are

$$\alpha = \frac{1.28255}{\sigma} \text{ and } \beta = \mu - 0.45005\sigma$$

If a and b are the estimates of the parameters α and β respectively, then they are computed from the following equations.

$$a = \frac{1.28255}{s_x} \qquad \text{...(3.26)}$$

$$b = \overline{x} - 0.45005\, s_x \qquad \text{...(3.27)}$$

where $\overline{x}$, and s_x are the sample mean and sample standard deviation of the observed sample which are computed from Eq. (3.23) and Eq. (3.24) and which are the estimates of μ and σ of the distribution respectively. When the sample size is small (say $n < 200$), Gumbel has suggested some modifications to the values of a and b obtained from Eq. (3.26) and Eq. (3.27). These modifications are discussed later in chapter 14.

Example 3.6. *Following are the 20 observed values of a random variable which constitute a sample. Find the mean, standard deviation and the skewness coefficient of the sample.*

1340 ; 1380, 1450, 618, 523, 508, 1220, 1780, 1060, 412

184, 1480, 876, 113, 516, 1090, 944, 397, 282, 353

Assuming the extreme value distribution as the proper choice to describe the random variable, estimate its parameters and hence find the probability for the variable to exceed 2000.

Solution.

$$\overline{x} = \frac{x_1 + x_2 + \ldots.. + x_n}{n}$$

$$= \frac{1340 + 1380 + \ldots + 282 + 353}{20} = 826.3$$

$$s_x^2 = \frac{(x_i - \overline{x})^2 + (x_2 - \overline{x})^2 + \ldots\ldots + (x_n - \overline{x})^2}{(n-1)}$$

$$= \frac{(1340 - 826.3)^2 + (1380 - 826.3)^2 + \ldots.. + (353 - 826.3)^2}{(20 - 1)}$$

$$= \frac{4722362.2}{19} = 248545.38$$

$$s_x = 498.54$$

$$\sum (x_i - \bar{x})^3 = (1340 - 826.3)^3 + (1380 - 826.3)^3 + \ldots\ldots + (353 - 826.3)^3$$

$$= 0.6741 \times 10^9$$

$$g_1 = \frac{n}{(n-1)(n-2)s_x^3} \sum (x_i - \bar{x})^3$$

$$= \frac{20 \times 0.6741 \times 10^9}{19 \times 18 \times (498.54)^3} = 0.318$$

The parameters of the extreme value distribution are computed as

$$a = \frac{1.28255}{s_x} = \frac{1.28255}{498.54} = \frac{1}{388.71}$$

$$b = \bar{x} - 0.45005\, s_x = 826.3 - 0.45005 \times 498.54$$

$$= 826.3 - 224.37 = 601.93$$

$$P[X \geq 2000] = 1 - P[X \leq 2000] = 1 - F(2000)$$

$$= 1 - e^{-e^{-y}}$$

$$y = a(x - b) = \frac{1}{388.71}(2000 - 601.93) = 3.597$$

$$P[X \geq 2000] = 1 - e^{-e^{-3.597}} = 1 - 0.973$$

$$= 0.027.$$

3.5 FREQUENCY ANALYSIS

The estimation of parameters and the selection of the distribution becomes unreliable when the observed data happens to be very small. In such cases, the probabilities can be obtained from the frequency analysis without bothering about which distribution the random variable follows.

Let $x'_1, x'_2, \ldots\ldots, x'_n$ be the original independent observations that are recorded in the same sequence of their occurrence. Let $x_1, x_2, \ldots\ldots, x_m, \ldots\ldots, x_n$ be the same observations arranged in the descending order of magnitude such that $x_1 \geq x_2 \geq x_3 \geq, \ldots\ldots, \geq x_m \geq, \ldots, \geq x_n$. The suffix m in the ordered sequence of observations is called the *rank* or the *order* of the observation. This, then, means that the highest value among the observations gets a rank of 1, the next highest gets a rank of 2, and so on while the observation with the least magnitude gets a rank n which is the size of the sample.

Let $P(x_m)$ denote the probability with which the value x_m is equalled or exceeded. That is

$$P(x_m) = P[X \geq x_m] \quad \text{...(3.28)}$$

In other words, $P(x_m)$ denotes the exceedence probability. In the ordered sequence if we consider the observation with rank m, this value x_m is equalled or exceeded m time out of n times, and therefore from the frequency interpretation of probability, we can write

$$P[X \geq x_m] = \frac{m}{n} = P(x_m) \quad \text{...(3.29)}$$

If the variable under consideration is an annual event such as the annual peak discharge, or annual runoff or annual precipitation etc., we can say that random variable takes a value equal to or more than x_m on the average once in $\frac{n}{m}$ years, which is variously known as the *recurrence interval, return period* or simply the *frequency*. The return period is usually denoted by T_r. Thus we find that the exceedence probability and the return period are reciprocals to each other.

$$T_r = \frac{n}{m} = \frac{1}{P(x_m)} \quad \text{...(3.30)}$$

The return period, therefore, indicates the average number of years within which a given event will be equalled or exceeded. Eq. (3.30) is called the *California formula*. The disadvantage of this formula is that for $m = n$, it gives an exceedence probability of 1 which should not be the case with a finite sample, The *Hazen's formula* gives the return period as

$$T_r = \frac{2n}{2m-1} \quad \text{...(3.31)}$$

No, doubt, the Hazen's formula gives an exceedence probability of less than 1 for $m = n$, but it gives a return period which is double the size of the sample for $m = 1$. The *Weibul's Formula* which avoids these difficulties and which is most commonly used in Hydrology is as given below

$$T_r = \frac{n+1}{m} \quad \text{...(3.32)}$$

Eq. (3.32) only will be used to determine the return periods in the remaining part of this book.

Now if a plot is made between the value of the random variable as abscissa and the corresponding exceedence probability as the ordinate, the resulting graph is called a *distribution graph* or a *probability plot* or an *empirical distribution*. This graph may be extrapolated or interpolated to obtain the exceedence probability corresponding to a given magnitude of the variable or to obtain the magnitude of the variable corresponding to any specified exceedence probability, without bothering about which distribution the random variable under consideration follows. Results obtained from such an extrapolation are only approximate because the extrapolation is usually influenced by a few end points and the result may differ from person to person considerably.

Therefore one must exercise enough caution when obtaining the results through extrapolation from frequency analysis especially when the analysis is based on shorter length of data. Of course, the best thing would be to identify the appropriate distribution to describe the random variable under consideration, evaluate its parameters and use the distribution to compute the required probabilities, whenever sufficient data is available.

Alternatively a graph may be prepared between the return period in years as the abscissa and the corresponding magnitude of the random variable as the ordinate. This graph can be used to obtain the return period corresponding to any given magnitude or the magnitude of the variable corresponding to any given return period. The graph of return period *Vs* the magnitude is preferred when the variable analysed is the annual peak flood.

Suppose, from the probability plot we read the magnitude of the variable corresponding to an exceedence probability of p. Then this magnitude is called (100 p) per cent *dependable value* of the variable. When the frequency analysis is carried out on the annual rainfall, or the annual runoff, information such as 50% dependable rainfall or 75% dependable rainfall, and 50% dependable runoff yield or 75% dependable runoff yield can be obtained. The following examples demonstrate the application of the frequency analysis in hydrology.

Example 3.7. *The annual rainfalls at a place for a period of 19 years from 1970 to 1988 in mm are given below.*

520, 615, 420, 270, 305, 380, 705, 600, 350, 550

560, 400, 520, 435, 395, 290, 430, 1020, 900.

Construct the frequency curve and hence find the 75% and 50% dependable rainfall. What is the probability that a rainfall of 800 mm or more occurs in any year ?

Solution.

Table 3.2. Frequency Analysis of Example 3.7

Year	*Annual rainfall in mm*	*Annual rainfall in descending order* x_m	*Rank* m	*Return period* $T_r = \frac{n+1}{m} = \frac{20}{m}$	*Exceedence probability* $P(x_m) = \frac{1}{T_r} = \frac{m}{20}$
1970	520	1020	1	20.00	0.05
1971	615	900	2	10.00	0.10
1972	420	705	3	6.67	0.15
1973	270	615	4	5.00	0.20
1974	305	600	5	4.00	0.25
1975	380	560	6	3.33	0.30
1976	705	550	7	2.86	0.35
1977	600	520	—	—	—
1978	350	520	9	2.22	0.45

1979	550	435	10	2.00	0.50
1980	560	430	11	1.82	0.55
1981	400	420	12	1.67	0.60
1982	520	400	13	1.54	0.65
1983	435	395	14	1.43	0.70
1984	395	380	15	1.33	0.75
1985	290	350	16	1.25	0.80
1986	430	305	17	1.18	0.85
1987	1020	290	18	1.11	0.90
1988	900	270	19	1.05	0.95

The given data is entered in columns (1) and (2) of Table 3.2 with the year in column (1) and the annual rainfall in column (2). The annual rainfall values of column (2) are arranged in the descending order of magnitude in column (3). In column (4) the ranks are assigned with a rank of 1 to the highest value, a rank of 2 to the next higher value and so on. Whenever a particular value

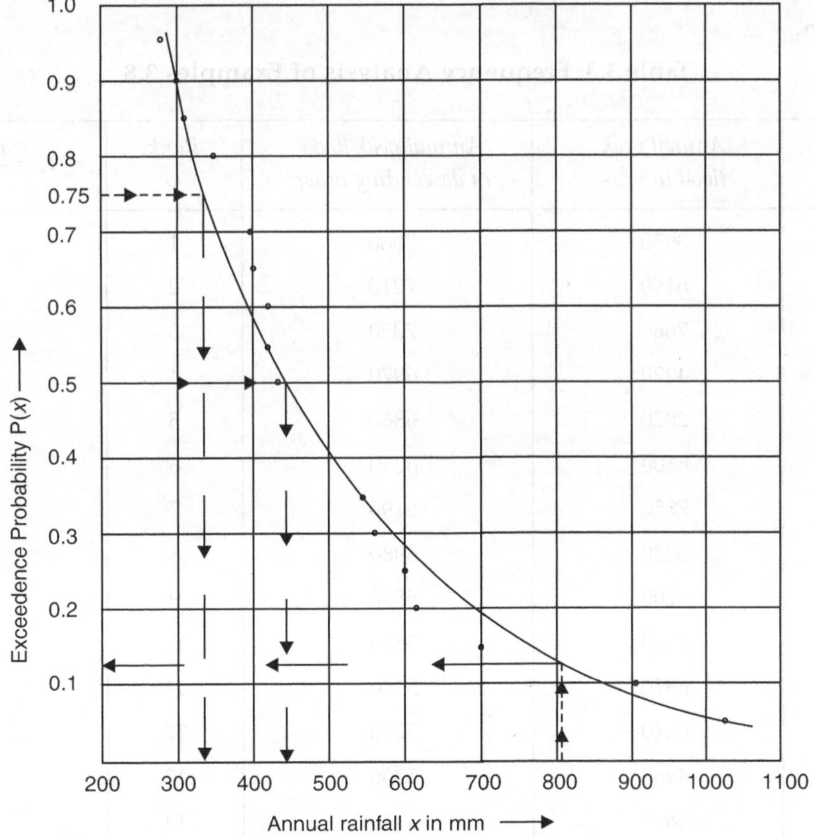

FIGURE 3.3 Probability plot of Example 3.7.

occurs more than once, all of them should have the same rank as they are all equal. In the present example there are two values with same magnitude of 520, which should get ranks of 8 and 9 in the normal course. But both should have same rank. Therefore 520 is assigned a rank of 9. In column (5) the return period of each of the ordered values is computed using the Weibul's formula. In column (6) the exceedence probability is computed as the reciprocal of the return period.

Now a graph is prepared between the annual rainfall of column (3) as the abscissa and the corresponding exceedence probability of column (6) as the ordinate. This graph is shown in Fig. 3.3. From this figure the annual rainfall corresponding to the exceedence probabilities of 0.75 and 0.50 are read as 335 and 450 mm respectively. Therefore the 75% and 50% dependable rainfalls are 335 and 450 mm. In other words, at the given place we can except a rainfall of 335 mm or more in 75 years out of 100 years, and a rainfall of 450 mm or more in 50 years out of 100 years. From the same figure the exceedence probability for a rainfall of 800 mm is read as 0.12.

Example 3.8. *The annual peak flood discharges recorded at a streamgauging site for 19 years during the period, 1961 to 1979 in m³/s are given below.*

3950, 6190, 7660, 4220, 2820, 5600, 7050, 5280, 5200, 4360, 6970, 6240,

4960, 5890, 5980, 3590, 6860, 7210, 5270.

Construct the frequency curve and hence find the flood peak with a return period of 50 years and 100 years.

Solution.

Table 3.3. Frequency Analysis of Examples 3.8

Year	*Annual peak flood in m³/s*	*Annual peak flood in descending order*	*Rank* m	$T_r = \frac{n+1}{m} = \frac{20}{m}$
1961	3950	7660	1	20.00
1962	6190	7210	2	10.00
1963	7660	7050	3	6.67
1964	4220	6970	4	5.00
1965	2820	6860	5	4.00
1966	5600	6240	6	3.33
1967	7050	6190	7	2.86
1968	5280	5980	8	2.50
1969	5200	5890	9	2.22
1970	4360	5600	10	2.00
1971	6970	5280	11	1.82
1972	6240	5270	12	1.67
1973	4960	5200	13	1.54
1974	5890	4960	14	1.43

1975	5980	4360	15	1.33
1976	3590	4220	16	1.25
1977	6860	3950	17	1.18
1978	7210	3590	18	1.11
1979	5270	2820	19	1.05

The required computations are shown in Table 3.3 adopting the same procedure as in Example 3.7. Now a graph is prepared between the return period in years as the abscissa and the annual flood peak in m^3/s as the ordinate as shown in Fig. 3.4. From this graph, the 50 years and 100 years annual flood peak discharges are read as 8200 m^3/s and 8650 m^3/s respectively.

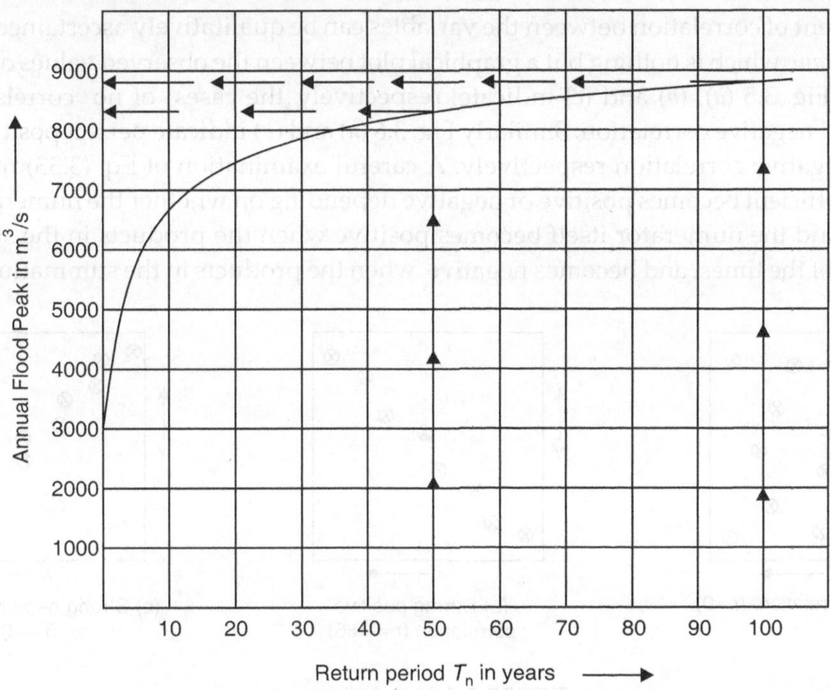

FIGURE 3.4 Frequency graph of Example 3.8.

3.6 CORRELATION

So far we have considered only one random variable. In practice, there are many problems involving two or more variables. If the variability in one random variable is accompanied by the variability in the other, these two variables are said to be *correlated*. For example, we can expect some dependence between annual rainfall over a basin and the corresponding annual runoff, or between the monthly evaporation and the mean monthly temperature etc. The degree of dependence between the variables is assessed through the correlation analysis. The measure of correlation, which is called the *correlation coefficient*, summarises in one number the direction and magnitude of correlation.

Let (x_i, y_i), $i = 1, 2, \ldots, n$ be the n pairs of observations of the random variables X and Y. Then the sample correlation coefficient of these two variables is given by

$$r = \frac{\frac{1}{n}\sum_{i=1}^{n}(x_i - \bar{x})(y_i - \bar{y})}{s_x s_y} \qquad \ldots(3.33)$$

where $\bar{x}$ and s_x are the mean and standard deviation of the variable X, and $\bar{y}$ and s_y are the mean and standard deviation of the variable Y. It can be shown that the value of r always lies between –1 and + 1. When the variables are independent, r approaches zero. When r equals + 1 the variables are said to be *perfectly positively correlated* and when r equals – 1, they are said to be *perfectly negatively correlated.*

The extent of correlation between the variables can be qualitatively ascertained by preparing the *scatter diagram*, which is nothing but a graphical plot between the observed values of the variables. For example, Fig. 3.5 (*a*), (*b*) and (*c*) indicate, respectively, the cases of no correlation, positive correlation and negative correlation. Similarly Fig. 3.6 (*a*) and (*b*) indicate perfect positive correlation and perfect negative correlation respectively. A careful examination of Eq. (3.33) reveals that the correlation coefficient becomes positive or negative depending on whether the numerator is positive or negative. And the numerator itself becomes positive when the products in the summation are positive most of the times, and becomes negative when the products in the summation are negative

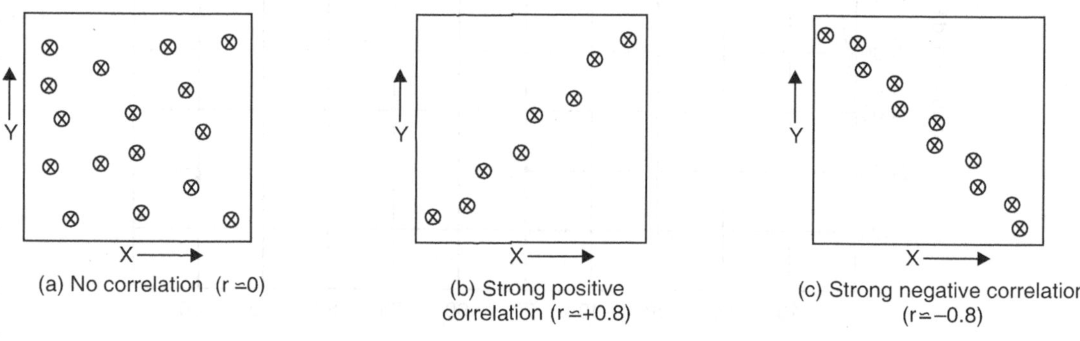

(a) No correlation (r =0)

(b) Strong positive correlation (r=+0.8)

(c) Strong negative correlation (r=–0.8)

FIGURE 3.5 Scatter diagrams.

most of the times. Any product in the summation is positive when x_i which is more than $\bar{x}$ is associated with y_i which is also more than $\bar{y}$ and when x_i which is less than $\bar{x}$ is associated with y_i which is also less than $\bar{y}$. This means that two variables are likely to be positively correlated when higher values of one variable are associated with the correspondingly higher values of the other variable, and lower values of one variable are associated with correspondingly lower values of the other variable. On the other hand, if higher values of one variable are associated with lower values of the other variable and *vice-versa*, the product and hence the numerator becomes negative leading to a negative correlation.

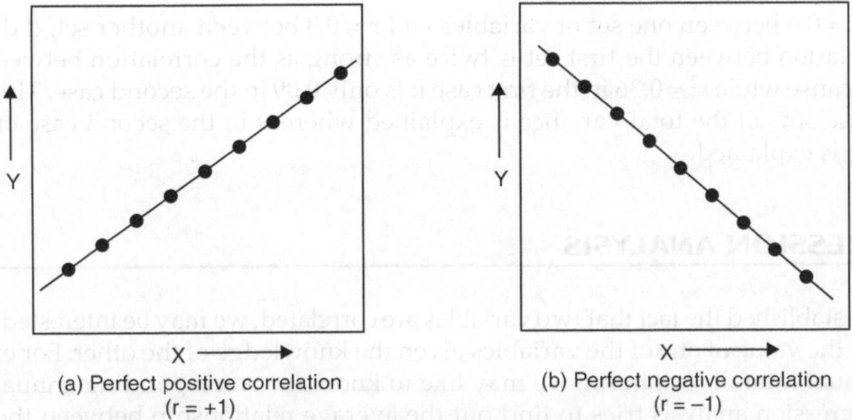

FIGURE 3.6 Perfect correlations.

It is to be noted here, however, that the correlation coefficient as defined by Eq. (3.33) is a measure of linear dependence only. In other words, even when the computed correlation coefficient is very small ($r \cong 0$), it is possible that some from of non-linear dependence may still exist between the variables. For example, consider the scatter diagrams in Fig. 3.7 (*a*) and Fig. 3.7 (*b*). If we compute the correlation coefficient for these cases, we may get a value which is almost equal to zero. But this does not mean at all that the variables are independent. We can only say there is no linear dependence. Thus a scatter diagram will give a better insight into the dependence among the variables than the blind computation of the correlation coefficient. The techniques of measuring the non-linear correlation are far more complicated.

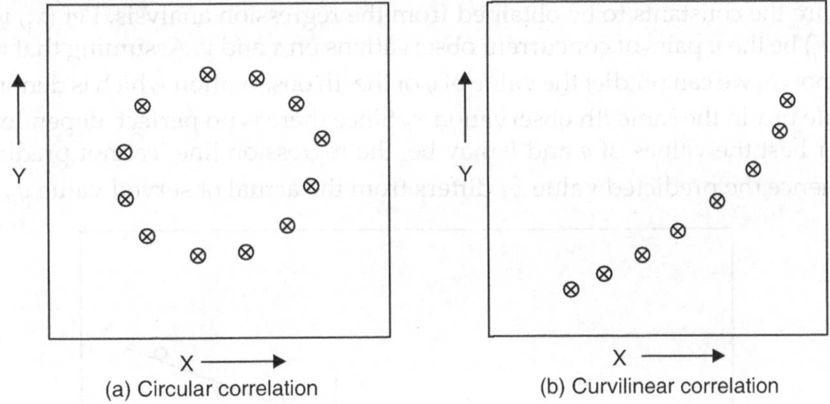

FIGURE 3.7 Scatter diagrams of non-linear correlation.

The coefficient of determination D, which is the square of the correlation coefficient ($D = r^2$), is perhaps more convenient and useful in interpreting the linear dependence among the variables. Because it is a measure of the variance, explained in on variable by its dependence on the other variable. That is,

$$D = \frac{\text{Variance explained}}{\text{Total variance}} \qquad \text{...(3.34)}$$

So, if $r = 0.6$ between one set of variables and $r = 0.3$ between another set, it does not mean that the correlation between the first set is twice as strong as the correlation between the second set. This is because while $r^2 = 0.36$ in the first case it is only 0.09 in the second case. This implies that in the first case 36% of the total variance is explained whereas in the second case only 9% of the total variance is explained.

3.7 REGRESSION ANALYSIS

After having established the fact that two variables are correlated, we may be interested in estimating or predicting the value of one of the variables given the knowledge of the other. For example, if we know the annual rainfall on a basin, we may like to know the corresponding annual runoff from the basin. Regression analysis tries to find out the average relationship between the variables. It refers to the method by which estimates are made of the values of one variable from a knowledge of the values of one or more other variables. Correlation is merely a tool of ascertaining the degree of linear dependence or the degree of association between the variables and, therefore, does not say that one variable is the cause and the other variable is the effect. In the regression analysis, however, one variable is taken as the dependent variable and the other the independent variables thus making it possible to study the cause and effect relationship.

Simple Linear Regression. Linear regression lines are fitted between the variables based on the observed data if it is felt that the relation between the two variables is linear. Let x be the independent variable and y be the dependent variable. We would like to obtain an expression to the dependent variable y as a linear function of the independent variable x. We assume the relationship between y and x as

$$y = a + bx \qquad ...(3.35)$$

where a and b are the constants to be obtained from the regression analysis. Let (x_1, y_1), (x_2, y_2), ..., (x_i, y_i), ..., (x_n, y_n) be the n pairs of concurrent observations on x and y. Assuming that the best values of a and b are known, we can predict the value of y of the ith observation which is denoted by $\hat{y}_i$ from the known value of x in the same ith observation x_i. Since there is no perfect dependence between y and x, however best the values of a and b may be, the regression line cannot predict the value of y exactly and hence the predicted value $\hat{y}_i$ differs from the actual observed value y_i. From Fig. 3.8,

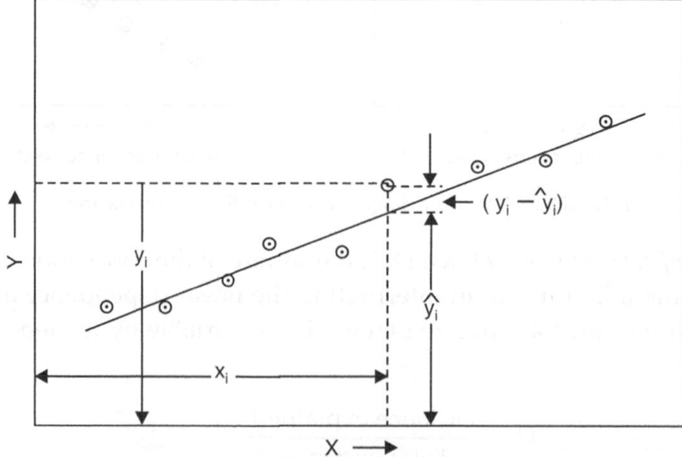

FIGURE 3.8 Regression line.

this deviation equals $(y_i - \hat{y}_i)$. But $\hat{y}_i = a + bx_i$. Hence the deviation becomes $(y_i - a - bx_i)$. Obviously the best regression line is the one which makes these deviations as small as possible. The criterion that is used to obtain the best values of a and b is that the sum of the squared deviations of all the observations shall be minimum.

Therefore, the criterion is

$$\text{Minimise } S = \Sigma\,(y_i - a - bx_i)^2 \quad \text{...(3.36)}$$

where the summation runs over all the observations in Eq. 3.36 and also in all the subsequent equations. The values of a and b which minimise S are given by

$$\frac{\partial S}{\partial a} = 2\,\Sigma\,(y_i - a - bx_i)(-1) = 0 \quad \text{...(3.37)}$$

$$\frac{\partial S}{\partial b} = 2\,\Sigma\,(y_i - a - bx_i)(-x_i) = 0 \quad \text{...(3.38)}$$

Eqns. (3.37) and (3.38) are called the normal equations. Eq. (3.37) gives the solution to a as

$$a = \bar{y} - b\bar{x} \quad \text{...(3.39)}$$

Substitution of Eq. (3.39) in Eq. (3.35) yields a very convenient form of regression line given by

$$(y - \bar{y}) = b(x - \bar{x}) \quad \text{...(3.40)}$$

Eq. (3.40) indicates that the regression line passes through the point of averages, that is through $(\bar{x}, \bar{y})$. The solution for b from Eqs. (3.37) and (3.38) is obtained as

$$b = r\,.\frac{s_y}{s_x} \quad \text{...(3.41)}$$

where r, s_x and s_y are as previously defined. Thus, the *regression coefficient b* retains the same sign as the correlation coefficient r.

Having obtained b from Eq. (3.41), the value of a can be computed either from Eq. (3.39) or Eq. (3.40).

It can be shown (left to the reader to show) that the values of the two regression constants a and b can also be determined is much simpler way from the following equations

$$b = \frac{\Sigma xy - (\Sigma x\,.\,\Sigma y)/n}{\Sigma x^2 - (\Sigma x)^2/n} \quad \text{...(3.42)}$$

and

$$a = \frac{\Sigma y - b\Sigma x}{n} \quad \text{...(3.43)}$$

If the analysis is carried out on the presumption that x is the dependent variable whose value is to be predicted based on the knowledge of the independent variable y, we get similar results as given below.

$$(x - \bar{x}) = b(y - \bar{y}) \quad \text{...(3.44)}$$

$$b = r\,.\frac{s_x}{s_y} \quad \text{...(3.45)}$$

As already pointed out , the regression line cannot make a perfect prediction unless $r = 1$. What is needed, then, is a measure which would indicate how precise is the prediction ? This measure is called the *standard error of estimate*, which is given by

$$S_{yx} = \sqrt{\frac{\sum (y_i - \hat{y}_i)^2}{n}} \quad ...(3.46)$$

which can be simplified to assume the form

$$S_{yx} = \sqrt{(1 - r^2)\, s_y^2} \quad ...(3.47)$$

This shows that higher the value of r^2, the smaller is the standard error of estimate. Since r^2 is defined as the coefficient of determination, the significance of r^2 rather than r can thus be properly appreciated.

Example 3.9. *The annual rainfall and runoff of a watershed, both expressed in centimetres, for a period of 17 years are given below. Do you suspect any dependence between rainfall and runoff ? If so, obtain the regression line to predict the annual runoff from annual rainfall. What is the standard error of estimate for the regression line ?*

Year	Rainfall (cm)	Runoff (cm)	Year	Rainfall (cm)	Runoff (cm)
1961	113	74	1970	119	78
1962	128	104	1971	152	109
1963	127	96	1972	137	96
1964	104	61	1973	165	124
1965	108	59	1974	151	103
1966	115	82	1975	160	134
1967	167	109	1976	130	87
1968	154	102	1977	149	106
1969	99	57	—	—	—

Solution. The computations required to determine the coefficient of correlation are given in Table 3.4 which is self explanatory. The letters x and y are used to denote the annual rainfall and annual runoff respectively.

Now

$$\bar{x} = \frac{1}{n}\sum x_i = \frac{2278}{17} = 134 \text{ cm}$$

$$\bar{y} = \frac{1}{n}\sum y_i = \frac{1581}{17} = 93 \text{ cm}$$

$$s_x = \sqrt{\frac{\sum (x_i - \bar{x})^2}{n}} = \sqrt{\frac{7902}{17}} = 21.56 \text{ cm}$$

$$s_y = \sqrt{\frac{\sum (y_i - \bar{y})^2}{n}} = \sqrt{\frac{7862}{17}} = 21.51 \text{ cm}$$

$$r = \frac{\frac{1}{n}\sum (x_i - \bar{x})(y_i - \bar{y})}{s_x \cdot s_y} = \frac{\frac{7271}{17}}{21.56 \times 21.51} = 0.922$$

$$D = r^2 = (0.922)^2 = 0.85$$

Since the coefficient of correlation is very high, there is every reason to believe that there is a strong linear dependence between the annual rainfall and the annual runoff. The required regression equation is

$$(y - \bar{y}) = b(x - \bar{x})$$

$$b = r \cdot \frac{s_y}{s_x} = 0.922 \times \frac{21.51}{21.56} = 0.92$$

Table 3.4. Regression Analysis of Example 3.9

i	*Rainfall* x_i	*Runoff* y_i	$(x_i - \bar{x})$	$(y_i - \bar{y})$	$(x_i - \bar{x})^2$	$(y_i - \bar{y})^2$	$(x_i - \bar{x}) \times (y_i - \bar{y})$
1	113	74	– 21	– 19	441	361	399
2	128	104	– 6	11	36	121	– 66
3	127	96	– 7	3	49	9	– 21
4	104	61	–30	– 32	900	1024	960
5	108	59	– 26	– 34	676	1156	884
6	115	82	– 19	–11	361	121	209
7	167	109	33	16	1089	256	528
8	154	102	20	9	400	81	180
9	99	57	– 35	– 36	1225	1296	1260
10	119	78	– 15	– 15	225	225	225
11	152	109	18	16	324	256	288
12	137	96	3	3	9	9	9
13	165	124	31	31	961	961	961
14	151	103	17	10	289	100	170
15	160	134	26	41	676	1681	1066
16	130	87	– 4	– 6	16	36	24
17	149	106	15	13	225	169	195
Σ	2278	1581	0	0	7902	7862	7271

$$(y - 93) = 0.92\ (x - 134)$$
$$y = 0.92x - 30.3$$

If we denote the rainfall and runoff by P and R as is done conventionally the required regression equation is

$$R = 0.92\ P - 30.3$$

The standard error of estimate is given by

$$S_{yx} = s_y \sqrt{1 - r^2}$$
$$= 21.51 \times \sqrt{(1 - 0.8464)} = 8.43 \text{ cm}$$

Multiple Linear Regression. The least square regression analysis can be extended to cases where there are more than one independent variables. Let y be the dependent variable and $x_1, x_2, ..., x_p$ be the independent variables. The multiple linear regression equation is then written as

$$y = a_0 + a_1\, x_1 + a_2\, x_2 + \text{..........} + a_p\, x_p. \quad ...(3.48)$$

where $a_0, a_1, a_2, ..., a_p$ are the regression constants to be determined. Denoting the observations on the variable as $(y_i, x_{1i}, x_{2i}, ..., x_{pi})$, the sum of the squared deviations is given by

$$S = \Sigma\ (y_i - a_0 - a_1\, x_{1i} - a_2 x_{2i} - .... - a_p x_{pi})^2 \quad ...(3.49)$$

where the summation runs from $i = 1$ to $i = n$, n being the number of observations. Equating the partial derivatives of this sum with respect to the regression constants to zero, we get the following normal equations.

$$\left.\begin{aligned}
\Sigma\, y_i &= \Sigma\, a_0 + a_1 \Sigma\, x_{1i} + a_2 \Sigma x_{2i} + \text{........} + a_p \Sigma x_{pi} \\
\Sigma\, y_i x_{1i} &= a_0\ \Sigma x_{1i} + a_1\ \Sigma x_{1i}{}^2 + a_2\ \Sigma x_{2i} x_{1i} + \text{......} + a_p\ \Sigma x_{pi} x_{1i} \\
&\vdots \\
\Sigma\, y_i x_{pi} &= a_0\ \Sigma x_{pi} + a_1\ \Sigma x_{1i} x_{pi} + a_2\ \Sigma x_{2i} x_{pi} + \text{......} + a_p\ \Sigma x^2{}_{pi}
\end{aligned}\right\} \quad ...(3.50)$$

The solution of Eq. (3.50) yields the values of $a_0, a_1, a_2, ..., a_p$.

Specifically, when we have only two independent variables (a case of $p = 2$), the equations reduce to the form

$$y = a_0 + a_1\, x_1 + a_2\, x_2. \quad ...(3.51)$$

$$\left.\begin{aligned}
\Sigma\, y_i &= n a_0 + a_1\, \Sigma x_{1i} + a_2\, \Sigma x_{2i} \\
\Sigma\, y_i x_{1i} &= a_o\, \Sigma x_{1i} + a_1\, \Sigma x^2{}_{1i} + a_2\, \Sigma x_{2i} x_{1i} \\
\Sigma\, y_i x_{2i} &= a_o\, \Sigma x_{2i} + a_1\, \Sigma x_{1i} x_{2i} + a_2\, \Sigma x^2{}_{2i}
\end{aligned}\right\} \quad ...(3.52)$$

The three equations in Eq. (3.52) provide the solution for a_0, a_1 and a_2.

Example 3.10. *The yearly fluctuation in the groundwater table is believed to depend on the annual rainfall and the volume of water pumped out from the basin. The data collected on these variables for a period of 10 years is given below. Establish the multiple linear regression equation among the variables.*

Year	*Water table fluctuation cm*	*Annual rainfall cm*	*Groundwater volume pumped out cm*
1971	*26.9*	*70.7*	*4.1*
1972	*– 35.8*	*42.7*	*5.2*
1973	*1.6*	*51.3*	*4.4*
1974	*– 15.4*	*58.0*	*5.6*
1975	*31.0*	*45.3*	*3.9*
1976	*58.8*	*74.7*	*4.3*
1977	*– 0.4*	*62.7*	*5.9*
1978	*–27.1*	*46.7*	*5.2*
1979	*0.7*	*79.3*	*4.4*
1980	*– 4.9*	*34.0*	*6.4*

Solution. Denoting the water table fluctuation by y, annual rainfall by x_1 and volume of water pumped out by x_2, the required regression equation is

$$y = a_0 + a_1 x_1 + a_2 x_2.$$

Table 3.5. Multiple Linear Regression Analysis of Example 3.10

i	y_i	x_{1i}	x_{2i}	$y_i x_{1i}$	$y_i x_{2i}$	$x_{1i} x_{2i}$	x^2_{1i}	x^2_{2i}
1	26.9	70.7	4.1	1901.83	110.29	289.87	4998.49	16.81
2	– 35.8	42.7	5.2	– 1528.66	– 186.16	222.04	1823.29	27.04
3	1.6	51.3	4.4	82.08	7.04	225.72	2631.69	19.36
4	–15.4	58.0	5.6	– 893.20	– 86.24	324.80	3364.00	31.36
5	31.0	45.3	3.9	1404.30	120.90	176.67	2052.09	15.21
6	58.8	74.7	4.3	4392.36	252.84	321.21	5580.09	18.49
7	– 0.4	62.7	5.9	– 25.08	– 2.36	369.93	3931.29	34.81
8	– 27.1	46.7	5.2	– 1265.57	– 140.92	242.84	2180.89	27.04
9	0.7	79.3	4.4	55.51	3.08	348.92	6288.49	19.36
10	– 4.9	34.0	6.4	– 166.60	– 31.36	217.60	1156.00	40.96
Σ	35.4	565.4	49.4	3956.97	47.11	2739.60	34006.3	250.44

The various summations required for the solution of a_0, a_1 and a_2 are obtained in Table 3.5. We have from this table

$$\Sigma y_i = 35.4\ ;\ \Sigma x_{1i} = 565.4\ ;\ \Sigma x_{2i} = 49.4\ ;\ \Sigma y_i x_{1i} = 3956.97$$
$$\Sigma y_i x_{2i} = 47.11\ ;\ \Sigma x_{1i} x_{2i} = 2739.60\ ;$$
$$\Sigma x^2_{1i} = 34006.32\ ;\ \Sigma x^2_{2i} = 250.44.$$

The normal equations are

$$10a_0 + 565.4a_1 + 49.4a_2 = 35.4$$
$$565.4a_0 + 34006.32a_1 + 2739.60a_2 = 3956.97$$
$$49.4a_0 + 2739.60a_1 + 250.44a_2 = 47.11$$

These equations give the solution as

$$a_0 = 47.5\ ;\ a_1 = 0.56 \text{ and } a_2 = -15.3$$

Therefore the required regression equation is

$$y = 47.5 + 0.56x_1 - 15.3x_2$$

Curvilinear Regression. A low value for the correlation coefficient between any two variables indicates the absence of linear dependence. However, there may exist a non-linear relationship between the variables. If the scatter diagram of the two variables shows a tendency for the points to be scattered around a curve rather than around a straight line, it is advisable to analyse the curvilinear relation between the variables. On an ordinary graph paper if the scatter diagram shows a curvilinear trend, in the next step the scatter diagram may be prepared between logarithms of the variables, that is between $\log y$ and $\log x$. If this approaches a straight line the required relation is of the form

$$y = ax^b \qquad \text{...(3.53)}$$

Because, the logarithmic transformation of Eq. (3.53) is a linear equation.

$$\log y = \log a + b \cdot \log x \qquad \text{...(3.54)}$$

Least square regression can now be performed on transformed variables to obtain the best values of $\log a$ and b.

If the scatter diagram of the dependent variable on logarithmic scale against the independent variables on ordinary scale approaches a straight line, then the required curvilinear relation is of the form

$$y = a\, e^{bx} \qquad \text{...(3.55)}$$

Because, the logarithmic transformation of Eq. (3.55) again assumes a linear form

$$\ln y = \ln a + bx \qquad \text{...(3.56)}$$

The linear regression analysis on $\ln y$ and x gives the best values of $\ln a$ and b.

Curvilinear data that do not approach a linear form under either a logarithmic transformation or a semilogarithmic transformation can be fitted by a polynomial relation given by

$$y = a_0 + a_1 x + a_2 x^2 + \ldots\ldots + a_p x^p \qquad \text{...(3.57)}$$

Applying the criterion that the sum of the squared deviations shall be minimum for the case of polynomial fitting, the following normal equations are obtained.

$$\Sigma y = na_0 + a_1 \Sigma x + a_2 \Sigma x^2 + \ldots\ldots\ldots a_p \Sigma x^p$$

$$\left.\begin{aligned} \Sigma xy &= a_0 \Sigma x + a_1 \Sigma x^2 + a_2 \Sigma x^3 + \ldots\ldots\ldots a_p \Sigma x^{p+1} \\ &\vdots \\ \Sigma x^p y &= a_0 \Sigma x^p + a_1 \Sigma x^{p+1} + a_2 \Sigma x^{p+2} + \ldots\ldots\ldots a_p \Sigma x^{2p} \end{aligned}\right\} \quad \ldots(3.58)$$

Eq. (3.58) provides the solution for the constants $a_0, a_1, a_2, \ldots\ldots\ldots, a_p$

The degree of the polynomial p may be determined by trial and error to achieve a stipulated level of accuracy in fitting.

If there are more than one independent variable then it becomes the case of multiple nonlinear regression. In many cases a simple logarithmic transformation may make the multiple curvilinear regression into a multiple linear regression. For example, a relation of the form

$$y = a_0 \, x_1^{a_1} \, x_2^{a_2} \ldots\ldots\ldots x_p^{a_p} \quad \ldots(3.59)$$

may be brought to a linear form after taking logarithms on both sides

$$\log y = \log a_0 + a_1 \log x_1 + \ldots\ldots + a_p \log x_p \quad \ldots(3.60)$$

Now the multiple linear regression analysis is carried out on the transformed variables to obtain the best values of log $a_0, a_1, a_2, \ldots\ldots\ldots, a_p$. As the number of variables increases in the multiple regression, the use of computer becomes inevitable.

REVIEW EXERCISES

Descriptive Questions

1. Differentiate between continuous and discrete random variables. Give few examples of each in hydrology.
2. Explain the role of the probability density function and the Cumulative Distribution Function in working out the probabilities of a random variable.
3. How is the kth moment about the origin of a distribution defined ? How is the kth central moment defined?
4. Define the mean, mode, median, standard deviation and skewness coefficient of a distribution. Sketch the probability density function of a random variable with positive, zero and negative skewness coefficients.
5. Under what conditions a random variable is likely to follow a normal distribution?
6. What is a standardised normal random variable ? What are its mean and variance? What is its probability density function?
7. How do you justify the Gumbel's extreme value distribution to describe the annual peak flood discharge?

8. Explain the step by step procedure involved in the frequency analysis to prepare an empirical probability plot from the sample of observed values.
9. Explain what do you understand by the *return period* ? Give few formulae which are used to determine the return period ?
10. What is meant by 75% dependable value of the random variable ? How is it obtained from frequency analysis ?
11. What is the difference between the correlation coefficient and the regression coefficient ? How are they related ?
12. What do you understand by the coefficient of determination and the standard error of estimate ?
13. A pth degree polynominal is to be established between the variables y and x. Derive the normal equations to obtain the coefficients of the polynominal using the least square regression.
14. What is the form of the regression equation to be tried if the scatter diagram on a semilog paper approaches a straight line ?

Analytical Questions

1. The probability density function of a random variable is given by $f(x) = kx^2$, for $0 \leq x \leq 3$. Find the value of k. What is the probability that the random variable takes a value of (*i*) less than or equal to 1, (*ii*) more than or equal to 2 and (*iii*) between 1.5 and 2.5 ?
2. A random variable is said to be exponentially distributed if its probability density function is given by

$$f(x) = \lambda e^{-\lambda x}, \text{ for } x \geq 0.$$

Find the mean, mode, median, standard deviation, coefficient of variation and the skewness coefficient for this distribution.

3. The probability density function of a random variable is given by

$$f(x) = kx, \text{ for } 0 \leq x \leq 9.$$

Evaluate k and find the mean, standard deviation and the skewness coefficient of the distribution.

4. The annual rainfall of a basin follows a normal distribution with a mean of 80 cm and a standard deviation of 10 cm. What is the probability that the annual rainfall exceeds 100 cm in any year ? What is the 75% dependable rainfall for the basin ? What is the annual rainfall with an exceedence probability of 0.05?
5. A random variable follows the extreme value distribution with a mean of 520 and a standard deviation of 280. What is the probability that the variable exceeds a value of 800 in any observation? What is the value of the variable with an exceedence probability of 0.01?
6. A random variable is estimated to equal or exceed a value of 240 with a probability of 0.02, and to equal or exceed a value of 280 with a probability of 0.01. Find the mean and standard deviation of the variable if it is assumed to follow the extreme value distribution. What is the value of the variable with an exceedence probability of 0.005?

7. The observed annual peak flood of a river in m^3/s for a period of 20 years from 1961 to 1980 are given below.

190, 155, 298, 136, 137, 131, 140, 124, 185, 104

91, 154, 109, 269, 164, 270, 142, 72, 130, 111.

Compute the mean, standard deviation and skewness coefficient for this sample.

8. For the data of annual peak flood discharges given in analytical question 7, prepare a graph of flood peak *versus* the return period and hence estimate the annual peak flood with a return period of 30 years.

9. The concurrent observations on annual rainfall and annual runoff of a basin in centimetres for a period of 15 years are given below.

Year	*Rainfall*	*Runoff*	*Year*	*Rainfall*	*Runoff*
1971	106	48	1979	88	20
1972	96	30	1980	74	17
1973	95	33	1981	101	34
1974	100	29	1982	69	21
1975	66	14	1983	55	18
1976	45	9	1984	95	49
1977	83	22	1985	78	30
1978	83	21			

Determine the correlation coefficient between rainfall and runoff and fit a linear regression line between rainfall and runoff. What is the standard error of estimate ?

10. The duration of the storm D, the storm rainfall P and the corresponding runoff R are given in the table below for 10 storms which occurred on a basin. Obtain a multiple regression equation for the runoff R in terms of storm rainfall and storm duration.

Storm duration D in hrs	*Storm rainfall P in cm*	*Storm runoff R in cm*
12	5.2	3.1
18	5.0	2.7
9	3.8	1.2
12	4.6	0.8
12	2.7	0.3
12	2.9	0.8
15	5.0	2.5
20	3.3	1.4
15	4.0	1.4
9	2.7	0.8

11. A random variable X has a probability density function $f(x)$ with a mean and standard deviation of μ and σ respectively. Another random variable Z is defined by $Z = (X - \mu)/\sigma$. Show that the new random variable Z has a mean and standard deviation of 0 and 1 respectively.
12. The Cumulative Distribution Function of a random variable X is given by $F(x) = kx$, for $0 \le x \le 9$. Find its probability density function.

Objective Type Questions

1. Which of the following is not a continuous random variable ?
   (*a*) Annual rainfall (*b*) Annual runoff
   (*c*) Number of rainy days (*d*) Wind velocity.
2. Which of the following is not a discrete random variable ?
   (*a*) Number of rainy days in a year (*b*) Annual peak flood
   (*c*) Number of cyclones in a year (*d*) Number of floods.
3. The unbiased estimate of the standard deviation is given by

   (*a*) $\sqrt{\dfrac{\Sigma(x_i - \bar{x})^2}{(n-1)}}$ (*b*) $\sqrt{\dfrac{\Sigma(x_i - \bar{x})^2}{n}}$

   (*c*) $\sqrt{\dfrac{\Sigma x_i^2 - \bar{x}^2}{n}}$ (*d*) $\sqrt{\dfrac{\Sigma x_i^2}{(n-1)}}$
4. The most appropriate distribution to describe any annual hydrologic variable is
   (*a*) Gumbel's distribution (*b*) Chi-squared distribution
   (*c*) Gamma distribution (*d*) Normal distribution
5. Gaussian distribution is another name for
   (*a*) Gumbel's distribution (*b*) Gamma distribution
   (*c*) Normal distribution (*d*) Log-normal distribution.
6. In the linear regression line $x = a + by$, the regression constant b is given by (r being the correlation coefficient between x and y)

   (*a*) $r \cdot \dfrac{s_x}{s_y}$ (*b*) $r \cdot \dfrac{s_y}{s_x}$

   (*c*) $\dfrac{r}{s_x}$ (*d*) $\dfrac{r}{s_y}$.
7. For a distribution which is symmetrical about mean, the skewness coefficient is
   (*a*) positive (*b*) negative
   (*c*) zero (*d*) sometimes positive and sometimes negative.

8. The Weibul's formula for the return period is given by

(a) $T_r = \dfrac{m+1}{n}$ (b) $T_r = \dfrac{n+1}{m}$

(c) $T_r = \dfrac{n-1}{m}$ (d) $T_r = \dfrac{n}{m-1}$

9. If x_1, x_2 and x_3 have the return periods of 50, 75 and 100 years, which of the following statements is true

(a) $x_1 < x_2 < x_3$ (b) $x_1 > x_2 > x_3$

(c) $x_1 < x_2$ and $x_2 > x_3$ (d) $x_1 < x_2$ and $x_2 < x_3$.

ANSWERS

Analytical Questions

1. $k = \dfrac{1}{9}$; (i) $\dfrac{1}{27}$ (ii) $\dfrac{19}{27}$ (iii) $\dfrac{12.25}{27}$

2. Mean $= \dfrac{1}{\lambda}$; mode = 0 ; median $= \dfrac{\text{In } 2}{\lambda}$

standard deviation $= \dfrac{1}{\lambda}$; coefficient of variation = 1 ;

skewness coefficient = 2

3. $k = \dfrac{2}{81}$; mean = 6 ; mode = 9; median = 6.364 ;

standard deviation = 2.1213 ; skewness coefficient = – 0.5657

4. 0.023 ; 73.25 cm ; 96.45 cm
5. 0.218 ; 1498.0
6. $\mu = 49.487$; $\sigma = 73.49$; 319.87
7. 155.6 m^3/s ; 60.59 m^3/s; 1.223
8. 340 m^3/s (approximately)
9. $r = 0.7476$; $R = 0.4822\ P - 28.19$; 7.624 cm
10. $R = 0.7513\ P + 0.046\ D - 2.0615$
12. $f(x) = \dfrac{1}{9}$, for $0 \le x \le 9$.

Objective Type Questions

1. (c)	2. (b)	3. (a)	4. (d)	5. (c)
6. (b)	7. (c)	8. (b)	9. (a)	

Chapter 4 THE CATCHMENT

4.1 INTRODUCTION

Quantitative assessment of hydrological parameters like precipitation, evaporation, infiltration and runoff, and their use in water balance studies or in the problems of design and forecasting will be rational only when they are applied to an area with well defined boundaries. Though the precipitation, evaporation and infiltration take place everywhere on the land surface, the resulting runoff passes through but one point on the stream. Therefore, it is only reasonable to take this area such that all the runoff generated from it flows only through the point at its outlet. Accordingly, the *catchment* (also known as the *drainage basin* or the *river basin*) is defined as the area drained by a stream or a system of connecting streams such that all the surface runoff originating in this area leaves the area in a concentrated flow through a single outlet. In American literature, the term drainage basin is commonly used to mean the area enclosed by the boundaries of the surface runoff system while its British equivalent is the catchment. In this book the terms drainage basin and the catchment are used synonymously to mean the boundary of a surface runoff system.

The line which divides the surface runoff between two adjacent river basins is called the *topographic water divide*, or the *watershed divide*, or simply the *divide*. The divide follows the ridge line around the basin crossing the stream only at the outlet point. It marks the highest points between the basins, but isolated peaks within a basin may reach higher elevations than any point on the divide. When a detailed topographic map is available, the basin area for any outlet point on the river can be very easily fixed by searching for the ridge points. Alternatively, the drainage area, especially for large river basins, can also be delineated by taking the boundary curve approximately midway between the contributing streams as shown in Fig. 4.1.

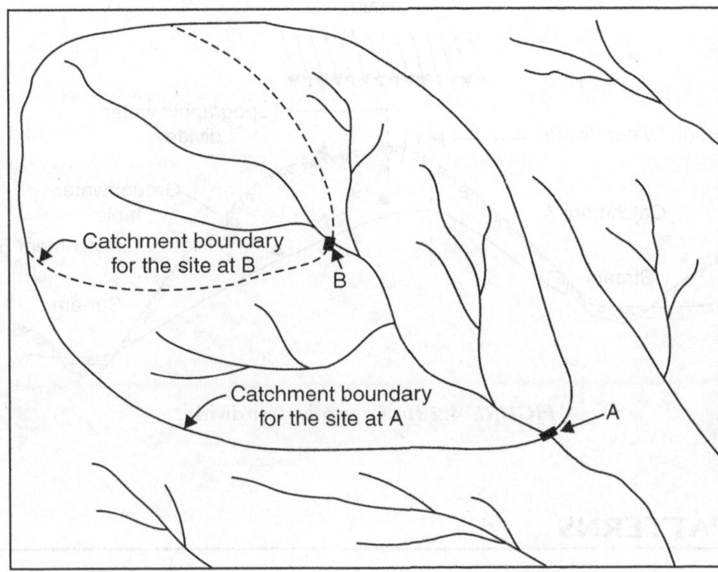

FIGURE 4.1 Topographic map showing location of a divide.

Sometimes, owing to the underlying geology, the runoff measured at the outlet of a particular basin may contain some contribution belonging to the precipitation fallen on a neighbouring basin by way of subsurface runoff as shown in Fig. 4.2 (*a*). Then the *watershed leakage* is said to occur.

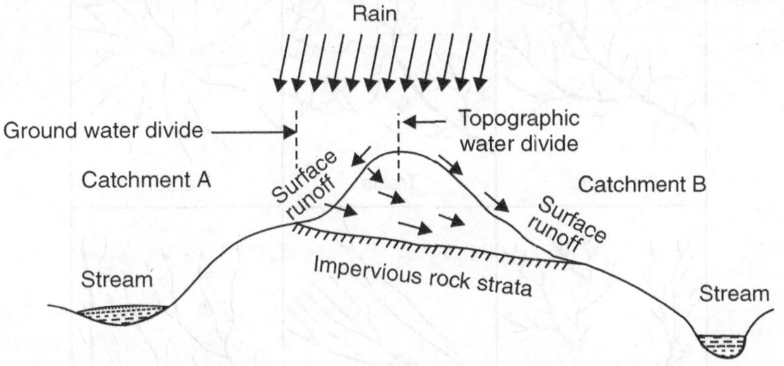

FIGURE 4.2 (*a*) Topographic divide.

Watershed leakage also occurs when the *groundwater divide* also known as *phreatic divide* and the topographic divide are not coincident in plan as shown in Fig. 4.2 (*b*). The subsurface runoff or groundwater runoff generated by the rain falling between the topographic divide and the groundwater divide of Fig. 4.2 (*a*) or Fig. 4.2 (*b*) will flow into the stream of catchment *B*, though it actually belongs to catchment *A*. It may, however, be noted here that it is difficult to fix the boundaries of the subsurface runoff system as the position of the groundwater table changes from time to time. In general, if the two streams flow more or less parallel, leakage is likely to occur from higher basin to lower basin. Even though the topographic divide and the groundwater divide do not coincide, the resulting discrepancy in the runoff measured at the basin outlet is generally not of a serious magnitude.

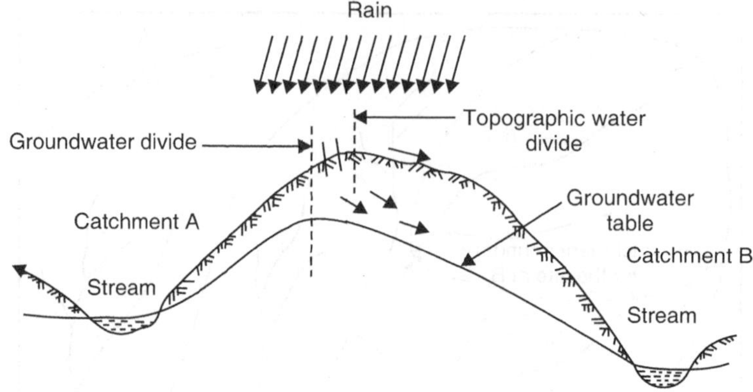

FIGURE 4.2 (*b*) Groundwater divide.

4.2 STREAM PATTERNS

The combined effects of climate and geology on the catchment topography yield an erosion pattern which is characterised by a network of channels or streams. Some of the most frequently observed stream patterns are shown in Fig. 4.3.

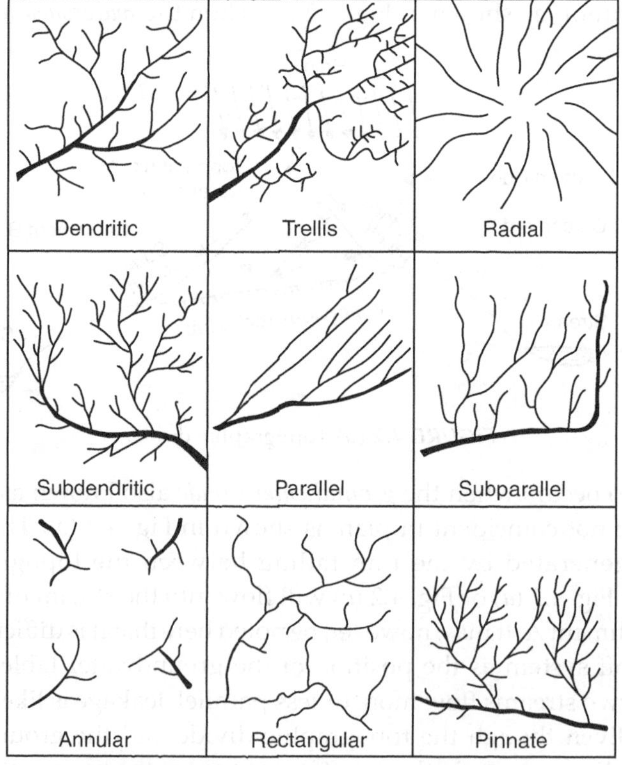

FIGURE 4.3 Commonly observed stream patterns.

When a region is homogeneous offering no variation in the resistance to the flow of water, the resulting streams run in all directions with no definite preference to any one particular direction. Then the stream pattern developed is called *dendritic* or *tree-like*.

The *trellis* drainage pattern develops when the underlying rock is strongly folded or sharply dipping. The longer streams will have preference to one particular orientation and the other tributaries will have an orientation at right angles to this.

The drainage pattern from dome mountains and volcanoes is of *radial* type where the streams emanate from a central focus and flow radially outward.

Owing to the Internal geologic structure of the land, sometimes, the *parallel* and *subparallel* drainage patterns are formed.

The streams which form in the weaker strata of a dome mountain indicate approximately circular or *annular* pattern. The annular pattern may be treated as a special form of the trellis pattern.

A region consisting of many rectangular joints and faults may produce a *rectangular* drainage pattern with streams meeting at right angles.

In a *pinnate* stream pattern, all the main streams run in one direction with the tributaries (resembling dendritic pattern) joining them at an oblique angle.

4.3 DESCRIPTION OF THE BASIN

The basin characteristics such as the area, land-surface topography and other morphological properties vary only with respect to geological time and thus may be treated as constant. The terminology with which the hydrologists describe a drainage basin is briefly presented here. For more detailed information the reader is referred to any good text book on geomorphology. There are many parameters used by the hydrologists to describe the drainage basin. The most common of them are discussed below.

Area. The area of the drainage basin is defined as the area of the closed curve which is obtained by projecting the catchment boundary on to a horizontal plane. It is usually determined by planimetering from reasonably large scale maps and is expressed in square kilometres. When the basin area is less than 25 km^2, it is expressed in hectares. The conversion to be used is 1 km^2 = 100 hectares.

Stream Order. The stream order is a classification reflecting the degree of branching or bifurcation of the stream channels within a basin. Assuming that one has a channel network map of the basin, the smallest fingertip tributaries are given the order 1. When two channels of order 1 join, a channel segment of order 2 is formed. When two channels of order 2 join, a channel segment of order 3 is formed and so on and so forth. This ordering of streams is shown in Fig. 4.4. The trunk of the stream through which all the discharge of water and sediment passes is therefore the stream segment of the highest order. Thus, the order of the main stream indicates the extent of branching in the basin.

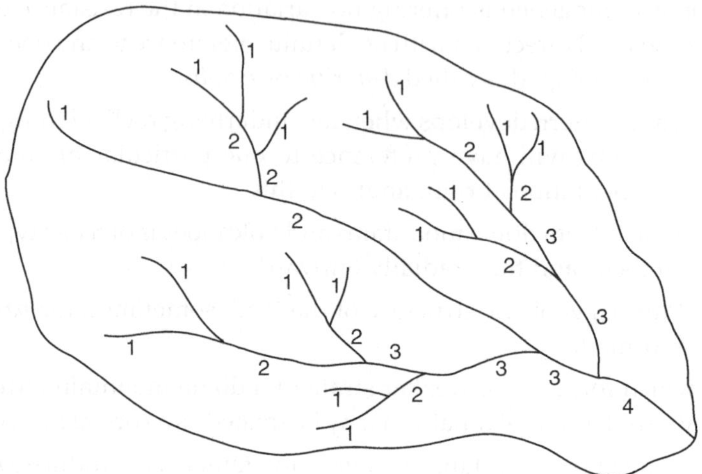

FIGURE 4.4 Ordering of streams.

Drainage Density. It is defined as the ratio of the total length of streams of all orders within the basin to the basin area.

$$D_d = \frac{\Sigma L}{A} \quad ...(4.1)$$

The drainage density may be thought of as a measure of the closeness of the spacing of the stream channels. Low drainage density reflects poor drainage conditions in the watershed. Low drainage densities are observed where soils are resistant to erosion or very permeable and where the slopes are small. Higher drainage densities are expected where soils are easily eroded or relatively impermeable, slopes are steep and vegetal cover is scanty.

The average length of *overland flow* $\overline{L_o}$ is taken to be half of the reciprocal of the drainage density. That is

$$\overline{L_o} = \frac{1}{2D_d} \quad ...(4.2)$$

Stream Density. The stream density D_s, also known as the *stream frequency*, is obtained as the ratio of the number of streams N_s to the area of the basin. That is

$$D_s = \frac{N_s}{A} \quad ...(4.3)$$

Length. It is the length measured along the main-stream from the catchment outlet to the remotest point on the catchment boundary. The main stream is distinguished by starting at the outlet and following the stream of highest order. When a bifurcation into two streams of equal order is reached, the branch having the larger catchment is followed.

Shape. The shape of the basin is quantitatively measured by various factors such as the *form factor, circularity ratio, elongation ratio* and *compactness coefficient*.

The form factor R_f is defined as the ratio of the basin area to the square of the basin length. That is

$$R_f = \frac{A}{L^2} \quad ...(4.4)$$

The circularity ratio R_c is defined as the ratio of the basin area to the area of the circle whose perimeter is equal to the perimeter of the basin. Thus, if P is the perimeter of the basin

$$R_c = \frac{4\pi A}{P^2} \quad ...(4.5)$$

The elongation ratio R_e is defined as the ratio of the diameter D_e of a circle whose area is same as the area of the basin to the length of the basin. That is

$$R_e = \frac{D_e}{L} = \frac{2}{L}\sqrt{\frac{A}{\pi}} \quad ...(4.6)$$

The compactness coefficient C_c is defined as the ratio of the perimeter of the basin to the perimeter of a circle whose area is equal to the area of the basin. That is

$$C_c = \frac{P}{\sqrt{4\pi A}} \quad ...(4.7)$$

It may be observed that the circularity ratio is nothing but the reciprocal of the square of the compactness coefficient.

$$R_c = \frac{1}{C_c^2} \quad ...(4.8)$$

Thus the circularity ratio and the compactness coefficient are not truly independent parameters.

Relief. The maximum basin relief H is the elevation difference between the basin outlet and the highest point on the basin perimeter usually expressed in metres.

Slope. There are many ways of defining the slope of a basin. A simple way of obtaining it is to divide the difference between the elevations of the highest point on the basin perimeter and the basin outlet by the distance between these two points.

The slope of the stream can also be obtained from the *stream profile*. A graph representing the horizontal distance along the mainstream *versus* the elevation is known as the stream profile. The slope of the stream may be taken as the total fall between the end points of the main stream divided by the length of the main stream. The slope of the line *ab* in Fig. 4.5 represents this. A more realistic definition of the stream slope is given by the line *cb*, which is drawn so as to have the same area under it as does the stream profile.

If the channel is divided into n equal segments with slopes of $s_1, s_2, s_3, \ldots\ldots\ldots, s_n$, the equivalent slope of the channel is obtained from the equation

$$s = \left[\frac{1}{n}\sum_{i=1}^{n} s_i^{1/2}\right]^2 \quad ...(4.9)$$

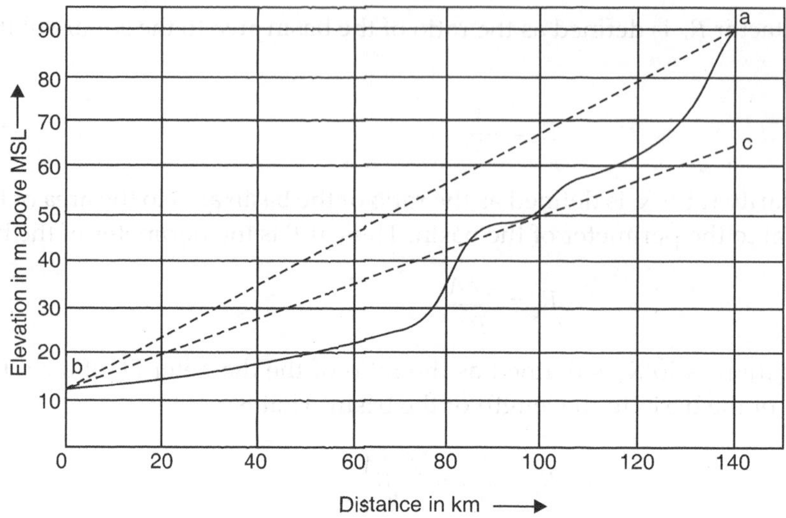

FIGURE 4.5 Stream profile.

Area-Length Relations. Data for a number of the larger rivers all over the world indicated the following relation between area and length.

$$L = 1.273\, A^{0.6} \qquad \ldots(4.10)$$

where L is the main channel length in km and A is the area of the basin in km^2. Eq. (4.10) suggests that a basin tends to elongate as it grows larger.

4.4 CLASSIFICATION OF WATERSHEDS

Based on area the watersheds are classified as small, medium and large. A watershed is considered to be small if its area is less than 250 km^2. A medium watershed will have the drainage area between 250 km^2 and 2500 km^2. When area of the watershed exceeds 2500 km^2, it is considered to be large. These limits should be considered as rough guidelines and may vary from one geographical area to another.

Depending on the land use the watershed may be referred to as urban, agricultural, mountainous, forest, desert, coastal or marsh watershed or a combination of two or more of them. The hydrological behaviour of these watershed may be so distinctly different that they have given rise to different branches of hydrology such as urban hydrology, forest hydrology, desert hydrology etc.

4.5 CLASSIFICATION OF STREAMS

Whenever it occurs, the rainfall over a catchment area will be usually confined to few hours to few days. But the flow in the streams will be found for much longer periods. This is mainly due to groundwater runoff and in some cases due to melting of snow. Based on the constancy of flow in the streams, they are classified into three types namely perennial or continuous, intermittent and ephemeral.

In a perennial stream, the flow in the stream is found the year round except during the periods of extreme drought. The flow is maintained by the ground water flow to the stream. The bottom of the channel is below the ground water table and as a result the groundwater moves to the channel. A perennial stream gains discharge from groundwater flow and it is therefore called a gaining stream. (See also Article 9.2)

In an intermittent stream the channel bed is intermittently below and above groundwater table. When the channel bed is below it gains discharge from the groundwater flow and when the channel bed is above it looses discharge to the groundwater and becomes dry. Therefore an intermittent stream behaves alternatively as gaining stream and a loosing stream.

An ephemeral stream is one which flows only during runoff producing storm events. The bed of an ephemeral stream is always above the groundwater table. Because of this condition, water in the stream infiltrates from the stream bed during the flow events and recharges the groundwater below. An ephemeral stream looses water as it flows.

A large stream may change from ephemeral to perennial as it moves toward the basin outlet. Because of the fluctuations in the groundwater table, the classifications of streams, it may be noted, is transitory.

REVIEW EXERCISES

1. From the topographic maps of your region locate the catchment boundary for one of the major streams.
2. For the drainage basin obtained in exercise 1, determine the following parameters : Area, stream order, stream density, drainage density, elongation ratio and form factor.
3. For the drainage basin of exercise 1, construct the stream profile and determine the average slope of the stream.
4. Collect the data on area and channel length for at least 10 streams and verify whether this data conform to Eq. (4.10).
5. What do you understand by watershed leakage ? Under what circumstances can you expect the watershed leakage ?
6. Define the catchment and the topographic divide. Explain how the catchment boundary can be obtained from the topographic maps.
7. Discuss briefly the various stream patterns.
8. A drainage basin is found to have the following data : Area of the basin = 2500 km^2 ; Length of the main stream = 110 km, the perimeter of the basin = 367 km. Determine the form factor, circularity ratio, elongation ratio and the compactness coefficient for the basin.
9. Define drainage density and various shape factors.
10. Explain with a neat sketch how the stream order of a channel network of a basin is determined.

ANSWER

8. 0.207, 0.233, 0.513, 20.7.

Chapter 5 PRECIPITATION

5.1 INTRODUCTION

From the hydrological viewpoint any form of moisture reaching the earth's surface from the atmosphere is called the precipitation. The types of precipitation, the forms of precipitation and the mechanisms required for the precipitation to occur have all been discussed in detail in Chapter 2 on Hydrometeorology. In a tropical country like India, major portion of the precipitation occurs in the form of rain only. For this reason the words precipitation and rainfall are used rather synonymously in this book.

Precipitation data is of utmost importance to hydrologists as it forms the basis of many hydrological studies. It is the variations in the rainfall distribution over space and time which create serious hydrological problems such as floods and droughts. This chapter presents the techniques of measurement of precipitation and analysis of precipitation data. A variety of instruments and techniques have been developed for measuring various aspects of precipitation such as the amount, intensity, rain drop size distribution etc. However, the discussion here is limited to only those which are used to measure the amount and intensity of precipitation.

5.2 MEASUREMENT OF PRECIPITATION

Precipitation was probably the first hydrological phenomenon to have been recorded by man. There is evidence that rainfall records were kept in India in the 4th century B.C. All forms of precipitation are essentially measured on the basis of the vertical depth of water that would accumulate on a level surface if the precipitation is retained where it fell. Precipitation is usually measured in millimetres and tenths of millimetres. If it is less than 1 mm it is recorded as *trace*. Rainfall with an intensity of 2.5 mm/h is called *light rain*, between 2.5 mm/h and 7.5 mm/h it is termed as *moderate rain* and when it exceeds 7.5 mm/h it is termed as *heavy rain*. A small surface area is taken for the purpose of measurement and the volume of precipitation water collected over that area is divided by the area to give the depth of precipitation. There are certain difficulties which come in the way of accurate measurement of precipitation. They are : (*i*) Any object used as a precipitation gauge must be above the ground surface in order to avoid the splash from the surroundings, flooding and accidents. As such it cannot measure the amount of precipitation exactly at the level of ground surface since the object extending above the surface of earth can create eddy currents which may affect the amount of catch. (*ii*) The measurement is never subjected to check by repetition and seldom by duplication. (*iii*) In urban localities it is difficult to locate the sites to be free from the effects of wind and at the same time clear of obstructions. (*iv*) The sample is unbelievably small. For example, the precipitation gauges with a receiving area of 20 cm diameter located at 30 km apart provide a sample of one part in 30 billion.

The prevailing wind at the time of precipitation would change the trajectory of the falling precipitation from vertical and more will fall on a given area normal to the trajectory than on the same area in a horizontal plane. However, since a catchment area is computed as the projection upon a horizontal plane the gauge with its receiving area in a horizontal plane gives a representative measurement. The precipitation is measured by an instrument called a *raingauge*. Raingauge is also variously known as *hyetometer*, *ombrometer* or *pulviometer*. The raingauges are of two types : Non-recording type or ordinary raingauges and Recording type or automatic raingauges.

5.3 NON-RECORDING RAINGAUGES

Standard non-recording raingauge prescribed by the IMD is the *Symon's gauge* the details of which are shown in Fig. 5.1 and Fig. 5.2. The gauge consists of a funnel with a sharp edged rim of 127 mm diameter, a cylindrical body, a receiver with a narrow neck and handle and a splayed base which is fixed in the ground. The receiver should have a narrow neck and should be sufficiently protected from radiation to minimise the loss of water from the receiver by evaporation. To prevent rain from splashing in and out, the vertical wall of the sharp edged rim is made deep enough and the slope of funnel steep enough (at least 45°). The rain falling into the funnel is collected in the receiver kept inside the body and is measured by means of a special measure glass (supplied along with the gauge) which is graduated in mm. The receiver has a capacity of 175 mm of rain. In regions of heavy rainfall, raingauges with receivers of 375 mm or 1000 mm capacity may be used.

The measure glass has a capacity of 25 mm and can be read to nearest 0.1 mm. The gauge is fixed on a masonry or concrete foundation of size 60 cm × 60 cm × 60 cm which is sunk into the ground. Into this foundation the base of the gauge is cemented as shown in Fig. 5.2 so that the rim of the gauge is exactly 30 cm above the ground level. The top of the gauge should be perfectly horizontal.

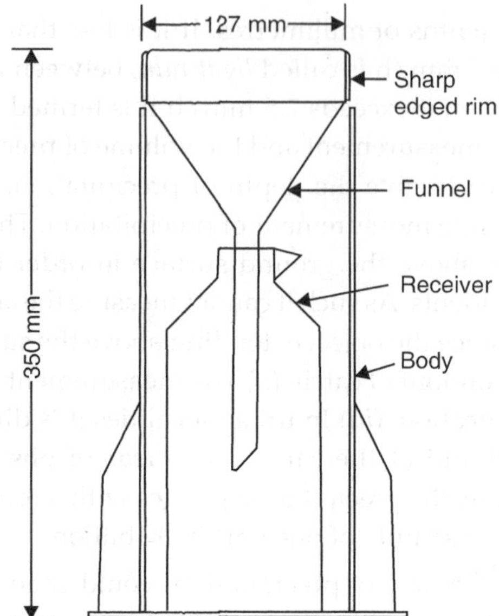

FIGURE 5.1 Symon's raingauge.

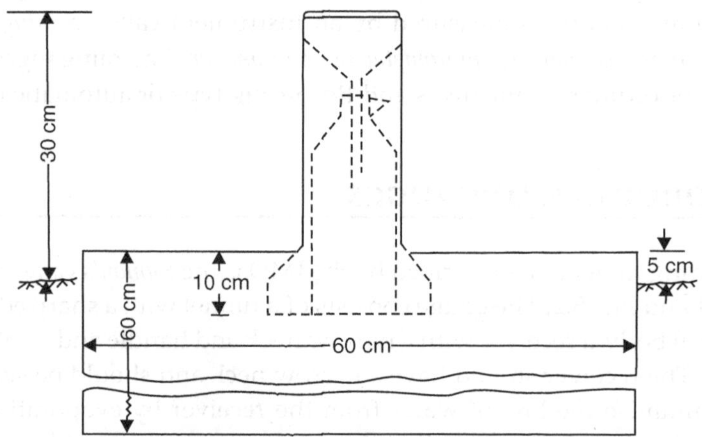

FIGURE 5.2 Installation of non-recording raingauge.

Recently, the IMD has changed over to the use of fibreglass reinforced polyester raingauges which are an improved version of the Symon's gauge. These gauges are available in different combinations of collector areas (100 cm² and 200 cm²) and receiver bottle capacities (2 to 10 litres). They have the capacity to measure rainfall depths of 100 mm to 1000 mm. They conform to the Indian standards IS : 5225–1969. Fig. 5.3 gives the details of the collector of 100 cm² area and the receiver for I.S. gauge. The details of other I.S. gauges can be obtained from IS : 5225.

At the routine time of observation the funnel is removed, the receiver is taken out and the rain water collected in the receiver is carefully poured into the measure glass and read without any parallax error. When the rainfall exceeds 25 mm the measure glass will be used as many times as required. The measured rainfall in the 24 hours ending with 8.30 A.M. is recorded as the rainfall of

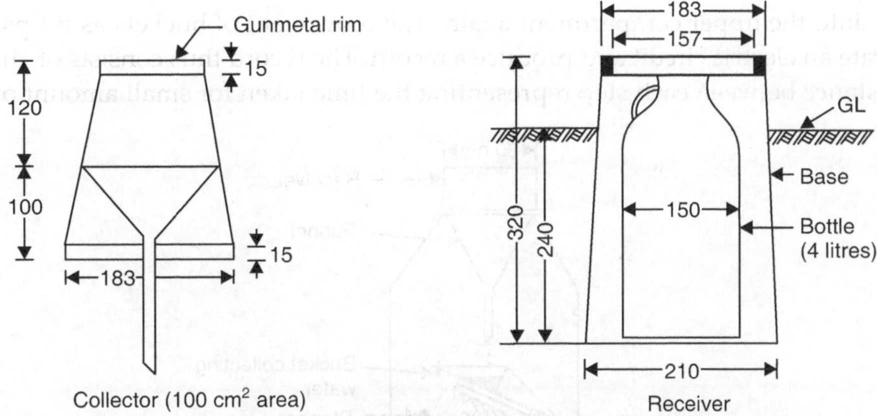

FIGURE 5.3 Indian standard raingauge.

the day on which 8.30 A.M. observation is taken. In regions of heavy rainfall, if it is suspected that the receiver (though of larger capacity) may not hold the entire rainfall of the day the measurements must be done more frequently with the last measurement being taken at 8.30 A.M. The sum of all the readings taken in the last 24 hours is recorded as the rainfall of that day.

5.4 RECORDING RAINGAUGES

Non-recording raingauges give the amount of rainfall only. They cannot provide the information regarding when exactly the rain commenced, when the rain ended, what is the intensity of rainfall and how the intensity of rainfall varies within the duration of the storm. In order to record the beginning and end of the rain and to measure the intensity of rainfall, a continuous record of rainfall with time is required. For this purpose we have to use the recording raingauges. Recording raingauges usually work by having a clock-driven drum carrying a graph on which a pen records the cumulative depth of rainfall continuously. Although there are different types of recording raingauges only three have gained widespread use. They are:

(*i*) Tipping (or tilting) bucket type

(*ii*) Weighing bucket type

(*iii*) Float type (with siphon arrangement).

5.4.1 Tipping Bucket Raingauge

The principle involved in this type of gauge is very simple. A container is divided vertically into two compartments and is balanced in an unstable equilibrium about a horizontal axis. In its normal position it rests against one of the two stoppers which prevent it from tipping over completely as shown in Fig. 5.4. The rain is led from a conventional collecting funnel into the uppermost compartment and after a predetermined rain (usually 0.25 mm) has fallen the bucket becomes unstable in its present position and tips over to its other position of rest. The compartments of the container are so shaped that water can now flow out of the lower one and leave it empty; meanwhile the rain falls into the upper compartment again. The movement of buckets as it tips over can be used to operate an electric circuit and produce a record. The record thus consists of discontinuous steps, the distance between each step representing the time taken for small amount of rain to fall.

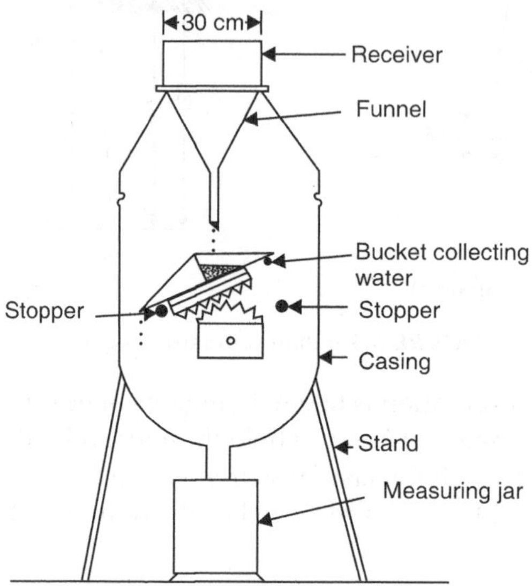

FIGURE 5.4 Tipping bucket raingauge.

The disadvantages of this type of gauge are as follows. If the buckets are designed to tip at a convenient frequency for a particular intensity of rainfall, they will tip either too soon or too late for other intensities. As a result both the intensity and amount of rainfall recorded will be in error except during a storm which has the same intensity for which the buckets are designed. The record obtained from this gauge is not in a convenient form. For higher intensities the bucket tips so rapidly that the jogs in the record tend to overlap and blend into one broad solid line making it difficult to read. The determination of the intensity for any given period requires the counting of jogs in the record during that period. The bucket takes a small but finite time to tip over and during the first half of its motion the rain is being led into the compartment already containing the calculated amount of rainfall. This error is appreciable in heavy rainfalls. For example, in a rainfall of intensity 15 cm/h the bucket tips every 6 seconds. About 0.3 s is required to complete the tip. This makes the intensity of rainfall recorded by the gauge less by 5%. Therefore the total rainfall of the day is always measured (as is done in the case of non-recording gauge) and this data may be used to correct the error. Owing to the discontinuous nature of the record, the instrument is not satisfactory for use in light drizzle or very light rain. The time of beginning and ending of rainfall cannot be determined accurately. This gauge is not suitable for measuring snow without heating the collector.

The biggest advantage of the tipping bucket gauge is that it is the only recording raingauge which can be used in remote places by installing the recorder at a convenient and easily accessible location.

5.4.2 Weighing-Bucket Raingauge

In this type of gauge the rain falling on the receiving area is collected by the funnel and is led into a storage bucket which rests on a weighing platform. The weight of the rainfall received since the recording began is recorded continuously by transmitting the movement of the platform through a system of links and levers to a pen which makes a trace on a suitably graduated chart secured around a drum as shown in Fig. 5.5. The drum is driven mechanically by a spring clock. The drum may be made to revolve once a day, once a week or once in any other desired period.

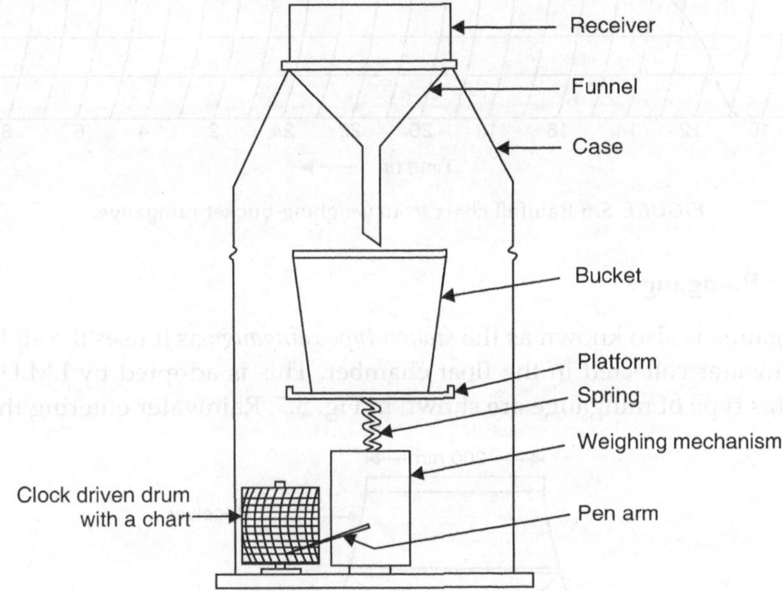

FIGURE 5.5 Weighing-bucket raingauge.

This type of gauge normally has no provision for emptying itself. To overcome this difficulty the mechanism may be arranged to reverse the travel of the pen after certain amount of precipitation has accumulated and reverse again after another equal amount so that the gauge may operate unattended for a week at a time except in regions of very intense precipitation where the precipitation may exceed the capacity of the gauge. A typical chart of rainfall record of a day from the weighing-bucket raingauge with reverse mechanism in action is shown in Fig. 5.6. In this chart the pen had reversed its path of travel (from upward movement to downward movement) at 3.40 h of the next day after recording 100 mm of rainfall. Therefore, for instance, the cumulative rainfall recorded by the gauge at, say, 5 h of the next day is 100 + 20 = 120 mm. The bucket is emptied and the pen is set to zero whenever the chart is changed.

The main usefulness of this type of gauge is that it can record snow, hail and mixture of rain and snow. All forms of precipitation are weighed and recorded in equivalent depth automatically. The disadvantages quoted against this gauge include: (*i*) the effects of temperature and friction on weighing mechanism may introduce errors in the record, (*ii*) shrinkage and expansion of the chart paper caused by changes in humidity may distort the time and the scale of rainfall, and (*iii*) failure of reverse mechanism results in the loss of record. The last difficulty may be eliminated by using only a single traverse of the pen but by reducing the scale of record graph.

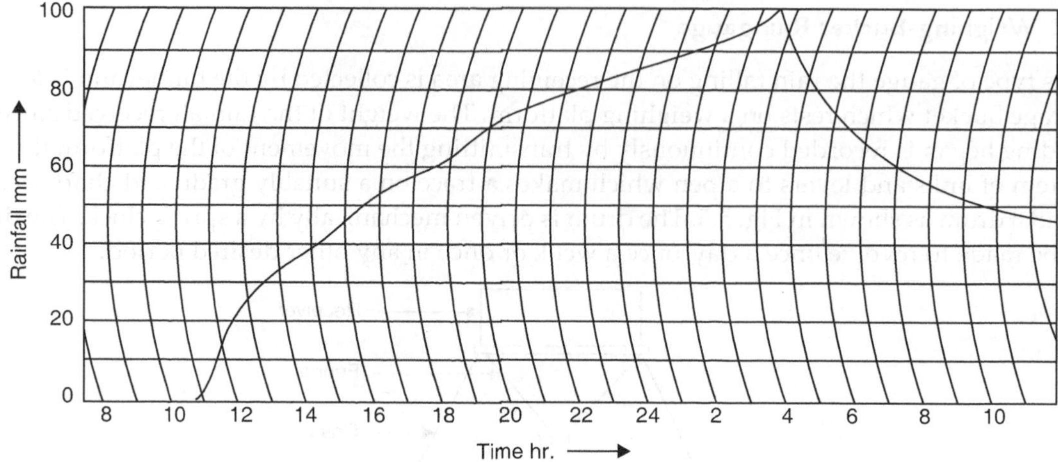

FIGURE 5.6 Rainfall chart from weighing-bucket raingauge.

5.4.3 Float Type Raingauge

This type of raingauge is also known as the *siphon type raingauge* as it uses the siphon mechanism to empty the rainwater collected in the float chamber. This is adopted by I.M.D. The details of construction of this type of raingauge are shown in Fig. 5.7. Rainwater entering the gauge at the

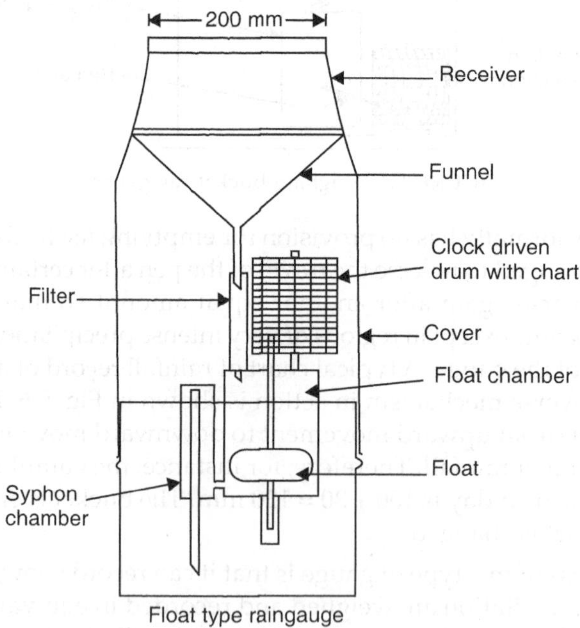

FIGURE 5.7 Float type raingauge.

top is led into the float chamber through a funnel and filter. The purpose of the filter is to prevent dust and other particles from entering the float chamber which may hinder the siphon mechanism. The float chamber consists of a float with a vertical stem protruding outside, to the top of which a pen is mounted. This pen rest on a chart secured around a clock driven drum. There is a small compartment by the side of the float chamber which is connected to the float chamber through a small opening at the bottom. This is called the siphon chamber which houses a small vertical pipe with bottom end open and the top end almost touching the top of the chamber. During the storm the rainwater collected in the float chamber raises the water surface in it and along with the water surface the float also rises enabling the pen to make a trace of cumulative depth of rainfall on the chart. When the float chamber is completely filled with water, the pen reaches the top of the chart. At this instant the siphoning occurs automatically through the pipe in the siphon chamber, the float chamber is emptied and the pen is brought to zero on the chart again. As the rain continues the pen rises again from the zero of the chart. The complete siphoning should be over in less than 15 seconds of time. This gauge cannot record precipitation in the form other than rain unless some sort of heating device is provided inside the gauge. The float may be damaged if the rainfall catch freezes.

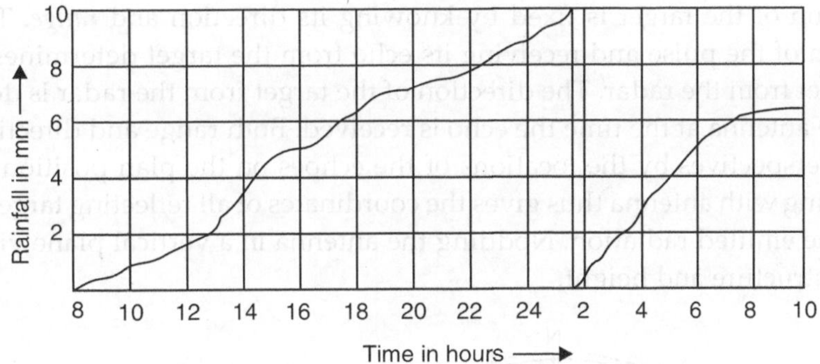

FIGURE 5.8 Rainfall chart from float type raingauge.

A chart from a float type raingauge with siphoning taking place during the storm is shown in Fig. 5.8. This chart indicates that the gauge has siphoned once at 1 : 30 h of the next day. Thus, the cumulative depth of rainfall recorded by the gauge at 5 h of the next day, for example, is 10 + 4 = 14 mm. If the rainfall is of large intensity, the siphoning may occur more than once during the period of the chart.

5.5 RADAR MEASUREMENT OF RAINFALL

Radar means *radio detecting and ranging*. It was originally designed to detect the aircrafts. But its use is increasingly found in the measurement of rainfall also. The principle involved in a radar is shown in Fig. 5.9 and Fig. 5.10. Electromagnetic waves (known as *pulses*) are produced by a transmitter and are radiated by a narrow beam antenna. The reflections of these waves from the targets in their range are known as *echoes*. The echoes are again intercepted by the same antenna. A receiver detects these echoes, amplifies and transforms them into video form on an indicator called the *Plan Position Indicator*. The screen of the indicator is illuminated dimly where there is no target and a bright spot occurs where there is a small target and a bright patch where there is an extended object such as a rain shower.

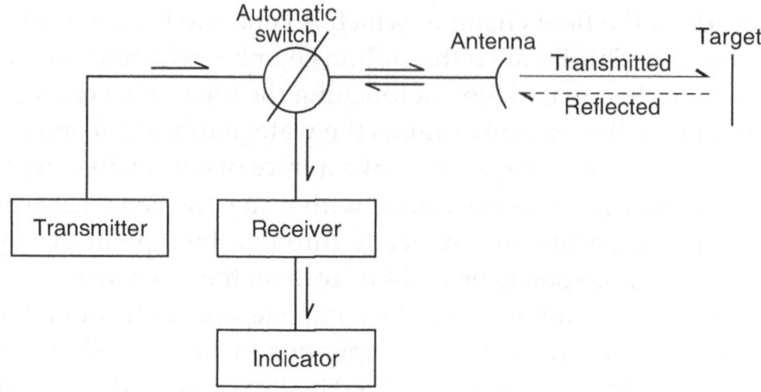

FIGURE 5.9 Working of a radar.

The location of the target is fixed by knowing its direction and *range*. The time interval between emission of the pulse and receiving its echo from the target determines the distance, or range, of the target from the radar. The direction of the target from the radar is determined by the orientation of the antenna at the time the echo is received. Both range and direction are displayed in their proper perspectives by the locations of the echoes on the plan position indicator, or the *radarscope*. Scanning with antenna thus gives the coordinates of all reflecting targets such as storms in the range of the emitted radiation. Nodding the antenna in a vertical plane yields information on precipitation structure and height.

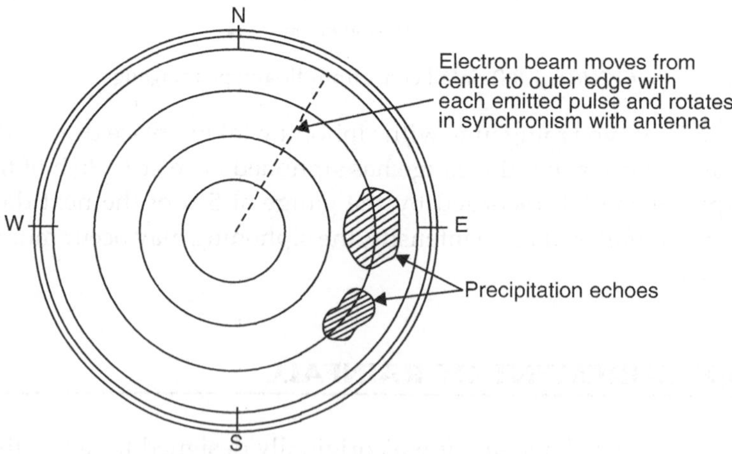

FIGURE 5.10 Plan position indicator.

The photographic exposures of echoes taken at 5 minute intervals are integrated and the resulting average echo intensities are calibrated with the observed rainfall intensities. It has been shown that the average echo intensity $\overline{P}_r$ can be expressed as

$$\overline{P}_r = \frac{C}{r^2} Z \qquad \text{...(5.1)}$$

where C is a constant, r is the distance to the target and Z is known as the radar-echo factor which is related to the rainfall intensity I, through the equation

$$Z = aI^b \qquad \text{...(5.2)}$$

where a and b are constants. Once such calibrations are available (that is fairly reliable estimates of a and b are obtained), the radar operation will yield fairly good quantitative analysis of the rainfall distribution in its range. When I is expressed in mm/h the values of a and b may be taken as 200 and 1.6 respectively.

The hydrological range of a radar is defined as the maximum range within which the relationship between the radar echo-intensity and the rainfall intensity as measured by raingauges remains reasonably valid. This range is about 200 km for some radars.

Satellites in Estimating Rainfall. When water balance studies are taken up on a global scale, it needs the information on precipitation over all places including oceans. The raingauge network may be inadequate for the purpose or it may be non existent, for example on oceans. In such cases, satellites may be of help. Satellites cannot measure precipitation directly. They provide photographs of the clouds. The rainfall intensities are obtained indirectly through the relationship between the brightness of the cloud photographs and the rainfall intensity. The relation between brightness of the cloud and rainfall intensity is usually determined by calibration with raingauge measurements and radar estimates. Except for rainfall estimates from a few weather radars installed on shipboards, the above calibration procedures are necessarily based on data over land surfaces. Such calibrations when used to estimate the rainfall over seas, their reliability depends on the precipitation processes over land surfaces and oceans being similar.

A major problem in estimating precipitation by satellites is that the photographs often do not reveal rainfall-producing clouds because of overlying cloud layers. Future developments in instrumentation and techniques may eventually lead to reasonably accurate estimates of rainfall.

5.6 ERRORS IN MEASUREMENT OF RAINFALL

Precipitation measurements are subject to various errors. Most of the errors lead to the measurements that are lower than the true values. As already pointed out earlier, the raingauge cannot accurately measure the amount of rainfall which would have fallen at the gauge site had the gauge not been there. In general the raingauge underestimates the rainfall that would have fallen on the ground in the absence of the gauge. The magnitude of the underestimate depends on the type of the gauge, the height of the rim of the gauge above the ground surface, the intensity of rainfall, the wind speed etc. Special investigations are required to assess the errors in rainfall measurement due to these reasons.

Some rainfall will be utilised to wet the receiving surface which is initially dry. This rainfall goes unrecorded as it evaporates back into the atmosphere, resulting in an underestimate of rainfall. It is estimated that 0.25 mm of rain is required to moisten the funnel and inside surfaces of a raingauges which is initially dry. This loss can easily add up to about 25 mm per year in some areas. This loss can be reduced to some extent by ensuring that the receiving surface of the gauge is smooth so that rain drops do not adhere to it.

Dents in the collector rim of the gauge may change its receiving area again resulting in underestimate of rainfall.

The main likely sources of error in measuring the catch of the non-recording gauges are the improper reading of measuring cylinder, the spilling of some of the water when transferring it to the measuring jar and the inability to transfer all the water from the receiver to the measuring jar. Proper training to the observer may eliminate most of these errors.

5.7 LOCATION OF RAINGAUGE

The amount of rainfall collected by a raingauge depends on its exposure conditions and therefore great care must be exercised in selecting a suitable site for its location. According to Indian Standards the following precautions must be strictly observed while selecting a site for a raingauge.

(*i*) The gauge shall be placed on a level ground, not upon a slope or a terrace and never on a wall or roof.

(*ii*) On no account the raingauge shall be placed on a slope such that the ground falls away steeply in the direction of prevailing wind.

(*iii*) The distance of the raingauge from any object shall not be less than twice the height of the object above the rim of the gauge.

(*iv*) Great care shall be taken at mountain and coast stations so that the gauges are not unduly exposed to the sweep of the wind. A belt of trees or a wall on the side of the prevailing wind at a distance exceeding twice its height shall form an efficient shelter.

(*v*) In hills where it is difficult to find a level space, the site for the gauge shall be chosen where it is best shielded from high winds and where the wind does not cause eddies.

The location of the gauge should not be changed without taking suitable precautions. Description of the site and surroundings should be made a matter of record.

5.8 ANALYSIS AND INTERPRETATION OF RAINFALL DATA

The precipitation process is essentially random in nature. We cannot predict with certainity what would be the rainfall for any given period in future. The rainfall magnitudes can be estimated only with some probability attached to them. Therefore the analysis of the rainfall data obtained over a long period in the past would help the hydrologist to make reasonable probabilistic estimates of rainfall to be used in various designs. The method of analysis and which aspects of rainfall are to be analysed depend upon the purpose of analysis.

The rainfall obtained from a single raingauge station is known as the *point rainfall* or *station rainfall*. The daily rainfalls recorded at a station may be totalled to yield the rainfalls for longer periods such as monthly rainfall, seasonal rainfall (for example the rainfall in the southwest monsoon season, *i.e.*, from Ist June to 30th Sept.), and annual rainfall. If the data at the station covers a period of more than 30 years, the *normal annual rainfall*, or *the normal monthly rainfall* for any month can be computed. For example, the normal June rainfall of the station is computed as the arithmetic average of the rainfalls in the month of June in the last 30 years. Similarly the normal annual rainfall of the station is obtained as the arithmetic average of the annual rainfalls in the last 30 years. The standard deviation computed for the same 30 years period is a measure of the variability of rainfall. The normal rainfalls are revised every decade by deleting the oldest 10 years data from the record and adding the most recent 10 years data to the record. If the observed rainfall in any year is less than the normal annual rainfall then it is called as *deficient year* or a *dry year* and in the converse it is called a *surplus year* or a *wet year*.

The data of annual rainfall can be graphically presented in many ways such as the *chronological chart*, *bar diagram* and the *ordinate graph*.

Chronological Chart. In this presentation the annual rainfall is plotted as the ordinate against the year in which it is recorded as the abscissa. The points are then joined by straight lines in between as shown in Fig. 5.11.

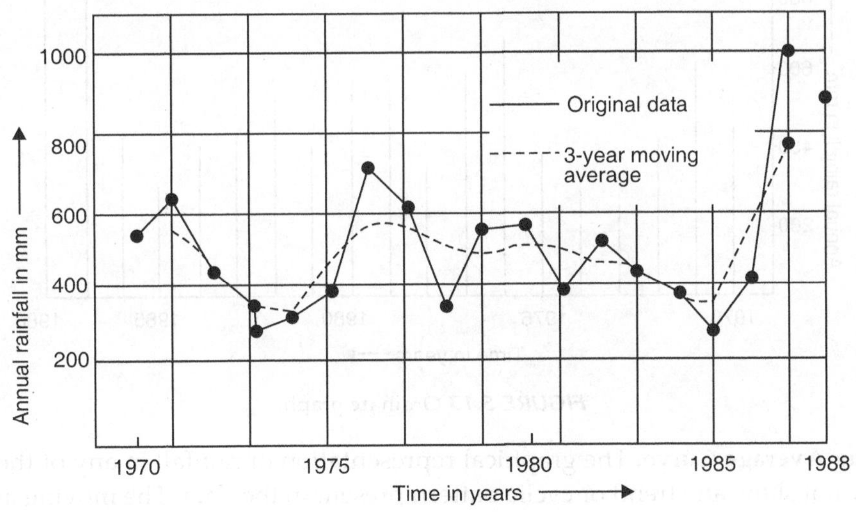

FIGURE 5.11 Chronological chart of rainfall.

Bar Diagram. In a bar diagram, the rainfall is represented as a rectangular bar whose height denotes the magnitude of the rainfall to some scale. A typical bar diagram is shown in Fig. 5.12.

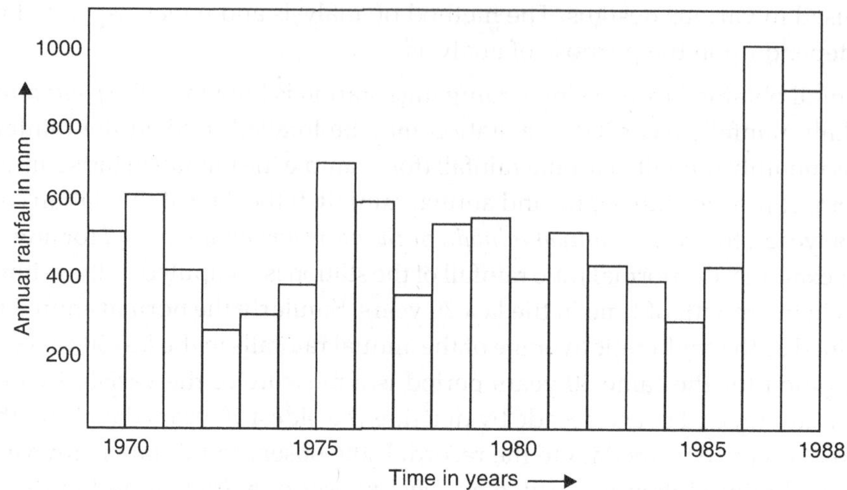

FIGURE 5.12 Bar diagram of rainfall.

Ordinate Graph. The ordinate graph presents the rainfall in any year as an ordinate line drawn to some scale at the corresponding year. Fig. 5.13 is an example of an ordinate graph.

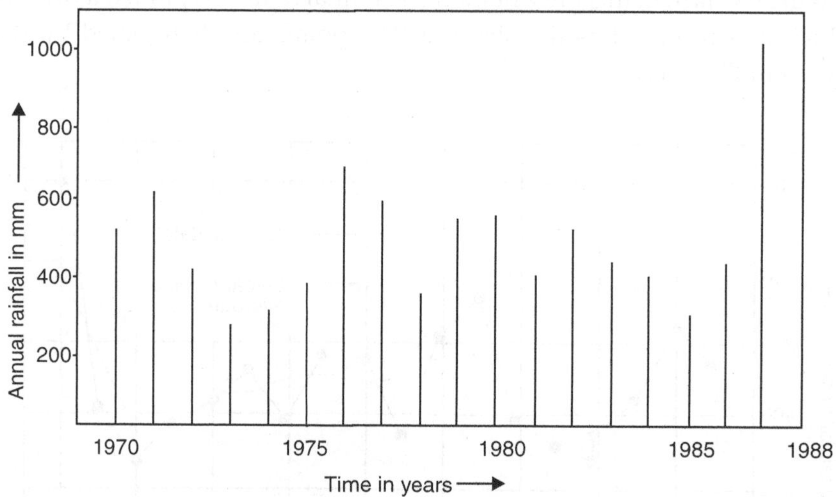

FIGURE 5.13 Ordinate graph.

Moving Average Curve. The graphical representation of rainfall in any of the above three methods may not show any trend or cyclic pattern present in the data. The moving average curve smoothens out the extreme variations and indicate the trend or cyclic pattern, if any, more clearly. It is also known as the *moving mean* curve. The procedure to construct the moving average curve is as follows. The moving average curve is constructed with a moving period of m year where m is generally taken to be 3 or 5 years. Let $x_1, x_2, \ldots\ldots, x_n$ be the sequence of given annual rainfall in the chronological order. Let y_i denote the ordinate of the moving average curve for the ith year. Then for $m = 3$, y_i is computed from

$$\left.\begin{aligned} y_2 &= \frac{x_1 + x_2 + x_3}{3} \\ y_3 &= \frac{x_2 + x_3 + x_4}{3} \\ &\vdots \\ y_i &= \frac{x_{i-1} + x_i + x_{i+1}}{3} \\ &\vdots \\ y_{n-1} &= \frac{x_{n-2} + x_{n-1} + x_n}{3} \end{aligned}\right\} \qquad \text{...(5.3)}$$

As can be seen from Eq. (5.3) the computed value of y correspond in time to the middle value of the x's being averaged and therefore it is convenient to use odd values of m. A moving average of m applied to a sequence of n values yields a sequence of $(n - 2k)$ values, where $k = (m - 1)/2$. For any general m, the y terms can be expressed as

$$y_i = \frac{1}{m} \sum_{j=i-k}^{i+k} x_j \text{ ; for } i = k + 1, k+2, \text{..........}, (n - k) \qquad \text{...(5.4)}$$

Although it is possible to use moving averages with any m, it is necessary that m be small compared to n. The moving average technique can be applied to other hydrological parameters as well such as temperature, wind speed etc. A 3-year moving average curve superimposed over the original sequence is shown in Fig. 5.11.

Generally no persistent regular cycles can be expected in the annual rainfall data. However, annual or seasonal cycles may be noticed when the moving average curve is constructed for monthly rainfall data.

Frequency Analysis. The data on annual rainfall may be subjected to frequency analysis to predict the *dependable rainfall* for any given probability level. (The reader is advised to refer to chapter 3 for frequency analysis and specifically to go through Example 3.7 for the details of obtaining the dependable rainfalls). Alternatively, an appropriate probability density function can be fitted to the rainfall data, if it is large, and the dependable rainfall values can be obtained from this probability function. The central limit theorem suggests that Normal distribution is an appropriate choice for the annual rainfall data.

Station Year Method. As mentioned earlier, the rainfall data obtained from a single rain gauge is known as point rainfall data. Such point rainfall data may be used in various applications like frequency analysis or development of intensity-frequency-duration curves (to be explained in Article 5.11). For such applications the rainfall record may not be of sufficient length. It may then be necessary to extend the point rainfall data. One simple and popular method commonly used for this purpose is known as station-year method.

In this method the records of number of raingauges including the raingauge whose record is to be extended are combined into a single composite record. The records may belong to the same or different periods. Obviously the length of the combined composite record is equal to the sum of the lengths of individual records. For instance, when the rainfall data of 10 raingauges (in which 5 gauges have 15 years record and the remaining 5 gauges have 20 years record) are combined they produce a combined record of length 175 years (5 × 15 + 5 × 20). Then according to the station-year method any of these ten gauges can be assumed to have an equivalent record of 175 years and this composite record can be used for intended purpose. The composite record is presumed to have the same characteristics as a 175 year record would have if it were recorded at one of the gauges.

The following conditions are to be satisfied before the station-year method can be applied :

(*i*) The individual raingauge records must be independent. This implies that the raingauges are located sufficiently far apart and the raingauges do not duplicate recording of the same rainfall values at the same time.

(*ii*) The area in which the gauges are located is meteorologically homogeneous. This means that the rainfall at all of the gauges is generated by the same types of storms and sources. One rough indication of homogenity is that the annual rainfall of all the gauges is approximately the same.

(*iii*) The rainfall frequency distribution is nearly the same for all the gauges, at least in the long run.

Example 5.1. *The annual rainfalls in mm recorded at a rainfall station for a period of 19 years from 1970 to 1988 are given below.*

520, 615, 420, 270, 305, 380, 705, 600, 350, 550

560, 400, 520, 435, 395, 290, 430, 1020, 900.

(*i*) *Find the average and standard deviation of the annual rainfall.*

(*ii*) *Plot the rainfall data as chronological chart, bar diagram and ordinate graph.*

Solution.

$$\bar{x} = \frac{x_1 + x_2 + \ldots\ldots + x_n}{n}$$

$$= \frac{520 + 615 + \ldots\ldots + 1020 + 900}{19}$$

$$= \frac{9665}{19} = 508.68 \text{ mm}$$

$$s_x^2 = \frac{(x_1 - \bar{x})^2 + (x_2 - \bar{x})^2 + \ldots.. + (x_n - \bar{x})^2}{(n-1)}$$

$$= \frac{(520 - 508.68)^2 + \ldots\ldots\ldots + (900 - 508.68)^2}{(19-1)}$$

$$= \frac{709592.11}{18} = 39421.78$$

$$s_x = \sqrt{39421.78} = 198.55 \text{ mm}$$

∴ The mean and standard deviation of the given rainfall data are 508.68 mm and 198.55 mm respectively.

The given rainfall data is presented as a chronological chart, bar diagram and originate graph in Fig. 5.11, Fig. 5.12 and Fig. 5.13 respectively.

Example 5.2. *For the rainfall data of Example 5.1 construct a 3-year moving average curve. Are there any apparent cycles or trends ? Discuss.*

Solution. The computations required to find the ordinates of the moving average curve are shown in Table 5.1.

Table 5.1. Computations for Moving Averages

i	*Year*	*Annual Rainfall* x_i	*Totals for moving average* $= x_{i-1} + x_i + x_{i+1}$	*Moving average* y_i = *col (4) ÷ 3*
(1)	(2)	(3)	(4)	(5)
1	1970	520		
2	1971	615	520 + 615 + 420 = 1555	518.3
3	1972	420	615 + 420 + 270 = 1305	435.0
4	1973	270	420 + 270 + 305 = 995	331.7
5	1974	305	270 + 305 + 380 = 955	318.3
6	1975	380	305 + 380 + 705 = 1390	463.3
7	1976	705	380 + 705 + 600 = 1685	561.7
8	1977	600	705 + 600 + 350 = 1655	551.7
9	1978	350	600 + 350 + 550 = 1500	500.0
10	1979	550	350 + 550 + 560 = 1460	486.7
11	1980	560	550 + 560 + 400 = 1510	503.3
12	1981	400	560 + 400 + 520 = 1480	493.3
13	1982	520	400 + 520 + 435 = 1355	451.7
14	1983	435	520 + 435 + 395 = 1350	450.0
15	1984	395	435 + 395 + 290 = 1120	373.3
16	1985	290	395 + 290 + 430 = 1115	371.7
17	1986	430	290 + 430 + 1020 = 1740	580.0
18	1987	1020	430 + 1020 + 900 =2350	783.3
19	1988	900		

The moving average curve is obtained by plotting the values of y_i in col. (5) against the corresponding year appearing in col. (2). The moving average curve is shown in Fig. 5.11, where it is superimposed over the original rainfall series. Though the variations in the original data are smoothened out to some extent in the moving average curve, no apparent trend or cyclicity is visible in the moving average curve.

Example 5.3. *Assuming that normal distribution is a good fit for the rainfall data of Example 5.1, determine the 50% and 75% dependable rainfalls. Also determine the probability for a rainfall of 800 mm or more to occur in any year. Noting that this data is same as that of Example 3.7, compare your answers with those of Example 3.7 which are determined using frequency analysis.*

Solution. The mean and standard deviation of the rainfall are already computed in Example 5.1 as 508.68 mm and 198.55 mm respectively.

Determination of 50% dependable rainfall : From the cumulative normal distribution function given in Table 3.1, the value of the standardised normal variable z which satisfies the condition that

$$P\,[Z \geq z] = 0.5$$

is obtained as 0. But $z = \dfrac{x - \overline{x}}{s_x}$ or $x = \overline{x} = 508.68$ mm.

∴ The 50% dependable rainfall is 508.68 mm. The corresponding value from frequency analysis in Example 3.7 is 450.0 mm.

Determination of 75% dependable rainfall : Again from Table 3.1, the value of z which satisfies the condition that

$$P\,[Z \geq z] = 0.75$$

is obtained as – 0.675

$$-0.675 = \frac{x - \overline{x}}{s_x}$$

or $$x = \overline{x} - 0.675\, s_x = 508.68 - 0.675 \times 198.55 = 374.66 \text{ mm.}$$

The corresponding value from frequency analysis in Example 3.7 is 335 mm

When $$x = 800 \text{ mm},\ z = \frac{800 - 508.68}{198.55} = 1.47$$

$$P\,[Z \geq 1.47] = 1 - P\,[Z \leq 1.47] = 1 - F\,(1.47)$$

$$1 - 0.9288 = 0.0712$$

∴ The annual rainfall in any year exceeds 800 mm with a probability of 0.0712. The corresponding value from frequency analysis in Example 3.7 is 0.12.

5.9 RAINFALL MASS CURVE AND RAINFALL HYETOGRAPH

For many hydrological designs the analysis of rainfall as given in the previous section is not adequate and the characteristics of individual storms may be needed. Suppose an engineer is involved in the hydraulic design of a culvert or a bridge. He has determined the catchment area at the site concerned and estimated that it may take about 8 hours for the water particle falling on the remotest point of the catchment to reach the catchment outlet. That means, if a storm lasts for a period of 8 h or more every portion of the catchment would be contributing to the flow at the outlet and that is the situation which produces the maximum flood flow. If the life span of the bridge is fixed at 50 years, the engineer would like to know what is the maximum rainfall depth (or maximum rainfall intensity) for a storm of 8 h duration that can be expected in the next 50 years so that he can predict the corresponding peak flood and design the bridge to withstand such a flood. If the area of catchment is large and the maximum rainfall intensity is likely to vary over the area, the engineer needs an areal average of the maximum intensity which is applicable for the whole area. Thus we see that the important parameters of storm rainfall required in such design problems are intensity, duration, frequency and areal distribution. Information regarding rainfall intensities is also required in the unit hydrograph analysis, infiltration studies and runoff simulation models. The rainfall intensity during a storm can be determined only from the rainfall data obtained from a recording raingauge. The techniques of analysis for determining the above characteristics are now presented.

The rate at which the rainfall is accumulating at any given instant of time is called the *rainfall intensity* at that time. The instantaneous rainfall intensity is usually denoted by i and expressed in mm/h. A graph showing the variation of rainfall intensity with time is called a *rainfall hyetograph*. A graph showing the cumulative depth of rainfall against time is known as the *rainfall mass curve*. As seen from the principle of its working, a float type or weighing type recording raingauge provides the information on rainfall mass curve. The cumulative rainfall is usually denoted by P, and expressed in mm. A typical rainfall mass curve is shown in Fig. 5.14. The difference in ordinates of

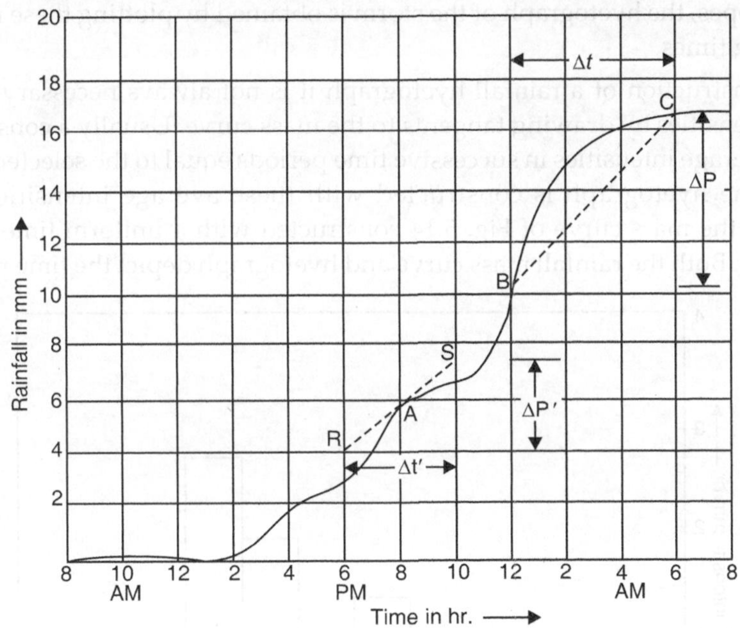

FIGURE 5.14 Determining rainfall intensity from rainfall mass curve.

the rainfall mass curve, $\Delta P = (P_2 - P_1)$, at two times t_2 and t_1 corresponding to the two points B and C on the rainfall mass curve gives the amount of rainfall occurring in the time interval $\Delta t = (t_2 - t_1)$. Thus the quantity $\dfrac{\Delta P}{\Delta t}$ is nothing but the average rainfall intensity in the time period of Δt. In other words if any two points on the rainfall mass curve are joined by a straight line the slope of that line gives the average intensity of rainfall for the time period between those two points. The average rainfall intensity is denoted by I. In the limit when the two points become nearer and nearer or as $\Delta t \to 0$, the line becomes a tangent to the mass curve and hence the slope of a tangent drawn to the rainfall mass curve at any time point gives the instantaneous rainfall intensity at that time.

$$i = \underset{\Delta t \to 0}{\text{Lt}} \frac{\Delta P}{\Delta t} = \frac{dP}{dt} \qquad ...(5.5)$$

Obviously if the slope of the mass curve is zero for any period, *i.e.*, when the mass curve is horizontal, then the intensity for that period is zero.

In the rainfall mass curve of Fig. 5.14, a tangent *RS* is drawn to it at *A*, corresponding to a time of 8.00 P.M. The slope of this tangent measured as $\frac{\Delta P'}{\Delta t'}$ gives the instantaneous intensity at 8.00 P.M. as $\frac{3.6 \text{ mm}}{4 \text{ h}}$ = 0.9 mm/h. Again the slope of the line joining *B* and *C* would give the average intensity of rainfall in the period 12.00 M.N. to 6.00 A.M. as $\frac{6.3 \text{ mm}}{6 \text{ h}}$ = 1.05 mm/h. However, it is to be noted that the intensity is not the same throughout the period from *B* to *C* since the mass curve is not having a uniform slope between *B* and *C*. If the instantaneous intensities are calculated by drawing tangents to the mass curve at different and numerous time points and measuring the corresponding slopes, the hyetograph of the storm is obtained by plotting these intensities against the corresponding times.

For the construction of a rainfall hyetograph it is not always necessary to determine the instantaneous intensities by drawing tangents to the mass curve. Usually a constant time interval is selected and average intensities in successive time periods equal to the selected time interval are computed and the hyetograph is constructed with these average intensities. A hyetograph corresponding to the mass curve of Fig. 5.14 constructed with a uniform time interval of 2 h is shown in Fig. 5.15. Both the rainfall mass curve and hyetograph depict the time pattern of a storm.

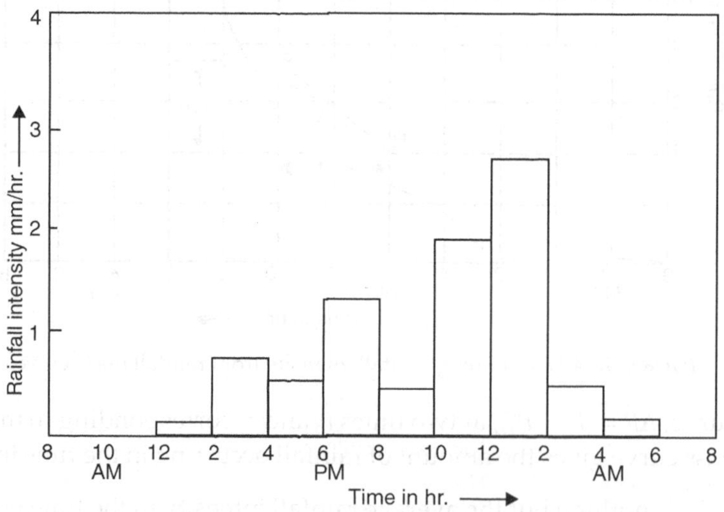

FIGURE 5.15 Rainfall hyetograph.

The interval used for the construction of a hyetograph depends on the sensitivity required. When the greatest possible sensitivity is required the smallest interval over which the rainfall rate is constant is selected. This interval may be as small as 5 minutes depending on the sensitivity of the instrument used. It is often more convenient, however, to use a uniform time interval suited to the purpose and the storm type. In many hydrological problems such as the estimation of runoff or the derivation of unit hydrographs, the minor fluctuations in rainfall intensity are not significant. Small area storms are frequently analysed at one hour interval and over large areas 6 hour intervals are often sufficient.

The hyetograph has the advantage that it uses a larger vertical scale than the mass curve and is appropriately used and compared with the corresponding hydrograph of runoff. The area under the hyetograph of a storm represents the total rainfall received during the period of that storm.

5.10 INTENSITY DURATION ANALYSIS

It can be generally observed that the most intense storms last for very short durations. As the duration of storm increases the maximum average intensity of the storm decreases. If the observed maximum rainfall intensities at a place for various durations such as 5 min, 10 min, 15 min, ..., 1 h, 2 h, 5 h, etc., are plotted against the respective durations, a graph known as the *intensity-duration graph* is obtained. A typical intensity-duration graph is shown in Fig. 5.17. A simple method of finding the maximum intensity for a given duration in any storm is to use a transparent scale with two vertical lines drawn on it at a distance equal to the required duration and to measure the maximum vertical intercept of the rainfall mass curve when the transparent scale is slided across the mass curve. Alternatively, the maximum intensity for any duration can also be found out analytically as described below. The rainfall mass curve is first tabulated at equal time intervals, the interval being equal to or less than the smallest duration for which the maximum intensity is required. Then the rainfall amount in every possible time interval equal to the required duration is found out by taking the difference of the ordinates of the rainfall mass curve at the corresponding times. Suppose the rainfall mass curve is tabulated at 15 minute interval and suppose we want the maximum rainfall in the storm for a duration of 45 mts. Then we find the rainfalls that have occurred between 10.00 to 10.45, between 10.15 to 11.00, between 10.30 to 11.15 etc., and pick up the largest of these values as the maximum rainfall for the 45 minute durations. Then the maximum intensity is obtained by dividing the maximum rainfall depth by this duration. Similarly the maximum intensity for any other duration can be obtained. This procedure is illustrated in Example 5.4.

The maximum intensity varies inversely with the duration and generally an equation of the form

$$I = \frac{C}{(t + a)^b} \quad ...(5.6)$$

is assumed between I and t. The values of C, a and b are obtained through regression analysis. In Eq. (5.6), while t is generally expressed in minutes, I may be expressed in mm/h or cm/h. The regression constants are different for different locations. Taking logarithms on both sides transforms Eq. (5.6) into a linear form.

$$\log I = \log C - b \log(t + a)$$

The best value of C, a and b are those for which the sum of the squared deviations is minimum.

i.e.,
$$S = \Sigma [\log I - \{\log C - b \log (t + a)\}]^2$$

should be minimum. Partial differentiation of S with respect to C and b yields

$$\Sigma \log I = n \log C - b \Sigma \log (t + a) \qquad ...(5.7)$$

$$\Sigma [\log I \log (t + a)] = \log C \Sigma \log (t + a) - b \Sigma[\log (t + a)]^2 \qquad ...(5.8)$$

where n is the number of observations and all the summations are over all the observed values. After obtaining the required summations, Eqs. (5.7) and (5.8) may be solved to provide the best values of C and b for any assumed value of a and the best value of a itself has to be found only by trial and error. This procedure is demonstrated in Example 5.5.

Example 5.4. *A storm commenced at 7.00 hours. The ordinates of the rainfall mass curve of this storm in mm as recorded by a recording rain gauge at 15 minute intervals are 0, 9.5, 17.0, 27.0, 40.5, 49.0, 63.0, 84.0, 95.0, 102.0, 110.0, 112.0 and 112.0. Construct the hyetograph of this storm for a uniform interval of 15 minutes.*

Solution. The ordinates of the rainfall mass curve are entered in Cols. (1) and (2) of Table 5.2. Rainfall depths in successive 15 minute intervals are computed and entered in Col. (3) while Col. (4) gives the corresponding rainfall intensities.

Table 5.2. Computations for Rainfall Hyetograph

Time	*Rainfall in mm*	*Rainfall in successive 15 minutes intervals*	*Rainfall intensity in mm/h*
7.00	0.0	—	—
7.15	9.5	9.5	38
7.30	17.0	7.5	30
7.45	27.0	10.0	40
8.00	40.5	13.5	54
8.15	49.0	8.5	34
8.30	63.0	14.0	56
8.45	84.0	21.0	84
9.00	95.0	11.0	44
9.15	102.0	7.0	28
9.30	110.0	8.0	32
9.45	112.0	2.0	8
10.00	112.0	0.0	0

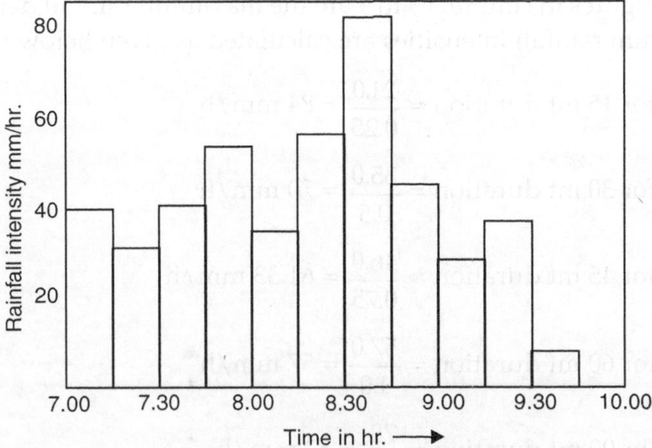

FIGURE 5.16 Rainfall hyetograph of Example 5.4.

The rainfall intensities in successive 15 minute interval of Col. (4) are plotted in Fig. 5.16 to give the hyetograph.

Example 5.5. *For the storm of Example 5.4 compute the maximum rainfall intensities for durations of 15, 30, 45, 60, 90, 120 and 180 minutes and plot the intensity duration graph.*

Solution. The necessary computations to obtain the maximum intensities are shown in Table 5.3.

Table 5.3. Computations for Determining Maximum Intensities

Time	*Cumulative Rainfall mm*	*Rainfall in any possible time interval equal to*						
		15 mt	*30 mt*	*45 mt*	*60 mt*	*90 mt*	*120 mt*	*180 mt*
7.00	0.0							
7.15	9.5	9.5						
7.30	17.0	7.5	17.0					
7.45	27.0	10.0	17.5	27.0				
8.00	40.5	13.5	23.5	31.0	40.5			
8.15	49.0	8.5	22.0	32.0	39.5			
8.30	63.0	14.0	22.5	36.0	46.0	63.0		
8.45	84.0	**21.0**	**35.0**	43.5	**57.0**	74.5		
9.00	95.0	11.0	32.0	**46.0**	54.5	**78.0**	**95.0**	
9.15	102.0	7.0	18.0	39.0	53.0	75.0	92.5	
9.30	110.0	8.0	15.0	26.0	47.0	69.5	93.0	
9.45	112.0	2.0	10.0	17.0	28.0	63.0	85.0	
10.00	112.0	0.0	2.0	10.0	17.0	49.0	71.5	**112.0**

The bold faced figures in columns 3 to 9 are the maximum rainfall depths for the respective durations. The maximum rainfall intensities are calculated as given below.

Max. intensity for 15 mt duration $= \dfrac{21.0}{0.25} = 84$ mm/h

Max. intensity for 30 mt duration $= \dfrac{35.0}{0.5} = 70$ mm/h

Max. intensity for 45 mt duration $= \dfrac{46.0}{0.75} = 61.33$ mm/h

Max. intensity for 60 mt duration $= \dfrac{57.0}{1.0} = 57$ mm/h

Max. intensity for 90 mt duration $= \dfrac{78}{1.5} = 52$ mm/h

Max. intensity for 120 mt duration $= \dfrac{95}{2.00} = 47.5$ mm/h

Max. intensity for 180 mt duration $= \dfrac{112}{3} = 37.33$ mm/h

The intensity-duration graph for this storm is shown in Fig. 5.17.

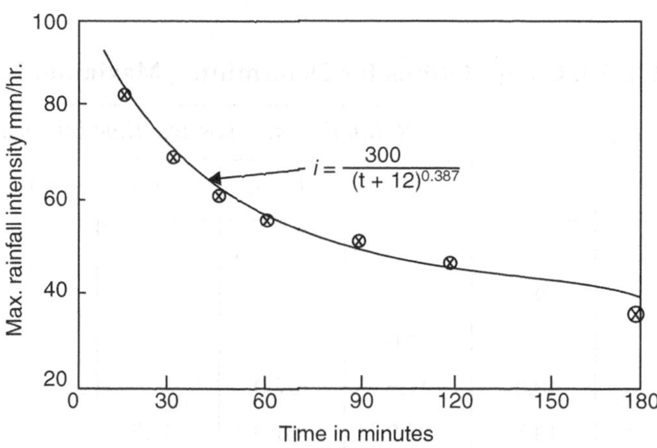

FIGURE 5.17 Intensity-duration graph.

Example 5.6. *For the intensity-duration graph of Example 5.5 fit an appropriate regression equation.*

Solution. The relationship between the maximum intensity and duration is assumed to be of the form given by Eq. (5.6). The required summations for determining the best values of C and b are shown in Table 5.4 for one trial value of a which is taken to be 8. The number of observations in the present case is 7. On substitution of the relevant summations in Eq. (5.7) and Eq. (5.8), we get

$$12.2777 = 7 \log C - 12.8709\, b$$

$$22.3604 = 12.8709 \log C - 24.2585\, b.$$

The solution of the above two simultaneous equations is

$$b = 0.362 \text{ and } \log C = 2.4196 \text{ or } C = 262.76$$

Therefore the required regression equation with $a = 8$ is

$$\hat{I} = \frac{262.76}{(t+8)^{0.362}}.$$

Table 5.4 Computations for Regression Analysis (trial with $a = 8$)

Duration *t* *minutes*	*Intensity* *I* *mm/h*	*log I*	*log (t + a)*	*log I.log(t + a)*	*{log (t + a)}²*	$\hat{I}$	$\Delta I = (I - \hat{I})$	ΔI^2
15	84.00	1.9243	1.3617	2.6203	1.8542	84.4527	– 0.4527	0.2049
30	70.00	1.8451	1.5798	2.9149	2.4958	70.4170	– 0.4170	0.1739
45	61.33	1.7877	1.7243	3.0825	2.9732	62.4268	– 1.0968	1.2030
60	57.00	1.7559	1.8325	3.2177	3.3581	57.0415	– 0.0415	0.0017
90	52.00	1.7160	1.9912	3.4169	3.9649	49.9730	2.027	4.1087
120	47.50	1.6767	2.1072	3.5331	4.4403	45.3680	2.132	4.5454
180	37.33	1.5720	2.2742	3.5750	5.1720	39.4743	– 2.1443	4.5980
	Σ	12.2777	12.8709	22.3604	24.2585			14.8356

Using the above equation the intensities for various durations are determined and entered under column $\hat{I}$. The sum of the squared deviations for this case is 14.8356. The procedure is repeated for other values of a and the results are as given under. For want of space the details of the regression analysis for the other trials are not presented here.

Trial	*a*	C	*b*	$\Sigma \Delta I^2$
1	8	262.76	0.362	14.836
2	10	281.33	0.375	14.475
3	12	300.00	0.387	**14.412**
4	14	320.00	0.399	14.636

It is thus seen that the sum of the squared deviations is the least when $a = 12$. Therefore the desired regression equation for the data under consideration is

$$\hat{I} = \frac{300}{(t+12)^{0.387}}.$$

5.11 INTENSITY-FREQUENCY-DURATION ANALYSIS

If the rainfall data from a recording raingauge is available for a longer period such as 40 or 50 years, the frequency of occurrence of a given maximum intensity can also be determined. Then we obtain the intensity-frequency-duration relationships. The procedure to obtain such relationships is now described.

Every storm in a year is analysed to find the maximum intensities for various durations as described in the previous section. Thus each storm gives one value of maximum intensity for a given duration. The largest of all such values is taken to be the maximum intensity in that year for that duration. Likewise the annual maximum intensity is obtained for all the durations. Similar analysis yields the annual maximum intensities for various durations in different years. It will then be observed that the annual maximum intensity for any given duration is not the same every year but it varies from year to year. In other words it behaves as a random variable. So, if 50 years of record is available then there will be 50 values of the maximum intensity of any duration which constitute a sample of the random variable. These 50 values of any duration can be subjected to the frequency analysis as explained in Chapter 3 and the maximum intensity of rainfall for various return periods can be obtained. Similar frequency analysis is carried out for other durations. Then from the results of this analyses graphs of maximum rainfall intensity against the return period for various durations such as those shown in Fig. 5.18 can be developed. The same results can be depicted in a different but more convenient form as shown in Fig. 5.19. From the curves of Fig. 5.19 the maximum intensity of rainfall for any duration and for any return period can be read out.

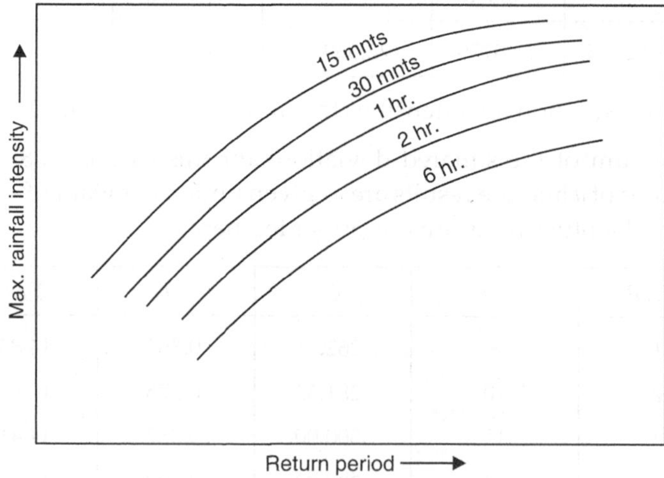

FIGURE 5.18 Intensity-frequency curves.

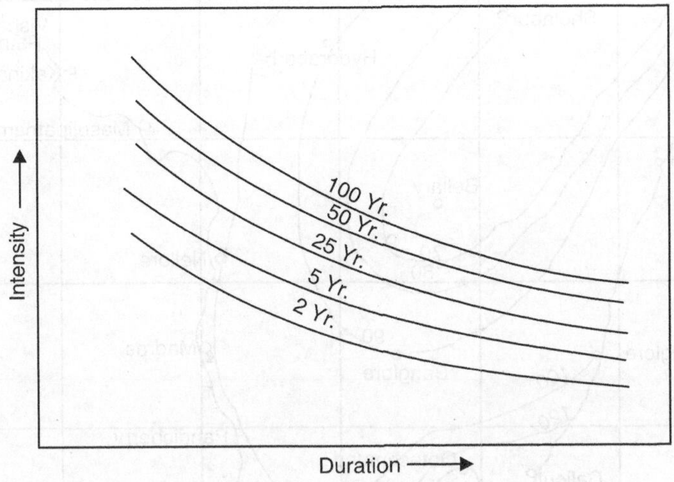

FIGURE 5.19 Intensity-frequency-duration curves.

Alternatively, for any constant return period an equation of the form given by Eq. (5.6) can be fitted between the maximum intensity and duration. That is

$$I = \frac{C_1}{(t + a_1)^{b_1}};\ I = \frac{C_2}{(t + a_2)^{b_2}};\ \text{etc.}$$

where C_1, a_1 and b_1 are applicable for one return period T_{r_1} and C_2, a_2 and b_2 are applicable for another return period T_{r2} etc. Generally it will be observed that the constants a and b are approximately same for all the return periods and only C is different for different return periods. In such a case one general equation may be developed for all the return periods as given by

$$I = \frac{KT_r^d}{(t + a)^b} \qquad ...(5.9)$$

where T_r is the return period in years and K and d are the regression constants for a given location. If a and b are not same for all the return periods, then an individual equation for each return period may be used.

When the intensity-frequency-duration analysis is carried out for a number of locations in a region, the relationships may be given in the form of Eq. (5.9) with a different set of regression constants for each location, or alternatively they may be presented in the form of maps (with each map depicting maximum rainfall depths for different combinations of one return period and one duration) which can be more conveniently used especially when one is dealing with large areas. Such maps are called the *isopluvial* maps. A map showing the maximum rainfall depths for a duration of one hour which can be expected with a frequency of once in 50 years over the south India is given in Fig. 5.20.

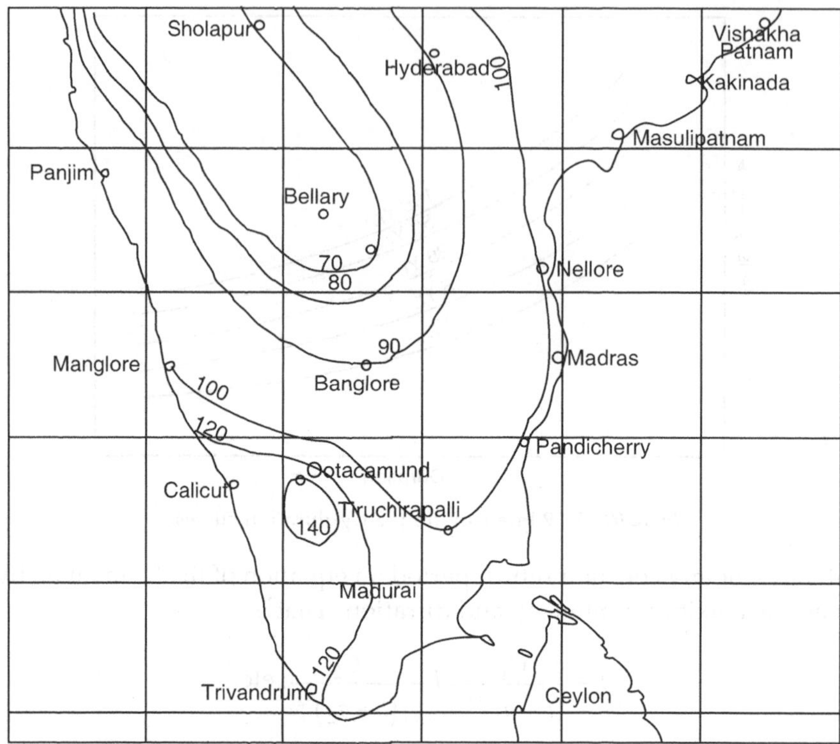

FIGURE 5.20 Isopluvial map of 50 years 1 hour rainfall over South India.

5.12 AVERAGE DEPTH OF PRECIPITATION OVER AN AREA

Since most hydrological problems require a knowledge of the average depth of rainfall over a significant area such as a river basin, some procedure must be used to connect the rainfall measured at the individual raingauges to the areal averages. The average depth of rainfall is also termed as *equivalent uniform depth of rainfall.* It is never possible to determine the exact average depth of rainfall. Because the raingauges give, at best, a very small sample. There are three methods of treating the raingauge records to arrive at an approximate answer and in general the three methods give three different approximations. They are : (*i*) *Arithmetic mean method* (*ii*) *Thiessen Polygon method* and (*iii*) *Isohyetal method.* The first two methods are purely mechanical processes requiring no special skill or judgement. On the other hand, the results obtained by the third method, which perhaps should be the most accurate, depend for their accuracy upon good judgement of the person making the computations.

5.12.1 Arithmetic Mean Method

This is the simplest of the three methods. As the name implies the result is obtained by dividing the sum of the rainfall amounts recorded at all the raingauge stations which are located within the area under consideration by the number of stations. That is

$$P = \frac{P_1 + P_2 + \ldots\ldots + P_n}{n} = \frac{1}{n}\sum_{i=1}^{n} P_i \qquad \ldots(5.10)$$

where P is the average depth of rainfall and $P_1, P_2, \ldots\ldots, P_n$ are the rainfalls recorded at stations 1, 2, ... etc. and n is the number of raingauge stations within the area. This method is also known as the *unweighted mean method* since the same weightage is given to rainfall recorded at all the gauges irrespective of their locations.

If the area is flat and the gauges are distributed uniformly over the area and if the variation of individual gauge record from the mean is not too large, this method is probably as accurate as any other method. Even in hilly terrains this method will yield fairly satisfactory results if the orographic influences on rainfall are considered in selecting the gauge sites. The method is very rapid even by manual computations and has got excellent adaptability to computer processing.

5.12.2 Thiessen Polygon Method

This method takes its name after A.M. Thiessen who suggested it first in 1911. It attempts to make allowance for irregularities in gauge locations by weighing the record of each gauge in proportion to the area which is closer to that gauge than to any other gauge. These areas for a given network of gauge stations are determined graphically as described below and as shown in Fig. 5.21.

(*i*) The basin area is plotted to some scale and on the same map the locations of the raingauge stations both within the area and also outside the area but nearby are indicated.

(*ii*) Straight lines are then drawn joining the adjacent raingauge locations to form triangles. Whenever it is required to divide a quadrilateral into two triangles, the shorter diagonal is preferred.

(*iii*) Perpendicular bisectors are now drawn to each side of all the triangles. These bisectors define a set of polygons one for each gauge, and each polygon contains only points that are closer to the gauge at its centre than to any other gauge.

(*iv*) The polygonal areas around each raingauge station within the basin boundary are then measured.

(*v*) The average depth of rainfall P is computed as

$$P = \frac{A_1 P_1 + A_2 P_2 + \ldots\ldots + A_n P_n}{A_1 + A_2 + \ldots\ldots + A_n} \qquad \ldots(5.11)$$

where $P_1, P_2, \ldots\ldots\ldots\ldots, P_n$ are the rainfalls recorded at raingauge stations with $A_1, A_2, \ldots\ldots\ldots, A_n$ as the polygonal areas around them. Since $A_1 + A_2 + \ldots\ldots\ldots\ldots + A_n = A$, the area of the basin, Eq. (5.11) can be written in the form

$$P = \frac{A_1}{A} P_1 + \frac{A_2}{A} P_2 + \ldots\ldots\ldots + \frac{A_n}{A} P_n \qquad \ldots(5.12)$$

The factors $\frac{A_1}{A}, \frac{A_2}{A}, ..., \frac{A_n}{A}$, which add up to unity are called the *Thiessen weights*. Thus the average depth of rainfall obtained by this method may be interpreted as the sum of the rainfalls multiplied by the respective Thiessen weights. For this reason this method is also known as the *weighted mean method.*

This method is fast when once the weights are known. It makes use of the rainfall record of stations located at a short distance beyond the boundary of the basin and their influence diminishes as their distance from the boundary increases. This is as it should be. On the other hand , in the arithmetic mean method every station is given equal weightage regardless of its location.

This method does not make any allowance for orographic influences in the basin. Every-time a new gauge is added or whenever a gauge is abandoned or fails to report, the network changes and the weights have to be determined for the new network. This inflexibility is the greatest limitation to this method.

5.12.3 Isohyetal Method

Perhaps the most accurate method of computing the average depth of rainfall is the isohyetal method. An *isohyet* may be defined as a line joining points of equal rainfall. In other words all the places along an isohyet experience the same amount of rainfall. The basin area along with the raingauge station locations in and around the area is drawn to a suitable scale. The rainfall amounts recorded by each station are indicated at the station locations. The isohyets are then drawn on the same map. First the points of a required isohyet are located by linear interpolation between the adjacent gauges. Then between the points thus located, the isohyets are faired by eye to produce smooth curves giving reasonable consideration to the evidence of nearby gauges not involved in the interpolation. The isohyets constructed using this procedure are shown in Fig. 5.22. The volume of rainfall between any two successive isohyets is computed as the product of the area between those isohyets and the average of the two isohyetal values. All such volumes in the basin area are totalled and divided by the basin area to yield the average depth of rainfall.

The isohyetal method utilises all the data available in the interpretation and it is well adapted to display and discussion. In constructing an isohyetal map the analyst can make full use of his knowledge of orographic effects and storm morphology. For example, while constructing the isohyets for a rugged hilly terrain the interpolation may be based on the difference in elevations of the adjacent raingauge stations rather than on the distance between them. The isohyets may be constructed at closer intervals around the local heavy rainfall centres.

The accuracy of the method is highly dependant upon the skill of the analyst. The method is very slow compared to the previous two. Its adaptability to computer processing is very poor.

Example 5.7. *Compute the average depth of rainfall for the catchment shown in Fig. 5.21 with an area of 200 km², using all the three methods. The rainfall amounts are indicated in mm at the respective gauge sites. Also compute the volume of rain water in m³ received by the catchment area in each case.*

Solution. The average depth of rainfall is determined by all the three methods as given below.

(*i*) *Arithmetic mean method*

$$P = \frac{P_1 + P_2 + \ldots\ldots + P_n}{n}$$

There are only 4 raingauge stations within the catchment area. Therefore

$$P = \frac{37 + 49 + 38 + 48}{4} = \frac{172}{4} = 43 \text{ mm.}$$

Volume of rainwater = Area × depth of rainfall = $A \times P$

$$= 200 \times 43 \text{ km}^2\text{-mm}$$

$$= (200 \times 1000 \times 1000) \times \left(\frac{43}{1000}\right) = 86 \times 10^5 \text{ m}^3.$$

(*ii*) *Thiessen Polygon method*

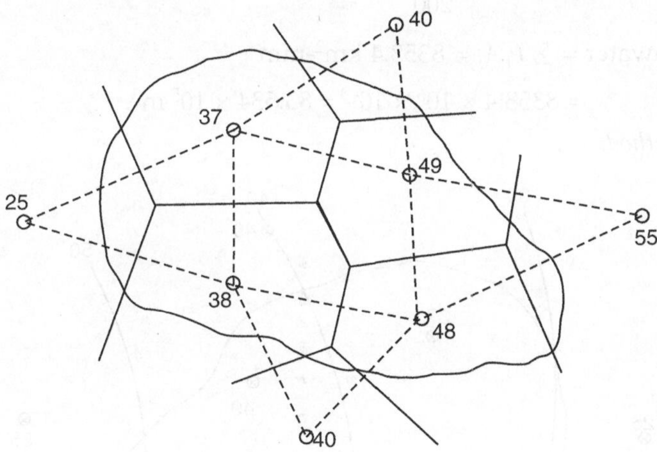

FIGURE 5.21 Thiessen polygon.

The construction of Thiessen polygons for the given raingauge network in the catchment is shown in Fig. 5.21. The computations involved in the determination of average depth of rainfall are given in Table 5.5.

Table 5.5. Computations for Thiessen Polygon Method

Station No. i	*Rainfall in mm* P_i	*Thiessen Polygon area within catchment boundary km*^{2} A_i	*Rainfall Volume in km*^{2}*-mm* $A_i \times P_i$
1	40	7.2	288.0
2	25	10.4	260.0
3	37	49.8	1842.6
4	49	35.8	1754.2
5	55	6.6	363.0
6	38	47.2	1793.6
7	48	41.5	1992.0
8	40	1.5	60.0
	Totals	200.0	8353.4

Average depth of rainfall $P = \dfrac{\Sigma P_i A_i}{A}$

$$= \frac{8353.4}{200} = 41.77 \text{ mm}$$

Volume of rainwater $= \Sigma P_i A_i = 8353.4$ km^2-mm

$$= 8353.4 \times 10^6 \times 10^{-3} = 83.534 \times 10^5 \text{ m}^3.$$

(*iii*) *Isohyetal method.*

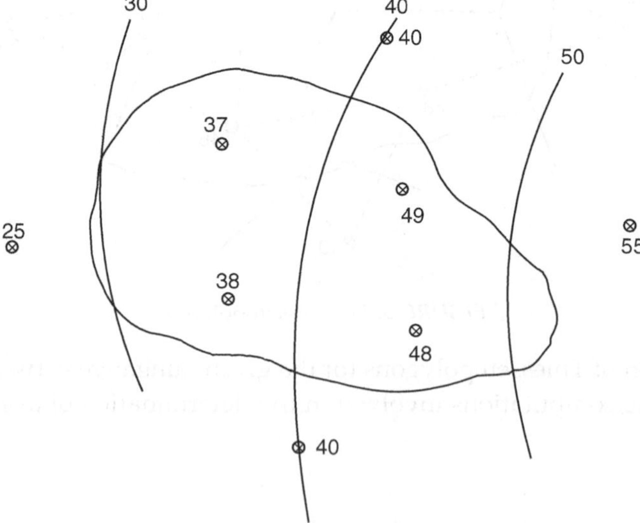

FIGURE 5.22 Isohyetal map.

The isohyets of 30, 40 and 50 mm rainfalls are drawn on the catchment area map and are shown in Fig. 5.22. The areas between successive isohyets are measured and tabulated in Table 5.6 to facilitate the determination of the average depth of rainfall.

Table 5.6. Computations for Isohyetal Method

Isohyetal value mm	*Enclosed area km^2*	*Net area km^2*	*Average rainfall mm*	*Rainfall volume km^2-mm*
50	14.0	14.0	53	742
40	103.2	89.2	45	4014
30	199.0	95.8	35	3353
< 30	200.0	1.0	29	29
				Total = 8138

$$\text{Average depth of rainfall } P = \frac{8138}{200} = 40.69 \text{ mm}$$

$$\text{Volume of rain water} = 8138 \text{ km}^2\text{-mm}$$
$$= 81.38 \times 10^5 \text{ m}^3.$$

Example 5.8. *Thiessen polygons constructed for a network of 10 raingauges in a river basin yielded Thiessen weights of 0.10, 0.16, 0.12, 0.11, 0.09, 0.08, 0.07, 0.11, 0.06 and 0.10. If the rainfalls recorded at these gauges during a cyclonic storm are 132, 114, 162, 138, 207, 156, 135, 158, 168 and 150 mm respectively determine the average depth of rainfall by Thiessen mean and arithmetic mean methods. Also determine the volume of surface runoff at the basin outlet if 35% of the rainfall is lost as infiltration. Take the area of the basin as 5800 km^2 and express your answer in million cubic metres.*

Solution. By arithmetic mean method, the average depth of rainfall is given by

$$P = \frac{132 + 114 + 162 + 138 + 207 + 156 + 135 + 158 + 168 + 150}{10}$$

$$= \frac{1520}{10} = 152 \text{ mm}$$

By Thiessen Polygon method, the average depth of rainfall is given by

$$P = \Sigma\, w_i P_i$$

where w_i is the Thiessen weight of ith raingauge whose rainfall is P_i

$$P = [0.10 \times 132 + 0.16 \times 114 + 0.12 \times 162 + 0.11 \times 138 + 0.09 \times 207 + 0.08 \times 156 + 0.07 \times 135 + 0.11 \times 158 + 0.06 \times 168 + 0.10 \times 150]$$

$$= [13.20 + 18.24 + 19.44 + 15.18 + 18.63 + 12.48 + 9.45 + 17.38 + 10.08 + 15.00]$$

$$= 149.08 \text{ mm}$$

Runoff depth = Rainfall – Loss

$$= 149.08 - 0.35 \times 149.08$$

$$= 96.9 \text{ mm.}$$

Volume of runoff = Area × runoff depth

$$= 5800 \times 96.9 \text{ km}^2\text{-mm}$$

$$= 562020 \times 10^3 \text{ m}^3$$

$$= 562.02 \text{ million cubic metres.}$$

Example 5.9. *A semicircle of diameter 20 km with an equilater triangle of side 20 km below its diameter is a close approximation to a river basin. The position coordinates of 5 raingauge stations A, B, C, D, and E located within the basin with respect to a coordinate axes system whose x-axis and origin are coincident with diameter and centre of the circle are (5, 5), (–5, 5), (–5, –5), (5, –5) and (0, 0) km respectively. If the rainfall recorded at these raingauge are 85, 92, 77, 80, and 105 mm respectively determine the average depth of rainfall using Thiessen polygon method.*

Solution. The area of the basin is shown in Fig. 5.23 with the locations of the raingauge stations A, B, C, D and E marked on it

The area of the basin A = area of semicircle + area of triangle

$$\text{Area of semicircle} = \frac{\pi \times 10^2}{2} = 157.08 \text{ km}^2$$

$$\text{Area of triangle} = \frac{1}{2} \times \text{base} \times \text{height}$$

$$\text{Height} = \sqrt{(20^2 - 10^2)} = \sqrt{300} \text{ km}$$

$$\therefore \quad \text{Area of triangle} = \frac{1}{2} \times 20 \times \sqrt{300} = 173.20 \text{ km}^2$$

$$\therefore \quad A = 157.08 + 173.2 = 330.28 \text{ km}^2$$

It can be seen from the Thiessen polygons constructed for the raingauge stations that the polygon of station E is a square of side $5\sqrt{2}$ km.

$\therefore$ Area of polygon for station $E = A_e = 5\sqrt{2} \times 5\sqrt{2} = 50 \text{ km}^2$.

Area of the polygons for A and B are equal and each is given by

$$\left[\frac{1}{2}(\text{area of semicircle}) - \frac{1}{4}(\text{area of square})\right]$$

$$\therefore \quad A_a = A_b = \frac{157.08}{2} - \frac{50}{4} = 66.04 \text{ km}^2.$$

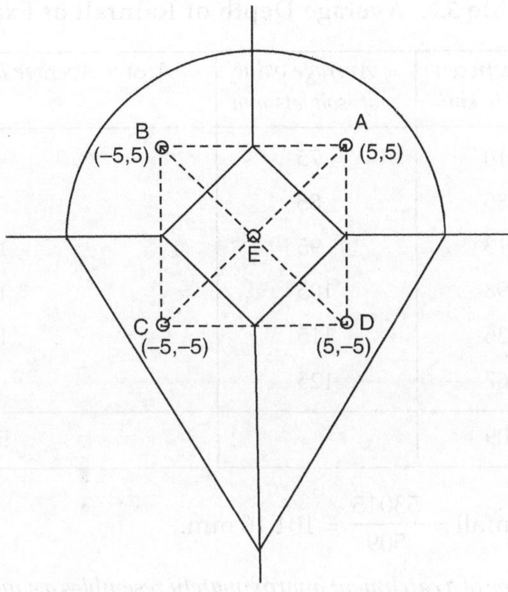

FIGURE 5.23 River basin of Example 5.9.

Again, area of the polygons for C and D are equal and each is given by

$$\left[\frac{1}{2}(\text{area of triangle}) - \frac{1}{4}(\text{area of square})\right]$$

$$\therefore \qquad A_c = A_d = \frac{173.2}{2} - \frac{50}{4} = 74.1 \text{ km}^2.$$

Now, the average depth of rainfall P is given by

$$P = \frac{P_a \times A_a + P_b \times A_b + P_c \times A_c + P_d \times A_d + P_e \times A_e}{A}$$

$$= \frac{85 \times 66.04 + 92 \times 66.04 + 77 \times 74.1 + 80 \times 74.1 + 105 \times 50}{330.28}$$

$$= \frac{28572.78}{330.28} = 86.51 \text{ mm}.$$

Example 5.10. *The analysis of a storm yielded the following information regarding isohyets. Calculate the average depth of rainfall.*

Isohyet interval in mm	*70–80*	*80–90*	*90–100*	*100–110*	*110–120*	*120–130*
Area in km²	*10*	*85*	*113*	*98*	*136*	*67*

Solution. The working is shown in Table 5.7.

Table 5.7. Average Depth of Rainfall of Example 5.10

Isohyetal interval mm	*Area between isohyets km²*	*Average value of isohyets mm*	*Area × Average isohyetal value km²-mm*
70–80	10	75	750
80– 90	85	85	7225
90–100	113	95	10735
100–110	98	105	10290
110–120	136	115	15640
120–130	67	125	8375
Σ	509		53015

$$\text{Average depth of rainfall} = \frac{53015}{509} = 104.16 \text{ mm.}$$

Example 5.11. *The shape of a catchment approximately resembles a square of side 8 km. With reference to an x-y coordinate frame whose origin is coinciding with one of the corners of the catchment the locations of the four corners of the catchment are (0, 0), (8, 0), (8, 8) and (0, 8). There are four raingauges A, B, C and D within this catchment whose positions with reference to the same coordinate frame are (2, 2), (6, 2), (6, 6) and (2, 6) respectively. The rainfall recorded by the raingauges A, B, C and D during a storm are 8, 6, 9 and 11 cm respectively. Determine the average depths of rainfall over the catchment by Arithmetic mean and Thiessen methods.*

Solution. By arithmetic mean method

$$\overline{P} = \frac{P_A + P_B + P_C + P_D}{4}$$

$$= \frac{8 + 6 + 9 + 11}{4} = \frac{34}{4} = 8.5 \text{ cm}$$

To determine $\overline{P}$ by the Thiessen polygon method, first the catchment is drawn to scale and the raingauges A, B, C and D are located on the same area as shown in Fig. 5.24 using the given information on the positional coordinates. The raingauge stations are connected by dashed lines and perpendicular bisectors to these dashed lines are then drawn as solid lines. It can be observed that the Thiessen polygons formed around all the 4 raingauge stations are squares of same size 4 km × 4 km = 16 km². The area of the catchment is A = 8 km × 8 km = 64 km².

$$\overline{P} = \frac{A_A}{A} P_A + \frac{A_B}{A} P_B + \frac{A_C}{A} P_C + \frac{A_D}{A} P_D$$

$$= \frac{16}{64} \times 8 + \frac{16}{64} \times 6 + \frac{16}{64} \times 9 + \frac{16}{64} \times 11 = \frac{1}{4} (8 + 6 + 9 + 11)$$

$$= \frac{34}{4} = 8.5 \text{ cm.}$$

Both the methods give same answer. This is because the Thiessen weights are same for all the raingauge stations.

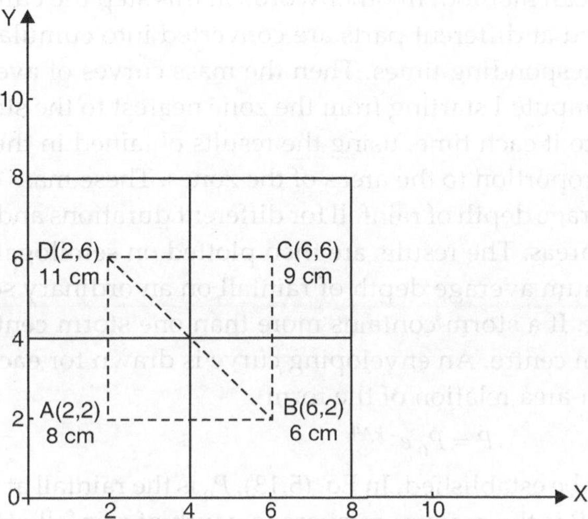

FIGURE 5.24 Thiessen polygon for Example 5.10.

5.13 DEPTH-AREA-DURATION ANALYSIS

In the previous sections, the variation of storm rainfall within its duration and over the area of its occurrence have been discussed separately. But in most of the design applications the maximum depth of rainfall that is likely to occur over a given area for a given duration is required. Where possible, the frequency (or the return period) of that rainfall should also be known. For example, the knowledge of maximum depth of rainfall occurring on areas of various sizes for storms of different durations is of interest in many hydrological design problems such as the design of culverts, design of minor irrigation schemes etc.

A storm of given duration over a particular area rarely produces uniform rainfall depth over the entire area. The storm usually has a centre where the rainfall P_0, is maximum and this is always larger than the average depth of rainfall P, applicable for the area as a whole. Generally the difference between these two values, that is $(P_0 - P)$, increases with increase in area and decreases with increase in the duration. Also the difference is more for convective and orographic precipitation than for cyclonic. To obtain a quantitative relationship between P_0 and P, series of storms with the data obtained from recording raingauges are to be analysed. The analysis of a typical storm is described below followed by a worked out example.

The storm rainfall data is plotted on the basin map and the isohyets are drawn. These isohyets divide the area into various zones. On the same map the Thiessen polygons are also constructed for all the raingauge stations. The polygon of a raingauge station may lie in different zones. Thus each zone will be influenced by a certain number of gauges, whose polygonal areas lie either fully or partially in that zone. The gauges which influence each zone along with their influencing areas are noted. Next for each zone the cumulative average depth of rainfall (areal average) is computed at various times using the data of rainfall mass curve at the gauges influencing the zone and applying the Thiessen weighted mean method. In other words in this step the cumulative depths of rainfall at different times recorded at different parts are converted into cumulative depths of rainfall for the zonal area at the corresponding times. Then the mass curves of average depth of rainfall for accumulated area are computed starting from the zone nearest to the storm centre and by adding one more zone adjacent to it each time, using the results obtained in the previous step and using the Thiessen weight in proportion to the areas of the zones. These mass curves are now examined to find the maximum average depth of rainfall for different durations and for various progressively increasing accumulated areas. The results are then plotted on semilogarithmic paper. That is, for each duration the maximum average depth of rainfall on an ordinary scale is plotted against the area on logarithmic scale. If a storm contains more than one storm centres, the above analysis is carried out for each storm centre. An enveloping curve is drawn for each duration. Alternatively, for each duration a depth-area relation of the form

$$P = P_0\, e^{-kA^n} \qquad ...(5.13)$$

proposed by Horton may be established. In Eq. (5.13), P_0 is the rainfall at the centre of the storm for any any given duration, P is the maximum average depth of rainfall over an area A for the same duration, and k and n are the constants to be obtained from the regression analysis. Another form of the equation that is suggested which may be fitted to the depth-area data is

$$\frac{100P}{P_0} = 100 - C\sqrt{A} \qquad ...(5.14)$$

where P_0, P and A retain the same meaning as in Eq. (5.13) and C is again a constant for a given duration. The following numerical example will illustrate the method just described.

Example 5.12. *The rainfall data obtained from 7 raingauges located in and around the basin area shown in Fig. 5.24 is as given below. The basin has an area of 2790 km². Analyse this data and develop the depth-area-duration curves for the storm.*

Time	*Cumulative rainfall in mm*						
in hours	A	B	C	D	E	F	G
4	0	0	0	0	0	0	0
6	12	0	0	0	0	0	0
8	18	15	0	0	0	6	0
10	27	24	0	0	9	15	6
12	36	36	18	6	24	24	9
14	42	45	36	18	36	33	15
16	51	51	51	36	42	36	18
18	51	63	66	51	60	39	18
20	51	72	87	66	66	42	18
22	51	72	96	81	66	42	18
24	51	72	96	81	66	42	18

Solution. The total rainfalls of 51, 72, 96, 81, 66, 42 and 18 mm are indicated at the respective raingauge stations *A*, *B*, *C*, *D*, *E*, *F* and *G* on the map. The isohyets with the values of 30, 45, 60 and 75 mm are constructed. These isohyets divide the basin area into five zones with areas as given below:

Zone I	415 km^2
Zone II	640 km^2
Zone III	1015 km^2
Zone IV	525 km^2
Zone V	195 km^2
Total:	2790 km^2

The Thiessen polygons are then constructed for the given raingauge network on the same map. From the Thiessen polygons and the isohyets the following information is noted :

Zone I is influenced by 3 stations *B*, *C* and *D* with areas of 105, 57 and 253 km^2 respectively.

Zone II by 4 stations *A*, *B*, *D* and *E* with areas of 37, 283, 20 and 300 km^2.

Zone III by 4 stations *A*, *B*, *E* and *F* with areas of 640, 20, 185 and 170 km^2.

Zone IV by 3 stations *A*, *F* and *G* with areas of 202, 275 and 48 km^2.

Zone V by stations *F* and *G* with areas of 37 and 158 km^2.

As can be seen from Fig. 5.25, Zone I is the nearest to storm centre while Zone V is the farthest.

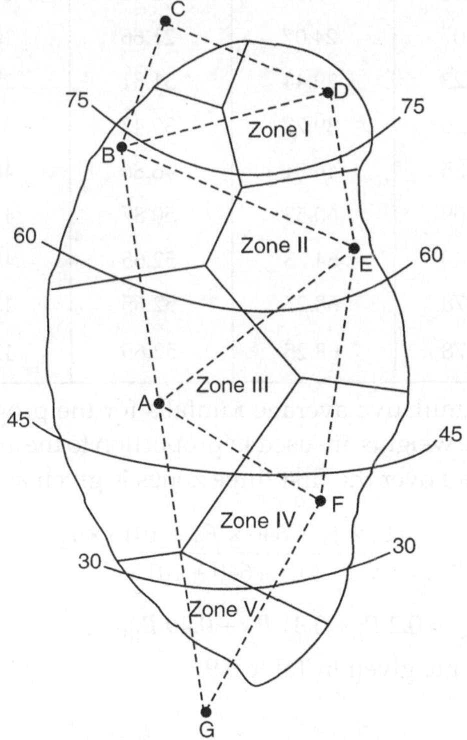

FIGURE 5.25 Depth-area-duration analysis.

The cumulative average depth of rainfall for each zone can be calculated using the data at stations *A, B, C, D, E, F* and *G* and the corresponding Thiessen weights. For example, the average depth of rainfall in Zone I at any time, P_I is computed from

$$P_I = \frac{105 \times P_B + 57 \times P_C + 253 \times P_D}{(105 + 57 + 253)}$$

where P_B, P_C and P_D are the cumulative rainfalls at stations *B, C* and *D* at the same time. That is

$$P_I = 0.253\ P_B + 0.137\ P_C + 0.61\ P_D$$

Similarly for zone II, we have

$$P_{II} = \frac{37 \times P_A + 283 \times P_B + 20 \times P_D + 300 \times P_E}{(37 + 283 + 20 + 300)}$$

or

$$P_{II} = 0.058\ P_A + 0.442\ P_B + 0.031\ P_D + 0.469\ P_E$$

These results are shown in Table 5.8.

Table 5.8. Cumulative Average Rainfalls in Various Zones in mm

Time	*Zone I*	*Zone II*	*Zone III*	*Zone IV*	*Zone V*
4	0	0	0	0	0
6	0	0.70	7.60	4.62	0
8	3.80	7.67	12.66	10.07	1.14
10	6.07	24.07	21.66	18.80	7.71
12	15.23	29.44	31.81	27.26	11.85
14	27.30	39.77	39.47	34.83	18.42
16	41.85	46.31	46.86	40.14	21.42
18	56.09	60.53	50.87	41.71	21.99
20	70.40	64.78	52.65	43.28	22.56
22	80.78	68.25	52.65	43.28	22.56
24	80.78	68.25	52.65	43.28	22.56

In the next step the cumulative average rainfalls for the progressively accumulated areas are worked out. Here again the weights are used in proportion to the areas of the zones. For example, the cumulative average rainfall over the first three zones is given as

$$P_{\mathrm{I+II+III}} = \frac{415 \times P_{\mathrm{I}} + 640 \times P_{\mathrm{II}} + 1015 \times P_{\mathrm{III}}}{415 + 640 + 1015}$$

$$= 0.2\ P_{\mathrm{I}} + 0.31\ P_{\mathrm{II}} + 0.49\ P_{\mathrm{III}}$$

The results of this step are given in Table 5.9

Table 5.9. Cumulative Average Rainfalls for Accumulated Areas in mm

Time hours	*I 330 km²*	*I + II 1085 km²*	*I + II + III 2280 km²*	*I + II + III + IV 2715 km²*	*I + II + III + IV + V 2790 km²*
4	0	0	0	0	0
6	0	0.43	3.94	4.08	3.79
8	3.80	6.15	9.34	9.49	8.91
10	6.07	17.00	19.28	19.18	18.38
12	15.23	23.86	27.76	27.66	26.55
14	27.30	34.87	37.12	36.66	35.38
16	41.85	44.56	45.69	44.57	42.95
18	56.09	58.79	54.91	52.24	50.12
20	70.40	66.99	59.96	56.59	54.21
22	80.78	73.17	63.12	59.11	56.55
24	80.78	73.17	63.12	59.11	56.55

Now for any zone the maximum average depths of rainfall for various durations of 4, 8, 12, 16 and 20 h can be obtained from Table 5.9 using the procedure explained in Example 5.5. These results are tabulated in Table 5.10.

Table 5.10. Maximum Depths of Rainfall for Accumulated Areas

Duration in hours	*Maximum average depths of rainfall in mm*				
	330 km²	*1085 km²*	*2280 km²*	*2715 km²*	*2790 km²*
4	28.79	23.92	18.42	18.17	17.64
8	55.17	43.13	36.35	35.08	34.04
12	74.71	60.84	50.97	48.16	46.33
16	80.78	72.74	59.18	56.59	54.21
20	80.78	73.17	63.12	59.11	56.55

For each duration, the maximum depths of rainfalls are plotted against the area on logarithmic scale as shown in Fig. 5.26.

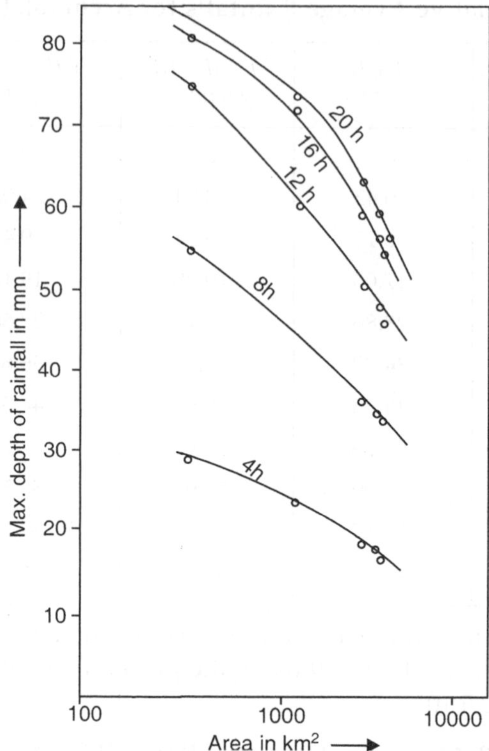

FIGURE 5.26 Depth-area-duration curves.

5.14 ESTIMATION OF MISSING PRECIPITATION RECORDS

The record at many raingauge stations may consist of short breaks due to several reasons such as the absence of the observer, instrumental failures etc. It is better to estimate these missing records and fill the gaps rather than to leave them. This applies especially when data is processed with an automatic equipment like an electronic computer.

Interpolation from Isohyetal Maps. Missing precipitation records may be interpolated from the isohyetal maps which cover the period in question. Skillfully drawn map takes local rainfall characteristics into account including the average patterns, distance from the sea, topography and storm paths. Interpolation of missing data from such maps will be reasonably accurate.

Arithmetic Average Method. If the normal annual precipitations at the adjacent stations are within 10% of the normal rainfall of the station under consideration, then the missing rainfall data may be estimated as a simple arithmetic average of the rainfalls at the adjacent gauges. Thus if the missing precipitation at station X is P_x, and $P_1, P_2, \ldots\ldots, P_m$ are the rainfalls at the m surrounding raingauges,

$$P_x = \frac{1}{m}(P_1 + P_2 + \ldots\ldots + P_m) \qquad \ldots(5.15)$$

Usually the data from three surroundings gauges will give good results.

Normal Ratio Method. If the normal annual rainfalls at the surrounding gauges differ from the normal annual rainfall of the station in question by more than 10%, the normal ratio method is preferred. In this method the rainfall values at the surrounding stations are weighed by the ratio of the normal annual rainfalls. That is

$$P_x = \frac{1}{m}\left[\frac{N_x}{N_1}P_1 + \frac{N_x}{N_2}P_2 + \ldots\ldots\ldots\ldots + \frac{N_x}{N_m}P_m\right] \qquad \text{...(5.16)}$$

where N_x is the normal annual rainfall at station X and $N_1, N_2, \ldots\ldots\ldots\ldots, N_m$ are the normal annual rainfalls at the m surrounding gauges respectively. A minimum of 3 surrounding stations are generally used in the normal-ratio method.

Example 5.13. *Raingauge station X did not function for a part of a month during which a storm occurred. The storm produced rainfalls of 84, 70 and 96 mm at three surrounding stations A, B and C respectively. The normal annual rainfalls at the stations X, A, B and C are respectively 770, 882, 736 and 944 mm. Estimate the missing storm rainfall at station X.*

Solution. The missing rainfall is given by

$$P_x = \frac{1}{3}\left[\frac{N_x}{N_a}P_a + \frac{N_x}{N_b}P_b + \frac{N_x}{N_c}P_c\right]$$

Here $N_x = 770$ mm and $P_a = 84$ mm

$N_a = 882$ mm $\quad P_b = 70$ mm

$N_b = 736$ mm $\quad P_b = 96$ mm

$N_c = 944$ mm

$$\therefore \quad P_x = \frac{1}{3}\left[\frac{770}{882}\times 84 + \frac{770}{736}\times 70 + \frac{770}{944}\times 96\right]$$

$$= \frac{1}{3}[73.333 + 73.234 + 78.305]$$

$$= \frac{1}{3}\times 224.872 = 74.957 \approx 75 \text{ mm.}$$

$\therefore$ The missing storm rainfall at raingauge station X may be taken as 75 mm.

Regression Method. In the regression method a multiple linear relationship is assumed between the precipitation at station X, where the record is missing and the precipitation at the nearby stations. That is

$$P_x = a_0 + a_1 P_1 + a_2 P_2 + a_3 P_3 + \ldots\ldots\ldots + a_m P_m \qquad \text{...(5.17)}$$

Usually a minimum of 3 stations are included in the analysis. That is, in Eq. (5.17) $m \geq 3$. The constants $a_0, a_1, a_2, \ldots\ldots\ldots, a_m$ are obtained through multiple regression analysis of the previously recorded data on $P_x, P_1, P_2, \ldots\ldots, P_m$ as explained in Example 3.10.

The advantage of regression method is that it accounts for, to some extent, the deficiencies in the Normal-ratio method. Also, if large amount of data is required to be estimated a random error term $t.\ s_y$ can be added to Eq. (5.17) where t is the normal random number with mean zero and standard deviation unity and s_y is the standard error of estimate. The normal random number can be generated using random number generator technique or can be taken from the table of normal random numbers available in some of the standard books on mathematics. The standard error of estimate s_y can be computed from the equation

$$s_y = \sqrt{\frac{1}{n}\sum_{i=1}^{n}(\hat{y}_i - y_i)^2} \quad \text{...(5.18)}$$

in which y_i is the observed value P_x in the ith observation and $\hat{y}_i$ is the value computed from the multiple regression equation using the observed values of $P_1, P_2, ..., P_m$ in the same i^{th} observation and n is the number of observations. Adding the term $t\, s_y$ to the estimates preserves the standard deviation which will be close to the observed standard deviation.

The above procedure is demonstrated in Example 5.14.

Example 5.14. *The recorded rainfall at 4 raingauge stations in an area for a period of 8 years is given below*

Year	*Annual rainfall in cm at the 4 raingauge stations*			
	X	A	B	C
1991	—	*57.7*	*63.0*	*44.2*
1992	*121.0*	*108.3*	*100.9*	*95.1*
1993	*95.0*	*75.3*	*90.5*	*60.4*
1994	*69.0*	*67.5*	*58.6*	*62.3*
1995	*86.0*	*55.7*	*69.6*	*80.3*
1996	*59.9*	*52.3*	*75.4*	*85.1*
1997	*50.3*	*46.1*	*62.5*	*40.3*
1998	*34.8*	*47.6*	*52.0*	*30.1*

Estimate the missing rainfall at station X in the year 1991 using (i) arithmetic mean method (ii) normal-ratio method and (iii) multiple regression method.

Solution. First we find the normal annual rainfalls at stations X, A, B and C

$$N_x = \frac{121.0 + 95.0 + ... + 50.3 + 34.8}{7} = \frac{516}{7} = 73.71 \text{ cm}$$

$$N_A = \frac{57.7 + 108.3 + ... + 46.1 + 47.6}{8} = \frac{510.5}{8} = 63.81 \text{ cm}$$

$$N_B = \frac{63 + 100.9 + ... + 62.5 + 52.0}{8} = \frac{572.5}{8} = 71.56 \text{ cm}$$

$$N_C = \frac{44.2 + 95.1 + ... + 40.3 + 30.1}{8} = \frac{497.8}{8} = 62.23 \text{ cm}$$

Arithmetic mean method: If the normal annual rainfall at A, B and C are to be within 10% of the normal annual rainfall at X, they must lie within 0.9 N_x to 1.1 N_x. That is between 66.34 and 81.08 cm. As we notice, N_A and N_B are not within this range. Therefore this method cannot be applied to this problem.

Normal ratio method: According to this method, the missing precipitation P_x is given by

$$P_x = \frac{1}{3}\left[P_A \frac{N_x}{N_A} + P_B \frac{N_x}{N_B} + P_C \frac{N_x}{N_C}\right]$$

$$= \frac{1}{3}\left[57.7 \times \frac{73.71}{63.81} + 63 \times \frac{73.71}{71.56} + 44.2 \times \frac{73.71}{62.23}\right]$$

$$= \frac{1}{3}[66.65 + 64.89 + 52.35] = \frac{183.89}{3} = 61.3 \text{ cm.}$$

Regression method: We shall use the record of 1992 to 1998 in the regression analysis. Let us denote P_x by the dependent variable y and denote P_A, P_B and P_C by the independent variables x_1, x_2 and x_3 respectively. Then the multiple regression equation is written as

$$y = a_0 + a_1 x_1 + a_2 x_2 + a_3 x_3$$

Then according to Eq. (3.50) the normal equation to provide solutions for a_0, a_1, a_2 and a_3 are given by

$$\left.\begin{aligned} \Sigma y_i &= n\, a_0 + a_1 \Sigma x_{1i} + a_2 \Sigma x_{2i} + a_3 \Sigma x_{3i} \\ \Sigma y_i\, x_{1i} &= a_0 \Sigma x_{1i} + a_1 \Sigma x_{1i}^2 + a_2 \Sigma x_{1i}\, x_{2i} + a_3 \Sigma x_{1i}\, x_{3i} \\ \Sigma y_i\, x_{2i} &= a_0 \Sigma x_{2i} + a_1 \Sigma x_{2i}\, x_{1i} + a_2 \Sigma x_{2i}^2 + a_3 \Sigma x_{3i}\, x_{2i} \\ \Sigma y_i\, x_{3i} &= a_0 \Sigma x_{3i} + a_1 \Sigma x_{1i}\, x_{3i} + a_2 \Sigma x_{2i}\, x_{3i} + a_3 \Sigma x_{3i}^2 \end{aligned}\right\} \quad \text{...(5.19)}$$

The various sums required for solving the above equations are computed as shown in Table 5.11. Substituting the summations, we get

$$516.0 = 7\, a_0 + 452.8\, a_1 + 509.5\, a_2 + 453.6\, a_3$$

$$36813.6 = 452.8\, a_0 + 32184.1\, a_1 + 34874.3\, a_2 + 31266.7\, a_3$$

$$40305.3 = 509.5\, a_0 + 34874.3\, a_1 + 38944.8\, a_2 + 34802.0\, a_3$$

$$36621.7 = 453.6\, a_0 + 31266.7\, a_1 + 34802.0\, a_2 + 32793.7\, a_3$$

When the above 4 simultaneous equations are solved, we get

$$a_0 = -25.9702\,;\ a_1 = 0.6314\,;\ a_2 = 0.5478\,;\ \text{and } a_3 = 0.2926$$

Table 5.11. Computation of sums for Multiple Regression Analysis of Example 5.14

i	y_i	x_{1i}	x_{2i}	x_{3i}	x_{1i}^2	x_{2i}^2	x_{3i}^2	$x_{1i}x_{2i}$	$x_{2i}x_{3i}$	$x_{3i}x_{1i}$	y_ix_{1i}	y_ix_{2i}	y_ix_{3i}	$\hat{y}_i$	$(y_i-\hat{y}_i)$	$(y_i-\hat{y}_i)^2$
1	121.0	108.3	100.9	95.1	11728.9	10180.8	9044.0	10927.5	9595.6	10299.3	13104.3	12208.9	11507.1	125.5	– 4.5	20.25
2	95.0	75.3	90.5	60.4	5670.1	8190.3	3648.2	6814.7	5466.2	4548.1	7153.5	8597.5	5738.0	88.8	6.2	38.44
3	69.0	67.5	58.6	62.3	4556.3	3434.0	3881.3	3955.5	3650.8	4205.3	4657.5	4043.4	4298.7	67.0	2.0	4.0
4	86.0	55.7	69.6	80.3	3102.5	4844.2	6448.1	3876.7	5588.9	4472.7	4790.2	5985.6	6905.8	71.1	14.9	222.01
5	59.9	52.3	75.4	85.1	2735.3	5685.2	7242.0	3943.4	6416.5	4450.7	3132.8	4516.5	5097.5	73.3	– 13.4	179.56
6	50.3	46.1	62.5	40.3	2125.2	3906.3	1624.1	2881.3	2818.8	1857.8	2318.8	3143.8	2027.1	49.2	1.1	1.21
7	34.8	47.6	52.0	30.1	2265.8	2704.0	906.0	2475.2	1565.2	1432.8	1656.5	1809.6	1047.5	41.4	– 6.6	43.56
Σ	516.0	452.8	509.5	453.6	32184.1	38944.8	32793.7	34874.3	34802.0	31266.7	36813.6	40305.3	36621.7			509.03

$$s_y = \sqrt{\sum_{i=1}^{7} \frac{(y_i - \hat{y}_i)^2}{7}} = \sqrt{\frac{509.03}{7}} = 8.53 \text{ cm}$$

Thus, the multiple regression is given by

$$y = -25.9702 + 0.6314\, x_1 + 0.5478\, x_2 + 0.2926\, x_3$$

or
$$P_x = -25.9702 + 0.6314\, P_A + 0.5478\, P_B + 0.2926\, P_C$$

Now the missing precipitation at X in the year 1991 is computed as

$$P_x = -25.9702 + 0.6314 \times 57.7 + 0.5478 \times 63.0 + 0.2926 \times 44.2$$
$$= -25.9702 + 36.4318 + 34.5114 + 12.9329$$
$$= 57.906 \text{ cm.}$$

The estimated value $\hat{y}_i$ for any i can be calculated using the multiple regression equation. For example for $i = 5$, we have

$$\hat{y}_5 = -25.9702 + 0.6314 \times 52.3 + 0.5478 \times 75.4 + 0.2926 \times 85.1$$
$$= 73.3 \text{ cm}$$

Similarly $\hat{y}_i$ for all i are computed. Then the standard error of estimate s_y is computed as

$$s_y = \sqrt{\frac{1}{n}\sum_{i=1}^{n}(y_i - \hat{y}_i)^2} = \sqrt{\frac{509.03}{7}} = 8.53 \text{ cm}$$

Inverse Distance Method. This method is used by U.S. National weather service in the studies of river forecasting. In this method four quadrants are delineated by north-south and east-west lines passing through the raingauge station where the missing rainfall is to be estimated. One raingauge station in each quadrant, which is the nearest to the raingauge station under question in that quadrant, is selected. The weight applicable to each of these four stations is computed as the reciprocal of the square of the distance between this station and the origin of the quadrants. Then the rainfalls recorded at the four stations in the four quadrants are multiplied by their respective weights and added. The resulting sum is divided by the sum of the weights to yield the missing rainfall. If one or more quadrants contain no raingauge stations, as for example for stations in a coastal area, then the estimation involves only the remaining quadrants. That is,

$$P_x = \frac{\left[\frac{1}{r_1^2} \times P_1 + \frac{1}{r_2^2} \times P_2 + \frac{1}{r_3^2} \times P_3 + \frac{1}{r_4^2} \times P_4\right]}{\left[\frac{1}{r_1^2} + \frac{1}{r_2^2} + \frac{1}{r_3^2} + \frac{1}{r_4^2}\right]} \quad ...(5.20)$$

where P_x is the missing rainfall at station X, P_1, P_2, P_3 and P_4 are the rainfalls recorded at the raingauge stations in the four quadrants which are nearest to station X in the respective quadrants, and which are at distances of r_1, r_2, r_3 and r_4 from station X respectively.

Example 5.15. *Station X failed to report the rainfall recorded during a storm. With respect to east-west and north-south axes set up at station X, the coordinates of 4 surrounding gauges, which are the nearest to station X in the respective quadrants, are (10, 15), (– 8, 5), (– 12, – 9) and (5, – 15) km respectively. Determine the missing rainfall at X, if the storm rainfalls at the four surrounding gauges are 73, 89, 68 and 57 mm respectively.*

Solution. $P_1 = 73$ mm ; $P_2 = 89$ mm ; $P_3 = 68$ mm ; $P_4 = 57$ mm.

Let r_1 be the distance between the station in the first quadrant and the station X. Then we have

$$r_1^2 = 10^2 + 15^2 = 325.$$

The weightage factor for this station $= \frac{1}{r_1^2} = \frac{1}{325} = 0.00308$

Similarly, we obtain

$$r_2^2 = (-8)^2 + 5^2 = 89 \; ; \; \frac{1}{r_2^2} = 0.01124$$

$$r_3^2 = (-12)^2 + (-9)^2 = 225 \; ; \; \frac{1}{r_3^2} = 0.00444$$

$$r_4^2 = (5)^2 + (-15)^2 = 250 \; ; \; \frac{1}{r_4^2} = 0.004$$

Substituting these values in Eq. (5.20), we get

$$P_x = \frac{(0.00308 \times 73 + 0.01124 \times 89 + 0.00444 \times 68 + 0.004 \times 57)}{(0.00308 + 0.01124 + 0.00444 + 0.004)}$$

$$= \frac{1.75512}{0.02276} = 77.11 \text{ mm.}$$

∴ The missing rainfall at X may be taken as 77.11 mm.

Estimates of missing precipitation data are generally more reliable for general type storms over flat terrain or over relatively smooth windward mountain slopes. If the terrain is rugged or if the storms are affected by severe and spotty convective activity, the reliability gets lessened. In general estimates for long intervals like month or a year are more reliable than for short interval such as day.

5.15 DOUBLE MASS CURVE

The double mass curve technique is used to test the consistency of rainfall record at any raingauge station which is suspected to contain certain discrepancies. The inconsistencies in the rainfall data of a station may be due to various reasons. The raingauge might have been installed at different sites in the past. That is, though there is a long and continuous rainfall record, the entire data is not homogeneous with respect to the present location of the gauge. The exposure conditions of the gauge may have undergone a significant change due to the growth of trees or the construction of tall buildings in the proximity of the gauge site. There may have been a change in the instrument, say from 125 mm to 200 mm raingauge. The raingauge may have been faulty for a part of the period of record, or the method and accuracy of measurement may have been suspected. Although a note is usually made in the station registers of the known changes of site and instruments, their effect or other unknown factors throw doubt on the accuracy and consistency of the record during the period covered. In all such cases it is better to test the record of any gauge which is suspect and adjust the data if required. The double mass curve analysis tests the record for its consistency and accuracy and provides a correction factor to assure that the data is reasonably homogeneous throughout its length and is related to a known site. It enables the data to be estimated for the missing periods or to be extrapolated beyond the existing length of record. The principle of double mass curve analysis is as follows.

A certain number of stations (usually 5 as minimum) with reliable data of approximately the same length and in the same climatic region are selected as *base stations*. It is necessary that the base stations themselves should not have any inconsistencies. The average of the annual rainfalls at these base stations is computed for every year and is cumulatively added beginning with the latest year. It is convenient to start from the latest year because we would like to adjust the data (if found to be inconsistent) of station X which is under consideration so that after adjustment it will be consistent and homogeneous with respect to its present location. Also, sometimes a recently established station record may need to be extrapolated to give estimates for earlier years. A similar cumulative addition of annual rainfall of station X is computed. Now a graph is plotted between the cumulative rainfall of the base stations as abscissa and the cumulative rainfall of station X as the ordinate as shown in Fig. 5.27. The resulting plot is called the double mass curve. The year to which a plotted point belongs is indicated for few points on the curve at regular intervals of 5 years.

If the data of station X is homogeneous and consistent with the data of the base stations the double mass curve is nearly a straight line. If there is a change in slope of the double mass curve at a certain point of time (for example, at point C in Fig. 5.27 pertaining to year 1953), the data of station X from this time into past is either inaccurate or non-homogeneous. A change in regional meteorological or weather conditions would not have had any influence on the slope of the double mass curve because the stations as well as its surrounding base stations would have been equally affected. That portion of the data which is found to be inconsistent can be adjusted by extending the mass curve line with original slope and reading the adjusted values from this extended line as shown in Fig. 5.27. Alternatively the data found to be inconsistent can also be adjusted by multiplying it with a correction factor which is nothing but the ratio of the slope of the adjusted mass curve to the slope of the unadjusted mass curve. Again referring to Fig. 5.27 the data of station X is inconsistent during the period 1941 to 1953. The adjusted rainfall of station X in any year during this period, P′ can be obtained as

$$P' = \frac{\tan \alpha'}{\tan \alpha} P \qquad \text{...(5.21)}$$

where P is the observed rainfall at station X in the same year, $\tan \alpha'$ is the slope of the adjusted mass curve (slope of CB') and $\tan \alpha$ is the slope of the unadjusted mass curve (slope of CB).

Considerable caution should be exercised in applying the double mass curve technique. A change in slope should not be considered significant unless it persists for at least 5 years and there is a collaborating evidence of a change in location or exposure or some other change. There is a regional consistency in precipitation pattern for long periods of time but this consistency becomes less pronounced for shorter periods. Therefore, the double mass curve technique is not generally recommended for the adjustment of daily or storm rainfalls.

Example 5.16. *The annual rainfall at station X, and the average of annual rainfalls at 25 surrounding base stations in cm are given below for a period of 36 years starting from 1941.*

Year	*Rainfall at X*	*Average rainfall of base stations*	*Year*	*Rainfall at X*	*Average rainfall of base stations*
1941	*163*	*135*	*1959*	*112*	*123*
1942	*119*	*111*	*1960*	*95*	*142*
1943	*121*	*124*	*1961*	*106*	*92*
1944	*129*	*111*	*1962*	*81*	*91*
1945	*126*	*123*	*1963*	*116*	*131*
1946	*120*	*90*	*1964*	*112*	*104*
1947	*153*	*138*	*1965*	*80*	*97*
1948	*172*	*119*	*1966*	*88*	*111*
1949	*127*	*108*	*1967*	*85*	*114*
1950	*108*	*107*	*1968*	*90*	*92*
1951	*126*	*111*	*1969*	*120*	*146*
1952	*190*	*142*	*1970*	*72*	*93*
1953	*112*	*112*	*1971*	*113*	*138*
1954	*97*	*99*	*1972*	*82*	*112*
1955	*86*	*93*	*1973*	*116*	*117*
1956	*111*	*131*	*1974*	*122*	*152*
1957	*68*	*92*	*1975*	*73*	*90*
1958	*88*	*142*	*1976*	*74*	*104*

(*i*) *Check whether the data of station X is consistent.*

(*ii*) *In which year a change in regime indicated ?*

(*iii*) *Compute the mean annual rainfall for station X at its present site for the given 36 years period first without adjustment and secondly with the data adjusted for the change in regime.*

(*iv*) *Compute the adjusted annual rainfalls at station X for the affected period.*

Solution. The computations to obtain the cumulative totals required for plotting the double mass curve are set out in Table 5.12. As seen from this table the data is entered starting from the most recent year. The annual rainfall at X, the average of annual rainfalls at base stations and the year in which these rainfalls are recorded are entered in Col. (2), Col. (4) and Col. (1) respectively. The cumulative annual rainfalls of station X and the cumulative of the average of the annual rainfalls of base stations are respectively shown in Col. (3) and Col. (5). A graph plotted between the cumulative of base stations as the abscissa and the cumulative of station X as the ordinate is shown in Fig. 5.27. Also shown in this figure are the years to which certain plotted points belong. From this double mass curve the following answers are obtained.

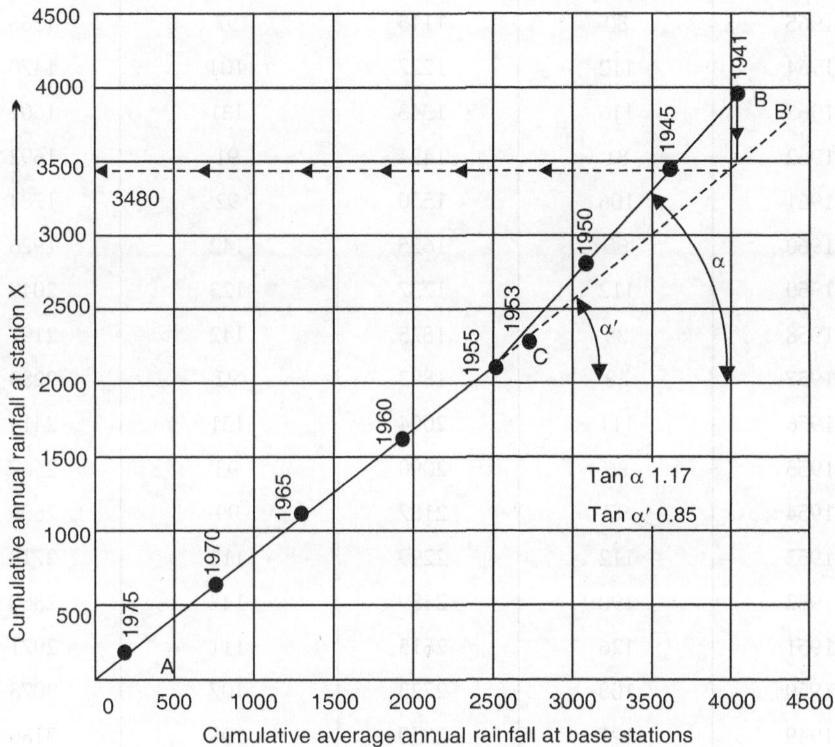

FIGURE 5.27 Double mass curve.

Table 5.12. Computations of Double Mass Curve

Year	*Annual rainfall at X cm*	*Cumulative rainfall at X cm*	*Average of annual rainfalls at 25 base stations cm*	*Cumulative rainfall of base stations cm*
1976	74	74	104	104
1975	73	147	90	194
1974	122	269	152	346
1973	116	385	117	463
1972	82	467	112	575
1971	113	580	138	713
1970	72	652	93	806
1969	120	772	146	952

1968	90	862	92	1044
1967	85	947	114	1158
1966	88	1035	111	1269
1965	80	1115	97	1366
1964	112	1227	104	1470
1963	116	1343	131	1601
1962	81	1424	91	1692
1961	106	1530	92	1784
1960	95	1625	142	1926
1959	112	1737	123	2049
1958	88	1825	142	2191
1957	68	1893	92	2283
1956	111	2004	131	2414
1955	86	2090	93	2507
1954	97	2187	99	2606
1953	112	2299	112	2718
1952	190	2489	142	2860
1951	126	2615	111	2971
1950	108	2723	107	3078
1949	127	2850	108	3186
1948	172	3022	119	3305
1947	153	3175	138	3443
1946	120	3295	90	3533
1945	126	3421	123	3656
1944	129	3550	111	3767
1943	121	3671	124	3891
1942	119	3790	111	4002
1941	163	3953	135	4137

(*i*) As the double mass curve is not having a uniform slope, it can be concluded that the record at station *X* is not consistent.

(*ii*) Since there is a break in the slope of the double mass curve in the year 1953, it may be said that the data at station *X* prior to 1953 is inconsistent and needs adjustment.

(*iii*) From Col. (3) of Table 5.12 the total of 36 years rainfalls at station *X* without adjustment = 3953 cm.

∴ The mean annual rainfall at X without adjustment

$$= \frac{3953}{36} = 109.8 \text{ cm.}$$

From the graph of double mass curve in Fig. 5.27, the cumulative rainfall at station X for 36 years after correction = 3480 cm.

∴ The mean annual rainfall at X after adjustment

$$= \frac{3480}{36} = 96.7 \text{ cm.}$$

(*iv*) The slope of the adjusted mass curve for the affected period of 1941 to 1953, that is, the slope of the line CB' = 0.85. The slope of the unadjusted mass curve, that is, the slope of the line CB = 1.17.

Therefore the correction factor for adjustment = $\frac{0.85}{1.17}$ = 0.7265. The adjusted annual rainfalls at station X for the period 1941 to 1952 are given below.

Year	*Observed rainfall at X*	*Adjusted rainfall at X*
	P	*P′ = 0.7265 P*
1941	163	118.4
1942	119	86.5
1943	121	87.9
1944	129	93.7
1945	126	91.5
1946	120	87.2
1947	153	111.1
1948	172	125.0
1949	127	92.3
1950	108	78.5
1951	126	91.5
1952	190	138.0

5.16 PROBABLE MAXIMUM PRECIPITATION (PMP)

The greatest depth of rainfall that can occur in a given duration at a given location is known as the *possible maximum*, or the *probable maximum* precipitation, abbreviated as PMP. It is the upper limit of the physically possible precipitation depth. It has been defined as that depth of precipitation which for a given area and duration can be reached but not exceeded under known meteorological conditions. If PMP for a given catchment is estimated, then it can be used to provide an estimate of the *probable maximum flood*, abbreviated as PMF, after appropriate adjustments for infiltration losses etc. The PMF has virtually no risk of being exceeded. There are two approaches for estimating the

PMP. The first approach uses the *maximisation* and *transposition* of real or model storms while the second approach utilises the statistical analysis of the extreme rainfall.

In the storm maximisation and transposition technique, the upper limits of the amount of moisture available in the air above the basin, the rate at which the wind may carry moist air into the basin, and the fraction of the inflowing water vapour that can precipitate are estimated from the available meteorological data. This indeed requires a large amount of data. Then assuming that the maximum air moisture and the maximum air inflow occur simultaneously the maximum precipitation that may occur is derived. In the absence of suitable data over the area under consideration the storms from very large distances are transposed with modification for factors such as elevation, latitude and distance from moisture source. Storms are not transposed in mountainous regions because it is not possible to adjust precipitation accurately for orographic influences.

In the statistical approach the annual series of the observed maximum rainfall depths for given duration are analysed to yield the mean and standard deviation. PMP is then expressed as the mean of the series plus K times the standard deviation. That is

$$\text{PMP} = \overline{P} + K.\,\sigma \qquad \text{...(5.22)}$$

The value of K is empirically determined. It varies with rainfall duration and inversely with the mean of the series. Its range is between 5 and 30. The PMP value estimated for a location is then adjusted suitably through an appropriate depth-area-duration relationship such as given in Eq. (5.13) or Eq. (5.14) so that it can be made applicable over a given area.

5.17 RAINGAUGE NETWORK

As pointed out earlier, the sampling area provided by a raingauge is extremely small. In order to get a reasonably accurate estimate of the average depth of rainfall over an area it is desirable to have as a large number of gauges as possible. The ratio of total area of the catchment to the total number of gauges in the catchment is defined as the *raingauge density* or the *network density*. Thus, the raingauge density gives the average area served by each gauge. From the view point of accurate hydrological analysis and forecasting, therefore, the raingauge density should be as small as possible. The actual gauge density to be adopted, however, will be governed by many factors such as topography, climate, the state of development of the region and largely by the finances available to install and maintain the network. The gauge density is also influenced by the uses for which the data are intended. For example, to determine the annual average depth of rainfall over a large area of level terrain a relatively sparse network may be sufficient. A dense network is required to determine the rainfall patterns in a thunder storm. A network should be so planned that it yields a representative picture of the areal distribution of rainfall. Local knowledge of the rainfall distribution combined with personal judgment is used in locating the raingauges.

The World Meteorological Organisation, WMO, has laid down certain norms regarding the minimum network density as given below :

Region I : Flat regions of temperate, mediterranean and tropical zones :

Minimum = 1 gauge for 600 to 900 km^2

Tolerable = 1 gauge for 900 to 3000 km^2

Region II : Mountainous areas of temperate, mediterranean and tropical zones :

Minimum = 1 gauge for 100 to 250 km^2

Tolerable = 1 gauge for 250 to 1000 km^2

Region III : Arid zones :

Minimum = 1 gauge for 1500 to 10000 km^2

For small mountainous islands with irregular precipitation one gauge for every 25 km^2 is suggested. Ten per cent of these gauges should be of recording type to enable the determination of rainfall intensities.

The density of raingauge network varies widely from country to country and from region to region within the country. From practical consideration of the Indian conditions, the Indian Standards Institution, ISI (IS: 4987–1968) has recommended the following densities :

(*i*) One gauge per 520 km^2 in plain areas, with denser network for the areas lying in the path of low pressure systems.

(*ii*) One gauge per 260 to 390 km^2 in regions with an average elevation of 1000 m above the mean sea level.

(*iii*) One gauge per 130 km^2 in predominantly hilly regions with heavy rainfall, higher density being preferred wherever possible.

It is also recommended that at least 10 per cent of the gauges are of recording type. The network arrived at based on these recommendations may be so located that all the gauges will have more or less equal Thiessen weights. In actual execution these recommendations may be implemented subject to the importance and economic feasibility of the project.

The adequacy of the number of gauges in the existing network may be ascertained following the procedure given by Indian Standards. The optimum numbers of raingauges N is given by the equation

$$N = \left[\frac{C_v}{p}\right]^2 \qquad \text{...(5.23)}$$

where p is the allowable degree of error in estimating the average depth of rainfall over the area and C_v is the *coefficient of variation* of the rainfall recorded at the m existing raingauge stations given by $(s_x/\bar{x}) \times 100$. It may be noted here that both p and C_v in Eq. (5.23) should be expressed as percentages. If $N < m$, the existing network estimates the average depth of rainfall with an error less than the allowable value of p and no more gauges are required. If $N > m$, the additional gauges required is given by $(N - m)$, rounded off to the next highest integer. The additional gauges may be so located within the area that along with the existing gauges they are evenly distributed over the entire basin.

Example 5.17. *The average annual rainfalls in cm at 4 existing raingauge stations in a basin are 105, 79, 70 and 66. If the average depth of rainfall over the basin is to be estimated within 10% error, determine the additional number of gauges needed.*

Solution. Mean of the rainfalls at the existing gauges is given by

$$\bar{x} = \frac{1}{m} \Sigma x_i = \frac{105 + 79 + 70 + 66}{4}$$

$$= \frac{320}{4} = 80 \text{ cm.}$$

The standard deviation of the rainfalls is given by

$$s_x^2 = \frac{\Sigma (x_i - \bar{x})^2}{(m-1)}$$

$$s_x^2 = \frac{(105-80)^2 + (79-80)^2 + (70-80)^2 + (66-80)^2}{(4-1)}$$

$$= \frac{922}{3} = 307.33$$

or $$s_x = \sqrt{307.33} = 17.53 \text{ cm}$$

$$C_v = \frac{s_x}{\bar{x}} \times 100 = \frac{17.53}{80} \times 100 = 21.91$$

$$p = 10\%$$

∴ $$N = \left[\frac{C_v}{p}\right]^2 = \left(\frac{21.91}{10}\right)^2 = 4.80$$

Additional gauges required = $(N - m) = (4.8 - 4)$

$= 0.8$ say 1.

5.18 ANNUAL RAINFALL IN INDIA

Though the average depth of annual rainfall over India is the largest for any country of comparable size in the world, considerable areal variation exists with about one-third of the country's cultivated land almost facing scarcity conditions. The average annual rainfall varies from about 150 mm in extreme west Rajasthan to over 2000 mm in Assam and north-east parts and western ghats. Cherrapunji situated at an elevation of 1330 m has an annual average of more than 11000 mm and it may receive a rainfall of as high as 1000 mm in a single day. Any day on which a rainfall of more than 2.5 mm is recorded is called a *rainy day*. The number of rainy days varies from place to place with a range of less than 10 to more than 100. The areal distribution of number of rainy days follows more or less the same pattern as that of mean annual rainfall.

At a given place, considerable variation of rainfall in time from year to year is also noticed. The coefficient of variation (ratio of standard deviation to mean, expressed as percentage) of the rainfall varies from place to place and its value ranges from 15 to 70. The coefficient of variation is least in regions of high rainfall and largest in regions of scanty rainfall. Thus, in the regions of Rajasthan, Haryana and Gujarat where the mean annual rainfall is low, the rainfall is also less certain.

5.19 RECORD RAINFALLS

If the greatest observed rainfalls of various places over the world for various durations are plotted on a logarithmic scale the curve enveloping all the points will generally be a straight line which can be described by

$$P = 417.0\, D^{0.48} \qquad \text{...[(5.23(}a\text{)]}$$

where P is the maximum depth of rainfall in mm and D is the duration in hours. For instance for a duration of 4 days the maximum rainfall depth given by Eq. (5.23) is 3729.3 mm. The maximum 4 day rainfall recorded anywhere in the world is at Cherrapunji, India and it is 3721 mm which occurred during Sept. 12–15, 1971.

5.20 SNOWFALL

Water released by the melting of snow is a major contributor to the streamflow of some rivers, sometimes resulting in floods. In the design of structures located in cold regions snow loads become critical and they should be properly accounted for. The effect of snow on surface transportation systems is widely appreciated. In all such situations snow is of great importance to hydrologists.

Snow is the solid form of water. It is usually composed of ice crystals mainly of hexagonal system and also liquid water. The amount of liquid water present in the ice is called the *water content* of the snow. The depth of snowfall can be measured by a yardstick when the accumulated depth is not large. *Snow boards* are sometimes used to readily identify the freshly fallen snow. These boards, square in shape with a side of at least 40 cm, are laid on the previous accumulation of snow. Several snow boards should be used to obtain more representative data. In places where large accumulations of snow are expected permanent *snow stakes* are recommended. The snow stakes, usually made of wood, are 45 mm in square with angle iron supports and are properly graduated. They resemble levelling staff in design. In remote and inaccessible areas, special type of snow stakes are used which can be read from a low-flying aircraft. All stakes for measuring snow depth should be installed where they will be least affected by blowing or drifting snow.

For a hydrologist the *water equivalent* of snow pack rather than the depth of snow pack itself is of greater importance. The ratio of the volume of melt-water from a snow sample to the initial volume of the sample is known as the *snow density*. Thus if D_s is the depth of snowfall and d_s is its snow density then the depth of water equivalent P is given by

$$P = d_s \cdot D_s \qquad \text{...(5.24)}$$

It has been observed that the snow density varies from 0.004 for freshly fallen snow at high latitudes to 0.91 for compacted snow in glaciers. An average snow density of 0.10 is often assumed for freshly fallen snow.

Snow surveys are conducted in the area of interest and snow samples are collected at different points with a spacing of 15 to 30 m. These samples are used in the determination of snow density at the respective points. The predetermined line along which the samples are collected is known as a *snow course*. Measurements are required at number of points because of the highly variable snowpack resulting from drifting and non-uniform melting. The sampling tube of diameter 40 to 90 mm with a cutting edge is driven vertically into the snowpack by rotating it till it reaches the bottom of the snowpack. Extension tubes are added if required. The depth of snowfall is noted from the graduations on the outer surface of the tube. The tube and its snow content are then withdrawn and weighed to determine the water equivalent. Since the determination of water equivalent by melting is not practical in the snow fields, the water equivalent is customarily determined by weight of the snow at the time of sampling.

Special types of radioactive snow gauges are also available which can be used to determine the variation of snow density with depth within the snowpack.

REVIEW EXERCISES

Descriptive Questions

1. What is the basis for measurement of precipitation ? What are the difficulties that come in the way of accurate measurement of precipitation ?
2. Describe with a neat sketch the principle of working of Symon's non-recording raingauge. How does the Indian Standard raingauge differ from Symon's gauge ?
3. Describe the principle of working of a tipping bucket type recording raingauge with a neat sketch. What are its advantages and disadvantages ?
4. Describe the principle of working of a weighing bucket type recording raingauge with a neat sketch. Mention its advantages and disadvantages.
5. Describe the principle of working of a float type recording raingauge with a neat sketch. Discuss its advantages and disadvantages.
6. What are the possible sources of error in the measurement of rainfall ?
7. Describe the principle of rainfall measurement using a radar with neat sketches.
8. What are the precautions to be taken in selecting a site for the location of a raingauge ?
9. Define normal annual rainfall, wet year and dry year. In how many ways can the annual rainfall data be graphically represented ? Describe them.
10. What is a moving-average curve ? How is it constructed ? What are its uses ?
11. What do you understand by dependable rainfall ? How do you analyse the available data to obtain 75% dependable annual rainfall ?
12. What is a rainfall hyetograph ? How is it derived from a rainfall mass curve ?
13. Describe the step by step procedure involved in the analysis for developing intensity-frequency-duration relationships. Sketch a typical set of these curves.

14. Describe the three methods of determining the average depth of rainfall over an area. Bring out the merits and demerits of each method.
15. What are the design applications of the depth-area-duration relations ? Explain the procedure of developing these relations.
16. Discuss the various procedures available to estimate the missing precipitation records.
17. How is the double mass curve technique used to check the consistency and adjust the rainfall record at a suspicious station ?
18. What is meant by Probable Maximum Precipitation ? Describe the methods of estimating PMP. What are its design applications ?
19. Define raingauge density. Discuss the ISI norms for the raingauge density.
20. Bring out the importance of snow surveys. What do you understand by snow density and how is it determined in the field ?
21. Explain how the satellites can measure the precipitation.
22. List out the advantages and disadvantages of tipping bucket type and weighing bucket type recording raingauges.
23. What do you understand by point rainfall record ? How can the point rainfall record be extended using station-year method.
24. What are the conditions to be satisfied for applying station year method ?
25. Explain the principle of regression method for estimating the missing precipitation record. What are its advantages ?

Analytical Questions

1. The cumulative rainfall depth with time during a storm as obtained from a recording raingauge at a station is given below.

Time in hours	*Rainfall in mm*	*Time in hours*	*Rainfall in mm*
10.00	0		
10.30	6	13.30	51
11.00	11	14.00	57
11.30	16	14.30	61
12.00	24	15.00	66
12.30	29	15.30	67
13.00	38	16.00	67

(*i*) Construct the hyetograph of this storm using uniform time interval of 30 minutes and also 2 hours.

(*ii*) Compute the maximum average intensities of rainfall for durations of 30 minutes, 1 hour, 2 hours, 3 hours and 5 hours in this storm and plot the resulting intensity duration curve.

(*iii*) Fit an appropriate regression equation for the intensity duration curve obtained above. Assume the best value of *a* as 10.

2. The hyetograph of a storm of 6 h duration is constructed with varying time intervals: at 20 minutes interval for the first one hour, at 40 minutes interval for the next 2 hours and at one hour interval for the last 3 hours. The successive ordinate of the hyetograph in mm/h are 66, 75, 54, 48, 69, 51, 38, 47 and 25. Determine the total rainfall depth produced by the storm.
3. The annual rainfalls in cm at a station for a period of 21 years from 1960 to 1980 are

   97, 125, 103, 81, 101, 119, 103, 79, 102, 118, 98

   83, 105, 123, 100, 86, 99, 114, 91, 83 and 106

   (*i*) Represent this data in the form of chronological chart, bar diagram and ordinate graph.

   (*ii*) Construct the 5-year moving average curve and superimpose it on the chronological chart. Comment on the moving average curve.
4. For the data of analytical question 3, determine the 75% dependable rainfall from frequency analysis and also by using the normal distribution as the appropriate fit.

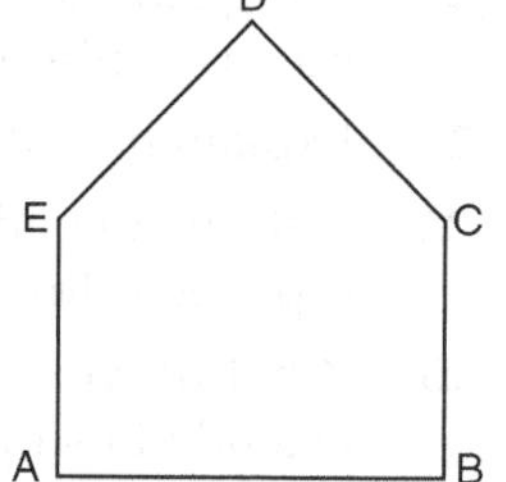

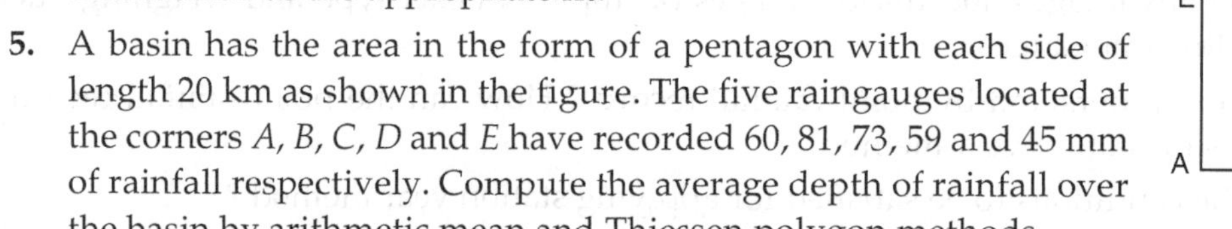

5. A basin has the area in the form of a pentagon with each side of length 20 km as shown in the figure. The five raingauges located at the corners *A*, *B*, *C*, *D* and *E* have recorded 60, 81, 73, 59 and 45 mm of rainfall respectively. Compute the average depth of rainfall over the basin by arithmetic mean and Thiessen polygon methods.
6. The network of 10 stations in and around a river basin have the Thiessen weights of 0.10, 0.06, 0.11, 0.07, 0.08, 0.09, 0.11, 0.12, 0.16 and 0.10 respectively. Stations 2, 4 and 5 lie outside the basin while the remaining are inside. If the rainfalls recorded at these gauges during a storm are 150, 168, 158, 135, 156, 207, 138, 162, 114 and 132 mm respectively. Determine the average depth of rainfall over the basin by arithmetic and Thiessen mean methods.
7. A major river basin is divided into four sub-basins with areas of 920, 705, 1075 and 1665 km^2. If the average annual rainfall on these sub-basins is 73, 85, 112 and 100 cm respectively. What is the average annual rainfall for the basin as a whole ?
8. The catchment area of a basin may be approximated as a semicircle of radius r km. With respect to the coordinate axis set up with its origin at the centre of the circle and the x-axis coincident with the diameter the area lies in the first and second quadrants and the position coordinates of the raingauge stations are (0, 0), $\left(\frac{r}{2}, \frac{r}{2}\right)$ and $\left(-\frac{r}{2}, \frac{r}{2}\right)$ km. Show that the Thiessen weights of the gauges are given by $0.5/\pi$, $(0.5 - 0.25/\pi)$ and $(0.5 - 0.25/\pi)$ respectively.
9. The isohyets drawn for a storm which occurred over a drainage basin of area 950 km^2 yielded the following information.

Isohyet interval in mm	85–75	75–65	65–55	55–45	45–35
Area between isohyets in km²	125	236	264	175	150

Determine the average depth of rainfall over the basin.

10. The shape of a catchment is in the form of a pentagon *ABCDE*. There are 4 raingauge stations *P*, *Q*, *R* and *S* inside the catchment. The position coordinates in km are *A*(0, 0), *B*(50, 75), *C*(100, 70), *D*(150, 0), *E*(75, –50), *P*(50, 25), *Q*(100, 25), *R*(100, –25) and *S*(50, –25). If the rainfalls recorded at *P*, *Q*, *R* and *S* are 88, 102, 112 and 116 mm respectively, determine the average depth of rainfall over the catchment using Thiessen polygon method.

11. During a month a raingauge went out of order while the other three gauges in the basin reported rainfalls of 107, 89 and 120 mm. If the normal annual rainfalls for these three gauges are 1120, 935 and 1200 mm respectively and the normal annual rainfall of the broken gauge is 978 mm, estimate the missing monthly rainfall at the broken gauge.

12. Neighbouring raingauge stations *A*, *B*, *C*, *D*, *E* and *F* have normal annual rainfalls of 610, 554, 468, 606, 563 and 382 mm respectively. During a storm, stations *B*, *C*, *D*, *E* and *F* have reported rainfalls of 22, 29, 35, 13 and 25 mm respectively and station *A* did not report as it was inoperative. Estimate the missing storm rainfall at *A* by the arithmetic average method and the normal-ratio method.

13. Station *A* failed to report the rainfall during a storm. With respect to east-west and north-south axes set up at station *A*, the co-ordinates of 4 surrounding gauges which are the nearest to *A* in the respective quadrants are (7.5, 7.5), (–10, 15), (–5, –6.5), (12, –8) km respectively. Determine the missing rainfall at *A*, if the storm rainfalls at the four surrounding gauges are 98, 67, 72 and 53 mm respectively.

14. Annual rainfall at station *X*, and the average of the annual rainfall at 15 nearby raingauge stations for a period of 35 years is given below.

Year	*Annual at X*	*Rainfall (mm) 15 stations average*	*Year*	*Annual at X*	*Rainfall (mm) 15 stations average*
1946	664	593	1964	534	562
1947	552	530	1965	491	481
1948	558	565	1966	519	484
1949	578	534	1967	456	481
1950	570	562	1968	545	582
1951	555	478	1969	534	514
1952	639	600	1970	453	496
1953	687	552	1971	474	532
1954	572	524	1972	466	540
1955	524	502	1973	478	484
1956	570	532	1974	554	621
1957	743	611	1975	433	486
1958	534	534	1976	437	600
1959	496	501	1977	458	434
1960	468	486	1978	545	547
1961	532	583	1979	560	636
1962	423	484	1980	435	478
1963	473	610			

(*i*) Examine the consistency of the data at station X.

(*ii*) In which year did a change in regime occur ? Discuss the possible reasons.

(*iii*) Determine the average annual rainfall at X for 35 years first without adjusting the data and later with adjustment for the change in regime.

15. The average annual rainfall of 5 raingauges in a basin are 89, 54, 45, 41 and 55 cm. If the error in the estimation of basin mean rainfall should not exceed 10%, how many additional gauges should be installed in the basin ?

16. The annual rainfalls at 7 raingauge stations in a basin are 58, 94, 60, 45, 20, 88 and 68 cm respectively. What is the percentage accuracy of the existing network in the estimation of the average depth of rainfall over the basin ? How many additional gauges are required if it is desired to limit the error to only 10% ?

17. During a snow survey the following information is collected.

The diameter of the sampling tube = 75 mm

Depth of snow pack = 286 mm

The weight of empty sampling tube = 5.25 kg

The weight of sampling tube + snowpack = 5.555 kg.

Determine the snow density and the depth of water equivalent of snowfall.

18. The coordinate distances in km of 5 raingauge station *X*, *A*, *B*, *C* and *D* are (0, 0), (4, 5), (–6, 8), (– 9, – 6) and (5, – 7) respectively. During July, 2005 station *X* was in operative and the other four stations *A*, *B*, *C* and *D* recorded rainfalls of 8.3, 10.1, 7.7 and 12.4 cm respectively. Calculate the missing July rainfall at X.

19. The shape of a watershed can be taken as square of each side 4 km. There are 5 raingauge stations *A*, *B*, *C*, *D* and *E* within this watershed. With reference to an *x-y* coordinate axes having its origin at one of the corners of the square, the locations of the other 3 corners and the 5 raingauge stations are (4, 0), (4, 4), (0, 4) and (1, 1), (3, 1), (3, 3), (1, 3) and (2, 2) respectively. Initially the four stations *A*, *B*, *C* and *D* have reported storm rainfalls of 55, 76, 68 and 40 mm respectively. Using the Thiessen polygon method compute the average depth of rainfall.

Later station *E* reported a rainfall of 63 mm. Compute the average depth of rainfall again by Thiessen polygon method considering the rainfall at *E* also. What is the corresponding value by arithmetic mean method.

20. The annual rainfalls at 4 raingauge stations for a period of 8 years are given below.

Year	X	A	B	C
1973	—	53.5	37.6	50.7
1974	103.0	85.8	80.9	85.9
1975	80.8	76.9	51.3	60.1
1976	58.7	49.8	52.9	48.7
1977	73.1	59.1	68.3	67.9
1978	50.9	64.0	72.3	60.7
1979	42.8	53.1	34.3	40.8
1980	29.6	44.2	25.6	30.7

Estimate the missing rainfall at X in 1973 using the multiple regression method. What is the standard error of estimate?

Objective Type Questions

1. The rainfall hyetograph is the graph drawn between
   (*a*) cumulative rainfall and time (*b*) rainfall intensity and time
   (*c*) rainfall depth and area (*d*) rainfall intensity and cumulative rainfall.
2. The rainfall mass curve shows the variation of
   (*a*) rainfall intensity with time (*b*) rainfall intensity with cumulative rainfall
   (*c*) rainfall excess with time (*d*) cumulative rainfall with time.
3. The nearest object from a raingauge should be at a minimum distance equal to
   (*a*) its height (*b*) twice its height
   (*c*) thrice its height (*d*) four times its height.
4. Double mass curve technique is used
   (*a*) to prepare the rainfall hyetograph from the rainfall mass curve
   (*b*) to prepare the rainfall mass curve from the rainfall hyetograph
   (*c*) to check the consistency of record at a suspected raingauge station
   (*d*) in developing isohyetal maps.
5. The maximum average rainfall intensity at a given location
   (*a*) increases with increase in duration (*b*) decreases with increase in duration
   (*c*) is independent of the duration of rainfall
   (*d*) sometimes increases and sometimes decreases with increase in duration.
6. The most accurate method of finding the average depth of rainfall over an area is
   (*a*) Isohyetal method (*b*) Arithmetic mean method
   (*c*) Thiessen polygon method (*d*) any of the above.
7. The Thiessen weights of 4 raingauges *A*, *B*, *C* and *D* covering a river basin are 0.15, 0.25, 0.30 and 0.30 respectively. If the average depth of rainfall for the basin is 5 cm, and the rainfalls recorded at *B*, *C* and *D* are 5 cm, 4 cm and 5 cm respectively. What is the rainfall at *A* ?
   (*a*) 5 cm (*b*) 6 cm
   (*c*) 7 cm (*d*) 8 cm.
8. For a given storm the average depth of rainfall over an area
   (*a*) decreases with increase in area (*b*) increases with increase in area
   (*c*) has no relation with area (*d*) none of the above.
9. An isohyet is a line joining points of
   (*a*) equal rainfall intensity (*b*) equal storm durations
   (*c*) equal rainfall excess (*d*) equal rainfall depth.
10. The chart removed from a recording type raingauge gives
    (*a*) a rainfall hyetograph (*b*) an isohyetal map
    (*c*) a rainfall mass curve (*d*) an intensity-duration curve.

11. The typical characteristics of convective showers is that they are of
    (*a*) high intensity and long duration
    (*b*) high intensity and short duration
    (*c*) low intensity and long duration
    (*d*) low intensity and short duration.
12. In which of the following the snow density would be maximum ?
    (*a*) fresh powder snow (*b*) virgin snow
    (*c*) `coarse snow (*d*) packed snow in glaciers.
13. In which of the following the snow density would be minimum ?
    (*a*) fresh powder snow (*b*) virgin snow
    (*c*) coarse snow (*d*) packed snow in glaciers.
14. Which of the following places is India records highest annual rainfall ?
    (*a*) Trivendrum (*b*) Bombay
    (*c*) Chirrapunji (*d*) Madras.
15. Any cyclic trend present in the rainfall data can be ascertained from
    (*a*) depth-area-duration curve (*b*) moving average curve
    (*c*) intensity-duration curve (*d*) double mass curve.

ANSWERS

Analytical Questions

1. (*i*) The ordinates of the hyetograph with a uniform interval of half an hour in mm/h are 12, 10, 10, 16, 10, 18, 26, 12, 8, 10, 2, 0

   The ordinates of the hyetograph with a uniform interval of 2 hours are 12, 16.5 and 5 mm/h.

   (*ii*) 26.0, 22.0, 17.5, 15.33 and 13.2 mm/h

   (*iii*) $I = \dfrac{(90.77)}{(t+10)^{0.337}}$

2. 287 mm
3. (*ii*) The ordinates of 5-year moving average curve are

   101.4, 105.8, 101.4, 96.6, 100.8, 104.2, 100.0, 96.0, 101.2,

   105.4, 101.8, 99.4, 102.6, 104.4, 98.0, 94.6, 98.6

   The moving average curve exhibits a cycle with a period of 5 years.
4. From frequency analysis 75% dependable rainfall = 88.5 cm

   From normal distribution 75% dependable rainfall = 91.485 cm
5. Arithmetic mean = 63.6 mm

   Thiessen mean = 63.01 mm

6. Arithmetic mean = 151.57 mm

Thiessen mean = 149.08 mm

7. 94.842 cm

9. 60.116 mm

10. 101.36 mm

11. 94.776 cm

12. 24.8 mm, 30.252 mm

13. 75.93 mm

14. (*i*) Data at X is not consistent

(*ii*) The change in regime occurred in 1959

(*iii*) Average without adjustment = 528.86 mm

Average with adjustment = 494.29 mm

15. 7 gauges

16. 15.395%, 10 gauges

17. 0.2414 gm/c.c., 69.04 mm

18. 9.51 cm

19. 59.75 cm ; 60.156 cm and 60.4 cm

20. $P_x = -12.9889 + 0.063618\, P_A - 0.954826\, P_B + 2.20482\, P_C$

Missing Precipitation in 1973 = 66.3 cm

Standard error of estimate = 5.9 cm.

Objective Type Questions

1. (*b*)	2. (*d*)	3. (*b*)	4. (*c*)	5. (*b*)
6. (*a*)	7. (*c*)	8. (*a*)	9. (*d*)	10. (*c*)
11. (*b*)	12. (*d*)	13. (*a*)	14. (*c*)	15. (*b*).

Chapter 6 STREAMFLOW MEASUREMENT

6.1 INTRODUCTION

The water which constitutes the flow in the surface stream is called *streamflow*. If the streamflow is unaffected by the artificial diversions, storage, or other works of man in or on the stream channels, then it is called as *runoff. In* other words runoff means the virgin streamflow. Streamflow forms the most important data for engineers and hydrologists since they are concerned mainly with estimating rates and volumes of the streamflow to be used in the design of water resources projects, or the changes in these values resulting from the man-made causes. It is, therefore, to be noted that whenever the streamflow data is used in the scientific investigations such as establishing the rainfall-runoff relations, unit hydrograph studies, design flood estimation etc., to be discussed in the later chapters, it has to be adjusted for artificial changes using suitable techniques till it resembles very nearly the virgin flow. In this chapter the techniques of streamflow measurement are presented.

It is rather difficult to measure the discharge of flow in the natural streams directly as it is done in the case of flow in pipes or laboratory flumes using the flow meters such as venturimeter, venturiflume, notches etc. But it is very easy to make a direct and continuous measurement of *stage* in the river which is nothing but the height of the water surface in the river above some arbitrary datum. Obviously the higher the stage in the river, the higher is the discharge. The general practice in the streamflow measurement is, therefore, to record the river stage and to convert the data on stage into the discharge data. This is accomplished through the *stage-discharge relationship* which is first established by actual measurement of discharge in the river at different stages. Once a stable stage-discharge relationship is established at a gauging site, the discharge measurement is discontinued and only the stage is recorded continuously.

6.2 MEASUREMENT OF STAGE

The river stage has been defined as the height of the water surface in the river at a given section above any arbitrary datum. It is usually expressed in metres. In many cases, the datum is taken as the mean sea level. Sometimes the datum may be selected at or slightly below the lowest point on the river bed.

The stage can be very easily measured by installing a vertical staff gauge which is nothing but a graduated scale such that a portion of it is always in the water at all times. It can be conveniently attached to a bridge pier or any other existing structure. It is read manually by noting the level of water surface in contact with it.

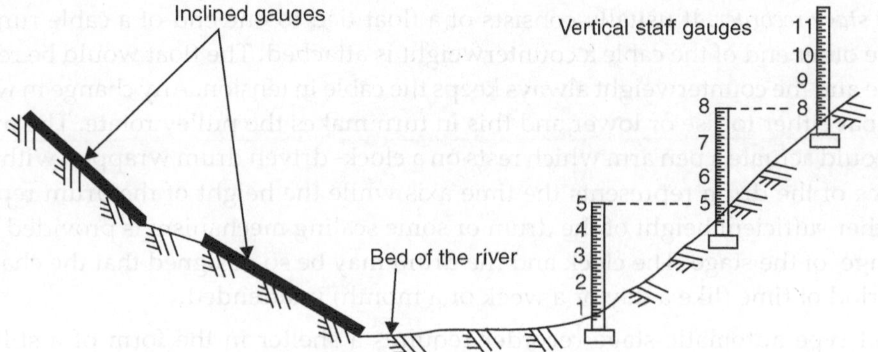

FIGURE 6.1 Measurement of stage in the river.

When the flow in the stream is subjected to large variations resulting in correspondingly large fluctuations in the stage, it may be beyond the range of a single vertical staff gauge to record the entire rise or fall in the water surface. In such situations it may be convenient to use several vertical staff gauges as shown in Fig. 6.1 covering the entire range, the graduations of all the gauges being reduced to the same datum. There should be a minimum overlap of 0.5 m between any two successive staff gauges.

Sometimes the gauge may be placed in an inclined position up the stream bank as shown in Fig 6.1. It must be properly anchored to the slope of the natural bank of the river channel. It is calibrated in the site by precise levelling and graduated accordingly. Inclined gauges may be constructed on one continuous slope or on more slopes. A flight of steps constructed alongside the inclined gauge proves to be convenient and facilitates taking of observations easily.

Proper care must be taken while installing the staff gauges to protect them from damage by boats, ice, or flood transported debris and to ensure that flow disturbances across the scale are at a minimum.

The river stage can also be measured manually by using another type of device called the *suspended weight gauge*. In this method a weight attached to a rope is lowered from a fixed reference point on a bridge or other overhead structure till it touches the water surface. By subtracting the length of the rope lowered from the reduced level of the fixed reference point the stage is obtained.

Though the manual gauges described above are simple and inexpensive they must be read frequently to get a continuous curve of the streamflow, especially when the stage is changing rapidly. Also, it is likely that the peak stage may be missed when it occurs in between the observations. *Recording type gauges* may be installed to overcome these difficulties. When the finances do not permit the installation of recording gauge, a *crest gauge* may be installed along with the manual gauge so that the peak stage is not missed. A crest gauge may consist of a small float which rises with the stage but is restrained at the maximum level attained by the water surface. A water soluble paint applied to a bridge pier and protected from rain may also be used as a crest gauge. The portion of the paint coming in contact with water is washed out indicating the peak stage attained during a flood. The data from the crest gauges are valuable in the establishment of flood profiles and in determining the slope for the investigation of the flood formulae.

Recording type gauge used to measure the stage continuously with time is also known as an *automatic stage recorder*. It usually consists of a float tied to one end of a cable running over a pulley. To the other end of the cable a counterweight is attached. The float would be resting on the water surface and the counterweight always keeps the cable in tension. Any change in water surface makes the float either to rise or lower and this in turn makes the pulley rotate. The movement of the pulley would actuate a pen arm which rests on a clock- driven drum wrapped with a chart. The circumference of the drum represents the time axis while the height of the drum represents the stage. So, either sufficient height of the drum or some scaling mechanism is provided to cover the expected range of the stage. The clock and the drum may be so designed that the chart runs for a specified period of time (like a day or a week or a month) unattended.

A float type automatic stage recorder requires a shelter in the form of a stilling well as shown is Fig 6.2. This stilling well gives protection to the float and counterweight from floating

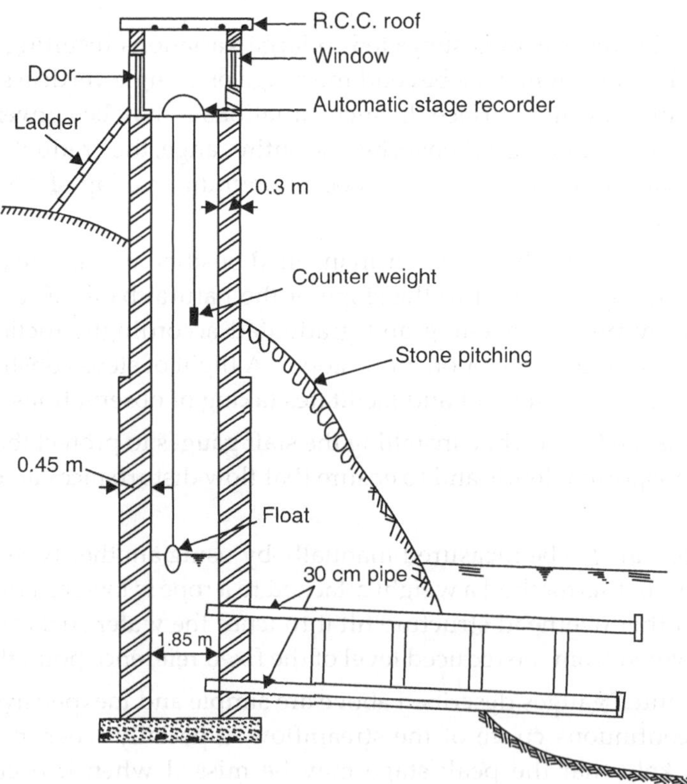

FIGURE 6.2 Shelter for automatic stage recorder.

debris and with proper design of intake pipes it suppresses the fluctuations resulting from surface waves in the river. Generally two or more intake pipes are placed to allow the water from the river into the well so that at least one will admit water at all the times. As the stilling well is likely to get filled with sediment it is necessary to make provision for the removal of silt from time to time. It is customary to install staff gauges inside and outside the well. These staff gauges serve to check the performance of the recorder and these are read each time the station is visited.

6.3 DISCHARGE MEASUREMENT BY AREA-VELOCITY METHOD

Though there are several methods of discharge measurement, perhaps the most accurate of all of them is the area-velocity method. In this method the discharge is obtained as the product of the cross-sectional area of flow in the river and the average velocity of flow. That is

$$Q = A \cdot V \qquad ...(6.1)$$

The measurement of discharge thus involves obtaining information at a number of points over the cross-section sufficient to determine the average velocity and also the area of flow. The number of points selected should be such that the measurements are completed within a reasonably quick time. This is essentially required when the stage in the river is changing rapidly at the time of discharge measurement. The procedure consists of dividing the flow area into a number of strips by means of *verticals* selected along the width of the river and measuring the depths at these verticals. The velocities are also measured at one or more points on the verticals to obtain the mean velocity for any vertical. The velocity at any point in the river is measured by inserting the *current meter* at that point facing the flow.

Current meter. There are mainly two types of current meters which are in common use, namely, *cup type* current meter (also known as the *price* current meter) and the *propeller type* current meter. The principle involved in both the meters is that the water flowing past the rotating element of the meter makes it revolve due to the unbalanced drag force acting on it and the speed of the rotating elements is directly proportional to the velocity of water. The cup type current meter consists of a wheel of six conical cups rotating about a vertical axis as shown in Fig. 6.3. The tail vanes (also known as the *fins*), will always align the meter along the direction of flow.

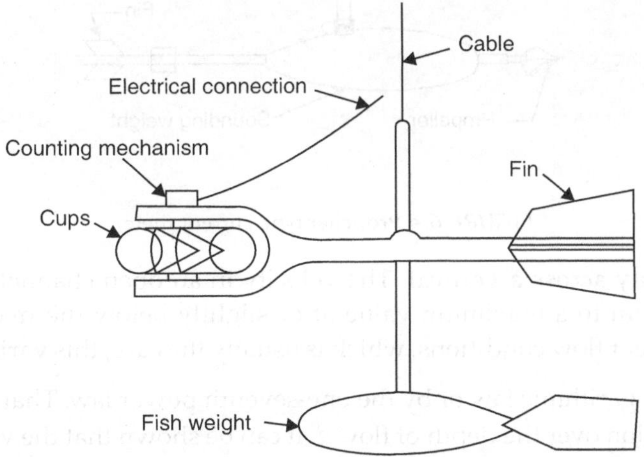

FIGURE 6.3 Cup type current meter.

The purpose of the fish weight with a streamlined shape at the bottom is to keep the meter cable as nearly vertical as possible. The usual weights recommended are 10, 15, 25, 35 and 50 kg. The recording unit consists of a headphone worn by the operator and an electric circuit with a battery and wire of the supporting cable. Each time the wheel of cups makes one revolution the electric circuit is closed and this causes a click in the headphone to be heard by the operator. The speed of the meter, after it is inserted at a point where velocity is to be measured, is obtained by starting a stop watch on a click and stopping it on another click after about 45 seconds.

The number of clicks counted (taking the first as zero) divided by the elapsed time gives the speed of the meter N in revolutions per second. Then the velocity measured by the meter is given by

$$v = a + bN \qquad \text{...(6.2)}$$

where a and b are the calibration constants of the meter supplied by the manufacturer. The normal range of velocities measured is 0.15 to 4.0 m/s. The principle of propeller type current meter (vide Fig. 6.4) is same as that of cup type current meter except that the rotating element is in the form of a propeller facing the direction of flow which rotates about a horizontal axis. The recording unit of some current meters consists of mechanical counters synchronised with an automatic timer avoiding the use of a headphone. In rivers of large depth special arrangements are to be made to support the meter overhead and to measure the cable length for lowering the meter. In shallow depths the meter is mounted on a rod and the observer wades through the stream.

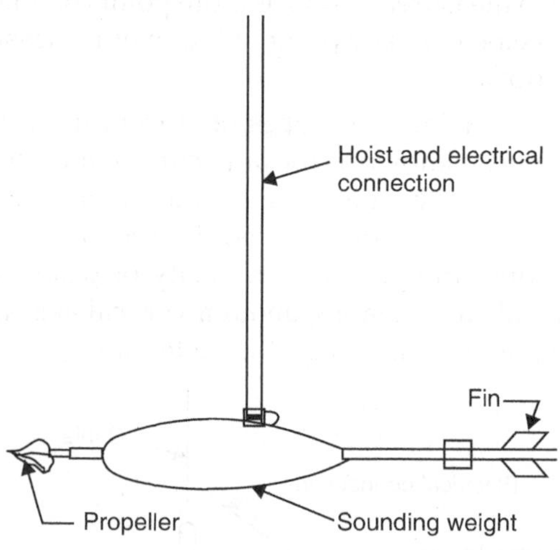

FIGURE 6.4 Propeller type current meter.

Average velocity across a vertical. The velocity in an open channel at any vertical varies from zero at the bottom to a maximum value at or slightly below the free surface as shown in Fig. 6.5. Under turbulent flow conditions, which is usually the case, this variation can be described closely either by the logarithmic law or by the one-seventh power law. That is, $v \propto \ln y$ or $v \propto y^{1/7}$. Integrating this variation over the depth of flow d, it can be shown that the velocity at a distance of 0.368 d from the bottom will be equal to the average velocity for the vertical in the case of logarithmic law and the velocity at a distance of 0.393 *d* from the bottom will be equal to the average velocity for the vertical in the case of one-seventh power law. Therefore the general practice is to take the velocity measured at a distance of 0.4 *d* from the bottom (or at a depth of 0.6 *d* from the free surface) as equal to the average velocity of the vertical. It can also be shown that (for both velocity distributions) the arithmetic average of the velocities at 0.2 *d* and 0.8 *d* is almost equal to the average velocity of the vertical.

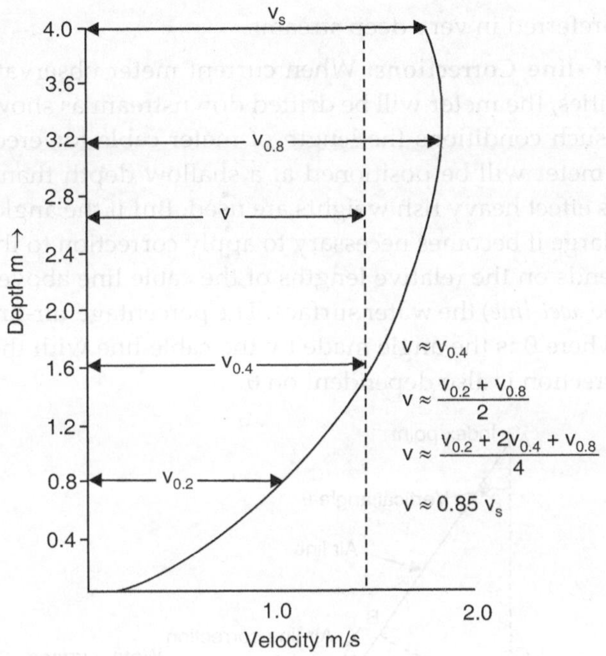

FIGURE 6.5 Velocity distribution at a vertical.

Thus the average velocity in a vertical can be obtained by any one of the following three methods. First the depth at the selected vertical is measured by lowering the cable along with the meter till it touches the bottom of the river bed. Next the meter is raised such that it occupies the position at 0.4 *d* from the bottom. Now the speed of the meter is recorded by noting the number of revolutions made by the meter and the corresponding time. Then the velocity is computed using Eq. (6.2) and this velocity itself is taken as the average velocity for the vertical. That is

$$V = v_{0.4} \qquad \text{...(6.3)}$$

This method is known as *single point method* or *six-tenth* method and is generally preferred for shallow depths.

In the *two point method*, after knowing the depth of flow *d* at the vertical the velocities are obtained at 0.2 *d* and 0.8 *d* from the bottom. The average of these two velocities is then taken as the mean velocity for the vertical. That is,

$$V = \frac{v_{0.2} + v_{0.8}}{2} \qquad \text{...(6.4)}$$

This method is supposed to give a better result of average velocity than the single point method.

In the *three point method*, the velocities are obtained at 0.2 *d*, 0.4 *d* and 0.8 *d* from the bottom and the average velocity for the vertical is given by

$$V = \frac{v_{0.2} + 2v_{0.4} + v_{0.8}}{4} \quad \text{...(6.5)}$$

This method is preferred in very deep streams.

Air-line and Wet -line Corrections. When current meter observations are made in swift waters with large velocities, the meter will be drifted downstream as shown in Fig. 6.6 and it will not hang vertically. In such conditions the length of meter cable lowered is more than the true vertical depth and the meter will be positioned at a shallow depth than indicated by the cable length. To minimise this effect heavy fish weights are used. But if the angle between the cable line and the true vertical is large it becomes necessary to apply correction to the measured depth. The manual correction depends on the relative lengths of the cable line above (known as the *air-line*) and below (known as the *wet-line*) the water surface. The percentage air-line correction is given by $k_a = (\sec\theta - 1) \times 100$, where θ is the angle made by the cable line with the vertical. Similarly the percentage wet-line correction is also dependent on θ.

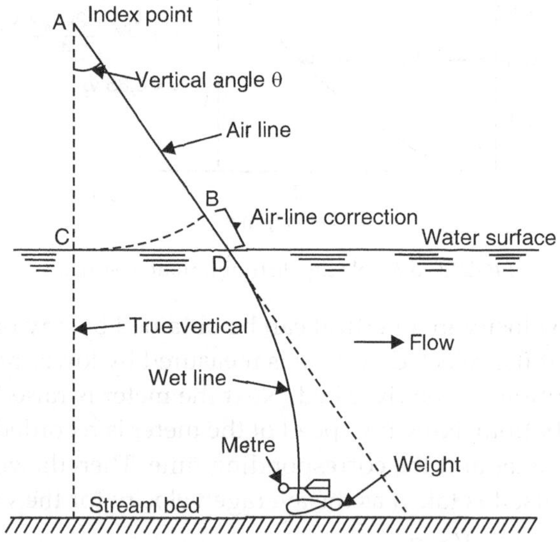

FIGURE 6.6 Position of current meter in swift water.

Thus if y is the vertical distance between the index point and the water surface and l_a is the air-line length of the cable, then we have

$$l_a = y\left(1 + \frac{k_a}{100}\right) \quad \text{...(6.6)}$$

Similarly if d is the true vertical depth of the meter below the water surface and l_w is the wet-line length of the cable, then

$$l_w = d\left(1+\frac{k_w}{100}\right) \qquad \text{...(6.7)}$$

where k_w is the percentage wet line correction. The percentage air-line and wet-line corrections are given for various cable line angle θ up to 30° in Table 6.1. These are as specified in IS : 2913–1964. Thus, for example, if the vertical distance between the index point and water surface is 3.5 m and

Table 6.1. Air-line and Wet-line corrections

Vertical angle θ *(degrees)*	*Air-line correction* k_a *(percentage)*	*Wet-line correction* k_w *(percentage)*
4	0.24	0.06
6	0.55	0.16
8	0.98	0.32
10	1.54	0.50
12	2.23	0.72
14	3.06	0.98
16	4.03	1.28
18	5.15	1.64
20	6.42	2.04
22	7.85	2.48
24	9.46	2.96
26	11.26	3.50
28	13.26	4.08
30	15.47	4.72

if we want to reach a point at a depth of 4.5 m below the water surface, then the total length of the cable to be lowered when the cable line makes an angle of 20° with the vertical is obtained as

$$l = l_a + l_w = \left(1+\frac{6.42}{100}\right)\times 3.5 + \left(1+\frac{2.04}{100}\right)\times 4.5$$

$$= 8.32 \text{ m}$$

Discharge Computation Procedure. After the mean velocities at number of verticals are obtained through current meter observations, the discharge in the river is determined by adding the incremental discharges in the various strips defined by the verticals. For example, in the strip *ABCD* in Fig. 6.7, the two velocities on the Vertical *EF* (which are at 0.2 *d* and 0.8 *d*) are averaged to get the mean velocity for the strip. The area of the strip is obtained as the measured depth of flow at the vertical *EF* multiplied by the width *AB* which is halfway to the adjacent verticals on either side of *EF*. Incremental discharges in the strips near the shores such as *GHI* and *G′H′I′* are ignored. No strip should include more than 10 per cent of the total flow. Thus about 20 to 30 verticals may be taken depending on the width of the stream. This procedure is known as the *mid-section* method and is used in Example 6.1.

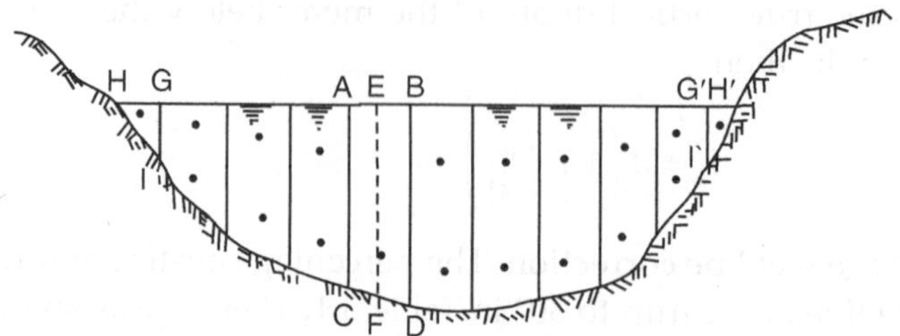

FIGURE 6.7 Current meter positions for discharge measurement.

Alternatively, after the point velocities are obtained from current meter observations, they are indicated at the corresponding points on a cross-section of the gauging site, preferably with an exaggerated vertical scale, and the contours of equal velocity known as the *isovels* are then drawn as shown in Fig. 6.8. The areas between successive isovels are next planimetered and are multiplied by the mean value of the isovels to give the incremental discharges. All these incremental discharges are then added to obtain the total flow. Though this method is more time consuming as it involves the construction of isovels, it is more accurate than the previous method.

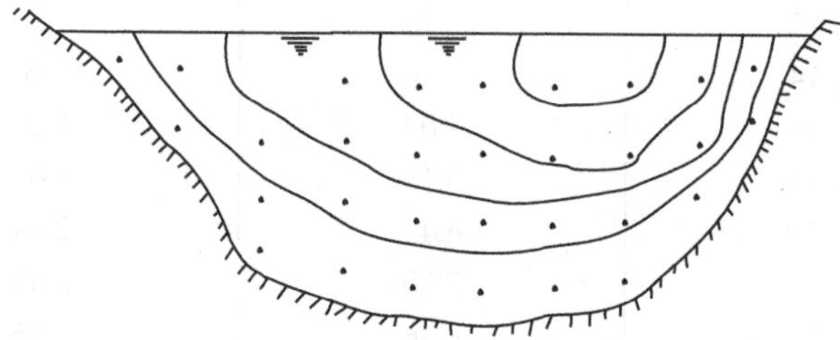

FIGURE 6.8 Isovel plotting for discharge computation.

Handling of current meter and cable is a difficult task especially when the depth of flow is large. Whenever possible an existing bridge is used as the overhead support for the meter cable. When no existing bridge is suitable, a cable way may be used. Where neither a bridge nor a cable way is practical, measurements may be made from a boat. But the discharge measurement using a boat is less satisfactory.

Example 6.1. *The current meter observations taken during the gauging of a stream are given in col. (1) to col. (5) of Table 6.2. The current meter rating may be taken as $v = 0.05 + 0.3\ N$, where v is in m/s and N is in rev/s. Compute the discharge in the stream.*

Solution. The data (in bold faced figures) as well as the computations are shown in Table 6.2. Col. (1) gives the distance from one bank where the vertical is taken. The measured depth of flow at this vertical is indicated in Col. (2) while Col. (3) gives the depth at which the current meter readings are taken. As seen from the given data, for the verticals near the two banks the velocity is measured only at 0.6 d, whereas at the other verticals the velocity is measured at 0.2 d and 0.8 d. Columns (4) and (5) give the revolutions counted and the corresponding time respectively. The speed of the current meter N in rev/s is obtained by dividing the number of revolutions given in Col. (4) by the time given in Col. (5) and this is entered in Col. (6). The velocity measured by the meter is then estimated as per the given rating equation using the value of N from Col. (6) and this is posted in Col. (7). In Col. (8) the mean velocity for the vertical is given. The area of the strip is computed by multiplying the depth of flow at the vertical by the width of the strip which is taken halfway to adjacent verticals on either side. For example, for the vertical which is at a distance of 2.4 m from the bank, the depth is 1.6 m, the width halfway to the left is $\frac{(2.4 - 1.6)}{2}$ = 0.4 m and the width halfway to the right is $\frac{(3.0 - 2.4)}{2}$ = 0.3 m. Hence the area of the strip = 1.6 × (0.4 + 0.3) = 1.12 m^2. The areas of the strips calculated thus are entered in Col. (9). The incremental discharge ΔQ for each strip is obtained as the product of the area of the strip and the mean velocity in the strip and is posted in Col. (10), the total of which gives the flow in the river at the time of gauging. That is $Q = \Sigma \Delta Q$. In the present case the discharge in the river is 1.656 m^3/s.

The reader may note that the condition that the discharge in any strip should not exceed 10 per cent of the total discharge is violated in this example. But, however, this example is framed only for the purpose of demonstrating the procedure involved and the violation of the condition is to reduce the computational effort.

Table 6.2 Discharge Measurement by Area-Velocity Method

Distance from bank m	*Depth of flow m*	*Meter depth m*	*No. of revolutions*	*Time sec*	*N rev/s*	*v m/s*	*Mean velocity V m/s*	*Area of the strip m²*	*Incremental discharge ΔQ m³/s*
0.8	0.5	0.3	12	48	0.250	0.125	0.125	0.40	0.050
1.6	1.0	0.8	23	52	0.442	0.183	0.223	0.80	0.178
–	–	0.2	36	51	0.706	0.262	–	–	–
2.4	1.6	1.28	27	54	0.500	0.200	0.228	1.12	0.225
–	–	0.32	41	60	0.683	0.255	–	–	–
3.0	2.0	1.60	33	58	0.569	0.221	0.245	1.20	0.294
–	–	0.40	45	62	0.726	0.268	–	–	–
3.6	2.0	1.60	32	58	0.552	0.216	0.243	1.20	0.292
–	–	0.44	44	60	0.733	0.270	–	–	–
4.2	1.8	1.44	28	53	0.528	0.208	0.238	1.26	0.300
–	–	0.36	42	58	0.724	0.267	–	–	–
5.0	1.2	0.96	24	50	0.480	0.194	0.227	0.96	0.218
–	–	0.24	35	50	0.700	0.260	–	–	–
5.8	0.6	0.36	14	45	0.311	0.143	0.143	0.48	0.069
6.6	0.0	–	–	–	–	–	–	–	–

Σ = 1.656

6.4 DISCHARGE MEASUREMENT BY MOVING-BOAT METHOD

The conventional area-velocity method of discharge measurement described in the previous section involves costly and tedius procedure and sometimes becomes impracticable especially during floods when facilities may be inundated or the site may be inaccessible, at site where no facilities exist, or at locations where unsteady flow conditions require that measurements be made as rapidly as possible. The moving-boat technique is a method applicable to the rapid measurement of discharge in large streams. It requires no fixed facilities and lends itself to the use of alternative sites if condition, make this desirable. This technique, the description of which now follows, has been developed by the United States Geological Survey in 1969.

The moving boat technique is similar to the conventional method as both use the area-velocity approach. The main difference is in the method of data collection. In the conventional method the data at each point in the cross-section are collected while the observer is in a stationary position. This may be called a *static* approach. In contrast, in the moving-boat method, the data are collected while the observer is on a boat which is rapidly traversing the cross-section which may be called a *dynamic* approach. The position of the current meter is fixed at a constant depth from the free surface unlike in the conventional method. Since the current meter moves with the same velocity as that of boat, the current meter records the resultant of boat and stream velocities which is the vectorial sum of these two. During a traverse of the boat across the stream, a sonic sounder records the depth of flow at different points which define the geometry of the cross-section. A continuously operating current meter measures the resultant of boat and stream velocities. A vertical vane aligns itself in a direction parallel to the movement of water past the current meter and an angle indicator attached to the vane assembly indicates the angle between the direction of the vane and the true course of the boat, that is, the angle between the resultant velocity and the boat velocity. The data from these instruments provide the information necessary for computing the discharge for the cross-section. The traverse is made without stopping along a preselected path that is normal to the streamflow, and the data are collected at regular intervals along the path, normally, at 30 to 40 observation points.

In the velocity triangle at any point of observation as shown in Fig. 6.9, V V_b and V_v stand for the component of stream velocity normal to the cross- section at the sampling point, the velocity of boat and the resultant velocity respectively. Stream velocity V which is perpendicular to the boat path can be determined as

$$V = V_v \sin \theta \qquad ...(6.8)$$

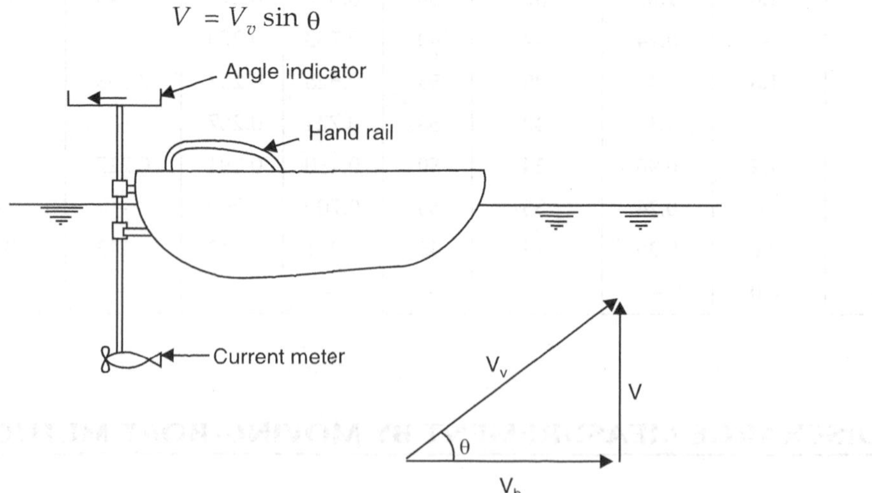

FIGURE 6.9 Moving-boat technique.

The solution of the above equation yields an answer that represents the component of stream velocity that is perpendicular to the true course, even though the direction of flow may not be perpendicular. This is the desired component. From the same velocity triangle, the distance that the boat has travelled along the true course between two consecutive observation points may be obtained as

$$L_b = \int V_v \cos\theta \, . \, dt \quad \text{...(6.9)}$$

If it is assumed that θ is approximately uniform over the relatively short distance in any one increment, then it may be treated as a constant. Alternatively, the average of the readings of θ at two consecutive points may be taken as the constant θ for the interval. Therefore Eq. (6.9) may be written as

$$L_b = \cos\theta \int V_v \, . \, dt \quad \text{...(6.10)}$$

Finally, d, the stream depth at each observations point is obtained from the sonic sounder chart. After determining V, L_b and d for each vertical, the mid-section method of computing discharge is used.

During a moving-boat discharge measurement the current meter is set at a predetermined fixed depth of about 1 m below the water surface. Strictly speaking the velocity obtained at this depth should be multiplied by a coefficient to adjust it to the mean velocity in its vertical. However, it is assumed that in the large streams where the moving-boat technique is applicable, these coefficients would be fairly uniform across a section thus permitting the use of an average cross-section coefficient to be applied to the total discharge. Information obtained from several vertical-velocity curves, well distributed across the measuring site, would be needed to determine a representative coefficient for the total cross-section. It is recommended that the results of at least six individual runs, each with 30 to 40 observation points, be averaged to obtain the discharge in uniform flow conditions. This is practicable because of the ease and speed with which extra runs can be made. For unsteady flow conditions, it will usually be desirable not to average the results from series of runs, but rather to keep them separate so as to better define the discharge variations with time. It is reported that the discharge measurements obtained by the moving boat technique agree within 4 per cent with those made by the conventional method. Moving-boat technique is not recommended in streams with water depth of less than 3 m.

To illustrate the above technique an example is presented here with all the computations involved shown in detail. The rating of the current meter is taken to be $V = 0.10 + 0.50\ N$. The average of the readings at the beginning and end of time interval is taken to be applicable for the mid-section. The computations proceed by finding the velocity recorded by the current meter V_v, the stream velocity V, the distance traversed by the boat L_b, the incremental area ΔA (= $L_b \, . \, d$) and finally the incremental discharge ΔQ (= $V . \Delta A$). The sum of all the incremental discharges would give the discharge in the stream at the time of gauging.

For convenience the computations are shown in a tabular statement in Table 6.3. In this example the sampling is done at equal time intervals. This leads to large incremental discharge in the middle of the stream. In practice the sampling time interval should be reduced for the observations in the central portion of the stream. Though the recommended number of sampling points is about 30 to 40, it is only 9 in this example. The smaller number of sampling points and the uniform sampling time interval are adopted in this example only to reduce the computations. It may also be noted that the final discharge of 489.45 m^3/s should be multiplied by a coefficient to get the true discharge.

Table 6.3. Discharge Measurement by Moving-boat Method

Time sec	*Current meter reading*	*Angle from angle indicator*	*Depth from sonic sounder m*	*N rev/s*	V_v *m/s*	*Average* θ	$L_b = V_v \cos\theta\, dt$ *m*	*Average depth d m*	$V = V_v \sin\theta$ *m/s*	$\Delta A = L_b \cdot d$	$\Delta Q = \Delta A \cdot V$
0	0	0	2.6								
30	26	21°	3.1	0.867	0.534	10.5	15.75	2.85	0.097	44.89	4.35
60	68	23°	4.6	1.400	0.800	22.0	22.25	3.95	0.300	85.66	25.70
90	130	27°	6.3	2.067	1.134	25.0	30.83	5.45	0.479	168.02	80.48
120	201	35°	8.1	2.367	1.284	31.0	33.02	7.2	0.661	237.74	157.15
150	265	31°	7.6	2.133	1.167	33.0	29.36	7.85	0.636	230.48	146.58
180	302	28°	6.5	1.233	0.717	29.5	18.72	7.05	0.353	131.98	46.59
210	335	20°	4.3	1.100	0.650	24.0	17.81	5.40	0.264	96.17	25.39
240	350	15°	1.8	0.500	0.350	17.5	10.01	3.05	0.105	30.53	3.21

Total : 489.45

6.5 DISCHARGE MEASUREMENT BY OTHER METHODS

There are several other procedures employed for the measurement of discharge in streams though none of them is comparable to the current meter method in reliability. The important of them are described here.

Slope-Area Method. In this method the flow in the channel is computed by one of the two open channel formulae, namely Chezy's or Manning's given by

$$Q = AC\sqrt{RS} \qquad \text{...(6.11)}$$

$$Q = A\frac{1}{n}R^{2/3}S^{1/2} \qquad \text{...(6.12)}$$

where A is the cross-sectional area, S is the energy slope, R is the hydraulic radius, C is the Chezy's coefficient and n is the Manning's rugosity coefficient.

A fairly straight reach of the river which is free from disturbances like backwater curve, drawdown curve or tributaries joining upstream or downstream is selected for the purpose. The length of the reach should be about 5 times the width of the river and not less than 300 m in any case. The drop in water surface in the reach selected must be at least 150 mm. The energy slope is computed as

$$S = \frac{(h_1 - h_2) + \left(\frac{V_1^2}{2g} - \frac{V_2^2}{2g}\right)}{l} \quad ...(6.13)$$

where h_1 and h_2 are the gauge reading at the *u/s* and *d/s* sections of the reach V_1 and V_2 are the average velocities in the stream at *u/s* and *d/s* sections of the reach and l is the length of the reach. From the field surveys the cross - sectional area of flow and the wetted perimeter are obtained at the beginning, middle and end of the reach. Then A and R to be used in Eq. (6.11) or Eq. (6.12) are determined as

$$A = \frac{A_1 + 2A_3 + A_2}{4} \quad ...(6.14)$$

$$P = \frac{P_1 + 2P_3 + P_2}{4} \quad ...(6.15)$$

$$R = \frac{A}{P} \quad ...(6.16)$$

where the suffix 3 stand for the mid-section of the reach. The value of C or n has to be carefully selected from the field data. As a first approximation the energy slope is taken as the water surface slope neglecting the velocity terms in Eq. (6.13) and with this S, the discharge Q is obtained from Eq. (6.11) In the next trial using the Q from pervious trial and A_1, and A_2, the velocities V_1 and V_2 are determined and the energy slope is revised with these velocities in Eq. (6.13). The discharge Q is computed using Eq. (6.11) and the revised energy slope is computed. The computations are made by successive approximations like this, till the discharge obtained in any two successive trials is nearly the same.

This method of discharge measurement is used mainly for the estimation of flood discharge when actual measurement for the area-velocity method are not possible for various reasons or when the estimation of discharge is to be made after the flood has passed the site. Further details of the method can be obtained from IS : 1912–1964.

This method can give only an approximate estimate of the discharge due to the following limitations : (*i*) Difficulty in proper selection of the rugosity coefficient n, (*ii*) The cross-sectional areas used in the discharge equation may be differing from those which existed during the time the flood passed the reach, (*iii*) The slope used in the discharge estimation cannot be precise as the water surface during the flood may be markedly warped.

Hydraulic Structures. On small streams the flow can be measured with the help of hydraulic flow measuring devices such as rectangular, triangular and Trapezoidal weirs, or flumes like venturi-flume and Parshall flume. The discharge in all these cases is expressed as a function of geometry of the structure and some reference head. The discharge expression will also contain an empirical coefficient. The details about the theory of these devices can be obtained from text-books on open channel flow.

An existing dam across a stream, a bridge opening or a causeway may also be used as a means of determining discharge indirectly. In the case of a dam the discharge can be expressed as a function of length of dam and head of the flow over the dam. In the case of bridge opening the discharge is a function of the area of flow at the constriction and the drop in water surface near the bridge. This method is known as the *contracted-opening method.* The flow over a causeway may be approximated as the flow over a broad-crested weir and the discharge estimated using the broad-crested weir formula.

Floats. In this method the floats are used to measure the velocity at various verticals across the width of the stream. After the mean velocity for the vertical is determined the discharge is computed adopting the same procedure as in area-velocity method. A float is allowed to travel on the water surface and the time taken by it to travel a stipulated distance is noted. This enables the determination of the surface velocity. The mean velocity in the vertical is taken as 0.85 times the velocity of the float at the surface.

$$V = 0.85\ v_s \qquad \text{...(6.17)}$$

Several floats introduced into the stream at various points give the mean velocities for several verticals. The distance travelled by the float must be large enough to give reasonably accurate velocity. In general it should be at least 180 m.

The main objections to this method are that the floats may not travel along straight paths between the sections and that the travel of the floats may be influenced by the prevailing wind.

Dilution Method. In this method a solution of a stable chemical such as common salt or sodium dichromate or a radioactive chemical, known as the *tracer,* is injected into the stream at either a constant rate or all at once. When the tracer is introduced into the stream at a constant rate, it is called the *plateau* method and when it it introduced all at once it is called the '*gulp*' method.

The solution will be diluted by the discharge of the stream. The concentration of the tracer at a cross-section downstream (known as the sampling section) of the injection point is determined. Let the tracer with a concentration of C_1 be injected into the stream at a constant rate of q m^3/s. Let Q be the discharge in the stream which already contains the same tracer with a concentration of C_0. If the concentration of the tracer at the downstream section C_2, then from the mass balance equation written for the tracer, we have

$$Q\,C_0 + q\,C_1 = (Q + q)\,C_2 \qquad \text{...(6.18)}$$

which on rearrangement takes the form

$$Q = q\frac{(C_1 - C_2)}{(C_2 - C_0)} \qquad \text{...(6.19)}$$

If the concentration of the tracer initially present in the stream is negligible, that is when $C_0 = 0$, the discharge in the stream is given by

$$Q = q\left(\frac{C_1}{C_2} - 1\right) \qquad \text{...(6.20)}$$

When the injection of the trace into the stream is sudden, the discharge is given by

$$Q = \frac{C_1\,\forall}{\int_{t_1}^{t_2}(C_2 - C_0)dt} \qquad \text{...(6.21)}$$

where ¥ is the volume of injected solution, $(t_2 - t_1)$ is the time in which all the tracer passes the sampling section, and C_2 is the concentration of the tracer at sampling section monitored continuously with time.

In this method, there is no need to measure velocity, depth, cross-sectional area, or any other hydraulic factors. But the accuracy of the method critically depends on two factors : (*i*) complete mixing of the injected solution throughout the stream cross-section before the sampling section is reached and (*ii*) absorption of the added tracer by the stream bottom materials, sediments, plants, organisms and the decomposition of the added tracer in the water of stream. For this reason, the concentration of the tracer should be determined at least at one more section downstream of the sampling section to verify that there is no systematic difference in the mean concentration from one section to another. Also, the tracer material should be so selected that :

(*i*) it dissolves readily in stream water at ordinary temperature,

(*ii*) it is absent in the stream water, or present only in negligible concentrations,

(*iii*) it is not decomposed in the stream water and is not retained or absorbed by sediments, plants or organisms,

(*iv*) it can be detected by simple methods,

(*v*) it is harmless to man and animals in the concentrations it assumes in the stream, and

(*vi*) it is not expensive.

Although radioactive elements are ideal tracers for this method, the health hazard may limit their use.

Electromagnetic Induction Method. Notable among the recent developments in the river gauging are the electromagnetic induction method and the ultrosonic methods. In the electromagnetic induction method a long conductor is burried at the bottom of the channel covering the entire width of the river which carries a current I to produce a controlled vertical magnetic field. As the water flowing in the stream cuts this magnetic field, it produces an electromagnetic force (emf) which is related to the discharge in the river, depth of flow, and the current in the conductor at the bottom of the river. Electrodes provided at the sides of the channel section measure the emf produced due to the flow in the river. It has been shown that the relationship among the variables of interest can be expressed as

$$Q = k_1 \left(\frac{Ed}{I} + k_2 \right)^n \quad ...(6.22)$$

where Q is the rate of flow in the river, d is the depth of flow, E is the emf induced, I is the current in the conductor and k_1, k_2, and n are the system constants to be obtained on calibration.

This method is particularly suitable for the rivers which undergo considerable change in their cross-sectional properties due to sedimentation, weed growth etc., and also in tidal channels where the flow changes rapidly with time both in magnitude and direction.

Ultrosonic Method. In this method certain reach of the river (of length at least equal to the width of the river) is made to have a rectangular lined cross-section. Two transducers which can emit as well as receive the ultrasonic signals are kept on either side of the reach such that the line joining these two is at 45° with the axis of the river as shown in Fig. 6.10.

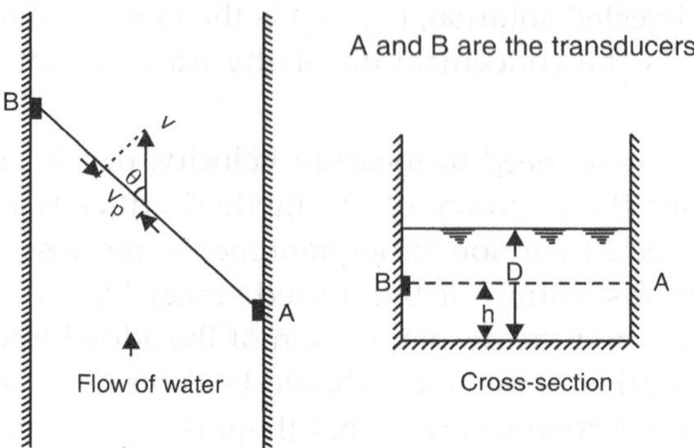

FIGURE 6.10 Ultrosonic method.

If u is the velocity of sound in water, the signal emitted at A would travel with a speed of $(u + v_p)$ while that at B with a speed of $(u - v_p)$ where v_p is the component of the flow velocity along the sound path. That is $v_p = v \cos \theta$. Therefore if t_1 is the time for the signal to travel from A to B, and t_2 is the time for the signal to travel from B to A, we can write

$$t_1 = \frac{L}{(u + v_p)} \text{ and } t_2 = \frac{L}{(u - v_p)}$$

where L is the length of the path from A to B. Thus we have

$$\frac{1}{t_1} - \frac{1}{t_2} = \frac{2v_p}{L} = \frac{2v \cos\theta}{L}$$

or

$$v = \frac{L}{2\cos\theta}\left(\frac{1}{t_1} - \frac{1}{t_2}\right) \quad ...(6.23)$$

The velocity obtained from Eq. (6.23) is the average velocity in the stream at a height h above the stream bed where the transducers are placed and not the average velocity for the entire cross-section. Therefore, when only one pair of transducers is used, the relation between $\frac{v}{V}$ and $\frac{h}{D}$ must be first established by calibration. This relation enables one to get the average velocity for the cross-section from the velocity measured by the pair of transducers located at h above the stream bed. Then the discharge is obtained as the product of the area of the flow and the average velocity. Estimation of discharge using only one pair of transducers and one signal path is called the *single-path gauging*. Alternatively for a given depth of flow, several pairs of transducers placed at different heights may be used to obtain the velocities at different heights which may be averaged to yield the average velocity for the cross-section. Then it is known as the *multiple-path gauging*.

Example 6.2 *In order to compute the flood discharge in a stream by the slope - area method the following data has been obtained.*

	u/s section	*middle section*	*d/s section*
Area	*108.6 m²*	*103.1 m²*	*99.8 m²*
Wetted perimeter	*65.3 m*	*60.7 m*	*59.4 m*
Gauge reading	*+ 316.8 m*	—	*+ 316.55 m*

Determine the flood discharge assuming Manning's n = 0.029 and length between u/s and d/s sections as 250 m.

Solution. Average area for the reach $A = \dfrac{A_1 + 2A_3 + A_2}{4}$

$$A = \frac{108.6 + 2 \times 103.1 + 99.8}{4} = 103.65 \text{ m}^2$$

Average wetted perimeter for the reach $P = \dfrac{P_1 + 2P_3 + P_2}{4}$

$$P = \frac{65.3 + 2 \times 60.7 + 59.4}{4} = 61.53 \text{ m}$$

Hydaulic radius $R = \dfrac{A}{P} = \dfrac{103.65}{61.53} = 1.685$ m

$$R^{2/3} = 1.416$$

$$Q = \frac{1}{n} A R^{2/3} S^{1/2} = \frac{1}{0.029} \times 103.65 \times 1.416 \times \sqrt{S}$$

or $$Q = 5060.98 \sqrt{S} \text{ m}^3/\text{s}.$$

Trial 1 : The slope is computed from Eq. (6.13) neglecting the velocity heads

$$S = \frac{h_1 - h_2}{l} = \frac{316.8 - 316.55}{250} = \frac{0.25}{250} = \frac{1}{1000}$$

$$\therefore \quad Q = 5060.98 \times \sqrt{\frac{1}{1000}} = 160.04 \text{ m}^3/\text{s}$$

Trial 2 : $$\frac{V_1^2}{2g} = \frac{\left(\dfrac{Q}{A_1}\right)^2}{2g} = \frac{1}{19.62} \times \left(\frac{160.04}{108.6}\right)^2 = 0.1107 \text{ m}$$

$$\frac{V_2^2}{2g} = \frac{\left(\dfrac{Q}{A_2}\right)^2}{2g} = \frac{1}{19.62} \times \left(\frac{160.04}{99.8}\right)^2 = 0.1311 \text{ m}$$

$$S = \frac{(h_1 - h_2) + \left(\dfrac{V_1^2}{2g} - \dfrac{V_2^2}{2g}\right)}{l} = \frac{0.25 + (0.1107 - 0.1311)}{250}$$

$$= \frac{0.2296}{250} = 9.184 \times 10^{-4}$$

$$Q = 5060.98 \times \sqrt{9.184 \times 10^{-4}} = 153.37 \text{ m}^3/\text{s}.$$

Trial 3 :

$$\frac{V_1^2}{2g} = \frac{1}{19.62} \times \left(\frac{153.37}{108.6}\right)^2 = 0.1017 \text{ m}$$

$$\frac{V_2^2}{2g} = \frac{1}{19.62} \times \left(\frac{153.37}{99.8}\right)^2 = 0.1204$$

$$S = \frac{0.25 + (0.1017 - 0.1204)}{250} = 9.252 \times 10^{-4}$$

$$Q = 5060.98 \times \sqrt{9.252 \times 10^{-4}} = 153.94 \text{ m}^3/\text{s}.$$

Since the discharges in trials 2 and 3 are nearly the same, the computations can be terminated here and the flood discharge may be taken as 153. 94 m^3/s.

Example 6.3. *In to a stream, with no trace of salt initially, a salt solution with a concentration of 20 mg/c.c. is introduced at a constant rate of 2 litres per minute. The samples collected at a downstream section sufficiently far away indicated an equilibrium salt concentration of 0.05 ppm. Determine the discharge in the stream from this data.*

Solution. From the given data, we have

$$C_0 = 0$$

$$C_1 = 20 \text{ mg/c.c.} = \frac{20}{1000} = 0.02$$

$$C_2 = 0.05 \text{ ppm} = \frac{0.05}{10^6} = 5 \times 10^{-8}$$

$$q = 2 \text{ liters/minute} = \frac{2 \times 10^{-3}}{60} = 3.33 \times 10^{-5} \text{ m}^3/\text{s}.$$

From Eq. (6.20),

$$Q = q\left(\frac{C_1}{C_2} - 1\right)$$

$$= 3.33 \times 10^{-5}\left(\frac{0.02}{5 \times 10^{-8}} - 1\right)$$

$$= 13.33 \text{ m}^3/\text{s}.$$

6.6 STAGE-DISCHARGE RELATIONS

After a sufficient number of discharge measurements have been made at a gauging station along with simultaneous stage observations, the results are plotted on an ordinary graph. It is customary to take the discharge as abscissa and the stage as the ordinate. Such a plot between the discharge and stage is known as the *stage-discharge relation* or the *rating curve* of the gauging station. Once a stable stage-discharge relation is established, it is only a matter of recording the stage continuously which can be readily converted into the discharge through the above relation. For most of the gauging stations a simple plot of stage *vs* discharge is satisfactory. Such a curve is approximately parabolic as shown Fig. 6.11. It may show some irregularities if the cross-section is irregular. In most of the cases the discharge measurements are made within a rather limited range around the average stage in the river. Both at very low stages and at very high stages during floods there may be no measurements. It is, therefore, necessary to extend the curve downward for the low stages and upward for the high stages. The methods of extending the stage-discharge curve are discussed in the next section. The stage-discharge curve has its greatest curvature in the low stages.

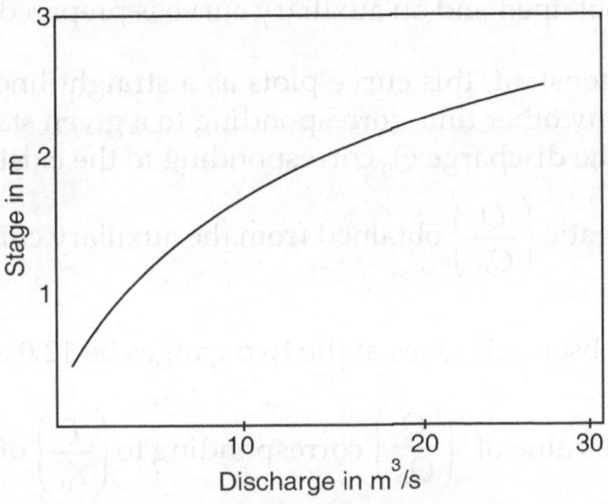

FIGURE 6.11 A simple stage-discharge relation.

Slope-stage-discharge relation. The stage-discharge data may not always lead to a well defined curve. Individual measurements may deviate considerably from the mean curve. This is because of the difference in water surface slope at the gauging station caused by varying backwater as a result of an obstruction downstream of the gauging site or as a result of high stages in a tributary joining the stream downstream of the gauging site. If either of these factors is present, the discharge in the stream may be different for different water surface slopes though the stage in the river is same. Under such conditions, therefore, the rating curve must include slope also as a parameter. Then it is called a slope-stage-discharge relation which is constructed as described below.

There will be two gauges in the stream recording the stage in the river at two sections ; one at the gauging station and the other, known as the *auxiliary* gauge, at a fixed distance, L, downstream. The distance L should be such that the recorded fall in the water surface between the two gauges should be at least 0.3 m. Let Q be the discharge in the river when the recorded fall is F, and let Q_0 be the discharge in the river when the recorded fall is F_0. While the stage in the river at the gauging site is same in both the cases. Then from open channel flow theory, we can write $Q \propto S^m$ and $Q_0 \propto S_0^m$, where S and S_0 are the corresponding energy slopes, since the cross-sectional area is same in both the cases. Assuming that energy slope may be taken to be the water surface slope we can replace S by $\left(\frac{F}{L}\right)$ and S_0 by $\left(\frac{F_0}{L}\right)$. This leads to the result

$$\frac{Q}{Q_0} = \left(\frac{S}{S_0}\right)^m = \left(\frac{F}{F_0}\right)^k \qquad ...(6.24)$$

While the exponent m is expected to be 0.5, k need not be equal to 0.5 since there is no assurance that the water surface between the two gauges is a straight line. Hence the value of k must be determined empirically. All the measurements on stage and discharge are plotted. The recorded fall for every measurement is indicated at the corresponding point. A simple stage-discharge relation is then fitted such that it passes through those points for which the recorded fall is equal to some standard fall F_0 (say 1.0 m). This curve is called Q_0 curve which gives the discharge when $F = F_0$. From the data of other points which are deviating from the Q_0 curve, the ratios $\left(\frac{Q}{Q_0}\right)$ and $\left(\frac{F}{F_0}\right)$ are obtained and an auxiliary curve is prepared as shown in Fig. 6.12. If the exponent k in Eq. (6.24) is constant, this curve plots as a straight line on logarithmic paper. The discharge in the stream at any other time corresponding to a given stage and a given fall F can be computed by first reading the discharge Q_0 corresponding to the existing stage from the Q_0 curve and multiplying this by the ratio $\left(\frac{Q}{Q_0}\right)$ obtained from the auxiliary curve from the known value of $\left(\frac{F}{F_0}\right)$. For example, let the observed stages at the two gauges be 12.0 and 11.3 m . $F = (12.0 - 11.3) = 0.7$ m ; $\frac{F}{F_0} = \frac{0.7}{1.0} = 0.7$. The value of $\left(\frac{Q}{Q_0}\right)$ corresponding to $\left(\frac{F}{F_0}\right)$ of 0.7 from the auxiliary chart of Fig. 6.12 is 0.73. The discharge corresponding to a stage of 12.0 m from the Q_0 curve is read as 45 m^3/s. Therefore the correct discharge is given by

$$Q = Q_0 \times \left(\frac{Q}{Q_0}\right) = 45 \times 0.73 = 32.85 \text{ m}^3/\text{s}.$$

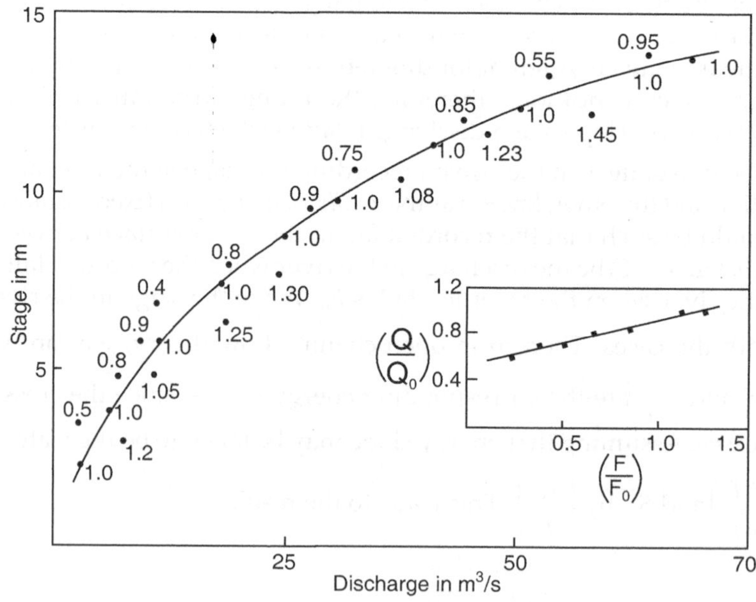

FIGURE 6.12 Slope-stage-discharge relation.

Stage-Discharge Relation Under Unsteady-Flow Conditions. Stream gauging stations not subjected to variable slope because of backwater may still be affected because of change in stage with time under unsteady flow conditions. During a rising stage of the river the velocity and discharge in the river are greater than they are for the same stage during steady flow conditions. Likewise, during a falling stage the discharge is less for any given stage than it is when the flow is steady. If Q_0 is the discharge under steady flow conditions for a given stage and Q is the discharge under unsteady flow conditions for the same stage but when the stage is changing at a rate of $\frac{dy}{dt}$, then it can be shown that

$$Q = Q_0 \sqrt{1 + \frac{1}{V_w S_0} \frac{dy}{dt}} \qquad \text{...(6.25)}$$

where V_w is the velocity of the flood wave and S_0 is the bed slope of the stream. For rising stages $\frac{dy}{dx}$ is positive and for falling stages (when the flood is receding) it is negative. Therefore, when plotted, the results of discharge measurement made during rising or falling stages do not fall on the true discharge curve but they lie to the right or left thereof, respectively. This descrepancy in the stage - discharge relationship caused by the variable slope because of changing stage under unsteady flow conditions can be treated in several ways.

One method is to develop a slope-stage-discharge relation as explained in the previous section, but this requires the installation and maintenance of the auxiliary gauge for ever.

The second method which assumes that the term $(V_w S_0)$ is a constant for the gauging site and which does not require an auxiliary gauge is as follows. A rating curve for steady flow conditions is first constructed using the discharge measurements made under fairly steady state conditions. This is the Q_0 curve. Making use of the discharge measurements taken during unsteady conditions a correction curve is constructed showing $\left(\frac{Q}{Q_0}\right)$ as a function of the rate of change of stage $\frac{dy}{dt}$ as shown in Fig. 6.13. The rating curve for steady state conditions along with the correction curve will facilitate the computation of discharge for any combination of stage and $\frac{dy}{dt}$. This is done by first reading Q_0 for the observed stage from steady state curve and then multiplying it by the correction factor indicated for the observed value of $\frac{dy}{dt}$.

For example, if the stage falls from 12.02 m to 11.98 m in an observation period of 1 hour, $\frac{dy}{dt} = \frac{(11.98 - 12.02)}{1.0} = -\ 0.04$ m/h and stage $= \left(\frac{12.02 + 11.98}{2}\right) = 12.0$ m. From the steady state rating curve of Fig. 6.13, Q_0 for a stage of 12 m is read as 70 m^3/s and from the correction curve for

$\frac{dy}{dt}$ = – 0.04 m/h, the value of $\left(\frac{Q}{Q_0}\right)$ is read as 0.875. Now the correct discharge = 70 × 0.875 = 61.25 m^3/s.

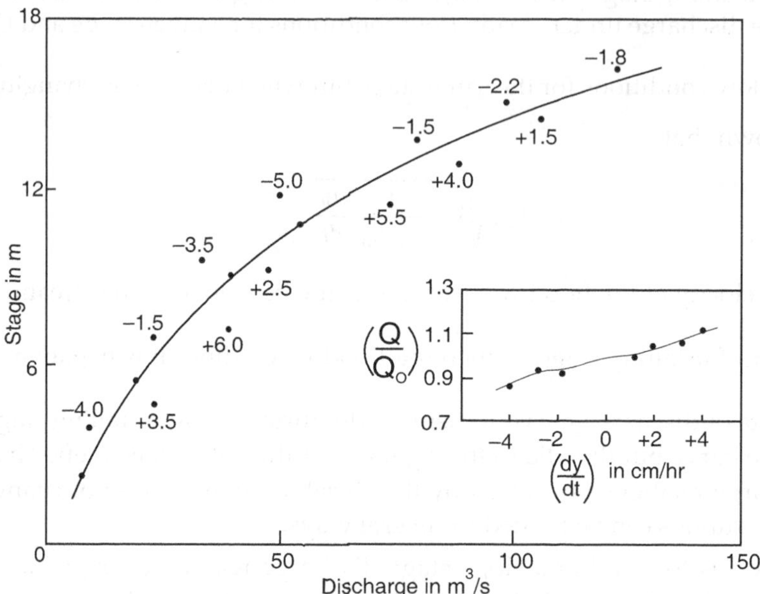

FIGURE 6.13 Stage-discharge relation for unsteady-flow conditions.

Alternatively from the measured discharges and $\frac{dy}{dt}$, and Q_0 from the steady state rating curve, the value of $\left(\frac{1}{V_w \cdot S_0}\right)$ may be determined from Eq. (6.25) and plotted against the corresponding stage which may be utilised as the correction curve instead of the curve of $\frac{dy}{dt}$ *vs* $\left(\frac{Q}{Q_0}\right)$. This is supposed to yield more accurate results since $(V_w \cdot S_0)$ is, strictly speaking, not a constant. For any observed stage the actual discharge is obtained by substituting Q_0 corresponding to the observed stage from the steady state curve, $\frac{1}{V_w \cdot S_0}$ from the correction curve and the observed value of $\frac{dy}{dt}$ in Eq. (6.25).

In these procedures, the rate of change of stage $\frac{dy}{dt}$ has to be obtained by drawing a tangent to the stage *vs* time graph (known as the *stage hydrograph*) at the time of observation. Since this is difficult to obtain, the practical method, however, consists of taking the change in stage during the hour preceding the gauge reading as $\frac{dy}{dt}$. Sometimes an increment of more than one hour may also be used.

Example 6.4. *The following data pertain to streamflow in a river at a gauging station. The rate of rise in water surface elevation = 10.0 cm/h. The normal discharge for the river stage obtained from the steady flow rating curve = 150 m³/s. Slope of the river bed = 3 × 10⁻⁴. Assume the velocity of the flood wave to be 2.5 cm/sec compute the river discharge.*

Solution. We have from Eq. (6.25)

$$Q = Q_0\sqrt{1 + \frac{1}{V_w S_0} \cdot \frac{dy}{dt}}$$

$$\frac{dy}{dt} = 10 \text{ cm/h} = \frac{10}{60 \times 60} \text{ cm/s} = 2.778 \times 10^{-3} \text{ cm/s}$$

$$S_0 = 3 \times 10^4 \,;\, V_w = 2.5 \text{ cm/s} \,;\, Q_0 = 150 \text{ m}^3/\text{s}$$

$$\therefore \quad Q = 150\sqrt{1 + \frac{1}{2.5 \times 3 \times 10^{-4}} \times 2.778 \times 10^{-3}}$$

$$Q = 150\sqrt{1 + 3.704} = 150 \times 2.1689 = 325.33 \text{ m}^3/\text{s}$$

6.7 EXTENSION OF THE RATING CURVE

The stage -discharge curve is usually established using the measurements of discharge taken around the mean stage in the river. During the floods or high stages the river may become inaccessible. Therefore, it is often necessary to have the stage-discharge curve extended beyond the highest measured discharges. For example, the magnitude of the design flood may be estimated using various techniques other than actual measurement. To know the stage in the river at the project site corresponding to the design flood we have to resort to only the stage discharge curve. Similarly the extension of rating curve becomes a necessity for river forecasting and other flood control problems. Several techniques have been developed as a result of experience and study of a large number of stage-discharge relations. In many cases these procedures will lead to a reasonably adequate extensions. However, it must be emphasized that any extension is open to possible serious error especially when the stream is of alluvial type subjected to aggradation and degradation and the results of extension should always be confirmed by alternative methods, and a field survey to establish the probable trend of the rating should be made if possible. These are many methods of extending the rating curve and only three methods that are widely used are described here.

Logarithmic Method. In this method an equation of the form

$$Q = k\,(y - a)^b \qquad \text{...(6.26)}$$

is assumed to be an appropriate fit between the discharge Q and the stage in the river y. The constants a, b and k are to be determined using the observed data on y and Q. The value of a is supposed to be the stage corresponding to zero discharge in the stream. Taking logarithms on both sides of Eq. (6.26),

$$\log Q = \log k + b \log (y - a) \qquad ...(6.27)$$

Eq. (6.27) suggests that a plot of Q *vs* $(y - a)$ would plot as a straight line on a log-log paper. This straight line can be extrapolated to determine the discharges corresponding to high stages. A trial value of a is assumed and the plot is made between Q and $(y - a)$ on log-log plot. If the plot is a straight line the correct value of a is obtained in the first trial itself. If the plot is a concave curve upward the value of a is decreased and if the plot is a concave curve downward the value of a is increased in the next trial and plot is remade. The trial and error procedure is continued till the plot is very nearly a straight line.

Alternatively, the least square regression analysis may be applied following the procedure of Example 5.6 to obtain the best values of a, b and k and Eq. (6.26) is then used for the purpose of extension. The criticism levelled against this method is that there is no assurance that Eq. (6.26) holds at high stages where there may be abrupt changes in the cross-section of the river.

The values of a, b and k may also be obtained from the following short-cut method without resorting to the rigorous regression analysis. Let (Q_1, y_1), (Q_2, y_2) and (Q_3, y_3) be three points on the existing rating curve such that $Q_2 = \sqrt{Q_1 Q_3}$. Then the value of a can be obtained from

$$a = \frac{y_1 y_3 - y_2^2}{y_1 + y_3 - 2y_2} \qquad ...(6.28)$$

Values of b and k can then be determined from Eq. (6.26). Since the constants thus obtained rarely provide a good fit throughout the entire range of the rating, the three points should be selected in the middle and upper flow range.

The value of 'a' may also be determined by the following graphical procedure which assumes that the rating curve is a parabola in the lower part. As shown in Fig 6.14, three points A, B and C are selected on the rating curve plotted on an ordinary graph such that $Q_B = \sqrt{Q_A \cdot Q_C}$. Vertical through B and a horizontal through C are drawn to meet at E and a vertical through A and horizontal through B are drawn to meet at D. Then lines AB and DE are extended to meet at F. The ordinate of the point F gives the desired value of a.

Efficiency of Stage-Discharge Curve. After fitting an appropriate stage-discharge curve, the standard error of estimate S_e is computed from the equation

$$S_e = \sqrt{\frac{1}{n} \Sigma \left(Q_0 - \hat{Q}\right)^2} \qquad ...(6.29)$$

where Q_0 is the observed discharge for a given stage and $\hat{Q}$ is the computed discharge from the equation fitted to the curve for the same stage and the summation is over all the n observations. Now the standard deviation of the observed discharges S_Q is calculated from the equation

$$S_Q = \sqrt{\frac{1}{n}\Sigma\left(Q_0 - \overline{Q}\right)^2} \qquad ...(6.30)$$

where $\overline{Q}$ is the average of the observed discharges. The efficiency of the fitted curve η is then calculated as

$$\eta = \frac{\text{Variance of observed discharges } S_e^2}{\text{Variance of observed discharges}}$$

$$\eta = \frac{S_Q^2 - S_e^2}{S_Q^2} \qquad ...(6.31)$$

Obviously closer is the value of η to 100, the better is the fitness of the curve. It is suggested that η should at least be 80% .

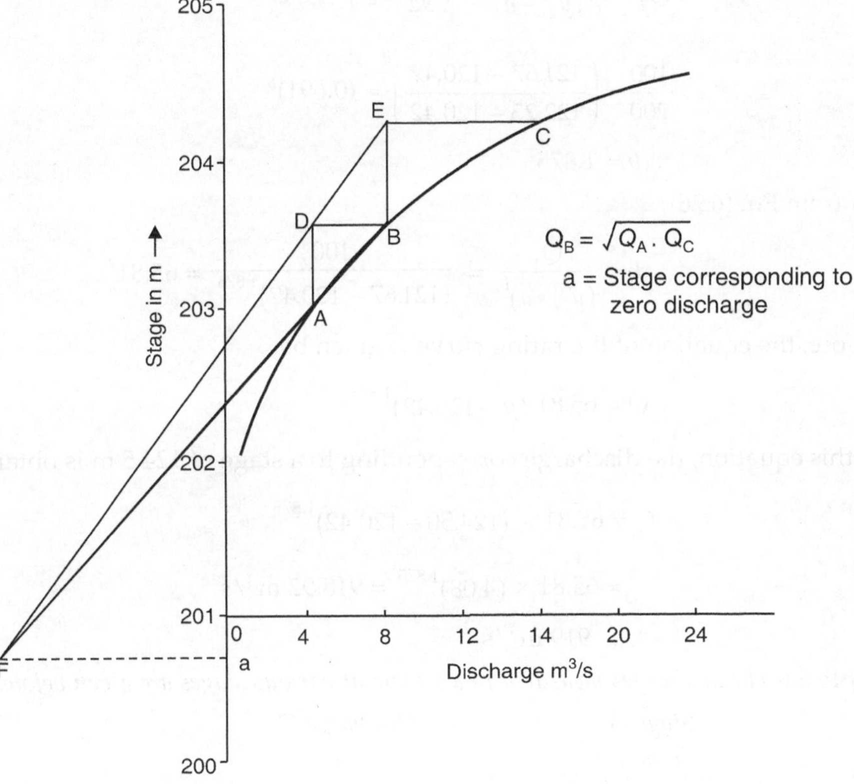

FIGURE 6.14 Determination of *a* by graphical method.

Example 6.5. *Three points on a rating curve of a stream gauging station obtained from an eye-fit for the stage discharge data have the following coordinates : (100 m³/s, 121.67 m), (200 m³/s, 122.23 m), and (400 m³/s, 123.04 m). Determine the equation of the rating curve and compute the discharge in the stream corresponding to a stage of 124.5 m.*

Solution.
$$y_1 = 121.67 \text{ m} \quad Q_1 = 100 \text{ m}^3/\text{s}$$
$$y_2 = 122.23 \text{ m} \quad Q_2 = 200 \text{ m}^3/\text{s}$$
$$y_3 = 123.04 \text{ m} \quad Q_3 = 400 \text{ m}^3/\text{s}$$

Since the given discharges satisfy the condition that $Q_2 = \sqrt{Q_1 \cdot Q_3}$, the constant a in the equation $Q = k\,(y - a)^b$ can be obtained as

$$a = \frac{y_1\, y_3 - y_2^2}{y_1 + y_3 - 2y_2}$$

$$= \frac{121.67 \times 123.04 - (122.23)^2}{121.67 + 123.04 - 2 \times 122.23} = 120.42.$$

Now,

$$\frac{Q_1}{Q_2} = \frac{k\,(y_1 - a)^b}{k\,(y_2 - a)^b} = \left(\frac{y_1 - a}{y_2 - a}\right)^b$$

or
$$\frac{100}{200} = \left(\frac{121.67 - 120.42}{122.23 - 120.42}\right)^b = (0.691)^b$$

or
$$b = 1.875$$

Again from Eq. (6.26),

$$k = \frac{Q_1}{(y_1 - a)^b} = \frac{100}{(121.67 - 120.42)^{1.875}} = 65.81.$$

Therefore, the equation of the rating curve is given by

$$Q = 65.81\,(y - 120.42)^{1.875}$$

From this equation, the discharge corresponding to a stage of 124.5 m is obtained as

$$Q = 65.81 \times (124.50 - 120.42)^{1.875}$$

$$= 65.81 \times (4.08)^{1.875} = 918.92 \text{ m}^3/\text{s}$$

$$\approx 919 \text{ m}^3/\text{s}.$$

Example 6.6 *The discharges measured in a stream at various stages are given below.*

Stage G in mm	*Discharge Q m³/s*
3.80	320.0
4.70	456.0
5.50	583.0
5.85	651.0
6.55	821.0

6.90	*906.0*
7.10	*985.0*
7.45	*1098.0*
7.85	*1296.0*

Plot the stage-discharge data on plain paper and fit a curve by eye. What is the stage corresponding to zero discharge ? Find the best fit equation to the curve and calculate the standard error of estimate and the efficiency of the curve.

Solution. The given stage-discharge data is plotted on a plain paper as shown in Fig. 6.15 and a smooth curve is drawn through the points by eye-fit. Extending this curve on lower side, it is observed that the intercept on vertical axis is 1.0 m. This corresponds to zero discharge and may be taken as first trial value for G_0. The equation to be fitted is taken as

$$Q = C\,(G - G_0)^m$$

where C and m are regression constants to be determined. Taking logarithm on both sides $\log Q = \log C + m \log (G - G_0)$ or $y = a + bx$ where $y = \log Q$, $a = \log C$, $x = \log (G - G_0)$ and $b = m$. The values of a and b can be determined from the following equations

$$b = \frac{\Sigma xy - (\Sigma x \,.\Sigma y)/n}{\Sigma x^2 - (\Sigma x)^2/n} \qquad ...(6.32)$$

and

$$a = \frac{\Sigma y - b\Sigma x}{n} \qquad ...(6.33)$$

where all the summations run for 1 to n, n being the no. of observations.

Table 6.4 Computation to Fitting the equation for Example 6.6 Trial with G_0 = 0.8 m

S.No.	G *m*	Q m^3/s	$G - G_0$	$x = \log (G - G_0)$	$y = \log Q$	xy	x^2	$\hat{Q}$	$(Q-\hat{Q})^2$	$(Q-\overline{Q})^2$
1	3.80	320	3.00	0.4771	2.5052	1.1953	0.2276	296.57	549.14	221530
2	4.70	456	3.90	0.5911	2.6590	1.5716	0.3494	453.88	4.51	112004
3	5.50	583	4.70	0.6721	2.7657	1.8588	0.4517	614.3	979.10	43127
4	5.85	651	5.05	0.7033	2.8136	1.9788	0.4946	690.19	1535.98	19508
5	6.55	821	5.75	0.7597	2.9143	2.2139	0.5771	851.95	957.69	920
6	6.90	906	6.10	0.7853	2.9571	2.3223	0.6167	937.69	1001.08	13301
7	7.10	985	6.3	0.7993	2.9934	2.3926	0.6389	988.01	9.06	37764
8	7.45	1098	6.65	0.8228	3.0406	2.5019	0.6770	1078.57	377.58	94452
9	7.85	1296	7.05	0.8482	3.1126	2.6401	0.7194	1185.75	12154.65	255358
Σ		7116		6.4589	25.7615	18.6753	4.7508		17568.79	797964

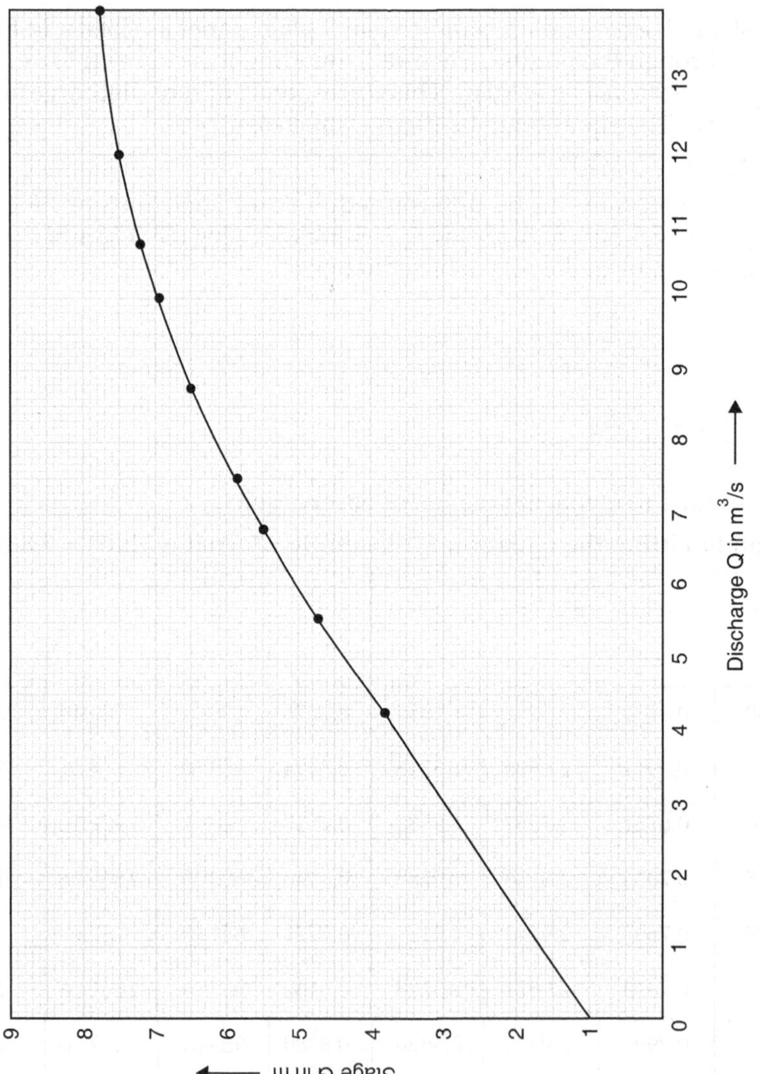

FIGURE 6.15 Stage discharge curve of Example 6.6.

All the required summations are computed as shown in Table 6.4 which is self explanatory. Though 5 trials are carried out with $G_0 = 1.0, 0.9, 0.8, 0.7$ and 1.1, the computation are shown only for $G_0 = 0.8$ in Table 6.4. For this trial

$$b = \frac{\Sigma xy - (\Sigma x . \Sigma y)/n}{\Sigma x^2 - (\Sigma x)^2/n} = \frac{18.6753 - (6.4589 \times 25.7615)/9}{4.7508 - (6.4589)^2/9} = 1.622$$

$$a = \frac{\Sigma y - b\Sigma x}{n} = \frac{25.7615 - 1.622 \times 6.4589}{9} = 1.698234$$

But $a = \log C \therefore C = 10^{1.698234} = 49.915$

$m = b = 1.622$. Therefore the required equation is

$Q = 49.915 (G - 0.8)^{1.622}$

$$\text{Standard error of estimate } S_e = \sqrt{\frac{\Sigma\left(Q - \hat{Q}\right)^2}{9}} = \sqrt{\frac{17568.79}{9}} = 44.18$$

$$\overline{Q} = \frac{\Sigma Q}{n} = \frac{7116}{9} = 790.67$$

$$\text{Standard deviation of the observed flow } S_Q = \sqrt{\frac{\Sigma\left(Q - \overline{Q}\right)^2}{n}}$$

$$S_Q = \sqrt{\frac{797964}{9}} = 297.76$$

$$\eta = \left(\frac{S_Q^2 - S_e^2}{S_Q^2}\right) \times 100 = \frac{(297.76)^2 - (44.18)^2}{(297.76)^2} = 97.8\%$$

The results of all the trials are summarised in the following table

Trial No.	G_0	C	*m*	S_e	*Q given by*
1	1.1	67.963	1.491	49.01	$Q = 67.963 (G - 1.1)^{1.491}$
2	1.0	62.355	1.525	48.30	$Q = 62.355 (G - 1.0)^{1.525}$
3	0.9	56.67	1.564	47.04	$Q = 56.67 (G - 0.9)^{1.569}$
4	0.8	49.915	1.622	44.18	$Q = 49.915 (G - 0.8)^{1.622}$
5	0.7	47.61	1.6313	47.61	$Q = 47.61 (G - 0.7)^{1.6313}$

Example 6.7. *The following measurements are available for the main gauge and auxiliary gauge which is downstream.*

Discharge m³/s	*Main gauge m*	*Auxiliary gauge m*
315	*132.5*	*132.0*
600	*132.5*	*131.0*

If the main gauge reading is 132.5 and the auxiliary gauge reading is 131.5 m. What is the discharge in the stream?

Solution. When $Q_0 = 315 \text{ m}^3/\text{s}$ $F_0 = 132.5 - 132.0 = 0.5 \text{ m}$

when $Q_1 = 600 \text{ m}^3/\text{s}$ $F_1 = 132.5 - 131.0 = 1.5 \text{ m}$

We can assume $Q_0 \propto F_0^m$ and $Q_1 \propto F_1^m$

$$\therefore \quad \frac{Q_0}{Q_1} = \left(\frac{F_0}{F_1}\right)^m \quad \text{or} \quad \frac{315}{600} = \left(\frac{0.5}{1.5}\right)^m$$

$$0.525 = (0.3333)^m$$

$$m = \frac{\log (0.525)}{\log (0.3333)} = \frac{-0.27984}{-0.47712} = 0.5865$$

Now $F_2 = 132.5 - 131.5 = 1.0 \text{ m}$

$$\frac{Q_0}{Q_2} = \left(\frac{F_0}{F_2}\right)^m = \left(\frac{0.5}{1.0}\right)^{0.5865} = 0.666$$

$$Q_2 = \frac{Q_0}{0.666} = \frac{315}{0.666} = 472.97 \text{ m}^3/\text{s}.$$

Steven's Method. This method is based on the Chezy's discharge formula given by Eq. (6.11). In this equation if $C\sqrt{S}$ is assumed constant for the given gauging site, and the mean depth of flow D is substituted for hydraulic radius R, then the discharge may be expressed as

$$Q = K A \sqrt{D} \qquad \text{...(6.34)}$$

where $K = C\sqrt{S}$. So, when known values of $A\sqrt{D}$ and Q are plotted on a graph, they represent nearly a straight line which can be easily extended. Values of $A\sqrt{D}$ for stages above the existing rating can be obtained by field measurements and used with the extended curve for estimates of Q corresponding to higher stages. The objection to this method is that $C\sqrt{S}$ may not be constant for all stages. However, as it takes the varying stream geometry into account it is a more rational procedure than the previous.

6.8 SELECTION OF A STREAMGAUGING SITE

A site selected on stream for the purpose of measuring discharge in the stream continuously is known as either a *streamgauging station* or a *hydrometric station*. The stage-discharge relation at streamgauging station is defined by the complex interaction of the channel characteristics including cross-sectional areas, shape, slope and roughness. The combination of these effects is called the *control*. A control is said to be a *permanent control* if the stage-discharge relationship which it defines does not change with time. On the contrary it is known as either an *impermanent control* or a *shifting control*. The shifting control may be due to the effects of changing channel scour and silt in alluvial streams, backwater, rapidly changing stage in the stream (unsteady flow conditions), variable channel storage, aquatic vegetation etc. After a gauging station has been established and sufficient discharge measurements have been made to determine the stage-discharge relation, it becomes necessary to make additional measurements at various stages from time to time to know just how rapidly or how slowly the control is changing. A control should never be assumed permanent until it has been proved so by a number of measurements made after the passage of floods.

There are four criteria in selecting a site for establishing a streamgauging station. They are accessibility, adequacy, stability and permanency. The station should be accessible under all conditions, particularly during floods. The station should be able to cover the full range of discharge which may occur. The stage-discharge relation at the site should change with time as little as possible. This aspect is very important both from the stand point of cost of operation and accuracy of streamflow computation. The station should never be located upstream of a junction with another stream near enough to be affected by backwater from that stream. It should also not be located within the influence of backwater from any project. The station should be so situated that the installation is almost permanent and is not likely to be disturbed. One of the most important features of streamflow record is its unbroken length. Also it should be located above points of flow diversion.

6.9 STREAMGAUGING NETWORK

The streamflow data is of paramount importance to both engineers and hydrologists and at the same time it is a costly affair. The limited budget should be utilised to obtain the reasonable estimates of the streamflow. The WMO has laid down the following norms for the streamgauging network.

Region I : Flat regions of temperate, mediterranean and tropical zones.

Minimum = 1 station per 1000 to 2500 km^2

Tolerable = 1 station per 3000 to 10000 km^2

Region II : Mountainous areas of temperate, mediterranean and tropical zones.

Minimum = 1 station per 300 to 1000 km^2

Tolerable = 1 station per 1000 to 5000 km^2

Region III : Arid zones.

Minimum : 1 station per 5000 to 20000 km^2.

For small mountainous islands with irregular rainfall the recommended density is one station per 140 to 300 km^2. The stations which record only the stage in the river should be installed at all major towns along the rivers and on all major rivers they should be installed in between the stream gauging stations. The following important points must be considered while locating the streamgauging stations :

(*i*) immediately below possible damsite, and other hydraulic structures.

(*ii*) at the inlet and outlet of natural lakes.

(*iii*) at the head of alluvial plains and above cities subject to flood damage.

(*iv*) at the limiting point just above the tidal influence and on the major tributary streams near their confluence with the main river.

6.10 STAGE AND DISCHARGE HYDROGRAPHS

A graph showing the variation of any property of water flow in the stream such as stage, discharge, velocity, sediment load etc., with time is known as a *hydrograph*. Specially if the stage is plotted against time it is called a *stage hydrograph* and likewise if the discharge is plotted against time then it is called a *discharge hydrograph* etc. However, the word hydrograph generally refers to discharge hydrograph unless specified otherwise. The discharge hydrograph at the gauging station is usually obtained from the observed stage hydrograph by converting the observed stages in the river into corresponding discharges through the stage-discharge relation of the gauging station. The hydrographs are plotted by computing the instantaneous discharges at sufficient number of points to indicate adequately all significant changes in the slope of hydrograph.

The volume of streamflow in any time interval may be obtained by integrating the hydrograph between the appropriate limits. Thus we have

$$\forall = \int_{t_1}^{t_2} Q \,.\, dt \qquad ...(6.35)$$

where $\forall$ denotes the volume of streamflow in the time interval $(t_2 - t_1)$. Since the discharge Q cannot be expressed as a well defined function of time t, the integration cannot be performed analytically. In practice the integration is done either graphically or numerically. In the graphical method the hydrograph is plotted on an ordinary graph paper with a scale of 1 cm = p hours on x-axis and 1 cm = q m^3/s on y-axis. Each square centimetre area of the graph then represents $pq \dfrac{\text{m}^3}{\text{s}}\text{h}$ or 3600 pq m^3. The area below the hydrograph between the time limits t_1, and t_2 measured in square centimetres multiplied by (3600 pq) therefore gives the volume of streamflow in the time interval $(t_2 - t_1)$. In the numerical method the ordinates of the hydrograph are measured at some uniform time interval of Δt hours. Let the time period $(t_2 - t_1)$ is divided into $(n - 1)$ equal intervals of Δt each and let $Q_1, Q_2, ..., Q_{n-1}$ and Q_n be the ordinates of the hydrograph in m^3/s at these intervals. Assuming linear variation of discharge over each time interval and applying the trapezoidal rule, the volume of the streamflow can be computed from the expression

$$\forall = 3600\ \Delta t \times \left[\sum_{i=1}^{n} Q_i - \frac{(Q_1 + Q_n)}{2} \right] \qquad ...(6.36)$$

The volume of streamflow during any day may be obtained by integrating the hydrograph between two successive mid-nights covering that particular day. If the volume of streamflow thus obtained is divided by a time period of 24 hours, the average discharge for that day, which is known as the *mean daily discharge*, is obtained.

Stramflow data are usually published in the form of mean daily flows. The picture presented by the mean daily flows is somewhat distorted as it averages out the peak flows. On large streams the maximum instantaneous discharge may only be slightly higher than the maximum mean daily flow as shown in Fig. 6.16 (*b*). On small streams the maximum instantaneous discharge is usually much greater than the highest mean daily flow as shown in Fig. 6.16 (*a*). Therefore, whenever possible, the hydrologist should secure and work with hydrographs of instantaneous flows when dealing with small basins.

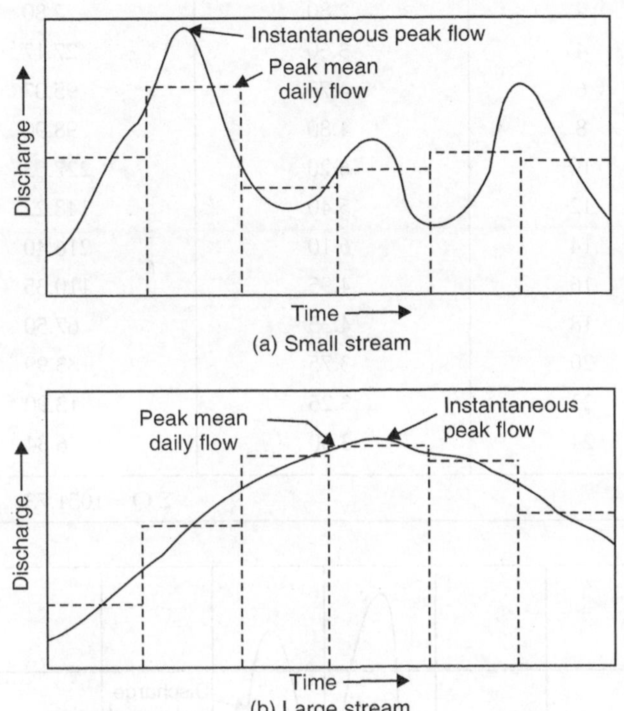

FIGURE 6.16 Instantaneous flow and mean daily flow.

Example 6.8. *The stage in the river is measured from 00 hours to 24 hours at a uniform interval of 2 hours and the recorded values in metres are :2.85, 2.80, 3.60, 4.75, 4.80, 6.20, 5.40, 6.10, 4.95, 4.35, 3.75, 3.25 and 3.00. The rating curve of this gauging site is given by* $Q = 23\,(y - 2.5)^{1.75}$*, where Q is in* m^3/s *and y is the stage in metres.*

Plot the given stage hydrograph and the corresponding discharge hydrograph. What is the volume of streamflow represented by the hydrograph in million m^3 *? What is the mean flow for the day ?*

Solution. The conversion of given stage into the corresponding discharge is shown in Table 6.5. For example, at 14 h, when the stage is 6.10 m, the corresponding discharge is computed as

$$Q = 23\,(6.10 - 2.50)^{1.75} = 216.40 \text{ m}^3/\text{s}$$

when the stage of column (2) is plotted as the ordinate against the time of column (1) as the abscissa the stage hydrograph is obtained. Likewise, when the discharge of column (3) is plotted against time the discharge hydrograph is obtained. The stage and discharge hydrographs are shown in Fig. 6.17.

Table 6.5 Conversion of Stage into Discharge

Time in hours	*Stage y in m*	*Discharge Q in m³/s*
0	2.85	3.66
2	2.80	2.80
4	3.60	27.17
6	4.75	95.07
8	4.80	98.90
10	6.20	227.03
12	5.40	148.22
14	6.10	216.40
16	4.95	110.35
18	4.35	67.50
20	3.75	33.99
22	3.25	13.90
24	3.00	6.84
		$\Sigma Q = 1051.73$

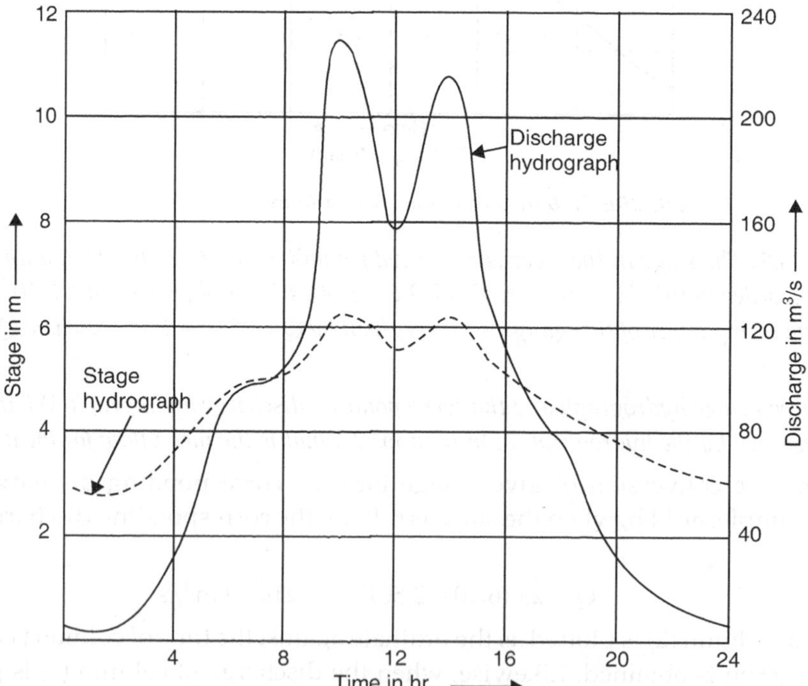

FIGURE 6.17 Stage and discharge hydrographs.

The volume of the streamflow can be computed from Eq. (6.36).

In the present case, $\Delta t = 2h$

$$\sum Q = 1051.73 \text{ m}^3/\text{s}$$

$$\left(\frac{Q_1 + Q_n}{2}\right) = \frac{3.66 + 6.84}{2} = 5.25 \text{ m}^3/\text{s}$$

$$\therefore \quad V = 3600\ \Delta t \times \left[\sum Q_i - \left(\frac{Q_1 + Q_n}{2}\right)\right]$$

$$= 3600 \times 2 \times [1051.73 - 5.25]$$

$$= 7534656 \text{ m}^3 = 7.535 \times 10^6 \text{ m}^3$$

$$= 7.535 \text{ million m}^3.$$

The mean daily flow $= \dfrac{7534656}{24 \times 3600} = 87.207 \text{ m}^3/\text{s}.$

6.11 UNITS OF STREAMFLOW

The basic unit of streamflow rate in the metric system is *cubic metre per second* (m^3/s) often written as *cumec*. Sometimes the flow rate may also be expressed in *m³/s/sq. km* which is nothing but the flow rate per unit area of the contributing catchment. This unit is particularly convenient when the flow rates of different streams with different catchment areas are to be compared. When the streamflow hydrograph is to be compared with the rainfall hyetograph which caused it, the flow rate may be expressed in *cm/h* or *mm/h* as the ordinate of the rainfall hyetograph has such units. A flow rate of 1 cm/h from a catchment of A sq. km is therefore equivalent to 2.78 A m^3/s. This unit must always be used in association with a specific drainage area.

Volume of flow may be expressed in cubic metres but this leads to very large numbers and therefore larger volume units are commonly used. The volume of water discharged in a period of 24 hours at a rate of 1 m^3/s is called one *cumec-day*. That is, one cumec-day is equal to $1\times 24 \times 60 \times 60 = 86400$ m^3. The other large units of volume are *cumec-hour* (3600 m^3), *cumec-month* (2592000 m^3) and *cumec-year* (31536000 m^3). *Hectare-metre* is also one of the commonly used volume unit. It represents the volume of water required to cover an area of one hectare to a depth of one metre. In other words, 1 *ha-m* = 10000 m^3. With a similar interpretation, sq. *km-cm* may also be used as a volume unit and again 1 sq. km - cm = 10000 m^3. The conversions among the various volume units just now mentioned may be listed as given below.

1 cumec-day = 86400 m^3 = 8.64 ha-m

1 cumec-month = 2592000 m^3 = 259.2 ha-m

1 million m^3 = 100 ha-m = 11.574 cumec-days

1 cumec-year = 31536000 m^3 = 31.536 million m^3 = 3153.6 ha-m

1 ha-m = 1 sq . km-cm = 10000 m^3 = 0.01 million m^3

Example 6.9. *The mean daily flows at a gauging station for a period of 7 days are 7, 27, 58, 41, 31, 20 and 13 m^3/s respectively. What is the total volume of streamflow at the site in cumec-days ? and in hectare-metres ? What is the mean flow rate for the week. If the drainage area at the site is 100 km^2 what is the runoff volume in cm ?*

Solution. Total volume of flow

$$= 7 + 27 + 58 + 41 + 31 + 20 + 13 = 197 \text{ cumec-days}$$

$$= 197 \times 86400 = 17020800 \text{ m}^3$$

$$= 1702.08 \text{ ha-m}$$

$$\text{Mean flow for the week} = \frac{197 \text{ cumec-days}}{7 \text{ days}}$$

$$= 28.143 \text{ m}^3/\text{s}$$

$$\text{Depth of runoff} = \frac{\text{Volume of runoff}}{\text{Area of the basin}}$$

$$= \frac{17020800}{100 \times 10^6} = 0.170208 \text{ m}$$

$$= 17.021 \text{ cm.}$$

Example 6.10. *What volume is represented by 57 mm of runoff depth from a basin of area 3300 km^2 ? Give the answer in m^3, cumec-days, ha-m and million cubic metres.*

Solution.

Volume of runoff = basin area × depth of runoff

$$= (3300 \times 10^6) \times (57 \times 10^{-3}) = 1881 \times 10^5 \text{ m}^3$$

$$= 188.1 \text{ million m}^3$$

$$= \frac{1881 \times 10^5}{10000} = 18810 \text{ ha-m}$$

$$= \frac{1881 \times 10^5}{86400} = 2177.08 \text{ cumec-days.}$$

Example 6.11. *A drainage basin has an area of 210 km^2. The average depth of rainfall received by it during a monsoon period is computed as 65 cm, while the runoff measured at its outlet during the same period is estimated to be 5.68×10^7 m^3. Compute the depth of runoff. What percentage of rainfall has become runoff ? If all this runoff volume is stored and used to irrigate a crop which requires 60 cm of water. How many hectares can be irrigated ?*

Solution. Depth of runoff $= \dfrac{\text{Volume of runoff}}{\text{Basin area}}$

$$= \frac{5.68 \times 10^7}{210 \times 10^6} = 0.2705 \text{ m}$$

$$= 27.05 \text{ cm.}$$

$$\% \text{ of the rainfall that became runoff} = \frac{27.05}{65} \times 100 = 41.62\%.$$

$$\text{Area of crop that can be irrigated} = \frac{\text{Volume of runoff}}{\text{Irrigation depth}}$$

$$= \frac{5.68 \times 10^7}{0.60} = 9.467 \times 10^7 \text{ m}^2$$

$$= 94.67 \text{ km}^2 = 9467 \text{ hectares.}$$

REVIEW EXERCISES

Descriptive Questions

1. Define the stage in a river. Describe with neat sketches how the stage can be measured with the help of a vertical staff gauge.
2. What do you understand by a crest gauge ? Explain the principle of working of any one type of crest gauge.
3. What is an automatic stage recorder ? Explain its principle of working. Describe the salient features of a shelter for an automatic stage recorder with a neat sketch.
4. Explain with a neat sketch the method of measuring the velocity at a point in a stream using a current meter.
5. Explain what is meant by the air-line and wet-line corrections ?
6. How is the average velocity over a vertical determined using single point, two point and three point methods ?
7. Explain the mid-section method of computing the discharge in a stream ? Show in a neat sketch the positions of velocity measurements over the cross-sectional area of the stream.
8. What is an isovel ? How is discharge in the stream computed using the isovels ?
9. Describe the moving boat method of measuring discharge in large streams.
10. What are the data to be obtained from field measurements to determine the discharge by slope-area method ? Explain how the discharge is computed ?
11. Describe the principle involved in the measurement of streamflow by the dilution method. What are the requisites of a good tracer used in the dilution method ?
12. Explain the principle of discharge measurement in the electromagnetic induction and ultrasonic methods.
13. What is a slope-stage discharge curve ? How is it constructed ?
14. How are the discharges adjusted when they are measured under unsteady flow conditions ?
15. Explain the logarithmic and Steven's methods of extending a stage discharge relationship listing out the merits and demerits of each.

16. Distinguish between
    (*i*) streamflow and runoff
    (*ii*) cup current meter and propeller current meter
    (*iii*) permanent control and shifting control.
17. What are the factors to be considered in locating a streamgauging site ?
18. Give some commonly used larger units of volume in metric system along with the interrelationship amongst them.
19. Define the efficiency of stage-discharge curve. Explain how it can be determined.
20. Explain how the best-fit equation can be obtained for the stage-discharge data. Describe how the stage corresponding to zero discharge in the stream is determined from the stage discharge data.

Analytical Questions

1. Assuming that the velocity distribution over a vertical follows one-seventh power law, given by $v = ky^{1/7}$, where v is the velocity at a point y metres above the stream bed and k is a constant. Find the mean velocity for the vertical with a depth of d metres. Also compute the point velocity at 0.4 d from the bed, and the average of the velocities at 0.2 d and 0.8 d from the bed. Compare these velocities with the mean for the vertical.
2. The depth of flow at a vertical in a stream is 3.5 m. The velocity of flow at various points on this vertical as measured by a current meter are given below.

Position of the current meter above the streambed in m	*Velocity in m/s*
0.7	2.62
1.4	2.94
2.1	3.16
2.8	3.28

The velocity measured at the surface by a float is 3.26 m/s. Determine the mean velocity for the vertical using

(*i*) single point
(*ii*) two point,
(*iii*) three point, and
(*iv*) surface float methods.

3. During the measurment of streamflow by a current meter, the depth of flow at a vertical and the vertical distance between the index point and the water surface were found to be 4.83 m and 5.26 m respectively. If the current meter is to occupy a position at 0.4 times the depth of flow from the streambed and if the cable line makes an angle of 26° with the vertical. What is the cable length to be lowered from the index point ?

4. Taking the rating of the current meter as $v = 0.05 + 0.8\,N$, where v is in m/s and N is in rev/s, compute the streamflow corresponding to the following observations.

Distance from bank, m	*Depth m*	*Current meter depth, m*	*Revolutions*	*Time s*
0.6	1.0	0.6	15	50
1.2	4.0	3.2	30	55
		0.8	48	53
2.0	5.5	4.4	40	46
		1.1	60	54
3.0	6.5	5.2	45	48
		1.3	67	52
3.8	4.5	3.6	33	54
		0.9	51	50
4.5	2.5	2.0	26	48
		0.5	44	55
5.0	1.0	0.6	20	47
5.6	0			

5. The data obtained during a stream gauging using a current meter with a rating $v = 0.04 + 0.65\,N$ is given below. The velocity at all the verticals is measured at 0.4 times the depth of flow from the streambed. Compute the discharge using the mid-section method.

Distance from bank, m	*Depth m*	*Revolutions*	*Time s*
2.2	0.8	18	44
4.4	1.8	30	52
6.5	3.2	40	42
10.0	3.7	65	51
13.0	3.0	54	54
16.5	2.8	51	50
19.6	2.6	46	49
21.7	2.4	44	53
24.0	2.0	20	41
25.0	0		

6. The data obtained during a streamgauging using the moving boat method is given below. The rating of the current meter is $V = 0.08 + 0.85\,N$.

Time from the start s	*Cumulative no. of revolutions*	*Angle of the resultant velocity relative to the boat velocity θ*	*Depth of flow from sonic sounder m*
0	0	0°	3.1
30	28	19°	3.4
60	65	22°	4.2
90	125	26°	5.3
110	183	36°	6.6
130	245	30°	6.9
150	298	28°	6.2
180	345	24°	4.9
210	386	20°	3.8
240	410	15°	2.7

Estimate the discharge in the stream.

7. The following field observations were obtained at two sections on a stream separated by a distance of 350 m along the stream.

	Area m^2	*Perimeter* *m*	*Gauge reading* *m*
u/s section	68.1	42.3	61.38
Middle section	72.3	45.7	—
d/s section	75.4	48.6	60.99

Determine the flood discharge assuming Manning's $n = 0.033$.

8. Sodium dichromate solution with a concentration of 25 mg/c.c. is introduced into a stream at a rate of 1.5 litres/minute. The samples collected at a downstream section sufficiently far away indicated an equilibrium concentration of 0.001 ppm. Determine the discharge in the stream. Assume no initial concentration of sodium dichromate in the stream.

9. A salt solution with a concentration of 85 mg/c.c. is discharged into a stream at a steady rate of 15 litres/s. The initial, salt concentration in the stream is found to be 0.65 ppm. The salt concentration at a downstream sampling section is measured as 3.0 ppm. Determine the discharge in the stream using the principle of dilution method.

10. In a gulp method of streamgauging by dilution, 45 litres of salt solution with a concentration of 250 mg/c.c. is introduced into the stream all at once. The stream did not contain any salt concentration initially. The continuous monitoring of salt concentration at a downstream section (sufficiently far away to ensure thorough missing) indicated that the entire concentration has passed the section in a period of 12 hours and the concentration curve can be approximated as a triangle with a maximum concentration of 0.005 ppm occurring at 4 hours from the beginning. Determine the discharge in the stream.

11. The stage in the river is recorded as 40.5 m at 8.00 A.M. and it is observed to fall by 0.15 m in the next one hour. The stage-discharge relationship for the steady state flow conditions at the gauging site is established as $Q = 56.85\,(y - 37.5)^{1.85}$, where Q is in m^3/s and y is in m. If the velocity of the flood wave is taken as 2.25 m/s and the bed slope of the stream at the gauging site is 1.43×10^{-4}. Determine the discharge in the stream at 8.00 AM.
12. The slope of the river at a gauging site is 1 in 4000. The normal discharge for an observed stage from a steady-flow rating curve is obtained as 225 m^3/s while the stage in the stream is observed to rise at a rate of 0.125 m/h. What is the actual discharge in the stream if the flood wave can be assumed to have a velocity of 2.5 m/s ?
13. Three points on a smooth curve drawn to best represent the stage-discharge data of a stream have the following coordinates.

Stage y in m	315.56	316.44	318.73
Discharge Q in m^3/s	50	250	1250

Develop an equation of the form $Q = k\,(y - a)^b$.

14. Prove equation (6.36).
15. The observed stage in a stream during a day is given below.

Time	*Stage m*	*Time*	*Stage m*
12 MN	212.85	2 PM	215.45
2 AM	212.80	4 PM	214.95
4 AM	213.60	6 PM	214.35
6 AM	214.75	8 PM	213.75
8 AM	218.4	10 AM	213.25
10 AM	217.4	12 MN	213.0
12 Noon	216.1		

The rating curve of the gauging station is $Q = 113.5\,(y - 212)^{1.68}$, where Q is in m^3/s and y is in m. Plot the given stage hydrograph and the corresponding discharge hydrograph. What is the volume of the streamflow represented by the hydrograph in million m^3 ? What is the mean flow for the day ?

16. The mean daily flows in m^3/s at a gauging station for a period of 7 days are, respectively, 31.9, 31.1, 30.2, 50.6, 44.4, 35.2 and 32.1. What is the total volume of the flow of the week in cumec-days and in hectare-metres ? What is the mean flow for the week in m^3/s ? If the drainage area at the site is 125 km^2. What is the streamflow volume in mm ?
17. What volume is represented by 83 mm of runoff depth from a catchment of area 5575 km^2 ? Give your answer in m^3, cumec-days, ha-m , and million m^3.
18. A drainage basin of area 500 km^2 received an average depth of monsoon rainfall of 73 cm and the corresponding runoff is measured to be 14×10^7 m^3. What is the depth of runoff ? What percentage of rainfall has become runoff ? If all this runoff volume were to be stored and used to irrigate a crop which required 60 cm of water, how many hectares can be irrigated ?

19. The stage discharge relation for a stream gauging site is given by

$$Q = 35.0\,(G - G_0)^{1.75}$$

where Q is the discharge in the stream in m^3/s, G is the stage in m and G_0 in the stage corresponding to zero discharge and is equal to 10.5 m. What is the stage in the river when the flow in the stream is 140 m^3/s and what is the discharge in the river corresponding to a stage of 13.8 m.

20. The following data pertain to a stream gauging station. The normal discharge from the steady flow rating curve is 187.0 m^3/s. The slope of the river bed is 6.25×10^{-4}. The rate of change of water surface elevation is 7.5 cm/h. Compute the river discharge if the velocity of the flood wave is 2.75 cm/s.

21. When the main gauge and auxiliary gauge (which is 200 m downstream of main gauge) readings are 156.3 m and 156 m respectively, the measured discharge in the stream is 98.6 m^3/s. And for readings of 156.3 m and 155.7 m the discharge is 139.5 m^3/s. What would be the discharge when main gauge and auxiliary gauge record 156.3 m and 155.85 m ?

22. Establish a stage-discharge rating curve from the following data obtained at a stream gauging site. Assume stage at zero discharge is 0.85 m

Stage m	*Discharge m^3/s*	*Stage m*	*Discharge m^3/s*
2.47	25.49	3.87	69.23
3.72	62.44	2.96	36.95
4.75	105.48	2.50	22.09
5.76	149.50	2.44	19.96
7.99	181.94	2.36	18.18
5.85	153.62	2.30	16.76
4.88	111.14	2.19	14.30

Compute the standard error of estimate and the efficiency of the rating curve

23. The data given below is obtained during the gauging of a stream.

Distance from left water edge m	*Depth d m*	*Velocity in m/s*	
		at 0.2 d	*at 0.8 d*
0	0.00	0.00	0.00
1	1.25	0.50	0.30
2	2.00	0.75	0.50
3	3.00	1.00	0.60
4	2.50	0.85	0.50
5	2.00	0.80	0.50
6	1.50	0.70	0.40
7	0.00	0.00	0.00

Compute the discharge in the stream.

24. The measured discharges, stage readings at main stage and at an auxiliary stage (600 m downstream of main stage) of a stream gauging site are given below. Develop a slope-stage-discharge relation from this data taking the standard fall F_0 as 0.3 m. Determine the exponent m in the relation $\frac{Q}{Q_0} = \left(\frac{F}{F_0}\right)^m$ by plotting Q/Q_0 vs F/F_0 on logarithm paper. Find the discharge for a stage of 7.6 m and a fall of 0.44 m.

Measured discharge	*Stage G in m*	
Q in m³/s	*Main stage*	*Auxiliary stage*
2659	3.57	3.29
2265	2.97	2.66
6853	4.91	4.60
765	1.83	1.52
15660	7.60	7.31
11814	6.41	6.09
15857	6.63	6.15
11921	7.17	6.94
6829	4.41	3.97
3191	3.28	2.90
10336	5.58	5.18
19878	8.30	7.95
16140	8.06	7.80
8427	6.58	6.39
3675	4.18	3.97

25. A drainage basin has an area of 1000 km^2. The average flows at the basin outlet for a period of 8 days in m^3/s are 10, 50, 80, 70, 60, 80, 30 and 20 respectively. Compute the 8-day mean flow. Compute the total volume of flow in m^3, hectare-metre and depth in mm. If this volume is used to meet the per-capita water demand of 150 liters/day of a town with a population of 1,10,000 for how many days it can meet the demand ?

Objective Type Questions

1. The stage in the river is defined as
   (*a*) the elevation of the water surface above an arbitrary datum
   (*b*) the average depth of flow in the stream
   (*c*) the hydraulic radius of the stream cross-section
   (*d*) the hydraulic depth of the stream cross-section.

2. The crest gauge is used to record
   (*a*) the lowest stage in the river.
   (*b*) the average stage in the river.
   (*c*) the peak stage in the river
   (*d*) the stage continuously with time.
3. If N is speed of the current meter in revolutions per second, the velocity measured by it is proportional to
   (*a*) $N^{1/2}$
   (*b*) N
   (*c*) $N^{3/2}$
   (*d*) N^2
4. In the two point method of finding the mean velocity across a vertical in a stream using a current meter, the velocities are measured above the stream bed at
   (*a*) 0.25 and 0.75 depths
   (*b*) 0.20 and 0.80 depths
   (*c*) 0.40 and 0.60 depths
   (*d*) 0.15 and 0.85 depths.
5. In the single point method of finding mean velocity across a vertical, the velocity is measured above the stream bed at
   (*a*) 0.4 d
   (*b*) 0.6 d
   (*c*) 0.7 d
   (*d*) 0.8 d.
6. The rating curve of a stream gauging station gives the variation of discharge in the stream with
   (*a*) the area of flow
   (*b*) the stage
   (*c*) the depth of flow
   (*d*) the velocity of flow.
7. The stage-discharge relationship of a gauging site, with Q denoting the discharge and y denoting the stage in the river, is usually expressed as
   (*a*) $Q = k\, y^b$
   (*b*) $Q = k\,(y - a)$
   (*c*) $Q = k\,(y - a)^b$
   (*d*) $Q = \dfrac{k}{(y - a)^b}$
8. A flood wave is passing through the gauging station. Let Q_i denote the discharge in the river when the stage is increasing and Q_d denote the discharge when the stage is decreasing with time while the stage in the river is same in both the cases. Then which of the following is true
   (*a*) $Q_i = Q_d$
   (*b*) $Q_i < Q_d$
   (*c*) $Q_i > Q_d$
   (*d*) $\dfrac{Q_d}{Q_i}$ = constant for all stages.
9. One km^2 -cm of volume is equal to
   (*a*) 1000 ha-m
   (*b*) 100 ha-m
   (*c*) 10 ha-m
   (*d*) 1 ha-m.

10. One cumec-day of volume is equal to
(a) 86400 m^3 (b) 8.64 ha-m
(c) 24 cumec-hours (d) all the above.

11. The dilution method of streamgauging is best suited for
(a) small streams with fairly steady turbulent flow
(b) meandering streams in alluvial plains
(c) streams with large depth and discharge
(d) any type of stream.

12. When the stage in the river is 4.8 m and the water surface slope is 1 in 1000 the discharge is measured to be 600 m^3/s. What would be the discharge carried by the river when the stage is same but the water surface slope is 1 in 1440 ?
(a) 300 m^3/s (b) 400 m^3/s
(c) 500 m^3/s (d) 600 m^3/s.

13. Which of the following is the largest unit of volume ?
(a) hectare-metre (b) km^2-mm
(c) million-m^3 (d) cumec-day.

14. The runoff from a drainage basin area 4320 km^2 is estimated as 10000 cumec-days. What is the depth of runoff ?
(a) 20 cm (b) 40 cm
(c) 43.2 (d) 21.6 cm.

15. If S_Q is the standard deviation of the measured discharges, S_e is the standard error of estimate, then the efficiency η of the curve fitted to stage discharge data is given by

(a) $\eta = \dfrac{S_e}{S_Q}$ (b) $\eta = \dfrac{S_Q - S_e}{S_Q}$

(c) $\eta = \dfrac{S_Q^2 - S_e^2}{S_e^2}$ (d) $\eta = \dfrac{S_Q^2 - S_e^2}{S_Q^2}$

ANSWERS

Analytical Questions

1. $V = 0.875\, kd^{1/7}$; $v_{0.4} = 0.8773\, kd^{1/7}$; $\dfrac{v_{0.2} + v_{0.8}}{2} = 0.8816\, kd^{1/7}$

2. (i) 2.94 m/s (ii) 2.95 m/s (iii) 2.945 m/s (iv) 2.771 m/s
3. 8.5 m
4. 14.97 m^3/s
5. 40.346 m^3/s
6. 1540 m^3/s
7. 101.45 m^3/s
8. 625 m^3/s
9. 542.534 m^3/s
10. 104.167 m^3/s
11. 404.84 m^3/s
12. 231.166 m^3/s
13. $Q = 136.84\,(y - 315.01)^{1.684}$
15. 66.96 million m^3; 775 m^3/s

16. 255.5 cumec-days ; 2207.52 ha-m ; 36.5 m^3/s ; 176.6 mm
17. 462.73×10^6 m^3 ; 5355.6 cumec-days ; 46273 ha-m ; 462.73 million m^3
18. 28 cm ; 38.36% ; 21538.5 hectares
19. 12.708 m, 282.8 m^3/s
20. 278.13 m^3/s
21. 120.76 m^3/s
22. $Q = 9.5\,(G - 0.85)^{1.73}$; $S_Q = 67.85$; $S_e = 10.03$; $\eta = 97.81\%$
23. 7.9625 m^3/s
25. 46.25 m^3/s ; 3.996×10^6 m^3 ; 399.6 ha-m ; 3.996 mm ; 242.2 days

Objective Type Questions

1. *(a)*	2. *(c)*	3. *(b)*	4. *(b)*	5. *(a)*
6. *(b)*	7. *(c)*	8. *(c)*	9. *(d)*	10. *(d)*
11. *(a)*	12. *(c)*	13. *(c)*	14. *(a)*	15. *(d)*

Chapter 7

EVAPORATION AND EVAPOTRANSPIRATION

7.1 INTRODUCTION

Evaporation and evapotranspiration are two of the most important and most complicated phases of the hydrologic cycle. These phases redistribute the heat energy between surfaces and atmosphere. Estimates of evaporation and evapotranspiration are required in the design of reservoirs, irrigation systems, scheduling and frequency of irrigation and water balance and simulation studies.

Evaporation is the process by which water from liquid or solid state passes into the vapour state and is diffused into atmosphere. When water is converted from solid state to vapour state without passing through liquid state then it is called *sublimation*. The process by which water passes from liquid to vapour state through plant metabolism is termed transpiration. The process by which water is evaporated from wet surfaces and transpired by plants together is called the *evapotranspiration*. These processes are now discussed in the same order.

7.2 EVAPORATION PROCESS

Consider a water body with a flat free surface. By supplying external thermal energy to the water body the kinetic energy of water molecules can be increased. When the molecules near the free surface attain enough kinetic energy they escape from the water body overcoming the intermolecular forces and eject themselves into the atmosphere. The amount of energy expended by a unit mass of water while passing from liquid to vapour state at constant temperature is called the *latent heat of evaporation* or *latent heat of vapourisation* denoted by L. When external energy is not made available, energy is removed from water body while evaporation is taking place and thus results in lowering of water temperature. Therefore evaporation is indirectly a cooling process. Out of the total atmospheric pressure on the free surface there will be some contribution by the vapour molecules present there. This partial pressure exerted by the vapour is called the vapour pressure. Continued supply of heat energy causes accumulation of more and more vapour molecules and a stage comes when the gaseous medium can no longer accommodate any more, and rejects the vapour molecules in the form of *condensation* at the same rate as vapourisation. At this stage the air is said to be saturated. At saturation, the partial pressure exerted by the water vapour is called the *saturation vapour pressure* and is denoted by e_s, which increases with temperature. Thus if the vapour pressure of air above the free surface of water is already equal to the saturation vapour pressure e_s, neither evaporation nor condensation takes place and then it is called an equilibrium state. In practice such an equilibrium can never exist under natural conditions of interest to hydrologists. Because the atmosphere into which evaporation takes place is infinite and various transport processes will also operate to transport the vapour in atmosphere both parallel and perpendicular to the liquid surface and prevent equilibrium from occurring.

Now it can be argued that for evaporation to occur it is necessary to have (*i*) a supply of water (*ii*) a source of heat and (*iii*) vapour pressure deficit, that is difference between the saturated vapour pressure of water e_s corresponding to water temperature and the actual vapour pressure of air above the free surface e_a. Under given conditions evaporation is proportional to the vapour pressure deficit and this fact was first recognised by Dalton in the year 1802 and since then it has been known as Dalton's law written in the form.

$$E = c\,(e_s - e_a) \quad ...(7.1)$$

where E is the evaporation and c is a co-efficient to account for the other factors which affect evaporation.

7.3 FACTORS AFFECTING EVAPORATION

A number of factors, both meteorological and physical, affect the rate of evaporation from a water body although it is difficult to assess quantitatively the relative importance of each of them. The meteorological factors may be listed as radiation, temperature, humidity, wind and atmospheric pressure, while the physical factors include water quality, and size and shape of the evaporating surface. Evaporation is expressed as the depth of water in mm escaped from a unit area in a given time. Thus it may be expressed as mm/h, mm/day, mm/month etc.

Radiation. Since the evaporation requires continuous supply of energy which is derived mainly from solar radiation, the radiation will be a factor of considerable importance. At a given place if the monthly evaporation depths and monthly solar radiation are plotted against the time in months these two curves show a very close relationship following the same trend. In the regions near equator where the annual solar radiation is large the annual evaporation is also large.

Temperature. The air and water temperatures are largely dependant on solar radiation. But the responses of these two to solar radiation are not in tune especially in water bodies with large depth. This is because a part of solar energy is stored in deep waters. The saturation vapour pressure of water increases with increase in temperature of water. Similarly at a constant atmospheric pressure, the partial pressure exerted by the water vapour present in the atmosphere also increases with temperature. As the evaporation is a function of the difference of these two vapour pressures, equal temperature increase of water and air may not increase the rate of evaporation. Since air temperatures are readily available evaporation is often compared with air temperature. However, evaporation is not dependent on air temperature alone. Thus, it may be seen that for two months having the same mean temperature the mean evaporation is not the same. In the case of deep water bodies for the same mean monthly temperature, the evaporation in a later (winter) month is more than the earlier (summer) month in the year, since a part of the heat received in the earlier month is utilised to increase the temperature of water at lower depths and the heat energy available for evaporation is reduced. During the later month though the incoming heat energy is less, additional heat energy is made available from storage for evaporation and therefore the evaporation is more. The difference in monthly evaporation of a winter month and a summer month in an year, the mean monthly temperature being same, decreases as the depth of water body decreases.

Wind. If the atmosphere above the water surface is still, it will soon get saturated with water vapour and the evaporation has to cease. But because of wind the water vapour is removed from time to time and fresh air capable of holding more vapour is brought over. Therefore higher wind velocities may cause higher evaporation. When the wind velocity is great enough to remove all the vapour as it is formed, a further increase in velocity will not increase evaporation appreciably. That means the relationship between wind speed and evaporation holds good only to a certain value and beyond this critical velocity no increase in evaporation may be noticed. This critical wind speed will vary with other conditions influencing evaporation.

Wind velocity will influence evaporation rate only when wind movement is turbulent since vapour removal is affected by turbulent diffusion and spreading. If the movement is laminar it is not likely to have any significant effect in the case of large water surfaces. Wind velocity observations are made at considerable heights from ground surface and it is the velocity at ground level that influences moisture removal. From observations of wind velocity at different heights the velocity just above the ground level can be obtained from Prandtl's logarithmic velocity distribution law.

The effect of wind on evaporation may be more pronounced over large bodies of water than over small areas. In the case of small evaporation pans a slight increase in wind velocity may be sufficient to remove the water vapour as fast as it is formed. Large water bodies like lakes and reservoirs may require large velocities.

Atmospheric Pressure. The decrease in atmospheric pressure would increase the rate of evaporation because the vapour pressure of air also decreases with atmospheric pressure and thus increases the vapour pressure deficit. The atmospheric pressure normally decreases with increase in altitude. As the altitude increases not only the atmospheric pressure but also the temperature decreases. Thus, the increase in evaporation due to decrease in atmospheric pressure may be nullified by the corresponding decrease in temperature. Therefore it may be difficult to investigate the effect of atmospheric pressure alone while the other conditions remain the same.

Quality of Water. The vapour pressure of pure water under given conditions is solely dependent on temperature. When a solute is dissolved in water, the vapour pressure of water is decreased. Since the evaporation is proportional to the difference between the vapour pressure of water and vapour pressure of air, the presence of solute in water reduces the evaporation. Evaporation decreases by one per cent for every one per cent increase in salinity and therefore the evaporation of sea water with an average salinity of 3.5 per cent is 2 to 3 per cent less than the evaporation from fresh water.

The turbidity of water, another aspect of water quality, probably has little direct effect upon evaporation although by affecting the albedo of water and consequently its heat budget it may have an indirect effect.

Size of Water Surface. The depth of evaporation from larger surface areas is less compared to evaporation from smaller surface areas, although of course the total volume of water evaporated will be more in the former. The reason for this phenomenon is that the air moving across a large lake or reservoir will gradually increase its water content and thus loses its water holding capacity. However, with other conditions remaining same the difference in evaporation rates from different sized surfaces will be considerably affected by the humidity of the incoming air. If this is initially high, then clearly it will be modified only slightly even after a lengthy passage across a large lake. There will, thus, be little difference in evaporation rates from large and small water surfaces.

Nature of Evaporating Surface. All surfaces exposed to precipitation are potentially evaporating surfaces. They may be broadly classified into three groups : land surfaces, snow surfaces and water bodies. The above discussion of factors affecting evaporation is mainly for the evaporation from free water surface, though it is applicable to other surfaces also in general.

The rate of evaporation from a saturated soil surface is approximately the same as that from an adjacent water surface of the same temperature. Evaporation decreases as the soil begins to dry. The rate of evaporation, therefore, is limited by the availability of water, or *evaporation opportunity*. The rate of evaporation from a soil surface is further dependent upon the presence or absence of vegetal cover. For the same evaporation opportunity, the evaporation rate decreases in the following order : bare ground, grass and crop lands, light forests and dense forests.

Evaporation from a surface covered with snow can occur only when the vapour pressure of the overlying air is less than the vapour pressure of the snow surface, that is, when the dew point of air is less than the temperature of the snow. For a given difference between surface and dew point temperatures evaporation from a snow surface is relatively small compared to that from a water surface.

7.4 ESTIMATION OF EVAPORATION

The estimation of evaporation from a water surface is frequently required in planning and design of many water resources projects. In providing a reservoir to cater to the needs of irrigation, watersupply or any other purpose its capacity should be determined giving due consideration to the evaporation losses from it. Similarly the design of a long canal in an arid or semi-arid region will be in serious error if allowance is not made for evaporation.

The methods of estimating evaporation may be divided into five categories. They are : (*i*) Empirical equations, (*ii*) Water balance method, (*iii*) Energy balance method, (*iv*) Mass transfer method, and (*v*) Combined energy balance and mass transfer method.

7.4.1 Empirical Equations

There are many empirical equations that are suggested to estimate evaporation from free-water surface. Most of them are based on Dalton's law, with modifications to account for some factors affecting evaporation such as wind velocity, temperature, atmospheric pressure etc. Some of the equations that are commonly used are given below.

Fitzgerald's equation

$$E = (0.4 + 0.124V)\,(e_s - e_a) \quad ...(7.2)$$

where E = Evaporation in mm/day

e_s = Saturated vapour pressure at the temperature of the water surface in mm of mercury

e_a = The actual vapour pressure of air in mm of mercury

V = Mean wind speed at the surface in km/h.

Horton's equation

$$E = 0.4\,(\psi\, e_s - e_a) \quad ...(7.3)$$

where ψ is a function of wind velocity given by $\psi = 2 - e^{-0.124\,V}$ and E, e_s, e_a and V have the same significance as in Eq. (7.2). For large areas, Horton suggested that the result of Eq. (7.3) be multiplied by the following area factor

$$(1 - P) + P\frac{(\psi - 1)}{(\psi - h)}$$

in which P is the fraction of the time during which the wind is turbulent and h is the relative humidity expressed as a fraction.

Meyer's equation

$$E = C\,(e_s - e_a)\,(1 + 0.06215\,V) \quad ...(7.4)$$

where E = Evaporation in mm/month

e_s = The saturation vapour pressure in mm of mercury corresponding to the mean monthly temperature of air

e_a = The actual vapour pressure in air based on mean monthly temperature and relative humidity

V = Monthly mean wind velocity in km/h at 10 m above ground.

C = 15 for small shallow ponds.

For large or deep water bodies the value of C is taken as 11 and also e_s is the saturation vapour pressure corresponding to mean monthly water temperature instead of air temperature and e_a is the actual vapour pressure in air about 10 m above the water surface.

Rohwer's equation

$$E = 0.771\,(1.465 - 0.000732\,P_a)\,(0.44 + 0.07334\,V)\,(e_s - e_a) \quad ...(7.5)$$

E = Evaporation in mm/day

P_a = Mean barometric reading in mm of mercury and e_s, e_a and V have the same meaning as in Eq. (7.2)

Lake Mead equation

$$E = 0.0331\,V\,(e_s - e_a)\,[1 - 0.03\,(T_a - T_w)] \quad ...(7.6)$$

where T_a = Average air temperature in °C

T_w = Average water surface temperature in °C

and e_s, e_a and V retain the same significance as in Eq. (7.2).

Though the empirical equations provide fast estimates of evaporation they must be used with caution. Their application becomes difficult if the data required in the equation is not available at a place of interest. Moreover, most of the meteorological quantities used in these equations are average daily values, whereas evaporation depends upon the total quantity of incoming solar radiation and the average may not represent the total.

Example 7.1. *The following meteorological data pertain to a large reservoir with a waterspread area of 15 km^2. The data represents the average values for the day.*

Water temperature = 24°C

Air temperature = 26°C

Atmospheric pressure = 752 mm of mercury

Wind speed at 0.5 m above ground level = 25.3 km/h

Relative humidity = 46%

Estimate the average daily evaporation from the reservoir and also the evaporation losses from the reservoir for a period of one week using different empirical formulae.

Solution. The saturation vapour pressure corresponding to the water temperature of 24° can be obtained from Table 2.1 as 29.831 mb or 29.831/1.33 = 22.43 mm of mercury. Similarly the saturation vapour pressure corresponding to air temperature of 26° is 33.608 mb or 33.608/1.33 = 25.27 mm of mercury. These values may also be obtained using Eq. (2.1).

$$\text{Relative humidity} = \frac{\text{Actual vapour pressure in air}}{\text{Saturation vapour pressure at air temperature}}$$

$$0.46 = \frac{e_a}{25.27} \quad \text{or} \quad e_a = 11.62 \text{ mm of mercury.}$$

The velocity at 0.5 m above ground level is itself taken to be the mean wind speed at the surface.

(*i*) *Fitzgerald's equation* :

$$\begin{aligned} E &= (0.4 + 0.124V)\,(e_s - e_a) \\ &= (0.4 + 0.124 \times 25.3)\,(22.43 - 11.62) \\ &= 3.5372 \times 10.81 = 38.24 \text{ mm/day} \end{aligned}$$

$$\text{Evaporation loss for one week} = 7 \times \frac{38.24}{1000} \times 15 \times 10^6$$

$$= 4.0152 \times 10^6 \text{ m}^3 \approx 4 \text{ million m}^3$$

(*ii*) *Horton's equation* :

$$E = 0.4\ (\psi e_s - e_a)$$

$$\psi = 2 - e^{-0.124\,V} = 2 - e^{-0.124 \times 25.3}$$

$$= 2 - 0.0434 = 1.9566$$

$$E = 0.4 \times (1.9566 \times 22.43 - 11.62) = 12.91 \text{ mm/day}$$

$$\text{Evaporation loss for one week} = 7 \times \frac{12.91}{1000} \times 15 \times 10^6$$

$$= 1.356 \times 10^6 \text{ m}^3 = 1.356 \text{ million m}^3$$

(iii) Meyer's equation :

$$E = C\,(e_s - e_a)\,(1 + 0.06215\,V)$$

Since it is a large water body and since the evaporation is estimated for one day the value of C may be taken as $\frac{11}{30}$. The velocity to be used in this equation is the velocity at 10 m above ground level. This can be obtained from the power law of Eq. (2.8) as shown below.

$$\frac{V_{10}}{V_{0.5}} = \left(\frac{10}{0.5}\right)^{0.15}$$

$$\therefore \quad V_{10} = 25.3 \times (20)^{0.15} = 39.653 \text{ km/h}$$

$$E = \frac{11}{30} \times (22.43 - 11.62) \times (1 + 0.06215 \times 39.653)$$

$$= 13.73 \text{ mm/day}$$

$$\text{Evaporation loss for one week} = 7 \times \frac{13.73}{1000} \times 15 \times 10^6$$

$$= 1.442 \times 10^6 \text{ m}^3 = 1.442 \text{ million m}^3$$

(iv) Rohwer's equation :

$$E = 0.771 \times (1.465 - 0.000732\,P_a)\,(0.44 + 0.07334\,V)\,(e_s - e_a)$$

$$= 0.771 \times (1.465 - 0.000732 \times 752)\,(0.44 + 0.07334 \times 25.3)\,(22.43 - 11.62)$$

$$= 17.5 \text{ mm/day.}$$

$$\text{Evaporation loss for one week} = 7 \times \frac{17.5}{1000} \times 15 \times 10^6$$

$$= 1.84 \times 10^6 \text{ m}^3 = 1.84 \text{ million m}^3$$

(v) Lake Mead equation :

$$E = 0.0331\,V\,(e_s - e_a)\,[1 - 0.03\,(T_a - T_w)\,]$$

$$= 0.0331 \times 25.3 \times (22.43 - 11.62)\,(1 - 0.03 \times 2)$$

$$= 8.51 \text{ mm/day}$$

$$\text{Evaporation loss for one week} = 7 \times \frac{8.51}{1000} \times 15 \times 10^6$$

$$= 0.894 \times 10^6 \text{ m}^3 = 0.894 \text{ million m}^3.$$

7.4.2 Water Balance Method

The ideal way of estimating evaporation from any water body such as a lake or reservoir would be to measure the various components of inflow, outflow and storage and to apply the mass balance equation. The general water balance equation for any time interval such as a day or a week may be written as

$$S_1 + P + \forall_{si} + \forall_{gi} - \forall_{so} - \forall_{go} - E - TR = S_2 \quad ...(7.7)$$

where S_1 is the storage at the beginning of the time interval

P is the precipitation falling into the water body

$\forall_{si}$ is the volume of surface inflow entering the water body

$\forall_{gi}$ is the volume of groundwater inflow entering the water body

$\forall_{so}$ is the volume of the surface outflow leaving the water body

$\forall_{go}$ is the volume of the groundwater outflow leaving the water body

E is the evaporation from the water body

TR is the transpiration losses from the water body, if any, and

S_2 is the storage at the end of the time interval.

All these quantities must be expressed in units of volume (m^3) or in equivalent depth (mm) over a reference area. Eq. (7.7) can be rearranged to take the form

$$E = (S_1 - S_2) + P + (\forall_{si} - \forall_{so}) + (\forall_{gi} - \forall_{go}) - TR \quad ...(7.8)$$

Practically, however, it is difficult to adopt this approach because of errors involved in measuring various terms, especially the quantities $\forall_{gi}$, $\forall_{go}$ and TR. The accuracy of the results may be improved to some extent by increasing the time interval for which the balance equation is applied. Studies under controlled conditions on some lakes yielded fairly accurate results of evaporation by this method.

7.4.3 Energy Balance Method

This method, also known as the energy budget method, is similar to the water balance method where the conservation equation is applied to energy instead of mass. For any given time interval on a unit surface area, say one square centimetre, the energy balance equation may be written as

$$Q_i + Q_a = Q_r + Q_b + Q_e + Q_h + Q_s \quad ...(7.9)$$

where Q_i = the incoming shortwave solar radiation both direct and diffused

Q_a = the net energy advected into the body of water by the sum of the processes such as rainfall, inflow, outflow, seepage etc.

Q_r = the reflected shortwave solar radiation

Q_b = the net outgoing longwave radiation

Q_e = the energy utilised in evaporation

Q_h = the energy conducted and convected from water surface to the atmosphere, called the *sensible heat*

Q_s = the energy stored in water.

The ratio of Q_h to Q_e is called the *Bowen's ratio* and is denoted by R. That is, Q_h may be replaced by $R \cdot Q_e$ in Eq. (7.9). If E is the evaporation in cm, ρ is the density of water in g/c.c and L_e is the latent heat of vapourisation of water, then Q_e is also equal to $\rho \cdot L_e \cdot E$. With these substitutions in Eq. (7.9) and some rearrangement, we obtain

$$E = \frac{(Q_i - Q_r - Q_b) + (Q_a - Q_s)}{\rho L_e (1 + R)} \quad ...(7.10)$$

where R is the Bowen's ratio given by

$$R = \frac{Q_h}{Q_e} = 0.00061\, P \frac{(T_s - T_a)}{(e_s - e_a)} \quad ...(7.11)$$

In Eq. (7.11), T_s and T_a are the water surface temperature and air temperature respectively in °C, e_s is the saturation vapour pressure corresponding to T_s, e_a is the actual vapour pressure of air where T_a is measured and P is the atmospheric pressure over water surface. The quantities e_s, e_a and P must all be expressed in same units such as mb or mm of mercury.

Estimation of evaporation by Eq. (7.10) is rather complicated because of the evaluation of various energy terms involved in it. It requires extensive instrumentation. *Radiometers* are available to measure the net radiation $(Q_i - Q_r - Q_b)$ directly. But the advected and stored energies Q_a and Q_s can be evaluated only from an approximate water budget and temperatures of the respective water volumes. For most lakes the advected energy Q_a and the stored energy Q_s tend to balance each other especially over long periods of time. Therefore they are frequently assumed to cancel when considering annual or mean annual evaporation.

7.4.4 Mass Transfer Method

In this method evaporation is considered as the turbulent transfer of vapour. Therefore this is also known as the *vapour flow approach* or the *aerodynamic approach*. The concepts of continuous mixing and the boundary layer theory are applied to obtain an expression for evaporation. Assuming an adiabatic atmosphere and logarithmic distribution for wind velocity, the equation is given as

$$E = \frac{46.08\,(e_1 - e_2)(V_2 - V_1)}{(T + 273)\ln\left(\frac{z_2}{z_1}\right)^2} \quad ...(7.12)$$

where E is evaporation in mm/h, z_1 and z_2 are the arbitrary lower and upper levels above the surface in metres, e_1 and e_2 are the vapour pressures measured at z_1 and z_2 in mm of mercury, V_1 and V_2 are the wind velocities measured at z_1 and z_2 in km/h and T is the average temperature of air in °C between z_1 and z_2.

Assuming z_1 to lie within the thin vapour saturated film above the water surface, e_1 equals e_s and with slight modification Eq. (7.12) may be written in the form of well known Dalton's equation

$$E = (a + b\,V)\,(e_s - e) \quad ...(7.13)$$

where e is the vapour pressure of air at which the wind velocity V is obtained. There are two cases that should be considered now. They are : (*i*) water temperature is same as the air temperature, and (*ii*) Water surface temperature is different from the air temperature. Case (*i*) occurs rarely and is emperically treated by the equation

$$E_a = C \,.\, \psi\,(V)\,(e_s - e_a) \qquad ...(7.14)$$

where C is an empirical coefficient

$\psi\,(V)$ is a function of wind velocity V

V is the wind speed measured at a stipulated height above the water surface

e_s is the saturation vapour pressure corresponding to air temperature

e_a is the actual vapour pressure in the air above

E_a is the evaporation from the water surface when water and air temperatures are equal.

Penman suggested the following equation for this case

$$E_a = 0.35\,(0.5 + 0.54\,V)\,(e_s - e_a) \qquad ...(7.15)$$

where E_a is in mm/day, e_s and e_a are in mm of mercury and V is in m/s measured at a height of 2 m above the free surface.

It is case (*ii*) which occurs normally and which is of practical interest. For this case also the equation should be of the form

$$E = C \,.\, \psi\,(V)\,(e_s' - e_a) \qquad ...(7.16)$$

where e'_s is the vapour pressure of the thin film of vapour that exists near the free surface between water and air, whose temperature is neither equal to water temperature nor to air temperature and is virtually impossible to measure.

7.4.5 Combined Energy-Budget and Mass Transfer Approach

The data requirements of either the mass balance or the energy balance methods make their use rather difficult. A brief investigation may give sufficient information for useful estimates of evaporation at a particular site concerned but the results cannot be applied for other locations. To obviate this difficulty Penman first utilised the best features of energy-budget and mass transfer approaches and derived an equation for evaporation which is for general application and which uses a limited number of meteorological parameters that can be measured fairly easily. A brief description of this method is now presented.

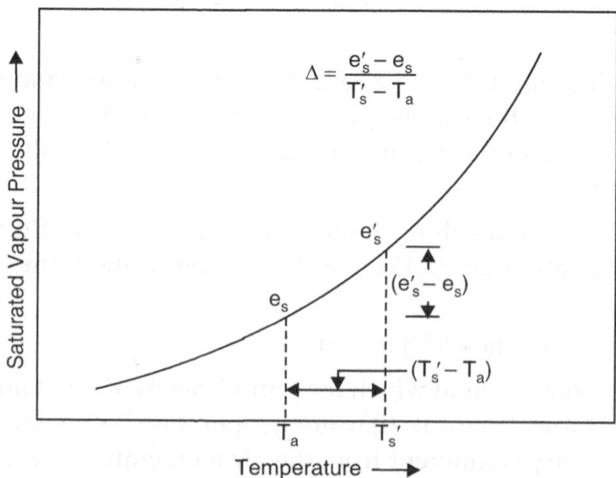

FIGURE 7.1 Saturated vapour pressure curve.

Assuming that the water and air temperatures are the same, the evaporation from water surface as per the mass transfer method is given by Eq. (7.14). Similarly, if the temperatures are not the same the evaporation from water surface, which is of interest, is obtained again from the mass transfer method from Eq. (7.16). Taking the ratio of these two, we get

$$\frac{E_a}{E} = \frac{(e_s - e_a)}{(e_s' - e_a)} \qquad ...(7.17)$$

From the energy-budget method, evaporation is given by

$$E = \frac{(Q_i - Q_r - Q_b) + (Q_a - Q_s)}{L_e\,(1+R)} = \frac{Q_n}{L_e\,(1+R)} \qquad ...(7.18)$$

where Q_n is the net solar energy used in evaporating water. Now from Eq. (7.11)

$$R = 0.00061\,P\,\frac{(T_s' - T_a)}{(e_s' - e_a)}$$

where T_s' is the temperature of the saturated vapour film which may be taken as the temperature of water itself. Even then the measurement of T_s' and e_s' is almost impossible. They are eliminated from the equation utilizing the slope of the saturated vapour pressure curve as shown in Fig. 7.1.

Let Δ be the slope of this curve at T_a. Then, assuming that T_s' and T_a are close to each other, we can write

$$\Delta = \frac{e_s' - e_s}{T_s' - T_a}$$

or

$$(T_s' - T_a) = \frac{(e_s' - e_s)}{\Delta} \qquad ...(7.19)$$

Substituting Eq. (7.19) in the expression for R, we get

$$R = 0.00061\,\frac{P}{\Delta}\,\frac{(e_s' - e_s)}{(e_s' - e_a)} \qquad ...(7.20)$$

Thus T_s' is eliminated from the equation. Now e_s' can be eliminated as shown below. Consider the terms $(e_s' - e_s)/(e_s' - e_a)$. This can be written as

$$\frac{e_s' - e_s}{e_s' - e_a} = \frac{e_s' - e_a + e_a - e_s}{e_s' - e_a}$$

$$= \frac{\left(e_s' - e_a\right) - \left(e_s - e_a\right)}{\left(e_s' - e_a\right)} = 1 - \frac{\left(e_s - e_a\right)}{\left(e_s' - e_a\right)}$$

Using the result of Eq. (7.17),

$$\frac{e_s' - e_s}{e_s' - e_a} = 1 - \frac{E_a}{E} = \frac{E - E_a}{E} \qquad \text{...(7.21)}$$

Substituting Eq. (7.21) in Eq. (7.20), we get

$$R = 0.00061 \frac{P}{\Delta} \cdot \frac{(E - E_a)}{E} \qquad \text{...(7.22)}$$

Thus we see that both T_s' and e_s' are eliminated from the equation. Now substituting Eq. (7.22) in Eq. (7.18), we obtain

$$E = \frac{Q_n}{L_e\left[1 + 0.00061 \frac{P}{\Delta} \cdot \frac{(E - E_a)}{E}\right]}$$

which, when brought to an explicit form, can be written as

$$E = \frac{\left(\frac{Q_n \Delta}{L_e} + 0.00061\, P.E_a\right)}{(\Delta + 0.00061\, P)} \qquad \text{...(7.23)}$$

Eq. (7.23) is called the Penman's equation for evaporation. The various terms in this equation are evaluated as explained here. First E_a is computed from Eq. (7.15). That is

$$E_a = 0.35\ (e_s - e_a)\ (0.5 + 0.54\ V) \qquad \text{...(7.15)}$$

where e_s is the saturated vapour pressure corresponding to the mean daily air temperature, e_a is the actual vapour pressure both expressed in mm of mercury and V is the wind speed in m/s measured at 2 m above the water surface. The value of e_s may be obtained either from Eq. (2.1) or from Table 2.1. Evaporation computed from Eq. (7.15) is in mm/day. This has to be converted into cm/day when it is used in Eq. (7.23).

Δ is the slope of the saturated vapour pressure curve corresponding to the air temperature in mm of mercury/°C. This can be obtained from the following approximate equation.*

$$\Delta = \frac{4098\, e_s}{(237.3 + t)^2} \qquad \text{...(7.24)}$$

where e_s is in mb and t is the air temperature in °C and Δ is in mb/°C.

P is the atmospheric pressure in mm of mercury. L_e is the latent heat of vapourisation of water in cal/g which varies with the temperature and can be obtained from

$$L_e = 597.3 - 0.564\ t \qquad \text{...(7.25)}$$

where t is the air temperature in °C.

---

* It may be noted that Eq. (7.24) is nothing but the derivative of Eq. (2.1).

Q_n is the net solar radiation used in evaporation in cal/sq. cm which is obtained as follows. For small time interval like a day the energy stored in water Q_s , and the energy advected into water body Q_a may be neglected without any appreciable error. The energy reflected back may be expressed as $r.Q_i$ where r is called the *reflection coefficient* or the albedo whose value may be taken as 0.05 for water surface. Then Q_n becomes

$$Q_n = Q_i\,(1 - r) - Q_b \qquad ...(7.26)$$

The incoming solar radiation Q_i is given by

$$Q_i = Q_0\left(a + 0.55\frac{n}{N}\right) \qquad ...(7.27)$$

$$a = 0.29 \cos \theta \qquad ...(7.28)$$

where Q_0 is the radiation in cal/cm²/day received at the top of the atmosphere, n is the actual number of sunshine hours in the day and N is the maximum possible hours of bright sunshine and θ is the latitude of the place. Both Q_0 and N are dependent on the latitude of the place and the month of the year as given in Table 7.1 and Table 7.2 respectively. The net outgoing radiation Q_b is given by

$$Q_b = \sigma\, T^4\,(0.56 - 0.09\,\sqrt{e_a}\,)\left(0.1 + 0.9\frac{n}{N}\right) \qquad ...(7.29)$$

where σ = *Stefan-Boltzman* constant = 118.944×10^{-9} cal/cm²/day/°K

T = mean daily temperature in °K

e_a = actual vapour pressure of air in mm of mercury

n, N = as defined in Eq. (7.27)

Nomographic solutions are also available for Eq. (7.23)

Thus the data required to estimate the evaporation using Penman's combined energy-budget and mass transfer approach given by Eq. (7.23) consist of

(*i*) the daily average atmospheric pressure

(*ii*) the daily average temperature

(*iii*) the daily average vapour pressure (or humidity)

(*iv*) the actual number of sunshine hours

(*v*) the daily average wind velocity at 2 m above the water surface.

These measurements are usually taken at all important meteorological observatories. The following example demonstrates the procedure of estimating evaporation using Penman's equation.

Example 7.2. *Estimate the daily evaporation from a water surface, using the Penman's approach, at a place of latitude 20°N on July 15th from the following data.*

Air temperature = 26°C

Vapour pressure of air = 13.25 mm of mercury

Number of sunshine hours = 7.6

Wind velocity at 2 m height = 2.5 m/s

Atmospheric pressure = 752 mm of mercury

Table 7.1. Total Daily Solar Radiation at top of the Atmosphere (cal/cm²/day)

Time / Latitude	Mar. 21	Apr. 13	May 6	May 29	June 22	July 15	Aug. 8	Aug. 31	Sep. 23	Oct. 16	Nov. 8	Nov. 30	Dec. 22	Jan. 13	Feb. 4	Feb. 26
N 90	—	436	795	1020	1109	1024	788	431	—	—	—	—	—	—	—	—
80	160	436	783	1014	1092	1009	777	431	158	7	—	—	—	—	—	7
70	316	541	771	967	1042	962	764	535	312	133	25	—	—	—	25	135
60	460	654	833	962	1008	957	825	648	455	281	150	74	50	75	150	284
50	592	754	893	987	1019	983	885	747	585	426	295	210	181	211	298	432
40	707	831	937	1001	1021	996	928	822	697	561	442	358	326	361	447	570
30	798	891	957	996	1004	989	949	882	788	683	581	507	480	509	585	690
20	866	921	951	963	963	958	943	911	856	783	706	646	623	649	712	792
10	908	921	920	907	899	903	913	914	897	860	813	770	755	775	819	870
0	922	899	862	828	814	824	855	889	912	913	898	877	868	881	904	923
S 10	908	849	783	728	708	725	776	838	897	937	954	959	961	964	964	949
20	866	773	680	611	584	608	674	763	856	934	987	1017	1029	1023	997	946
30	798	674	559	479	449	477	554	665	788	903	993	1051	1072	1056	1002	915
40	707	554	425	339	306	338	421	549	697	844	972	1058	1091	1063	982	853
50	592	420	284	199	170	198	282	416	585	765	928	1044	1088	1048	936	775
60	460	277	144	70	48	70	143	274	455	663	865	1013	1077	1022	872	672
70	316	131	24	—	—	—	24	130	312	548	801	1023	1113	1028	809	555
80	166	7	—	—	—	—	—	7	158	442	814	1072	1166	1077	820	447
90	—	—	—	—	—	—	—	—	—	442	825	1088	1183	1094	833	447

Table 7.2 Maximum Possible Bright Sunshine Hours

Time of year → *Latitude* *N or S* ↓	*N July* 17 *S Jan* 15	*Aug.* 19 *Feb* 15	*Sept.* 17 *March* 15	*Oct.* 18 *April* 15	*Nov.* 17 *May* 15	*Dec.* 16 *June* 15	*Jan.* 13 *July* 15	*Feb.* 11 *Aug* 15	*March* 13 *Sep.* 15	*April* 12 *Oct.* 15	*May* 13 *Nov.* 15	*June* 14 *Dec.* 15
0	12.1	12.1	12.1	12.1	12.1	12.1	12.1	12.1	12.1	12.1	12.1	12.1
10	12.6	12.4	12.2	11.9	11.7	11.5	11.6	11.8	12.0	12.3	12.6	12.7
20	13.2	12.7	12.2	11.7	11.2	10.9	11.0	11.4	12.0	12.6	13.1	13.3
30	13.9	13.1	12.3	11.4	10.6	10.2	10.4	11.0	11.9	12.8	13.6	14.1
40	14.6	13.6	12.4	11.1	9.9	9.3	9.6	10.5	11.8	13.2	14.3	15.0
50	15.9	14.2	12.5	10.6	9.0	8.1	8.5	9.9	11.7	13.6	15.3	16.3
60	18.0	15.3	12.7	10.0	7.4	5.9	6.6	8.7	11.5	14.3	17.0	18.7
70	24.0	17.6	13.2	8.8	3.6	0	0	6.7	11.3	15.2	21.3	24.0
80	24.0	24.0	14.6	4.2	0	0	0	0	11.2	20.2	24.0	24.0
90	24.0	24.0	24.0	0	0	0	0	0	0	24.0	24.0	24.0

Solution. From Table 2.1, e_s for a temperature of 26°C is read as 33.608 mb

$$e_s = 33.608 \text{ mb} = \frac{33.608}{1.33} = 25.27 \text{ mm of mercury}$$

$$e_a = 13.25 \text{ mm of mercury}$$

$$V = 2.5 \text{ m/s}$$

$$E_a = 0.35\,(e_s - e_a)\,(0.5 + 0.54\,V)$$

$$= 0.35\,(25.27 - 13.25)\,(0.5 + 0.54 \times 2.5)$$

$$= 7.783 \text{ mm/day} = 0.7783 \text{ cm/day}$$

From Table 7.1, for a latitude of 20°N on July 15th $Q_0 = 958$ cal/cm^2

From Table 7.2, for a latitude of 20°N on July 15th N = 13.2 h, n = 7.6 h ; $a = 0.29 \times \cos 20° = 0.2725$

$$Q_i = Q_0\left(a + 0.55\frac{n}{N}\right)$$

$$= 958\left(0.2725 + \frac{0.55 \times 7.6}{13.2}\right) = 564.42 \text{ cal/cm}^2$$

$$Q_b = \sigma\, T^4\,(0.56 - 0.09\sqrt{e_a})\left(0.1 + 0.9\frac{n}{N}\right)$$

$$T = 273 + 26 = 299\ K$$

$$Q_b = 118.944 \times 10^{-9} \times 299^4 \times (0.56 - 0.09\sqrt{13.25}) \times \left(0.1 + \frac{0.9 \times 7.6}{13.2}\right)$$

$$= 950.66 \times 0.2324 \times 0.6182$$

$$= 136.58 \text{ cal/cm}^2$$

$$Q_n = Q_i\,(1 - r) - Q_b = 564.42\,(1 - 0.05) - 136.58$$

$$= 399.62 \text{ cal/cm}^2$$

$$\Delta = \frac{4098\, e_s}{(237.3 + t)^2}$$

$$= \frac{4098 \times 33.608}{(237.3 + 26)^2} = 1.987 \text{ mb/°C}$$

$$= \frac{1.987}{1.33} = 1.494 \text{ mm of mercury/°C}$$

$$L_e = 597.3 - 0.564 \times t$$

$$= 597.3 - 0.564 \times 26 = 582.636 \text{ cal/g}$$

$$P = 752 \text{ mm of mercury}$$

$$E = \frac{\frac{Q_n \Delta}{L_e} + 0.00061\, P.E_a}{\Delta + 0.00061\, P}$$

$$= \frac{\frac{399.62 \times 1.494}{582.636} + 0.00061 \times 752 \times 0.7783}{1.494 + 0.00061 \times 752}$$

$$= \frac{1.0247 + 0.357}{1.494 + 0.4587} = 0.7076 \text{ cm}$$

or $$E = 7.076 \text{ mm/day.}$$

7.5 MEASUREMENT OF EVAPORATION

Various methods of estimating evaporation discussed in the previous section can be used to estimate the evaporation from lakes and reservoirs. But these methods are not directly applicable in design problems and also in many cases the required data may not be available though the Penman's method is increasingly used. However, most estimates of reservoir evaporation both for design and operation have been made using the measured evaporation. Evaporation measurements are also required to validate the empirical equations that are developed. Evaporation is usually measured either by *atmometers* or by the *evaporation pans* also known as the *evaporimeters.*

7.5.1 Atmometers

The atmometers are provided with some special surfaces which are kept wet and from these surfaces the evaporation takes place. There will be some source to continuously supply water and keep the evaporating surface wet. The water loss observed at the source is an indication of the evaporation.

Livingstone Atmometer. The livingstone atmometer consists of a 5 cm diameter spherical surface made of 2.5 mm thick porus material. It is filled with distilled water and connected to a supply reservoir by means of a capillary tube as shown in Fig. 7.2. Water is continuously drawn from the supply reservoir to meet the evaporation needs of the spherical surface. The decrease in the water level of the supply reservoir is taken as a measure of evaporation.

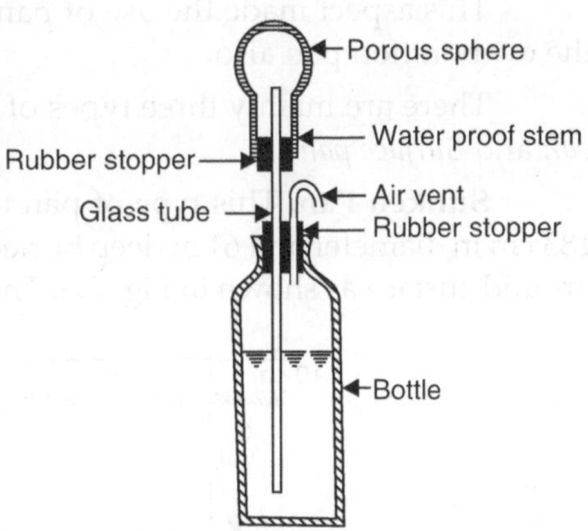

FIGURE 7.2. Livingstone atmometer.

Piche Atmometer. It consists of a graduated glass tube 1.5 cm in diameter and 30 cm long with one end open. It is filled with distilled water and then the open end is covered with a filter paper. The filter paper is held in position by a metal clip. Now the tube is kept in an inverted position so that the water wets the filter paper and evaporates. The water loss in the tube is recorded as a measure of evaporation.

The atmometers are not as common as the evaporation pans because they do not have the same exposure as large water surfaces. They tend to overestimate the wind effects and underestimate the radiation effects. The rate of evaporation recorded by atmometers is generally higher than the evaporation recorded by evaporation pans.

7.5.2 Evaporation Pans

The pans are commonly used as evaporation measuring devices because they are inexpensive and simple to instrument. In evaporation pans the depth of evaporation during any time interval is measured as the drop in water surface level in the pan in that interval corrected for precipitation, if any, obtained from an adjacent or a nearby standard raingauge. The observations are usually taken on daily basis and after taking the measurement on each day the water level in the pan is restored to a stipulated value by adding or removing required amount of water. Any cylindrical vessel of 1.2 to 1.8 m diameter, 0.3 m high with open top and made of galvanised iron sheet can be used as an evaporation pan. The water level within the pan is maintained between 5 to 7.5 cm below the rim of the pan.

Pan Coefficient. A criticism against the use of evaporation measurements obtained from pans in the designs may be levelled on the basis that the evaporation measured from a small pan will not be same as the evaporation from a large lake or reservoir since the exposure conditions are not identical in both cases. Specifically, the heat storing capacity and the heat transferred from the side and bottom of a pan are quite different from those of a large lake or reservoir. Also the height of the rim above the water surface in an evaporation pan affects the wind action over the surface. In addition it creates a shadow of variable magnitude over the water surface which affects the radiation incident to the water surface. Nevertheless, it has been observed that the ratio of reservoir evaporation to pan evaporation, called the *pan coefficient*, is constant on annual basis and does not vary excessively from region to region for a given type of pan.

$$\text{Pan coefficient} = \frac{\text{Actual evaporation from reservoir}}{\text{Measured evaporation from pan}}$$

This aspect made the use of pans popular and many meteorological observatories include the evaporation pan also.

There are mainly three types of pans that are in common use, namely *Sunken pan, Floating pan* and *Surface pan.*

Sunken Pan. This type of pan used by United States Bureau of Plant Industry (USBPI) is 183 cm in diameter and 61 m deep buried in the ground such that the top of its rim is 10 cm above the ground surface as shown in Fig. 7.3. The water surface in the pan is maintained within ±15 mm of

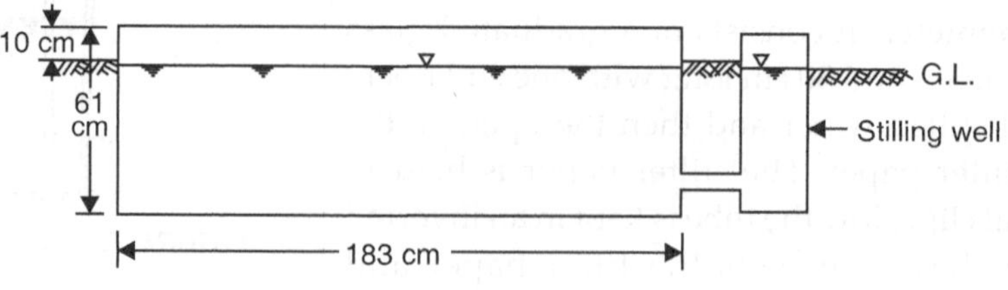

FIGURE 7.3. Sunken pan.

the ground level. Evaporation is measured by means of a pointer gauge in a stilling well outside the pan. The stilling well and the pan are interconnected by a small tube. The *colorado sunken pan* is square in shape of side 91.5 cm and 46 cm deep.

The sunken pan has the advantage that certain boundary effects such as the direct solar radiation on side walls and the heat exchange through side walls are negligible. The pan coefficient for this pan is understandably high and is about 0.95. The disadvantages of this pan are : (*i*) it tends to collect trash and difficult to maintain, (*ii*) it is difficult to instal this pan, (*iii*) it is difficult to detect any leak, and (*iv*) it is difficult to evaluate the amount of heat which is transferred directly through the walls of the pan to the surrounding soil which may be required in theoretical studies.

Floating Pan. This pan is supported by drum floats and is kept floating in a lake. The water level in the tank is maintained at the same level as the water surface in the surrounding lake while the rim of the pan is 7.5 cm above the water level. The floating pan used by United States Geological Survey (USGS) is square in plan with a side of 90 cm and 45 cm deep.

The use of these pans is not very common. There are many operational difficulties associated with this type of pan. For example, it is not easily accessible, and there may be splashing of water into or out of the pan due to wave action which is very difficult to quantify. The pan coefficient for this type of pan ranges from 0.70 to 0.82, but it is generally recommended as 0.80.

Surface Pan. The most common type of evaporation pan is the surface pan. It may also be called as a *land pan*. In most countries the surface pan conforms to the standard United States Weather Bureau (USWB) class A land pan which is shown in Fig. 7.4. It is 122 cm in diameter and 25.4 cm deep and is made of unpainted galvanised iron. It is installed 15 cm above the ground

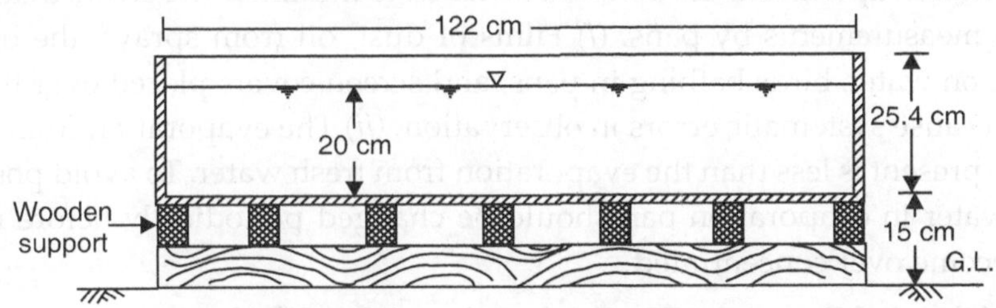

FIGURE 7.4 U.S. Weather bureau class a land pan.

level over a wooden frame. The space below the wooden frame permits the free circulation of air under the pan. It is filled to a depth of 20 cm. Water surface level is measured daily with a hook gauge in a stilling well. The stilling well is provided within the pan. The evaporation is computed as the difference between observed levels adjusted for precipitation, if any. The annual pan coefficient is 0.7. The true coefficient on seasonal basis may vary from 0.6 to 0.8. The advantages of this pan are : (*i*) stable pan coefficient, (*ii*) easy access for observation, (*iii*) more stability compared to floating pan, (*iv*) relative freedom from dirt and trash, and (*v*) reasonable cost of installation.

ISI Standard Pan. The standard pan, also known as the *modified class A pan*, specified by IS: 5973–1970 is made of 0.9 mm thick copper sheet. It is 122 cm in diameter and 25.5 cm deep. It is tinned inside and painted white outside. It is placed on a 10 cm high square platform of 122.5 cm width made of wood. A fixed point gauge is kept in the pan to indicate the water level. The top of the pan is covered fully with a hexagonal mesh of galvanised iron to protect it from birds. It is filled with water to a depth of 19 cm. A calibrated cylindrical measure is used to add or remove water in order to maintain the water level in the pan to a fixed mark indicated by the pointer gauge in the stilling well. The details of the ISI standard pan are given in Fig. 7.5. The pan coefficient for this pan ranges from 0.65 to 1.10 and an average value of 0.8 is generally recommended.

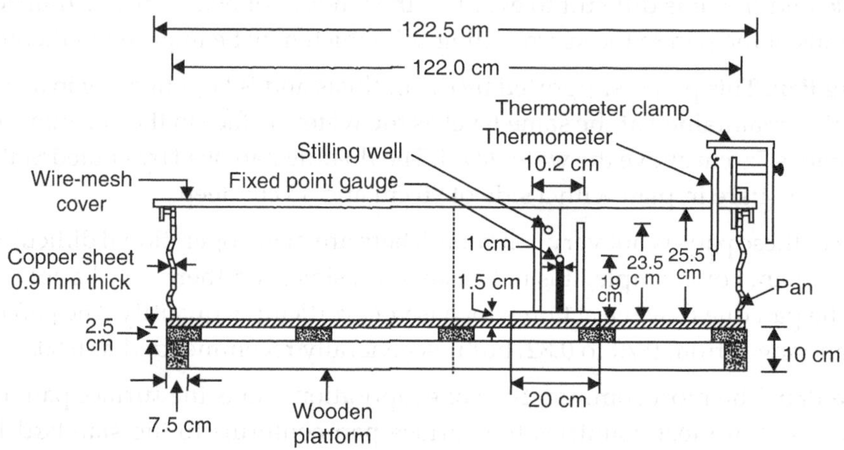

FIGURE 7.5. ISI standard pan.

Errors in Pan Evaporation. Care should be taken to minimise the errors that may creep into the evaporation measurements by pans. (*i*) Films of dust, oil from sprays, the oily secretion of insects alighting on water, birds bathing in pans, and screen covers placed over the pans to keep out the birds can cause systematic errors in observation. (*ii*) The evaporation from water in which some solutes are present is less than the evaporation from fresh water. To avoid possible error due to this reason, water in evaporation pan should be changed periodically before minerals in the supply water become over concentrated.

Effect of Advected Energy on Pan Evaporation. Advection by means of water added to the evaporation pan is usually insignificant. But advection of sensible heat through the pan walls is sufficient to produce moderate variation in the pan coefficient under varying climatic regimes. The measured evaporation from the pan E_p needs adjustment to account for the advected energy before it is used along with appropriate pan coefficient to compute the lake evaporation. This is especially true since the annual heat transfer through the bottom of the reservoir is essentially zero. Using the concept of Bowen's ratio, a function can be derived for estimating heat transfer through class *A* pan from observations of air and water temperatures, wind movement, and atmospheric pressure. Assuming that the pan coefficient is 0.7 when air and pan water temperatures are equal, the lake evaporation *E* can be computed from

$$E = 0.7\,[\,E_p \pm 0.0064\,p\,\alpha_p\,(\,0.37 + 0.00255\,v_p)\,\mid T_0 - T_a\mid^{0.88}\,] \qquad ...(7.30)$$

where evaporation is in mm/day, p is the atmospheric pressure in kilopascals, v_p is wind velocity in km/day, T_0 is the water surface temperature and T_a is air temperature in degrees Celsius. The sign following E_p is (+) when $T_0 > T_a$ and (–) when $T_0 < T_a$. The coefficient α_p can be evaluated using the following approximate equation.

$$\alpha_p = 0.34 + 0.0117\, T_0 - 3.5 \times 10^{-7}\, (T_0 + 17.8)^3 + 0.0135\, v_p^{\,0.36} \qquad ...(7.31)$$

where T_0 is in degrees Celsius and v_p is the wind velocity at 150 mm above the pan rim in km/day.

Lake evaporation computed from Eq. (7.30) should be adjusted if the net advection to lake is appreciable.

Evaporimeter Network. As pointed out earlier, the meteorological observatories usually include an evaporation pan also. Thus the daily evaporations are measured along with other meteorological parameters such as rainfall, temperature, atmospheric pressure, humidity etc. The WMO recommends the following minimum network of evaporimeters.

Region I : Arid zones — one station for every 30,000 km^2

Region II : Humid temperate regions — one station for every 50,000 km^2

Region III : Cold regions — one station for every 1,00,000 km^2.

Example 7.3. *The average water spread areas that are likely to be maintained during the operation of a reservoir after its completion and the observed monthly pan evaporations at a proposed reservoir site are given below.*

Month	*Average water spread area in hectares*	*Pan evaporation in cm*
January	872.0	10.2
February	797.0	15.3
March	754.5	25.4
April	739.5	30.5
May	726.0	28.0
June	717.5	17.8
July	735.0	15.3
August	765.0	14.0
September	807.5	14.0
October	850.0	15.3
November	891.0	12.7
December	917.0	10.2

Estimate the annual evaporation loss from the reservoir in million m^3. If 75% of this loss can be prevented and the water thus saved is utilised to irrigate a crop with a requirement of 57 cm of water, how much area can be irrigated ? Assume a pan coefficient of 0.70.

Solution. For the sake of convenience, the computations are shown in Table 7.3.

Table 7.3. Computation of Reservoir Evaporation Losses

Month ①	*Average water spread area in ha* ②	*Monthly pan evaporation in cm* ③	*Monthly lake evaporation in cm* ④ = 0.7 × ③	*Monthly evaporation loss in ha-cm* ⑤ = ② × ④
Jan.	872.0	10.2	7.14	6226.08
Feb.	797.0	15.3	10.71	8535.87
March	754.5	25.4	17.78	13415.01
April	739.5	30.5	21.35	15788.33
May	726.0	28.0	19.60	14229.60
June	717.5	17.8	12.46	8940.05
July	735.0	15.3	10.71	7871.85
Aug.	765.0	14.0	9.80	7497.00
Sept.	807.5	14.0	9.80	7913.50
Oct.	850.0	15.3	10.71	9103.50
Nov.	891.0	12.7	8.89	7920.99
Dec.	917.0	10.2	7.14	6547.38
				Σ = 113989.16

The lake evaporation in any month is obtained as the product of the monthly pan evaporation and the pan coefficient, that is 0.7. These values are given is column (4). The monthly evaporation loss is computed as the product of the average water spread area in that month and the lake evaporation. These values appear is column (5), the total of which gives the annual evaporation loss.

∴ Annual evaporation loss = 113989.16 ha-cm

$$= 113989.16 \times 10^4 \times \frac{1}{100} \text{ m}^3$$

$$= 11.4 \times 10^6 \text{ m}^3 = 11.4 \text{ million m}^3$$

Depth of irrigation = 57 cm = 0.57 m

$$\text{Area of crop that can be irrigated} = \frac{11.4 \times 10^6}{0.57} = 20 \times 10^6 \text{ m}^2$$

$$= \frac{20 \times 10^6}{10^4} = 2000 \text{ hectares.}$$

Example 7.4. *During a daily routine observation, 10.8 litres of water was added to bring the water surface in the evaporation pan to the stipulated level and the nearby raingauge measured 3.6 mm of rainfall. What was the evaporation recorded for the day if the diameter of the pan is 122 cm ?*

Solution. Diameter of the pan = 122 cm

Area of the pan $= \frac{\pi}{4} \times 122^2 = 11689.87 \text{ cm}^2$

Volume of water added = 10.8 litres = 10800 cm^3

Depth of water added $= \frac{\text{Volume}}{\text{Area}} = \frac{10800}{11689.87} = 0.924$ cm

Rainfall = 3.6 mm = 0.36 cm

Evaporation = depth of water added + rainfall

= 0.924 + 0.36 = 1.284 cm

= 12.84 mm

Example 7.5. *What is the evaporation, if 4.75 litres of water is removed from an evaporation pan of diameter 1.22 m and the simultaneous rainfall measurement is 8.8 mm ?*

Solution. Area of the pan $= \frac{\pi}{4} \times 122^2 = 11689.87 \text{ cm}^2$

Volume of water removed = 4.75 litres = 4750 cm^3

Depth of water removed $= \frac{4750}{11689.87} = 0.406$ cm = 4.06 mm

Evaporation = rainfall – depth of water removed

= 8.80 – 4.06 = 4.74 mm.

7.6 REDUCING EVAPORATION FROM WATER SURFACES

Although total prevention of evaporation is ruled out, many attempts have been made to reduce the evaporation with a varying degree of success. The reduction of evaporation has a great economic importance especially in arid climates since the evaporation from storage in these regions is very substantial where the available surface water resource itself is meagre. Evaporation control is a particularly important way to conserve water because it usually requires little new construction and additional water becomes available without construction delays. In many cases it may cost less to reduce evaporation than to collect and store an equivalent amount of water from other sources. Controlling only dry-season evaporation with short-lived methods could have economic significance in arid regions because evaporation is the greatest during the driest seasons, which are also the peak periods of water use. Suppression of evaporation in impounded waters also suppresses the increase in salt concentration that occurs with evaporation. If the evaporation is controlled by floating materials, it will cutoff the light, thereby reducing the growth of undesirable algae and submerged aquatic weeds.

The methods of reducing evaporation may be classified into three groups. Though the first two are limited in scope the third offers great potential and can definitely be tried if it proves to be economical.

Reducing Surface-Area. The volume of water lost as evaporation is directly proportional to the exposed waterspread area. Therefore the evaporation can be reduced by reducing the surface area. The suggested alternatives are such as

(*i*) Constructing reservoirs with minimum ratio of area to storage.

(*ii*) Storing water below ground

(*iii*) Storing water in one large reservoir instead of several small reservoirs.

(*iv*) Selecting proper reservoir sites, and

(*v*) Straightening stream channels and thus reducing meandering of surface areas of water.

Mechanical Covers. This method applies mostly to small reservoirs. The covers include roofs, floating rafts and wind breaks, floating wax blocks, floating light weight concrete blocks. Removal of aquatic plant life near the water surface may also save loss of water through transpiration.

Suppression of Evaporation by Surface Films. Certain chemicals known as *polar compounds* can act as evaporation suppressors. These substances possess molecules which have great affinity for water at one end and other end repels water. When these substances are in contact with water each molecule tends to align itself with one end in water and the other end projecting from the surface of the water. Ideally this would form a film of one molecule thick on the water surface which is called a *monolayer*. This layer will suppress the evaporation. The thickness of this film is less than a micro-cm. It is invisible and exhibits none of the interference patterns of colour associated with oil films. Its presence on the surface of water is characterised by the absence of tiny wavelets since these have been suppressed by the film. The film is penetrable by rain drops but will close again after being broken. It is flexible and moves with the motion of the water surface. It does not easily break up as thick films of oil. *Hexadeconal,* also known as *cetyl alcohol* is the polar compound that is most commonly tried. No apparent toxicity of hexadeconal to wild life and fish life have been observed.

The effectiveness of the mono-molecular layers depends on many factors such as wind action, method of application, and oxidation or degradation by micro organisms. Wind may sweep off the film and pile up the material on shore. This becomes a major problem, especially on large reservoirs. Effective application of material is also a big technical problem in the case of large reservoirs. The material may be applied in a solvent, such as kerosene, as an emulsion, or as a solid in the form of powder, pellets, or cakes.

The reduction of evaporation by use of monofilms varies with the kind of material and conditions under which it is used. As high as one-third reduction has been claimed by some investigators. Theoretically less than 25 grams of hexadeconal is sufficient to form a compact layer over a surface area of one hectare. In practice, however, considerably more material is required because it is removed by wind, birds (it is reported that when the material is used in the form of pellets it is eaten away by birds), insects and aquatic life.

Though the monolayers reduce evaporation by acting as physical barriers to vapour transfer, they do not reduce the amount of solar energy the water absorbs, and they decrease the amount of heat normally lost from the water because inhibiting evaporation also inhibits the cooling effects of vapourisation. Accordingly as evaporation is suppressed the water temperature increases. When the film is disrupted, the evaporation rate from the treated water surface is higher than the evaporation from an untreated surface. Furthermore, an intact alcohol barrier is impossible to maintain because of wind and wave action. These problems come in the way of using the monolayers. Attempts are now under way to circumvent the problem of maintaining a continuous film on the water surface using a plastic net to restrict the drift and disruption of alcohol layers. Some new type of monolayers with silvery surface which can reflect the energy and also act as a barrier for vapour transfer may be the answer to the problem of reducing the temperature of water body.

From a hydrologic view point, no side effects should be anticipated because of suppression of evaporation in the balance of hydrologic cycle. Evaporation from land areas forms only a very small component of water vapour in the atmosphere.

7.7 TRANSPIRATION

Transpiration is the process by which water vapour escapes from the living plant, principally through the leaves and enters the atmosphere. This process constitutes an important phase of the hydrologic cycle since it is the principal mechanism by which the precipitation falling on land areas is returned to the atmosphere. Though it is not practicable to separate evaporation and transpiration from an engineer's view point a knowledge of each process is required to understand the techniques employed to estimate them.

The water extracted by the plant roots from the soil is transferred through the plant to the *intercellular space* within the leaves. Air enters the leaf through the *stomata* openings in the leaf surface and the *chloroplasts* within the leaf use carbon dioxide from the air and a small portion of the available water to manufacture carbohydrates required for plant growth (*Photosynthesis*). As air enters the leaf, water escapes through the open stomata ; this is the process of transpiration. The ratio of water transpired to that used in forming plant matter is very large-upto 800 or more.

7.7.1 Factors Affecting Transpiration

The factors affecting transpiration may be considered under three groups : the plant factors, the soil factors and the climate factors.

The plant factors that affect transpiration are the extent and efficiency of the root systems in moisture absorption, the leaf area, leaf arrangement and structure and stomatal behaviour. Plants with fewer stomatal openings per unit area and less surface area exposed to radiation transpire relatively less water. Plants having root systems reaching to the water table will transpire water which is independent of moisture content in the zone of aeration. All plants can control stomatal opening to some extent and therefore have some ability to reduce the transpiration during periods of drought. However, the control is only to reduce the transpiration and the upper limit on transpiration is governed by available radiant and sensible energy. The rate of transpiration is largely independent of plant type, provided there is adequate soil water and the surface is entirely covered by vegetation.

The soil factors include those governing the water supply to the roots. The amount of water held in the soil after excess gravitational water has drained is defined as the *field capacity*. Moisture content at which permanent wilting of plants occurs is called the *wilting point*. The moisture in the soil between field capacity and wilting point is a measure of *available water*. Some investigators believe that transpiration is independent of available moisture until it has receded to the wilting point, while others assume that transpiration is roughly proportional to available moisture.

The climatic factors affecting transpiration are solar radiation, light intensity, atmospheric pressure, temperature and wind etc. Since photosynthesis is highly dependent on the radiation received, about 95 per cent of daily transpiration occurs during daylight hours. Plant growth stops when temperature drops to 4°C. During the nights transpiration virtually ceases due to stomatal closure. Thus, over a period of a month or longer, evaporation from a free water surface may exceed the transpiration from a crop. Over a few hours during the day, however, the crop may evaporate more than a water surface.

7.7.2 Measurement of Transpiration

It is not possible to measure transpiration loss from an appreciable area under natural conditions. Therefore the measurements are generally restricted to small samples under laboratory conditions.

In one method the transpiration is computed as the increase in moisture content of the space in a closed container in which one or more plants are grown.

Freshly cut leaves, twigs or stalks of a growing plant immediately after cutting are weighed at short time intervals till wilting commences. The method is based on the assumption that transpiration continues at approximately the same rate before and after cutting.

Most commonly, transpiration is measured using a *phytometer*. It is a large vessel filled with soil in which one or more plants are rooted. The soil surface is sealed after the plant is well-established so as to exclude precipitation and to prevent the escape of moisture except through the plant. Periodical measurements of the weight of phytometer along with the amounts of water added in the intervening intervals of time give an estimate of transpiration. Since it is not possible to simulate natural conditions the results of phytometer observation are useful only for theoretical studies.

Transpiration ratio which is defined as the amount of water transpired by a plant in its growth to produce a unit weight of dry matter exclusive of roots is sometimes used to estimate the transpiration loss.

7.8 EVAPOTRANSPIRATION

Evapotranspiration, denoted by *ET*, is the sum of the water used by plants in a given area in transpiration and the water evaporated from the adjacent soil in the area in any specified time. *Consumptive use*, *CU* is the evapotranspiration from an area plus the water used directly in the metabolic process of building plant tissue. The amount of water used in the actual metabolic process is insignificant compared to evapotranspiration and therefore the term *CU* is generally taken as equivalent to *ET*. In an area covered with vegetation it is difficult and also unnecessary from practical view point to separately evaluate evaporation and transpiration. It is more convenient to estimate the evapotranspiration directly. Only over those areas of earth's surface where no vegetation is present, for example ice and snow fields, bare rock slopes, some desert areas, and water surfaces, will purely evaporation occur. Evapotranspiration represents the most important aspect of water loss in the hydrologic cycle.

The *potential evapotranspiration, PET*, is defined as the evapotranspiration which would occur if there was always an adequate water supply available to a fully vegetated surface. It is the upper limit of *ET* for a crop in a given climate. If the water supply available is less than *PET*, the deficit will be drawn from soil moisture storage until the available soil moisture is reduced to about 50 per cent. Beyond this the actual evapotranspiration, *AET*, will be less than *PET*, until the wilting point is reached when evapotranspiration ceases. Seasonal consumptive use of a crop is the total quantity of water used by it in evapotranspiration during the entire growing season. A knowledge of *PET* and seasonal *CU* is useful in designing the irrigation systems.

7.8.1 Factors Affecting Evapotranspiration

In general the factors that govern the evaporation and transpiration also govern the evapotranspiration process. However, a distinction between *PET* and *AET* is important. Potential evapotranspiration is controlled essentially by meteorological factors, whereas the actual evapotranspiration is also considerably affected by plant and soil factors.

In general *PET* tends to increase as the temperature, sunshine and wind speed increase and as the humidity decreases. Radiation supplies the energy for the evapotranspiration process. A close relationship exists between the net solar radiation and evapotranspiration. The type of plant may also possibly influence the evapotranspiration in the sense that albedo of the vegetation surface may vary from plant to plant. A light coloured surface will obviously reflect away more (and therefore absorbs less) radiation than a dark- coloured surface. But most plant surfaces fall within a fairly narrow colour range. The albedo of most green agricultural crops is very nearly constant at about 0.25. The orientation of row crops also influences the evapotranspiration as the interception of solar energy is a function of orientation. The presence or the absence of crops in the surrounding lands may also influence evapotranspiration. Air masses passing over a cropped and irrigated area of the field in question are cooler and more humid than those coming from a bare ground. The evapotranspiration is therefore less in the former case than in the latter.

As already pointed out the *AET* is influenced by the soil and plant factors also. When the vegetation surface is dry and when the soil moisture content falls below optium value the *AET* rate will be limited by the resistance to water movement imposed by the soil pores and the plant. The density of vegetation cover which is defined as the ratio of leaf surface area to ground area (which is of the order of 5 to 10) will also influence *AET*. It has been observed that a reduction in the density of vegetation cover reduces the evapotranspiration rates. Removing the lower leaves, if properly timed, can reduce water lost by transpiration with little or no loss in grain yield. *AET* losses will also differ according to the main stages of the plant development. Thus, in the seedling stage the evapotranspiration is high, while after ripening of the grain crops the evapotranspiration may be very much less than *PET*.

7.8.2 Measurement of Evapotranspiration

There are mainly four methods of direct measurement of evapotranspiration. They are :

(*i*) Water-budget method,

(*ii*) Field experimental plots,

(*iii*) Soil moisture depletion studies, and

(*iv*) Lysimeter method.

Water-budget-method. This method, also known as inflow outflow method, is suitable for evaluating evapotranspiration of large area such as watersheds over relatively long periods of time. The evapotranspiration can be evaluated provided all other constituent items of the following water balance equation are known

$$\text{Water supply} = \text{Water disposal} \quad ...(7.32)$$

The items constituting the supply are : surface inflow, subsurface inflow, precipitation, and imported water. The items constituting the disposal are : Surface outflow, subsurface outflow, evapotranspiration, domestic, municipal and industrial use, exported water, increase in surface storage, increase in groundwater storage. It is evident that the above rigorous definition of water balance makes evaluation extremely difficult. However, the normal annual evapotranspiration can be readily computed as the difference between long-term average of precipitation and streamflow since the change in storage items over a long period of years is inconsequential.

Field Experimental Plots. In this method a field plot is chosen and the amounts of water added to the irrigation plot under observation by way of precipitation and irrigation are measured along with runoff. The moisture content in various layers of the soil within the root zone depth are measured both at beginning and end of the crop season. Then the evapotranspiration is computed as

$$ET = I - Q - \Delta S \quad ...(7.33)$$

where I is the total inflow in mm including precipitation and irrigation water, Q is the total surface runoff in mm and ΔS is the increase in soil moisture storage in mm which is computed from the following equation

$$\Delta S = \sum_{i=1}^{n} \frac{M_{bi} - M_{ei}}{100} \times G_i \, D_i \quad ...(7.34)$$

where M_{bi} = Moisture content in per cent at the beginning of the season in ith layer

M_{ei} = Moisture content in per cent at the end of the season in ith layer

G_i = The apparent specific gravity of the ith layer of the soil

D_i = Depth of ith layer of soil, in mm, within the root zone depth

n = The number of soil layers considered in the entire root zone.

Since it is difficult to evaluate the deep percolation losses, they are not included in the balance equation. Hence it is recommended that water is added to the plot only in small quantities to minimise this loss. Another precaution that is required is that the water table at the location of the plot should be deep, so that the plants do not derive any supply from this source.

If the soil moisture is maintained near field capacity throughout the season the evapotranspiration resulting from Eq. (7.33) represents potential evapotranspiration.

Soil moisture depletion studies. The studies involve measurement of soil moisture from various depths at as frequent intervals of time as possible throughout the growth period of a crop. The method gives satisfactory results for irrigated field crops grown on fairly uniform soils and where the soil moisture variations within the root zone are not influenced by groundwater. The evapotranspiration u, for any time period between two successive samplings is obtained as

$$u = \sum_{i=1}^{n} \frac{M_{1i} - M_{2i}}{100} \times G_i \, D_i \qquad \text{...(7.35)}$$

where M_{1i} = Soil moisture percentage at the time of first sampling in the ith layer

M_{2i} = Soil moisture percentage at the time of second sampling in the ith layer

The terms G_i, D_i and n have the same significance as in Eq. (7.34).

Summation of the evapotranspiration values of all sampling intervals gives the seasonal evapotranspiration or the seasonal consumptive use. Of course, due allowance should be made in the summation to account for the evapotranspiration that occurred between irrigation application and soil moisture sampling.

Lysimeter. A lysimeter (also known as *evapotranspirometer*) consists of a circular tank about 60 to 90 cm in diameter and 180 cm deep. Sometimes larger diameters upto 3 m with 3 m depth are also used. The lysimeter is filled with soil and individual crops or natural vegetation, for which the evapotranspiration is required, are grown. It is buried so that its top is in flush with the surrounding ground surface. The sides of the lysimeter are impervious whereas the bottom is pervious. Water passing through the soil column is collected at the bottom and conducted through a small tube to a measuring gauge in an adjacent pit. Soil moisture measurements may be obtained from moisture sampling, weighing, or hydraulic method. In weighing type, the tank is mounted on a self recording scale. In the hydraulic method the tank is floated on water or a suitable heavy liquid and changes in weight are recorded through pressure changes observed in a manometer. Evapotranspiration is then determined from the measurements of the amount of water added, outflow at the bottom and changes in the soil moisture. For better results the lysimeter may be surrounded by a guard area of at least 4 hectares of the same crop with similar moisture conditions.

The limitation of lysimeter method is that differences may exist between the lysimeter and natural conditions of soil profiles, soil moisture regime, plant rooting characteristics, methods of water application, water table, temperature and the net energy exchange.

7.8.3 Estimation of Evapotranspiration from Evaporation Data

The potential evapotranspiration and the evaporation from a thin free water surface are affected by the same meteorological factors : radiation, humidity, wind and temperature. Though some differences in surface roughness, albedo characteristics and other factors exist, a close relationship between the evaporation from a well located evaporation pan and evapotranspiration can be expected. The procedure then is, therefore, to take the readings of evaporation pan and multiply the measured evaporation with an appropriate *crop factor* to obtain the evapotranspiration. Extensive experimentations under varying climatic conditions and for different crops are required to establish the crop factor. The crop factor is observed to be almost unity, when it is based on the evaporation observed in a sunken pan.

7.8.4 Estimation of Evapotranspiration from Climatological Data

The difficulties in measurement of evapotranspiration in the field led to the development of empirical equations which relate the evapotranspiration with some readily available climatic data like temperature, humidity, wind speed, day length, and sunshine hours. Four of the most commonly used equations are explained here. These are *Blaney-Criddle, Thornthwaite, Christiansen,* and *Penman* equations. Evapotranspiration is also estimated using integration method, and the groundwater fluctuations.

Blaney-Criddle Equation. Blaney and Criddle developed a simple equation for estimating evapotranspiration. It is assumed that the evapotranspiration is closely correlated with the mean monthly temperatures and daylight hours. The monthly consumptive use factor f is defined as

$$f = \frac{p \,.\, T_m}{100} \qquad \text{...(7.36)}$$

where T_m is the monthly mean temperature in °F, p is the monthly daylight hours expressed as per cent of the daylight hours of the year and f is in inches.

In other words p is obtained from the expression

$$p = \frac{\text{possible sunshine hours for the particular month}}{\text{possible sunshine hours for the whole year}} \times 100$$

or

$$p = \frac{\text{possible sunshine hours for the particular month}}{(365 \times 12)} \times 100.$$

The value of p depends on the latitude of the place and the month of the year which can be obtained from Table 7.4.

Table 7.4. Monthly Percentage of Daytime hours in the Year

Latitude N	*Jan*	*Feb*	*Mar*	*Apr*	*May*	*Jun*	*Jul*	*Aug*	*Sep*	*Oct*	*Nov*	*Dec.*
0°	8.49	7.67	8.49	8.22	8.49	8.22	8.49	8.49	8.19	8.49	8.22	8.49
10°	8.11	7.40	8.44	8.43	8.81	8.57	8.84	8.74	8.26	8.29	7.89	8.08
20°	7.73	7.26	8.20	8.52	9.14	9.22	9.25	8.95	8.30	8.19	7.58	7.88
30°	7.30	7.03	8.38	8.72	9.53	9.49	9.67	9.22	8.34	7.99	7.19	7.14
40°	6.73	6.73	8.30	8.92	9.99	10.08	10.34	9.56	8.41	7.78	6.73	6.53
50°	5.98	6.32	8.25	9.25	10.69	10.93	10.90	10.00	8.44	7.43	6.07	5.65
60°	4.67	5.70	8.05	9.66	11.72	12.39	12.33	10.72	8.57	7.00	5.04	4.15

The monthly consumptive use is then obtained as

$$u = k \,.\, f \qquad \text{...(7.37)}$$

where k is an empirical crop coefficient. The monthly consumptive use u are added for all the months of the crop to yield the seasonal consumptive use or the total evapotranspiration in inches. The value of k depends on the month and the place. Average values of k for some selected crops are given in Table 7.5. These values may be used in the computations in the absence of more reliable values which can be obtained only from local experimental data.

Table 7.5. Values of the crop factor k for use in Blaney-Criddle Equation

Crop	*Jan*	*Feb*	*Mar*	*Apr*	*May*	*Jun*	*Jul*	*Aug*	*Sep*	*Oct*	*Nov*	*Dec*
Rice	—	—	—	0.85	1.00	1.15	1.30	1.25	1.10	0.90	—	—
Maize	—	—	—	0.50	0.60	0.70	0.80	0.8	0.6	0.5	—	—
Wheat	0.50	0.70	0.75	0.70	—	—	—	—	—	—	—	—
Sugarcane	0.75	0.80	0.85	0.85	0.90	0.95	1.00	1.00	0.95	0.90	0.85	0.75
Cotton	—	—	—	0.50	0.60	0.75	0.90	0.85	0.75	0.55	0.50	0.50
Vegetables	0.50	0.55	0.60	0.65	0.70	0.75	0.80	0.80	0.70	0.60	0.55	0.50

The Blaney-Criddle equation gives reasonably accurate estimates of evapotranspiration provided a locally developed crop coefficient is used. However it takes only temperature and daylight hours into account and the other important factors like humidity and wind are ignored.

Example 7.6. *The mean monthly temperature in May and November at a place of latitude 20°N are 28.4° and 20.2° centigrade respectively. Determine the evapotranspiration for sugarcane crop at this place for these two months.*

Solution. From Table 7.4 for a place of latitude of 20°N

p in the month of May = 9.14 and

p in the month of Nov. = 7.58.

From Table 7.5, the value of the crop coefficient k for sugarcane in the month of May is 0.90 and in the month of November it is 0.85.

Temperature in May = 28.4°C = 83.12°F

Temperature in November = 20.2°C = 68.36°F

$$\text{The consumptive use factor for May} = f = \frac{9.14}{100} \times 83.12$$

$$= 7.597 \text{ inches}$$

Evapotranspiration in May = 0.90 × 7.597 = 6.837 inches

= 17.37 cm

$$\text{The consumptive use factor for Nov.} = f = \frac{7.58}{100} \times 68.36 = 5.182 \text{ inches}$$

Evapotranspiration in Nov. = 5.182 × 0.85 = 4.404 inches
= 11.19 cm.

Thornthwaite Equation. Thornthwaite defined the heat index i_m for any month m as

$$i_m = \left(\frac{t_m}{5}\right)^{1.514} \qquad m = 1, 2, \ldots\ldots\ldots, 12 \qquad \ldots (7.38)$$

where t_m is the mean monthly temperature in °C. The twelve monthly heat indices are then added to obtain the annual heat index I. That is

$$I = \sum_{m=1}^{12} i_m \qquad \ldots(7.39)$$

The monthly potential evapotranspiration for any month is then calculated from the equation

$$PET = 1.6\, b \left(\frac{10\, t_m}{I}\right)^a \qquad \ldots(7.40)$$

where a is an exponent given by the equation

$$a = 6.75 \times 10^{-7} I^3 - 7.71 \times 10^{-5} I^2 + 1.792 \times 10^{-2} I + 0.49239 \qquad \ldots(7.41)$$

and b is a factor to correct for unequal day length between months. The value of b is obtained from the expression

$$b = \frac{\text{possible sunshine hours for the particular month}}{12 \times 30\,(\text{or } 31)}$$

The evapotranspiration given by Eq. (7.40) will be in cm/month.

The Thornthwaite equation suffers from the same limitations as that of Blaney-Criddle equation as it considers temperature and daylight hours as the only factors influencing the evapotranspiration. However, it is found to give reasonable estimates of *PET* in the temperate, continental climate of North America where it was originally developed because there the temperature and radiation are strongly correlated. In other parts of the world this equation has been less successful. Nevertheless, it is more widely used because it is simple.

Example 7.7. *The mean monthly temperatures in °C at a place of latitude 20°N from January to December are 16.6, 18.5, 23.3, 27.6, 28.4, 25.8, 24.4, 23.8, 23.5, 23.6, 20.2 and 17.1 respectively. Compute the evapotranspiration at this place for the months of May and November.*

Solution. The computation of monthly heat indices is shown in Table 7.6 from which the annual heat index I is found to be 120.072.

Table 7.6 Computation of monthly heat indices in Thornthwaite equation

Month	*Monthly mean temperature °C* t_m	*Monthly heat index* $i_m = \left(\frac{t_m}{5}\right)^{1.514}$
Jan	16.6	6.152
Feb	18.5	7.249
Mar	23.3	10.279
Apr	27.6	13.283
May	28.4	13.870
Jun	25.8	11.994
Jul	24.4	11.022
Aug	23.8	10.614
Sep	23.5	10.413
Oct	23.6	10.480
Nov	20.2	8.281
Dec	17.1	6.435
		$I = \Sigma\, i_m = 120.072$

$$a = 6.75 \times 10^{-7} (120.072)^3 - 7.71 \times 10^{-5} (120.072)^2 + 1.792 \times 10^{-2} (120.072) + 0.49239$$
$$= 1.1685 - 1.11157 + 2.15169 + 0.49239$$
$$= 2.701.$$

From Table 7.4, for May $p = 9.14$

$$\text{Possible sunshine hours in May} = \frac{9.14}{100} \times 365 \times 12 = 400.332$$

$$\therefore \quad \text{For May} \qquad b = \frac{400.332}{12 \times 31} = 1.076$$

$$PET \text{ in May} = 1.6 \times 1.076 \times \left(\frac{10 \times 28.4}{120.072}\right)^{2.701} = 17.61 \text{ cm.}$$

From Table 7.4 again, for November $p = 7.58$

$$\text{Possible sunshine hours in November} = \frac{7.58}{100} \times 365 \times 12 = 332.004$$

$$\therefore \quad \text{For November } b = \frac{332.004}{12 \times 30} = 0.9222$$

$$PET \text{ in November} = 1.6 \times 0.9222 \times \left(\frac{10 \times 20.2}{120.072}\right)^{2.701} = 6.0135 \text{ cm.}$$

Christiansen Equation. The christiansen equation for the estimation of the potential evapotranspiration is written as

$$PET = 0.473\, Q_0 \cdot C \qquad \text{...(7.42)}$$

where Q_0 is the daily solar radiation at the top of the atmosphere converted into equivalent depth of evaporation in mm, C is a coefficient derived from a series of climatic measurements, and PET is in mm/day. Specifically

$$C = C_T C_H\ C_W C_S\ C_E \qquad \text{...(7.43)}$$

The coefficients in Eq. (7.43) are evaluated using the meteorological data. The temperature coefficient C_T, the humidity coefficient C_H, the wind coefficient C_W, the sunshine coefficient C_S and the elevation coefficient C_E are given by the following equations.

$$C_T = 0.393 + 0.5592\left(\frac{T}{20}\right) + 0.0476\left(\frac{T}{20}\right)^2 \qquad \text{...(7.44)}$$

$$C_H = 1.25 - 0.348\left(\frac{H_n}{40}\right) + 0.12\left(\frac{H_n}{40}\right)^2 - 0.0218\left(\frac{H_n}{40}\right)^4 \qquad \text{...(7.45)}$$

$$C_W = 0.708 + 0.3276\left(\frac{W}{96.56}\right) - 0.036\left(\frac{W}{96.56}\right)^2 \qquad \text{...(7.46)}$$

$$C_S = 0.542 + 0.64\left(\frac{S}{80}\right) - 0.4992\left(\frac{S}{80}\right)^2 + 0.3174\left(\frac{S}{80}\right)^3 \qquad \text{...(7.47)}$$

$$C_E = 0.97 + 0.03\left(\frac{E}{304.8}\right) \qquad \text{...(7.48)}$$

in which T is the mean daily temperature in °C, H_n is the relative humidity measured at noon in per cent, W is the mean daily wind velocity in km/day at 0.6 m height, S is the per cent of bright sunshine hours given by $\left(\frac{n}{N} \times 100\right)$, and E is the elevation above mean sea level in m.

If the monthly potential evapotranspiration is computed for one month, directly from Eq. (7.42), then Q_0 is taken as the monthly solar radiation and all the coefficient must be evaluated using the respective mean monthly meteorological parameters. In addition the value thus obtained must be multiplied by another coefficient C_M whose value ranges from 0.90 to 1.10 depending on the latitude of the place.

Penman Equation. The Penman equation for estimating evaporation from a water surface, that is Eq. (7.23), is based on the most complete theoretical approach. The evaporation E obtained from Eq. (7.23) is multiplied by a coefficient K to yield the potential evapotranspiration.

$$PET = K \cdot E \qquad \text{...(7.49)}$$

Penman developed values of K for short grass which varied from 0.6 to 0.8 depending on the month of the year. For better estimates of PET it is essential that the values of K for particular crop be developed locally.

Alternatively, the Penman equation (7.23) can be used for direct evaluation of PET by using an appropriate value of albedo for the vegetation surface under consideration. For fresh green vegetations the value of albedo may be taken as 0.25.

Example 7.8. *Estimate the daily potential evapotranspiration using the data of Example 7.2 for a crop grown in that place. Take the albedo for the crop as 0.25.*

Solution.

$$Q_n = Q_i (1 - r) - Q_b$$

$$= 564.42 (1 - 0.25) - 136.58$$

$$= 286.735 \text{ cal/cm}^2$$

$$\Delta = 1.494 \text{ mm of mercury/°C}$$

$$E_a = 0.7783 \text{ cm/day}$$

$$L_e = 582.636 \text{ cal/gm}$$

$$P = 752 \text{ mm of mercury}$$

$$PET = \frac{\dfrac{Q_n \Delta}{L_e} + 0.00061\, P.E_a}{\Delta + 0.00061\, P}$$

$$= \frac{\dfrac{286.735 \times 1.494}{582.636} + 0.00061 \times 752 \times 0.7783}{1.494 + 0.00061 \times 752}$$

$$= \frac{0.73525 + 0.357}{1.9527} = 0.56 \text{ cm/day}$$

$$= 5.6 \text{ mm/day.}$$

Integration Method. In this method the total evapotranspiration from a given area is determined as the summation of evapotranspirations from the individual crops, natural vegetation, bare land, and water surface existing in that area. It requires the knowledge of evapotranspiration rates from the split up areas which may be obtained through the empirical equations described earlier.

Groundwater Fluctuations Method. Daily rise and fall of water table give an indication of evapotranspiration losses. This is especially true when vegetation obtains most of its moisture for growth from the capillary fringe above the water table. This method has been used by the United States Geological Survey and the empirical equation developed by them is

$$ET = Y_s (24a + b) \qquad \text{...(7.50)}$$

where ET is evapotranspiration in inches/day, Y_s is the specific yield of the soil, a is the rate of rise of water table from midnight to 4 A.M. in inches/h, and b is the net change in water table elevation during the day in inches.

7.8.5 Evapotranspiration Control

Evapotranspiration may possibly be reduced from a region through the following measures.

(*i*) By destroying the unwanted plants that transpire efficiently

(*ii*) By breeding plant varieties that transpire less

(*iii*) By reducing air movement over a crop by wind breaks of interplanted rows of taller plants

(*iv*) By applying chemical antitranspirants

(*v*) By placing watertight moisture barriers or water-retardant mulches on the soil surface

(*vi*) By appropriate land use changes in the catchment which increase streamflow.

The attempts of reducing evapotranspiration through land use changes should be undertaken only after thorough study of possible related side effects such as high rate of erosion and increased peak flow.

REVIEW EXERCISES

Descriptive Questions

1. Discuss the factors affecting evaporation.
2. Describe various methods of estimating evaporation from water bodies.
3. Derive the Penman's evaporation equation for open water surfaces. Explain how Penman's equation differs from the other empirical equations.
4. Define pan coefficient. Discuss the relative merits and demerits of sunken, floating and land pans.
5. Describe the ISI standard evaporation pan with a neat sketch. In what way it is different from USWB class A land pan ?
6. Describe how the evaporation is measured using atmometers.
7. Discuss various methods of reducing evaporation from a water body.
8. Evaporation is indirectly a cooling process. Justify the statement.
9. Visit the nearest meteorological observatory in your area and find out what type of evaporation pan is used there. Collect the evaporation data and compute the monthly and annual evaporations of that place.
10. Find out how the evaporation losses are accounted for in a major water resources project of your state.
11. Bring out the difference between evaporation, transpiration, evapotranspiration and consumptive use.

12. Discuss the various factors affecting evapotranspiration.
13. Distinguish between the potential evapotranspiration and the actual evapotranspiration.
14. Explain how the evapotranspiration can be estimated using the Blaney-Criddle and Thornthwaite equations.
15. How is transpiration measured from an experimental field plot ?
16. Explain how transpiration is measured with a phytometer.
17. Explain the use of a lysimeter in measuring evapotranspiration.
18. How is evapotranspiration estimated using Penman's equation ?
19. Explain the integration method of computing evapotranspiration.
20. How can evapotranspiration be reduced ?

Analytical Questions

1. Estimate the daily evaporation from a large reservoir using the empirical equations of Horton, Meyer and Rohwer from the following data :

   Water surface temperature = 28°C ; Air temperature = 30°C

   Atmospheric pressure = 766 mm of mercury ; Relative humidity = 42% ;

   Wind speed at 0.5 m above ground level = 7.5 m/s.
2. Estimate the evaporation from a free water surface on November 30th from the following data using Penman's equation :

   Latitude of the place : 18° 45 N ; Air temperature = 20°C

   Actual vapour pressure in air = 15.0 mm of mercury

   Actual sunshine hours : 5.5 ; Atmospheric pressure : 755 mm of mercury

   Wind velocity 2 m above ground : 1.8 m/s.
3. Compute the weekly evaporation from a reservoir using the water-budget method from the following data recorded during the week :

   Average inflow into the reservoir = 32.5 m^3/s

   Average outflow from the reservoir = 40.2 m^3/s

   Rainfall during the week = 73.6 mm

   Average waterspread area = 15.8 km^2

   Estimated seepage = 0.25 million m^3

   Storage at the beginning of the week = 9180 ha-m

   Storage at the end of the week = 8630 ha-m.
4. The main canal of a major irrigation project runs for a length of 38 km and has a lined trapezoidal section of bed width 8 m and side slopes 1 : 1. The depth of flow during the rabi season is maintained between 1.75 m and 2.5 m. The class A pan evaporation for the season is 61.1 cm. What is the conveyance loss in the canal during the season.

5. The average waterspread areas in the reservoirs and the pan evaporation at the reservoir site for twelve months are given below.

Month	*Water spread area in hectares*	*Pan evaporation in cm*
Jan	681	11.5
Feb	623	17.2
Mar	590	28.5
Apr	578	34.3
May	567	31.5
Jun	560	20.0
Jul	574	17.2
Aug	598	15.7
Sep	630	15.7
Oct	664	17.2
Nov	696	14.3
Dec	716	11.5

The pan coefficient may be assumed as 0.8. If the water in the reservoir fetches at the rate of Rs. 0.75 per m^3, how much additional revenue accrues to the project per year if the evaporation losses are reduced to 25% ?

6. If 8.75 litres of water is added to an evaporation pan of 1.22 m diameter to bring the water surface to the stipulated level and if a nearby raingauge measured a rainfall of 4.2 mm. What is the evaporation recorded for the day ?

7. What is the evaporation, if 4.2 litres of water is removed from an evaporation pan of diameter 1.22 m and the simultaneous rainfall measurement is 8.75 mm ?

8. The mean monthly temperatures during July to October at a place of latitude 10° N are 25.3, 24.1, 22.0 and 22.6 degrees C respectively. Determine the consumptive use of vegetables grown during this period using Blaney-Criddle formula.

9. The mean monthly temperatures at a place of latitude 10° N from January to December are 20.5, 22.8, 28.8, 34.0, 35.1, 31.9, 30.1, 29.4, 29.0, 29.1, 25.0 and 21.1 degrees C respectively. Compute the evapotranspiration at this place using the Thornthwaite equation for the months of April and December.

10. Taking the albedo of the crop as 0.25 and using the Penman's equation. Estimate the potential evapotranspiration for the data of Analytical Question 2.

Objective Type Questions

1. According to Dalton's law, evaporation is proportional to
   (*a*) the vapour pressure gradient
   (*b*) the difference between the saturation vapour pressure at 100°C and the actual vapour pressure
   (*c*) the difference between the actual vapour pressure and the saturation vapour pressure at 0°C
   (*d*) the difference between the saturation vapour pressure at a given temperature and the saturation vapour pressure at 0°C.
2. The salinity in water
   (*a*) reduces the evaporation (*b*) does not affect evaporation
   (*c*) increases the evaporation (*d*) difficult to say.
3. In terms of the saturation vapour pressure e_s corresponding to water surface temperature, and the actual vapour pressure in air e_a, the Dalton's evaporation law may be written as
   (*a*) $E = C\,(e_s + e_a)$ (*b*) $E = C\,(e_s - e_a)$
   (*c*) $E = C \cdot \frac{e_s}{e_a}$ (*d*) $E = C\,.\,e_s\,.\,e_a$
4. The pan evaporation is denoted by E_p and the actual evaporation by E. Then
   (*a*) $E_p > E$ (*b*) $E > E_p$
   (*c*) $E = E_p$ (*d*) $E \lesseqgtr E_p$.
5. The pan coefficient is defined as
   (*a*) $\frac{E}{E_p}$ (*b*) $\frac{E_p}{E}$
   (*c*) $(E_p - E)$ (*d*) $E_p . E$.
6. Which of the following has the largest pan coefficient
   (*a*) USWB class A land pan (*b*) Sunken pan
   (*c*) Floating pan (*d*) ISI standard pan.
7. The average pan coefficient of USWB class A land pan is equal to
   (*a*) 0.70 (*b*) 0.75
   (*c*) 0.85 (*d*) 0.95.
8. What is the average daily evaporation loss from a reservoir with an average water spread area of 15 km^2 in a month having lake evaporation of 20 cm.
   (*a*) 200000 m^3 (*b*) 300000 m^3
   (*c*) 100000 m^3 (*d*) 150000 m^3.

9. The wind velocity at a height of 2 m above the ground is 15 km/h. What would be the velocity at a height of 10 m above the ground ?
   (*a*) 19 km/h (*b*) 75 km/h
   (*c*) 3 km/h (*d*) 50 km/h.
10. The chemical compound which is generally used to reduce the evaporation from water surface is
   (*a*) D.D.T. (*b*) Alum
   (*c*) Cetyl alcohol (*d*) Potassium dichromate.
11. Atmometer is used to measure
   (*a*) evaporation (*b*) transpiration
   (*c*) evapotranspiration (*d*) all the above.
12. The evaporation through plants and from the surrounding soil together is called
   (*a*) hydration (*b*) vapourisation
   (*c*) transpiration (*d*) evapotranspiration.
13. Lysimeter is an instrument used to measure
   (*a*) evaporation (*b*) infiltration
   (*c*) evapotranspiration (*d*) transpiration.
14. The albedo of green vegetation surface is usually taken as
   (*a*) 0.05 (*b*) 0.15
   (*c*) 0.25 (*d*) 0.35.

ANSWERS

Analytical Questions

1. 16.98 mm/day, 15.23 mm/day, 19.3 mm/day
2. 3.52 mm/day
3. 0.1112 m/week or 1.588 cm/day
4. 0.2 million m^3
5. Rs. 64, 35, 360
6. 11.69 mm
7. 5.16 mm
8. 470.13 mm
9. 39.076 cm and 4.46 cm
10. 2.656 mm/day.

Objective Type Questions

1. (*a*)	2. (*a*)	3. (*b*)	4. (*a*)	5. (*a*)
6. (*b*)	7. (*b*)	8. (*c*)	9. (*a*)	10. (*c*)
11. (*a*)	12. (*d*)	13. (*c*)	14. (*c*).	

Chapter **8**

INFILTRATION

8.1 INTRODUCTION

As described in the hydrologic cycle, infiltration is one of the ways through which the precipitation reaching the earth's surface is disposed of. The other ways include evaporation and evapotranspiration, overland flow, and surface runoff. Infiltration is the only source for groundwater replenishment and thus forms an important arc of the hydrologic cycle.

Infiltration may be defined as the process by which water enters the surface strata of the earth. The infiltrated water first meets the soil moisture deficiency, if any, and thereafter the excess water moves vertically downwards to reach the groundwater table. This vertical movement is called *percolation.*

A soil under given conditions has an upper limit on its absorbing capacity. The *infiltration capacity* of a soil under given conditions is defined as the maximum rate at which it is capable of absorbing water and is denoted by f. The actual infiltration observed in a given soil, f_a, will be equal to or less than its infiltration capacity f depending on whether or not the rate of source supply is more or less than the infiltration capacity. Specifically if i denotes the rate of rainfall, then

$$f_a = f, \quad \text{if} \quad i \geq f \quad \text{and}$$

$$f_a = i, \quad \text{if} \quad i < f.$$

The information regarding the infiltration rates is required in the estimation of surface runoff and groundwater recharge. Unfortunately f is not a constant but varies within the duration of storm as well as seasonally. The factors affecting the infiltration capacity are now discussed.

8.2 FACTORS AFFECTING INFILTRATION CAPACITY

The variations in the infiltration capacity are large. The infiltration capacity is influenced by many factors. Some factors contribute to special variation while the others to temporal variation.

Depth of Surface Detention and Thickness of Saturated layer. Infiltration takes place due to combined influences of gravity and capillary forces. The infiltration of water through a soil surface may be visualised as a flow through a large number of tiny pipes as shown in Fig 8.1. As

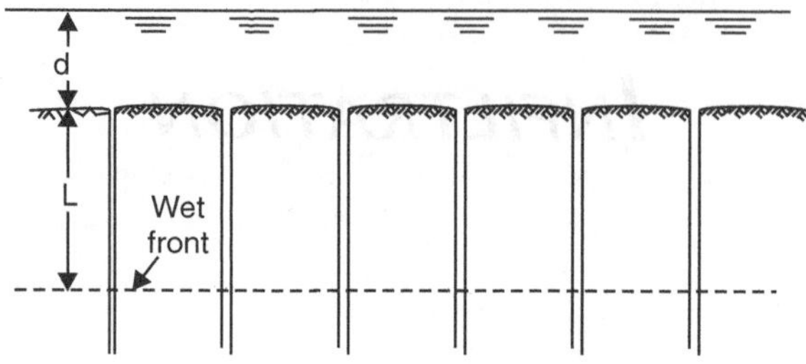

FIGURE 8.1 Infiltration process.

the infiltration continues the *wet front* will be travelling downwards. At any instant of time the resistance to flow is proportional to the thickness of the saturated layer up to wet front *L*, while the driving head is proportional to $(L + d)$, *d* being the depth of detention. At the beginning of the process when *L* is either less than *d* or of comparable magnitude with *d*, the resistance to flow is rather less and hence water enters rapidly. This is one of the reasons why *f* is large at the beginning of the process. When the thickness of saturated layer *L* becomes large and hence *d* becomes negligible compared to *L* any increase in *L* produces same effect both on resistance and driving head and so *f* remains constant. This suggests that the infiltration capacity should decrease with time in a continuous rain and become a constant, f_c, ultimately as shown in Fig. 8.2.

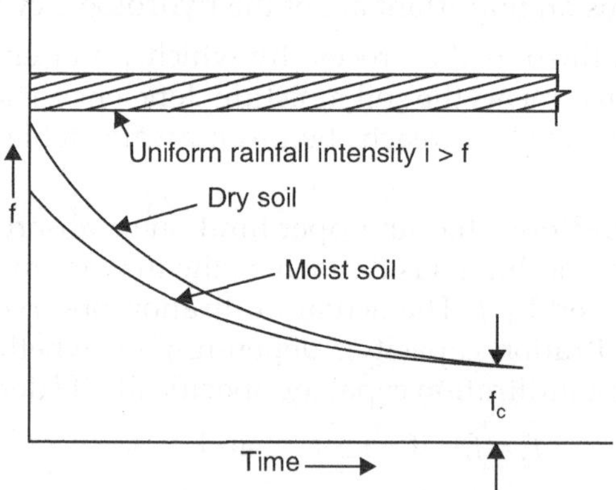

FIGURE 8.2 Variation of infiltration capacity with time.

Soil Moisture. If a soil is completely dry at the beginning of rain there is a strong capillary attraction for moisture in the subsurface layers that acts in the same direction as gravity and gives high initial value of infiltration. As water percolates down, the surface layers become semi-saturated and the capillary forces diminish and hence *f* also reduces. If the soil is moist to begin with, the infiltration starts with lesser capacity compared to that of a dry soil. These two trends are shown in Fig. 8.2. Another, aspect of soil moisture is that when the soil is subjected to wetting, very fine soil particles called *colloids* will swell slightly and reduce the size of voids. This is another reason why *f* reduces with time.

Compaction. The clay surfaced soils are compacted even by the impact of rain drops which reduce *f*. This effect is negligible on sandy soils. Compaction not only reduces the porosity but also the pore sizes. When the compaction is artificial due to man-made effects the initial infiltration capacity is very low and it is further reduced during the storm. Overgrazed pastures, playgrounds, and areas subjected to heavy vehicular traffic will have less infiltration capacities. Clay soils when exposed dry to the storm, apart from being compacted by the rain drop impact may become highly impermeable because of the in wash of fine particles along with rain water into the pores. This factor also reduces *f* during the storm. High infiltration capacities produced by ploughing and cultivation, burrowing animals and insects, and decay of vegetable matter such as roots are often reduced very rapidly by compaction due to rain.

Surface Cover Conditions. The nature of surface cover has also an important influence on the infiltration. The presence of a dense cover of vegetation on the surface increases *f*. The vegetative cover retards the movement of overland flow and causes high depths of detention. It reduces the effect of rain drop compaction. It also provides a layer of decaying organic matter which promotes the activity of burrowing insects and animals. Transpiration by vegetation tends to keep the soil moisture at low levels. All these factors tend to increase *f*.

Surfaces covered with snow, and paved urban areas will obviously have very low or zero infiltration capacity.

Temperature. The effect of temperature on infiltration is explained through viscosity. The flow through soil pores is by and large laminar for which the resistance is directly proportional to viscosity. At high temperatures since viscosity of water is low, high infiltration capacities are expected. In winter months when low temperatures prevail, the infiltration capacity is also low. This is one of the factors responsible for seasonal variations in *f*.

Others. The other factors which may marginally affect the infiltration capacity include the entrapped air in the pores, the quality of water, freezing, etc. The presence of entrapped air increases the resistance to flow and therefore reduces infiltration. The quality of water, particularly its turbidity in terms of clay and colloids it contains, and the salts it contains such as fertiliser residues from alkaline soils or other sources, may have diminishing influence on infiltration capacity.

8.3 ANNUAL AND SEASONAL CHANGES IN INFILTRATION CAPACITY

The infiltration capacity at a given location changes both seasonally and annually. Many factors influencing infiltration which are discussed earlier are not constant throughout the year. The seasonal variation in them, in turn, produces the seasonal variation in *f*. For example, soil moisture, temperature, soil cover produce high variability in *f*. Usually *f* will have higher values in summer because of low soil moisture and high temperatures.

The annual changes in *f* are mainly because of changes in land use pattern. Except for major changes in land use the variation occurs very slowly and it may not be noticeable. However, when an entire catchment or a large portion of it is subjected to sudden deforestation or intense agriculture the variation is very marked.

8.4 MEASUREMENT OF INFILTRATION

Infiltration rates are required in many hydrologic problems such as runoff estimation, soil moisture budgeting and in irrigation. There are two general approaches to the determination of infiltration rate. One of these approaches, usually referred to as hydrograph analysis method, analyses the observed rainfall hyetograph and the corresponding runoff hydrograph from a small plot or a natural watershed to estimate the infiltration rates. In the hydrograph analysis method the actual infiltration rate curve is obtained. But the derived estimate of infiltration is as accurate as the precision of the measurement of rainfall and runoff from the basin. The other method uses the *infiltrometers*. The infiltrometer always gives the infiltration capacity rate and the information from such infiltrometer tests at various locations in the basin may give a fairly satisfactory estimate of the average infiltration capacity rate for the entire basin as a whole.

8.5 INFILTROMETERS

Infiltrometers are of two types. They are (*i*) *flooding type infiltrometers* and (*ii*) *rainfall simulators*. In the flooding type of infiltrometers water is applied in the form of a sheet usually with a constant depth of flooding. They may use a single ring or two rings to delineate the sample area. In the former case it is known as a *simple infiltrometer* or a *tube infiltrometer*. In the latter case it is called a *double ring infiltrometer*. In the rainfall simulators water is applied by sprinkling at a uniform rate that is in excess of infiltration capacity.

Infiltrometers are used on small watersheds or on sample areas within a watershed. If there is a wide variation in soils and vegetation, the watershed is subdivided into relatively uniform sub-areas each having approximately a single soil-cover complex.

Tube Infiltrometer. This consists of a metal cylinder of diameter 25 to 30 cm and length 50 to 60 cm, with both ends open. It is driven into a level ground such that about 10 cm of the cylinder is above the ground as shown in Fig. 8.3. The surface of the soil inside the ring is protected by placing a perforated metallic dish over it. A pointer is set inside the ring to indicate the water level (corresponding to a stipulated depth of flooding of usually about 5 cm) to be maintained. The water is poured quickly into the ring till the water level touches the tip of the pointer and infiltration is allowed to proceed. As the water level goes down it is brought to the original level by adding required amount of water. The volumes of water added and the time intervals at which these volumes have been added are noted down to compute the infiltration rates. The observations are continued till almost uniform infiltration rate is obtained which may take 3 to 6 hours depending on the soil type.

A major drawback of the tube infiltrometer is that the infiltrated water percolates laterally at the bottom of the ring as shown in Fig. 8.3. Thus, the tube is not truly representing the area through which the infiltration is taking place. This drawback is rectified to a large extent in the double ring infiltrometer.

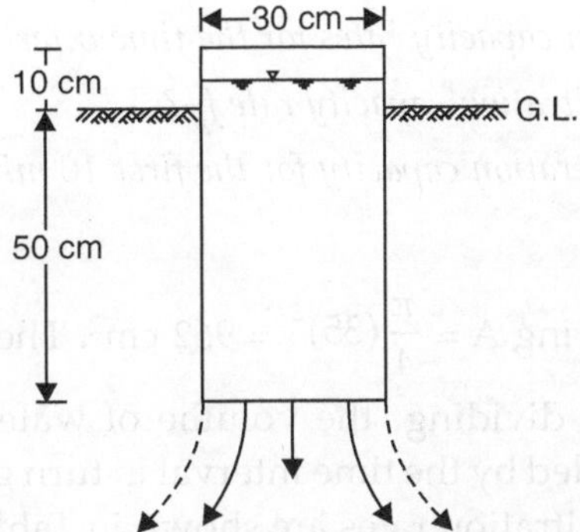

FIGURE 8.3 Tube infiltrometer.

Double Ring Infiltrometer. This is the most commonly used flooding type infiltrometer. It consists of two concentric rings driven into soil to a depth of about 15 cm uniformly without tilt and disturbing the soil to the least. The diameters of the rings may vary between 25 to 60 cm as shown in Fig. 8.4. Water is then applied in both the inner and outer rings to maintain a constant depth of about 5 cm. Water is replenished after the level falls by about 1 cm. The water depth in the inner and outer rings should be kept same during the observation period. Readings of volume of water added at successive time intervals to maintain constant depth of flooding in the inner ring are taken to facilitate the computation of infiltration capacity rate. As the purpose of outer ring is to suppress the lateral percolation of water from the inner ring, the water added to it need not be measured though water is added to maintain the same depth as in the inner ring. The experiment has to be carried out till a constant infiltration rate is observed. The surface of the soil may be protected by a perforated disk of sheet metal so that the turbidity of surface water is kept at minimum.

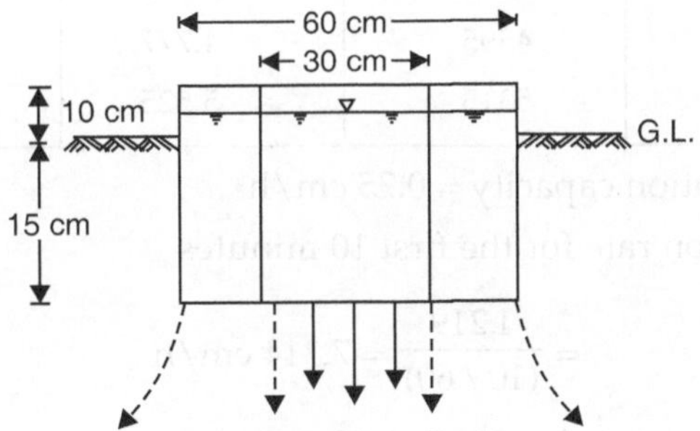

FIGURE 8.4 Double ring infiltrometers.

Example 8.1. *An infiltration test on a ring with 35 cm diameter yielded the following data.*

Times from the start, mts	0	2	5	10	20	30	60	90	150	210
Cumulative vol. of water added cm^3	0	278	658	1173	1924	2500	3345	3875	4595	5315

(a) *Determine the infiltration capacity rates for the time intervals in the experiment.*

(b) *What is the ultimate infiltration capacity rate* f_c *?*

(c) *What is the average infiltration capacity for the first 10 minutes and for the first 30 minutes of the experiment ?*

Solution. The area of the ring $A = \frac{\pi}{4}(35)^2 = 962$ cm^2. The depth of infiltration in any time interval is therefore obtained by dividing the volume of water added to the ring in that time interval by 962, which when divided by the time interval in turn gives the infiltration capacity rate. The computations to give the infiltration rates are shown in Table 8.1.

Table 8.1 Determination of Infiltration Rates

Time in minutes t	Δt *hours*	*Cumulative vol. of water added in cm*^{3} *V*	*Cumulative infiltration depth in cm* $F = V/A$	*Incremental infiltration* ΔF *in cm*	*Infiltration rate in cm/h* $f = \Delta F/\Delta t$
0		0			
2	0.033	278	0.289	0.289	8.76
5	0.050	658	0.684	0.395	7.90
10	0.083	1173	1.219	0.535	6.45
20	0.167	1924	2.000	0.781	4.68
30	0.167	2500	2.599	0.599	3.59
60	0.500	3345	3.477	0.878	1.76
90	0.500	3875	4.028	0.551	1.10
150	1.000	4595	4.777	0.749	0.75
210	1.000	5315	5.525	0.748	0.75

The minimum infiltration capacity = 0.75 cm/h

The average infiltration rate for the first 10 minutes

$$= \frac{1.219}{(10/60)} = 7.314 \text{ cm/h}$$

The average infiltration rate for the first 30 minutes

$$= \frac{2.599}{(30/60)} = 5.198 \text{ cm/h}$$

The infiltration rates are plotted against their time intervals as a bar diagram in Fig. 8.5. By joining the mid- points of the bars one would get the approximate instantaneous infiltration capacity curve. Alternatively the infiltration capacity curve can also be obtained by first plotting the cumulative infiltration of column (4) of Table 8.1 and measuring the slopes of the tangents drawn to it at different points of time as the infiltration capacity rates at the corresponding times.

Infiltrometer with adjustable depth of flooding. The depth of flooding in the infiltrometer can be maintained at any desired level and also the constant depth of flooding can be maintained automatically by having two feeder tanks as shown in Fig. 8.6 supplying water to the inner and outer rings. Each feeder tank has two outlet pipes one at the bottom called the *water pipe* and the other at the top called the *air pipe*. There is an opening at the top of the tank which is used to fill the tank and subsequently it is sealed with stop cock. The feeder tank is filled with water such that the level in the tank is just below the air pipe. The piezometric tube with a graduated scale attached to the tank indicates the water level inside. The diameter of feeder tank can be so selected that it results in a convenient magnification. For example, a tank with 15 cm diameter, when the diameter of inner ring is 30 cm, gives a fall of 4 cm corresponding to an infiltration depth of 1 cm. The open ends of the outlet pipes are kept in the respective rings such that they are almost in level with the water surface in the rings. The water pipe will be slightly dipping into water, while the air pipe is in level with the water surface the difference in elevation being of the order of 1 mm. The feeder tanks rest on platforms that can be lowered or raised by means of the screws fixed to their supporting stands. This arrangement facilitates the maintenance of any depth of flooding in the rings.

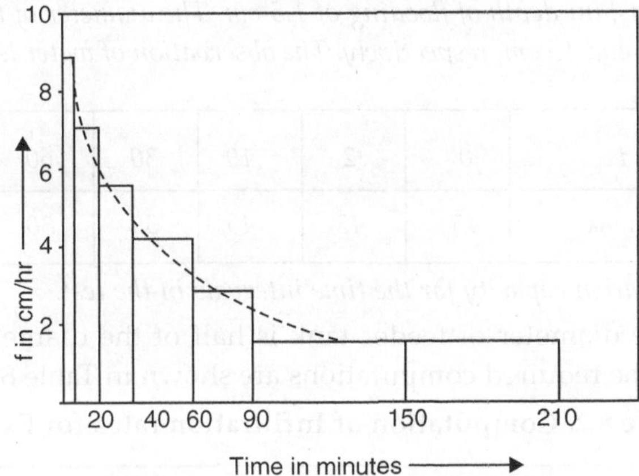

FIGURE 8.5 Infiltration capacity curve.

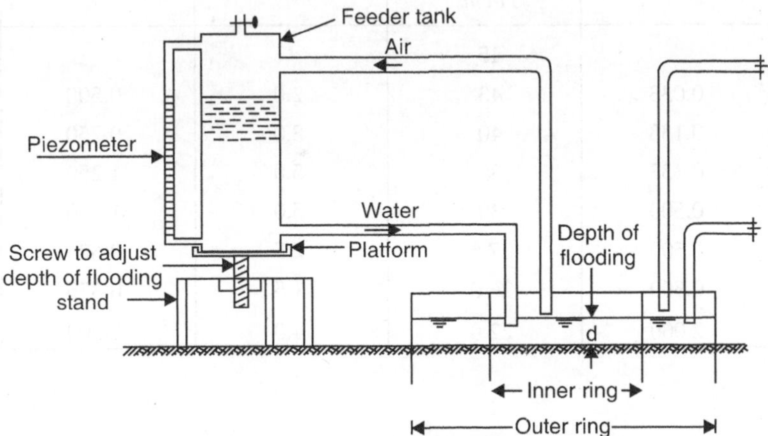

FIGURE 8.6 Double ring infiltrometer with automatic and adjustable depth of flooding.

The infiltration measurement is carried out as follows. At the site selected the ground is levelled and the rings are driven concentrically. The two feeder tanks are filled with water by closing the ends of the water pipes with fingers. Now the platforms of the tanks are adjusted so that the ends of air pipes are exactly at the required elevation corresponding to the stipulated depth of flooding. Water is now poured quickly into both the rings simultaneously till the water level just submerges the tips of the air pipes in both the rings. At this juncture the fingers which are closing the water pipes are removed. As the infiltration progresses ; the water level in the rings goes down exposing the end of the air pipe through which air rushes in and drives water through the waterpipes into the rings. So the water level in the rings starts increasing and when the level just touches the tip of the air pipe the path for the flow of air is cutoff and hence the flow of water from feeder tank stops. Again when the level in the rings falls because of infiltration the above mechanism comes into play feeding required water to the rings. Thus, the depth of flooding in the rings is automatically maintained constant always. When the process has started, one has to note down continuously at regular time intervals the level in the tank feeding water to the inner ring.

Example 8.2. *Infiltration test is conducted on an automatic adjustable depth of flooding type double ring infiltrometer with a constant depth of flooding of 1.5 cm. The diameter of the inner ring and the tank feeding water to it are 30 cm and 15 cm, respectively. The observation of water level in the tank are as given below :*

Time in minutes since start t	0	2	10	30	60	90	120	240
Water level in cm in feeder tank	45	43	40	35	30	27.5	26	22

Determine the infiltration capacity for the time intervals in the test.

Solution. Since the diameter of feeder tank is half of the diameter of the inner ring the magnification ratio is 4. The required computations are shown in Table 8.2.

Table 8.2. Computation of Infiltration rates for Example 8.2

Elapsed time t minutes	Δt *hours*	*Level in feeder tank H cm*	ΔH *cm*	$\Delta F = \Delta H/4$ *cm*	$f = \Delta F/\Delta t$ *cm/h*
0		45			
2	0.033	43	2.0	0.500	15.000
10	0.133	40	3.0	0.750	5.625
30	0.333	35	5.0	1.250	3.750
60	0.500	30	5.0	1.250	2.500
90	0.500	27.5	2.5	0.625	1.250
120	0.500	26.0	1.5	0.375	0.750
240	2.000	22.0	4.0	1.000	0.500

Rainfall Simulators. In this type of infiltrometers water is applied on to a plot of about 2 m × 4 m in the form of drops resembling rainfall through series of nozzles and the resulting runoff is continuously measured. The experiment consists of three stages namely *calibration run, test run* and *analytical run*. During a calibration run a plastic sheet or a metal pan is placed over the plot. Since infiltration is suppressed the total runoff during this run equals the total rainfall, which on division by the duration of rainfall yields the uniform rate of applied rainfall. In a test run the water is applied (with the same nozzle opening and with same pressure as in the calibration run) at uniform rate and it is continued till the observed rate of runoff becomes constant. Though the rainfall is stopped the runoff continues for some more time afterwards because of the water stored in detention. The analytical run is started just when the water disappears from the soil surface and continued till a constant rate of runoff is obtained again. A typical plot of rainfall and runoff rates during a test run and analytical run is shown in Fig. 8.7.

The water balance equation at any time during the test run can be written as

$$P = Q + F + V_d + D_a \qquad \text{...(8.1)}$$

where P is the cumulative rainfall, Q is the cumulative runoff, V_d is the depression storage, and D_a is the water stored in detention, all expressed in terms of depth over the plot. Alternatively, rearranging the terms, the expression for cumulative infiltration depth F, may be written as

$$F = P - Q - V_d - D_a \qquad \text{...(8.2)}$$

P and Q can be obtained by integrating the rainfall hyetograph and runoff hydrograph. But unless V_d and D_a are known Eq. (8.2) cannot be solved. The analytical run is utilised to obtain the values of V_d and D_a. Since the analytical run is taken on the plot when it is saturated the infiltration

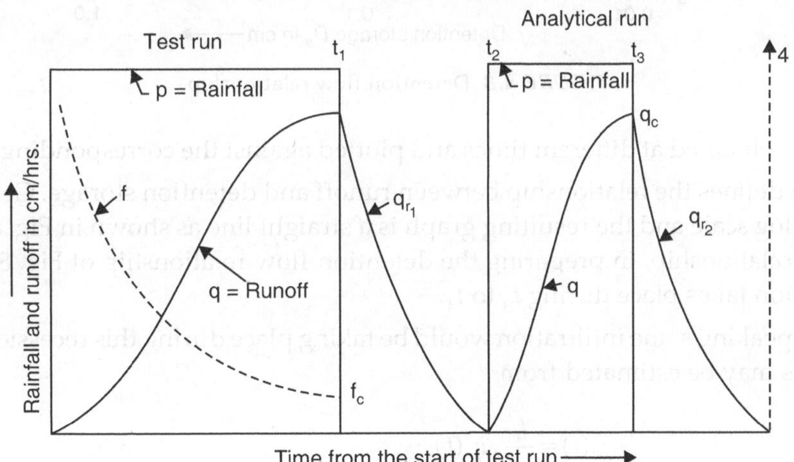

FIGURE 8.7 Hydrographs of test and analytical runs in rainfall simulator.

rate can be taken to be constant throughout this run and equal to $(p - q_c)$. Therefore for the period t_2 to t_3, the infiltration depth $F = (p - q_c)\,(t_3 - t_2)$. The cumulative rainfall $P = p\,(t_3 - t_2)$ and the cumulative runoff,

$$Q = \int_{t_2}^{t_3} q \, . \, dt$$

Since the rainfall has ceased at t_3, the runoff during t_3 to t_4 is essentially due to the detention storage, if infiltration is ignored during this period. Thus D_a at t_3 given by

$$D_a = \int_{t_3}^{t_4} q_{r2} \cdot dt$$

Then applying Eq. (8.2) for the analytical run at t_3, wet get

$$V_d = P - F - Q - D_a \qquad ...(8.3)$$

All the terms on the right hand side are known and hence the depression storage, which is a constant for a given plot, is evaluated. But the detention storage D_a will be varying with time. At any time t, let D_a (t) and q_r (t) be the detention storage and runoff rate respectively. Then

$$D_a(t) = \int_{t}^{t_4} q_r(t) \, . \, dt.$$

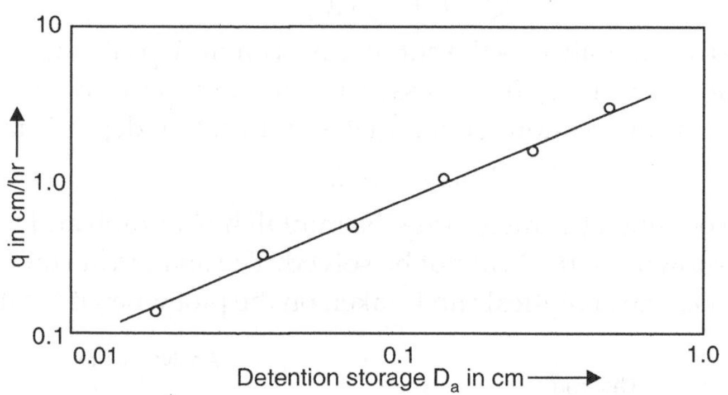

FIGURE 8.8 Detention flow relationship.

If $D_a(t)$ is obtained at different times and plotted against the corresponding runoff rates the resulting graph defines the relationship between runoff and detention storage. Generally this plot is made on log-log scale and the resulting graph is a straight line as shown in Fig. 8.8 and is called *Detention-Flow* relationship. In preparing the detention-flow relationship of Fig. 8.8 it is assumed that no infiltration takes place during t_3 to t_4.

Strictly speaking some infiltration would be taking place during this recession period. These infiltration rates may be estimated from

$$f_r(t) = \frac{f_c}{q_c} \, . \, q_r(t) \qquad ...(8.4)$$

The infiltration during recession is then added to the observed q_r to obtain revised estimate of D_a *Vs* q relationship. That is the graph is plotted between $D_a(t)$ *Vs* $[q_{r'}(t) + f_{r'}(t)]$ instead of $D_a(t)$ *Vs* $q_{r'}(t)$. Now Eq. (8.2) is applied successively at different times during the test run which would yield the graph between the cumulative depth of infiltration and time. The slope of a tangent drawn to this graph at any time would give the infiltration capacity rate at that time. The infiltration capacity rate curve for the test run is shown in Fig. 8.7.

8.6 HYDROGRAPH ANALYSIS

The infiltration rate curve as it occurs during a storm can be estimated through analysis of runoff hydrograph. Such analyses have distinct advantages over infiltrometer techniques because they consider variations in rainfall, vegetation, depression storage as they occur in nature. Several investigators have suggested various techniques of hydrograph analysis. All these techniques are essentially based on the mass balance Eq. (8.1) with some special procedures of obtaining the depression storage and detention storage components. Detailed discussion of these methods can be found in Handbook of Applied Hydrology by Ven Te Chow.

8.7 INFILTRATION EQUATIONS

The infiltration capacity rate curve as obtained from infiltrometer is essentially observed to be a decaying curve with a very high infiltration capacity rate at the beginning and decreasing in an exponential form to attain a minimum constant value. Various investigators have proposed mathematical expressions to describe the shape of this curve .Some of the most commonly used equations are given below.

Horton's equation. Horton developed a mathematical equation for defining the rate curve of infiltration capacity given by

$$f = f_c + (f_0 - f_c)\, e^{-kt} \qquad ...(8.5)$$

where f_0 is the initial infiltration capacity, t is the time since the start of rainfall, and k is a constant depending on type of soil and condition of vegetable cover. The effect of k on the rate curve is shown in Fig. 8.9.

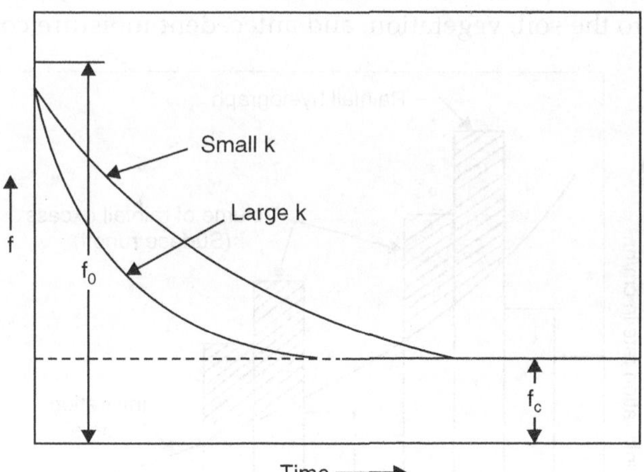

FIGURE 8.9 Effect of the parameter k on infiltration curve.

The minimum infiltration capacity f_c is related to the hydraulic conductivity of the soil. The parameters f_c and k may be regressed with physical characteristics of the watershed to make Eq. (8.5) more useful practically. The f_0 may be related to the *antecedent soil moisture.*

Philips equation. This is given by

$$f = A + \frac{S}{2} t^{-1/2} \qquad \text{...(8.6)}$$

where S and A are the parameters to be evaluated which are similar to f_0 and f_c of Horton's equation.

Kostiakov equation. Sometimes it is more convenient to present the infiltration data in the form of a cumulative curve than a rate curve. This curve may be called the *mass infiltration curve.* Kostiakov's equation for this curve is given by

$$F = a\, t^n \qquad \text{...(8.7)}$$

where F is the cumulative infiltration depth from the start of rainfall upto time t and a and n are the two parameters to be evaluated.

Holtan's equation. For large t, the cumulative infiltration F becomes a constant. It is called ultimate volume of infiltration and is denoted by F_c. The volume of potential infiltration F_p is given by $F_p = (F_c - F)$. The Holtan's infiltration equation is expressed as

$$f - f_c = a\, F_p^n \qquad \text{...(8.8)}$$

where a and n are constants. The value of a will vary from one watershed to the other but will be same for all the storms occurring on a given watershed. Its value will usually be in the range 0.2 and 0.8. Holtan used values of a = 0.62 and n = 1.387.

8.8 INFILTRATION INDICES

The infiltration capacity curves which are developed either from infiltrometer tests or the hydrograph analyses methods can be used to estimate the runoff from a given storm. The infiltration rate curve appropriate to the soil, vegetation, and antecedent moisture conditions existing at the

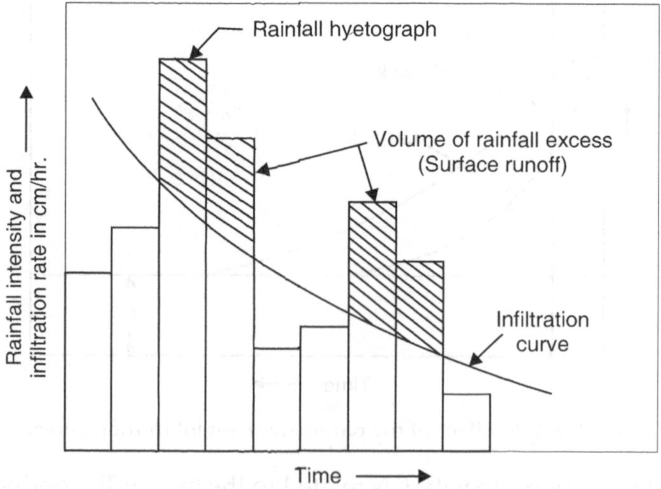

FIGURE 8.10 Relationship between rainfall, runoff and infiltration.

time of occurrence of storm is superimposed on the rainfall hyetograph with base lines coincident as shown in Fig 8.10. The area of the rainfall hyetograph above the infiltration curve would then represent the runoff volume whose time distribution may be obtained through the application of unit hydrograph principle. The rainfall volume below the infiltration curve represents the total depth of infiltration during the storm. The problem, however, lies in selecting the appropriate infiltration curve representative of the conditions existing at the start of the storm. Attempts have been made to correlate the parameters of the infiltration rate curve with soil types, land use patterns and the soil moisture. Soil moisture is often quantified through the *antecedent precipitation index*, which is discussed in Chapter 10.

Though this approach appears to be simple there are some difficulties. If the rainfall intensity is always more than the infiltration capacity the results are satisfactory. If the rainfall intensity fluctuates above and below the infiltration capacity rate curve the problem is complicated. The intensity of rainfall is rarely uniform both in time and space. If the rainfall is less than infiltration capacity the contribution to soil moisture is less than assumed and the drop in infiltration curve is correspondingly less. It is also likely that different infiltration capacity curves are required at different points in the basin depending on soils, vegetation and antecedent moisture.

The above difficulties led to the use of infiltration indices. These indices in general express the infiltration as an average rate throughout the storm. Since the infiltration capacity actually decreases with prolonged rainfall, the use of an average value assumes too little infiltration during the first part of the storm and too much near the end of it. Therefore infiltration indices are best suited for use in the analysis of major flood producing stroms occuring on wet soils, or stroms of such intensity and duration that the infiltration rate might be assumed to have reached a final constant rate prior to or early in the storm. Some of the commonly used infiltration indices are described below.

ϕ-index. The ϕ-index is an average rainfall intensity above which the rainfall volume equals the runoff volume. This is illustrated in Fig. 8.11. In this figure the rainfall hyetograph is plotted on a time base and a horizontal line is drawn such that the shaded area above the line exactly equals the measured runoff. Since the unshaded area below the line is also measured rainfall but did not appear as runoff it represents all the losses including depression storage, evaporation, interception as well as infiltration. However, infiltration is the largest loss compared to the other

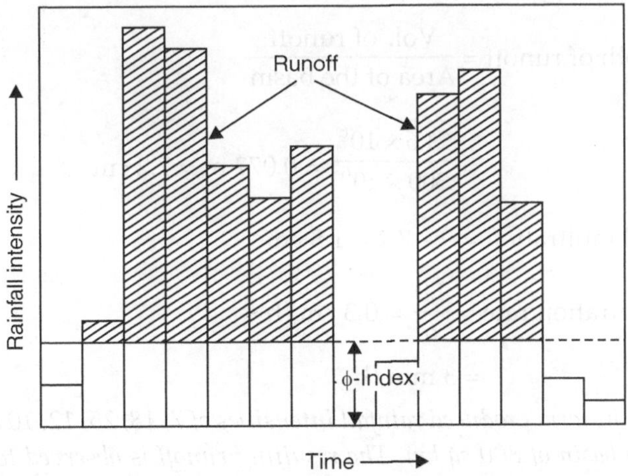

FIGURE 8.11 Definition sketch of ϕ-index.

losses. The ϕ-index can be determined for each flood event for which the runoff measurements are available. Then probably a relationship between the size of the flood, antecedent soil moisture and the ϕ-index could be developed. Since the data required to derive a precise infiltration rate curve for large watersheds is very large, the ϕ-index will be very useful in predicting the infiltration from a storm on large water - sheds. The ϕ-index is used in unit hydrograph studies to define the pattern of rainfall excess.

W-index. The W-index is a refined version of ϕ-index. It excludes the depression storage and interception from the total losses. It is the average infiltration rate during the time rainfall intensity exceeds the capacity rate. That is

$$W = \frac{F}{t} = \frac{(P - Q - S)}{t} \qquad ...(8.9)$$

where F is the total infiltration, t is time during which rainfall intensity exceeds infiltration capacity, P is total precipitation corresponding to t, Q is the total storm runoff and S is the volume of depression storage and interception. Thus W-index is essentially equal to ϕ-index minus the depression and interception storage. While the segregation of infiltration and retention seems to be a refinement, the task of estimating retention is so difficult that combining it with infiltration is probably equally satisfactory which explains the popularity of ϕ-index.

W_{min}-index. This is the lowest value of W-index which is observed under very wet initial conditions. Under these conditions since the retention rate is very low W-index and ϕ-index tend to be equal. This index is principally used in studies of maximum flood potential.

Example 8.3. *The total observed runoff volume during a 6 h storm with a uniform intensity of 1.5 cm/h is 21.6×10^6 m³. If the area of the basin is 300 km², find the average infiltration rate for the basin.*

Solution. Total rainfall = Intensity of rainfall × Duration

$$= 1.5 \times 6 = 9 \text{ cm}$$

$$\text{Volume of runoff} = 21.6 \times 10^6 \text{ m}^3$$

$$\text{Area of the basin} = 300 \text{ km}^2 = 300 \times 10^6 \text{ m}^2$$

$$\text{Depth of runoff} = \frac{\text{Vol. of runoff}}{\text{Area of the basin}}$$

$$= \frac{21.6 \times 10^6}{300 \times 10^6} = 0.072 \text{ m} = 7.2 \text{ cm}$$

$$\text{Total infiltration} = 9 - 7.2 = 1.8 \text{ cm}$$

$$\text{Average infiltration rate} = \frac{1.8}{6} = 0.3 \text{ cm/h}$$

$$= 3 \text{ mm/h.}$$

Example 8.4. *A 6 h storm produced rainfall intensities of 7, 18, 25, 12, 10 and 3 mm/h in successive one hour intervals over a basin of 800 sq km. The resulting runoff is observed to be 2640 hectare-metres. Determine ϕ-index for the basin.*

Solution. Total Vol. of runoff

$$= 2640 \text{ hectare-metres}$$
$$= 2640 \times 10000$$
$$= 264 \times 10^5 \text{ m}^3$$

Area of the basin = 800 km^2 = 800 × 10^6 m^2

$$\therefore \quad \text{The depth of runoff} = \frac{264 \times 10^5}{800 \times 10^6} = 0.033 \text{ m}$$
$$= 33 \text{ mm.}$$

The given rainfall hyetograph is shown in Fig. 8.12.

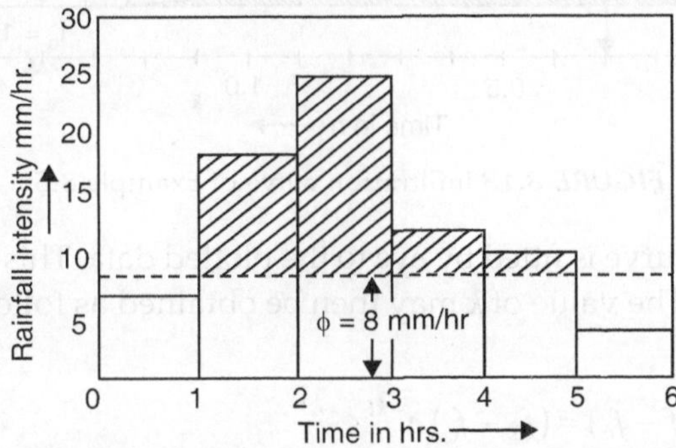

FIGURE 8.12 ϕ-index of Example 8.4.

By trial and error if ϕ-index is chosen to be 8 mm/h, the rainfall volume above the ϕ-index line is equal to

$$0 + (18 - 8) + (25 - 8) + (12 - 8) + (10 - 8) + 0$$
$$= 0 + 10 + 17 + 4 + 2 + 0 = 33 \text{ mm.}$$

This is exactly equal to the observed runoff.

Therefore ϕ = 8 mm/h

Example 8.5. *From a double ring infiltrometer the following infiltration rates were observed.*

Times in hours	0.0167	0.0583	0.125	0.25	0.50	0.75	1.25
f in cm/h	8.76	7.90	6.45	4.68	2.75	1.76	1.10

Determine the constants f_0, f_c and k of the Horton's equation to describe the above data.

Solution. The Horton's equation is given by

$$f = f_c + (f_0 - f_c)\, e^{-kt}$$

The constants f_0, f_c and k can be found out from the following two methods.

Method (*i*) : The given data is plotted on an ordinary graph as shown in Fig. 8.13

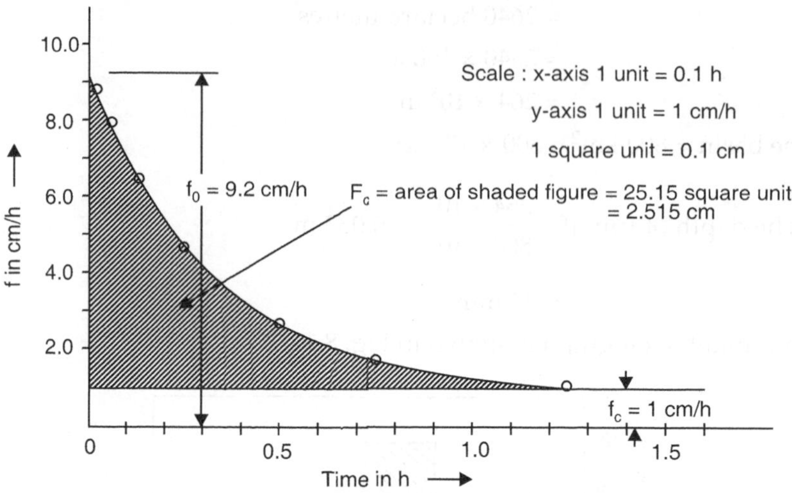

FIGURE 8.13 Infiltration curve of Example 8.5.

Then an exponential curve is fitted by eye to the plotted data. This yielded f_0 and f_c as 9.2 cm/h and 1.0 cm/h respectively. The value of k may then be obtained as follows. The Horton's equation is rewritten as

$$(f - f_c) = (f_0 - f_c)\, e^{-kt}$$

Integrating the above equation on both sides, we get

$$\text{L.H.S} = \int_0^{\infty} (f - f_c)\, dt = \text{the area of the shaded portion in Fig. 5.6}$$

$$= F_c$$

$$\text{R.H.S.} = \int_0^{\infty} (f_0 - f_c)\, e^{-kt} \,.\, dt = \frac{(f_0 - f_c)}{k}$$

From these two results we can determine k from the equation

$$k = \frac{f_0 - f_c}{F_c} = \frac{(9.2 - 1.0)}{2.515} = 3.26$$

The Horton's equation to describe the given data is therefore given by

$$f = 1.0 + 8.2\, e^{-3.26t}$$

Method (*ii*) : Taking logarithms to the natural base on both sides of the Horton's equation,

$$\ln (f - f_c) = \ln (f_0 - f_c) - kt$$

Letting y denote $\ln(f - f_c)$ and c denote $\ln(f_0 - f_c)$, the above equation assumes a linear form as

$$y = c - kt.$$

Now a graph between y and t may be plotted on an ordinary graph paper and a straight line is fitted by eye. The slope of this line defines the value of k while its intercept on y-axis gives the value of c from which f_0 can be evaluated. Alternatively a regression equation may be fitted between y and t using the procedure of Example 3.9 which is adopted in the present case. The value of f_c is taken to be 1 cm/h. The computations are set out in Table 8.3.

Table 8.3 Determination of constants in Horton's equation

t	f	$(f-f_c)$	$y = ln\,(f-f_c)$	$(y-\bar{y})^2$	$(t-\bar{t})^2$	$(y-\bar{y})(t-\bar{t})$
0.0167	8.76	7.76	2.0490	1.7961	0.1638	– 0.5424
0.0583	7.90	6.90	1.9315	1.4950	0.1318	– 0.4440
0.125	6.45	5.45	1.6956	0.9738	0.0879	– 0.2925
0.25	4.68	3.68	1.3029	0.3530	0.0294	– 0.1018
0.50	2.75	1.75	0.5596	0.0223	0.0062	–0.0117
0.75	1.76	0.76	– 0.2744	0.9667	0.1080	– 0.3231
1.25	1.10	0.10	– 2.3026	9.0685	0.6866	– 2.4952
Σ 2.95			4.9616	14.6764	1.2137	– 4.2167

$$\bar{t} = \frac{2.95}{7} = 0.4214$$

$$s_t = \sqrt{\frac{1.2137}{7}} = 0.4164$$

$$\bar{y} = \frac{4.9616}{7} = 0.7088$$

$$s_y = \sqrt{\frac{14.6764}{7}} = 1.448$$

$$r = \left(-\frac{4.2167}{7}\right)\Big/(0.4164 \times 1.448) = 0.999$$

$$b = \frac{r \,.\, s_y}{s_t} = -3.4739$$

$$(y - 0.7088) = -3.4739\ (t - 0.4214)$$

or
$$y = 2.1727 - 3.4739\ t$$

$$\ln\,(f_0 - f_c) = 2.1727 \qquad \therefore \qquad (f_0 - f_c) = 8.782$$

$$f_0 = 8.782 + 1 = 9.782 \text{ cm/h}$$

The Horton's equation is therefore given by

$$f = 1.0 + 8.782\ e^{-3.4739\,t}$$

Example 8.6. *The cumulative depth of infiltration in an experiment on a tube infiltrometer is observed to follow the equation $F = 0.165\ t^{0.65}$, where F is in cm and t is in minutes. Determine the equation for infiltration rate and the average infiltration rate.*

Solution. $F = 0.165\ t^{0.65}$

$$f = \frac{dF}{dt} = 0.165 \times 0.65 \times t^{-0.35} = 0.10725\ t^{-0.35}\ \text{cm/minute}$$

$$f = 6.435\ t^{-0.35}\ \text{cm/h, where } t \text{ is in minutes}$$

or

$$f = 1.5353\ t^{-0.35}\ \text{cm/h, where } t \text{ is in hours}$$

$$f_{av} = \frac{F}{t} = 0.165\ t^{-0.35}\ \text{cm/minute}$$

$$= 9.9\ t^{-0.35}\ \text{cm/h, where } t \text{ is in minutes}$$

$$= 2.362\ t^{-0.35}\ \text{cm/h, where } t \text{ is in hours.}$$

Example 8.7. *The Horton's infiltration equation for a basin is given by $f = 6 + 16\ e^{-2t}$ where f is in mm/h and t is in hours. What are the values of f_0, f_c and k ? If a storm occurs on this basin with an intensity of more than 22 mm/h determine the depth of infiltration for the first 45 minutes and the average infiltration rate for first 75 minutes.*

Solution. On comparison with the Horton's equation in the standard form one can observe that $f_c = 6$ mm/h, $f_0 = 22$ mm/h and $k = 2$/h.

Since the rainfall intensity is more than f_0, the infiltration takes place at the capacity rate throughout the storm. Hence the cumulative depth of infiltration for the first 45 minutes is given by

$$F = \int_0^{3/4} f.dt = \int_0^{0.75} \left(6 + 16e^{-2t}\right) dt$$

$$= \left[6t - 8e^{-2t}\right]_0^{0.75}$$

$$= [6 \times 0.75 - 8 \times e^{-1.5}] - [0 - 8]$$

$$= 12.5 - 8e^{-1.5} = 12.5 - 1.785$$

$$= 10.715\ \text{mm.}$$

The cumulative depth of infiltration for the first 75 minutes is given by

$$F = \int_0^{1.25} f.dt = \left[6t - 8e^{-2t}\right]_0^{1.25}$$

$$= [6 \times 1.25 - 8 \times e^{-2.5}] - [0 - 8]$$

$$= 15.5 - 0.657 = 14.843\ \text{mm.}$$

The average infiltration rate for the first 75 minutes is given by

$$f_{av} = \frac{F}{t} = \frac{14.843}{1.25} = 11.874\ \text{mm/h.}$$

8.9 INITIAL ABSTRACTIONS

Not all the precipitation reaching the ground surface becomes runoff. Apart from the evaporation and infiltration which are the major losses, there are two other processes known as *interception* and *depression storage* which reduce the volume of water available for runoff. Much of the rain falling during the initial period of the storm is stored on the vegetal cover as interception. Some more precipitation is requiring to fill up all the depressions before the overland flow commences. The interception and depression storage are together called the initial abstractions or the *initial losses*. The word *surface retention* is used to denote interception, depression storage and also the evaporation during the storm but it does not include water temporarily detained enroute to the streams.

Interception. The amount of precipitation intercepted by the vegetal cover depends on many factors such as density and composition of vegetation, storm characteristics and the prevailing wind at the time of storm. It is very difficult to exactly measure the interception. Interception is satisfied in the first part of the storm and it would cease after the vegetation is saturated. It is estimated that the annual interception by a well-developed forest may be anything between 10 to 20% of the annual rainfall. If the area experiences a large number of small stroms it can be even greater than 25 %. Fig. 8.14 shows the variation of interception loss expressed as per cent of rainfall depth with storm rainfalls. It can be seen from this figure that the interception loss is large for stroms with small rainfall and it approaches a small constant value for large storms.

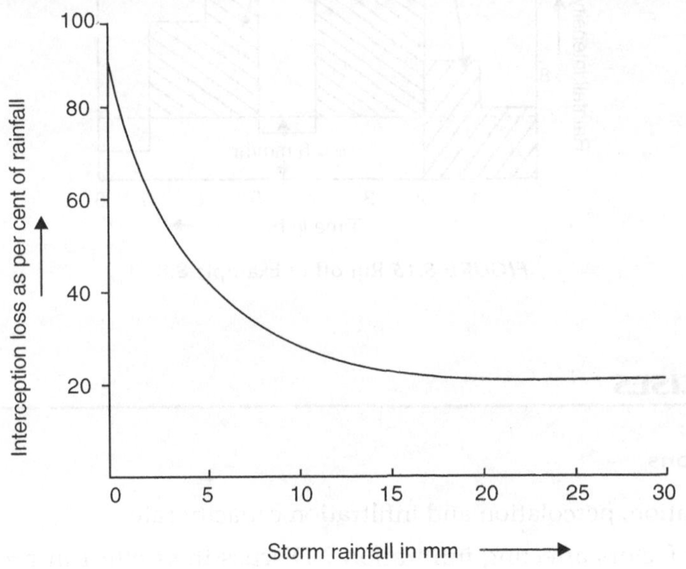

FIGURE 8.14 Interception loss curve.

Depression Storage. The water which goes to fill up the surface depressions is not available to runoff as it is eventually lost as either infiltration or evaporation. Thus depression storage is considered as loss from rainfall. The depression storage depends again on many factors such as the type and the nature of catchment, slope of the catchment and the antecedent precipitation that reflects the soil moisture. It is observed that the depression storage for most basins lie between 10 to 50 mm.

Example 8.8 *A 500 sq. km watershed received a 8 h storm which produced hourly intensities of 6, 9, 20, 16, 4, 14, 12 and 2 mm/h. If the initial abstractions are estimated to be 15 mm and ϕ-index is 5 mm/h, what would be the runoff volume produced by the storm ?*

Solution. The rainfall occurring in the first two hours is 15 mm which is same as the initial abstractions. Therefore no runoff is produced in this period. In the remaining 6 h period the rainfall excess is given by

$$(20 - 5) + (16 - 5) + 0 + (14 - 5) + (12 - 5) + 0 = 42 \text{ mm.}$$

$\therefore$ The runoff volume = Area of the basin × Runoff depth

$$= 500 \times 10^6 \times \frac{42}{1000} = 21 \times 10^6 \text{ m}^3$$

If the initial abstractions are neglected and ϕ-index is applied throughout the storm the runoff is overestimated by 5 mm resulting in an error of about 12%.

The basin produces surface runoff only from the end of 2nd hour.

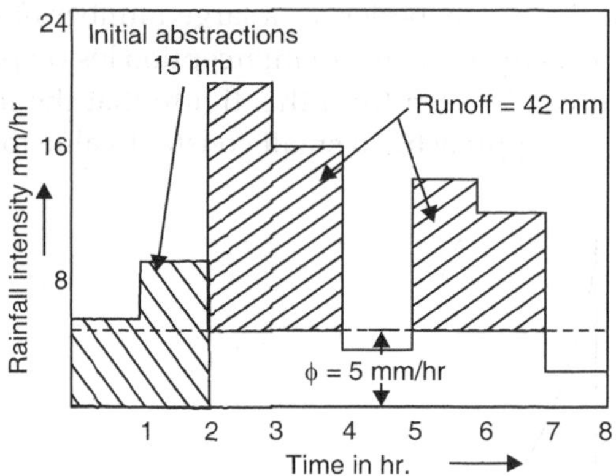

FIGURE 8.15 Runoff of Example 8.8

REVIEW EXERCISES

Descriptive Questions

1. Define infiltration, percolation and infiltration capacity rate.
2. What are the factors affecting infiltration ? Discuss their effect in producing variation in infiltration rate during a storm, and also in producing seasonal and spatial variations in infiltration rate.
3. Describe how infiltration capacity rate can be measured using a double ring infiltrometer. How is it better than a tube infiltrometer ?
4. Explain the working of a double ring infiltrometer with adjustable depth of flooding with the help of a neat sketch.

5. Describe the method of obtaining the infiltration capacity rate curve of an experimental plot using a rainfall simulator.
6. Discuss various infiltration equations. Explain how the constants f_c, f_0 and k in the Horton's equation can be obtained from the experimental data.
7. Define ϕ-index and W-index and bring out the difference between them. How is ϕ-index determined from the rainfall hyetograph ?
8. What are the initial losses ? How does the interception loss vary with the magnitude of storm rainfall ?
9. Show the effect of initial moisture and k on the shape of the infiltration curve described by Horton's equation with the help of sketches.
10. How is ϕ-index used in estimating the runoff from a storm ? How is the estimate of runoff affected by ignoring the initial loss ?

Analytical Questions

1. An infiltration test conducted on a double ring infiltrometer with an inner ring of diameter 30 cm yielded the following data

Time in minutes	0	5	10	30	60	120	180	240	300	360
Cumulative vol. of water add in cm³	0	46	90	246	435	662	842	1000	1154	1300

Determine the infiltration capacity rates for the time intervals in the test and plot the variation of infiltration capacity rate with time.

2. A seven hour storm produced the following rainfall intensities (in mm/h) at half an hour intervals over a basin of area 1830 km^2.

4, 9, 20, 18, 13, 11, 12, 2, 8, 16, 17, 13, 6 and 1.

If the corresponding observed runoff is 36.6 million m^3, estimate the ϕ-index for the storm.

3. A storm with a uniform intensity of 1.6 cm/h for a period of 8 hours occuring over a basin of area 650 km^2 produced a runoff estimated to be 57.2 million m^3. Find the average infiltration rate during the storm.
4. The ordinates of a rainfall mass curve of a storm over a basin of area 850 km^2 measured in mm at one hour interval are

0, 10, 22, 30, 39, 45.5, 50, 55.5, 60, 64 and 68

If the infiltration during this storm can be represented by Horotn's equation with $f_0 = 6.5$ mm/h, $f_c = 1.5$ mm/h and $k = 0.15$ /h, estimate the resulting runoff volume.

5. The cumulative depth of infiltration in an experiment on a tube infiltrometer is observed to follow the equation $F = 0.2\,t^{0.55}$, where F is in cm and t is in minutes. Determine the equation for infiltration rate and the average infiltration rate.

6. If the infiltration takes place at a capacity rate given by Horton's equation show that the average infiltration rate for the first t hours is given by

$$f_{av} = f_c + \frac{\left(1 - e^{-kt}\right)(f_0 - f_c)}{kt}.$$

7. The Horton's infiltration equation for a basin is given by $f = 10 + 30\, e^{-0.8t}$, where f is in mm/h and t is in hours. What are the values of f_0, f_c and k ? If a storm occurs on this basin with an intensity of more than 40 mm/h, determine the depth of infiltration for the first one hour and the average infiltration rate for the first two hours.

8. The infiltration rates observed during a test on a double ring infiltrometer are as given below.

t in hours	0.0417	0.125	0.333	0.75	1.5	2.5	3.5	4.5	5.5
f in cm/h	0.781	0.747	0.662	0.535	0.370	0.255	0.224	0.218	0.207

Determine the constants f_0, f_c and k of the Horton's equation which fits the above data.

9. A storm during a dry weather has rainfall intensities of 8, 12, 40, 38, 30, 26, 28, 5, 16, 32, 36, 24, 14 and 4 mm/h at half an hour intervals. What is the runoff volume from a basin area of 600 km^2 if the initial abstractions are 10 mm and the ϕ-index for the basin is 10 mm/h ?

What is the per cent error in runoff estimate if the initial abstractions are neglected ?

10. A basin has an area of 750 km^2. Forty per cent of it is impervious due to either rock outcrop or urbanisation. Thirty per cent is thick forest. The remaining is agriculture land. The ϕ-indices for these portions may be taken as 0, 15, and 7.5 mm/h respectively. An eight hour storm produced rainfall intensities of 6, 9, 20, 16, 4, 14, 12 and 2 mm/h in successive one hour intervals. Determine the corresponding runoff depth and runoff volume in million m^3.

Objective Type Questions

1. Infiltration capacity of the soil is defined as
   (*a*) the depth of water absorbed by the soil during the storm
   (*b*) the intensity of rainfall above which the rainfall volume equals the observed runoff volume
   (*c*) the maximum rate at which the soil absorbs water
   (*d*) the permeability of the soil in vertical direction.

2. In the standard notation, the Horton's infiltration equation is given by
   (*a*) $f = f_c + (f_0 - f_c)e^{-kt}$ (*b*) $f = f_0 + (f_c - f_0)\, e^{-kt}$
   (*c*) $f = f_c + (f_0 + f_c)\, e^{-kt}$ (*d*) $f = f_c + (f_0 - f_c)\, e^{kt}$

3. A 6 h storm with hourly intensities of 7, 18, 25, 12, 10, and 3 mm/h produced a runoff of 33 mm. Then the ϕ-index is
   (*a*) 7 mm/h (*b*) 3 mm/h
   (*c*) 10 mm/h (*d*) 8 mm/h.

4. ϕ-index is defined as
   (*a*) the difference between maximum and minimum infiltration capacities
   (*b*) the difference between the total rainfall and the total runoff divided by the duration of the storm
   (*c*) the rainfall intensity above which the rainfall volume equals the observed runoff volume
   (*d*) the minimum infiltration rate during the storm.
5. W-index will be always
   (*a*) equal to ϕ-index (*b*) more than ϕ-index
   (*c*) less than ϕ-index (*d*) a constant fraction of ϕ-index.
6. A 26 h storm with a uniform intensity of 1.5 cm/h produced a runoff depth of 72 mm. The average infiltration rate during this storm is
   (*a*) 3 mm/h (*b*) 6 mm/h
   (*c*) 9 mm/h (*d*) 12 mm/h.
7. A 6 h storm of uniform intensity and a rainfall of 8 cm produced a runoff of 5 cm. What is the runoff produced by a 12 h storm of uniform intensity and a total of 11 cm assuming ϕ-index to be same during both storms ?
   (*a*) 5 cm (*b*) 8 cm
   (*c*) 11 cm (*d*) 9 cm.
8. Under identical conditions if the infiltration capacity measured by double ring infiltrometer is f_d and that measured by a rainfall simulators is f_s, then which of the following is true ?
   (*a*) $f_s < f_d$ (*b*) $f_s > f_d$
   (*c*) $f_s = f_d$ (*d*) difficult to tell.
9. Interception loss when expressed as per cent of storm rainfall
   (*a*) increases with increase in storm rainfall
   (*b*) decreases with increase in storm rainfall
   (*c*) decreases with decrease in storm rainfall
   (*d*) is dependent of storm rainfall.
10. Which of the following has the highest infiltration capacity
   (*a*) forest land
   (*b*) grazed pasture
   (*c*) concrete pavement in air port
   (*d*) rock out crop.

ANSWERS

Analytical Questions

1. 7.812, 7.464, 6.621, 5.348, 3.211, 2.547, 2.235, 2.179 and 2.065 mm/h
2. 10 mm/h

3. 5 mm/h
4. 27.107 mm or 23.041 million m^3
5. $f = 0.11t^{-0.45}$ and $f_{av} = 0.2t^{-0.45}$
7. 30.65 mm ; 19.964 mm/h
8. $f_0 = 0.775$; $f_c = 0.20$ cm/h ; $k = 0.882$
9. 92 mm, + 1.09%
10. 45.05 mm or 33.79 million m^3

Objective Type Questions

1. (*c*) 2. (*a*) 3. (*d*) 4. (*c*) 5. (*c*)
6. (*a*) 7. (*a*) 8. (*a*) 9. (*b*) 10. (*a*).

Chapter 9

GROUNDWATER

9.1 INTRODUCTION

In general, the term *groundwater* or *subsurface water* refers to the water that occurs below the surface of earth. The main source of groundwater is infiltration. The infiltrated water after meeting the soil moisture deficiency percolates deeply and becomes groundwater. The groundwater is free from pollution and is very useful for domestic use in small towns and isolated farms. It can be made available at a small capital cost and also in least possible time. In arid regions, groundwater is often the only reliable source of water for irrigation. Groundwater has been an important water resource throughout the ages. Like any other natural resource, groundwater is also not unlimited. They must be wisely managed and protected against undue exploitation and contamination by pollutants or salt water. The subject of groundwater attracted the attention of agricultural engineers, civil engineers, geologists, geophysicists and scientists from various other disciplines. Excellent books are available on the subject. This chapter presents an elementary treatment of groundwater as a part of the hydrologic cycle.

The formation below the earth's surface is divided into two zones by an irregular surface called the *water table*. At all points on the water table the pressure is atmospheric. The zone between the ground surface and the water table is called the *unsaturated zone* or the *vadose zone*. In the vadose zone the soil pores may contain either air or water or both. Hence it is also called the zone of aeration. (vide Fig. 9.1). Water in this zone is held to the soil particles by capillary forces. Thus, while this water is able to move within the vadose zone, it cannot move out of the zone into wells or other places that are exposed to atmospheric pressure. In the zone below the water table all the soil pores are completely filled with water and hence it is called the *zone of saturation* or the *phreatic zone*. The phreatic zone may extend to considerable depth, but as the depth increases the weight of overburden tends to close the pore space and relatively little water is found at depths greater than 3 km. Since the water in the phreatic zone is under pressure more than atmospheric, if a hole is dug in this zone water begins to flow into the hole whereas it is not so in the vadose zone. It is this water in the phreatic zone which moves freely under the force of gravity that is actually called the groundwater, the occurrence and movement of which are the topics for discussion in this chapter. The water in the vadose zone is of interest to agronomists and agricultural engineers.

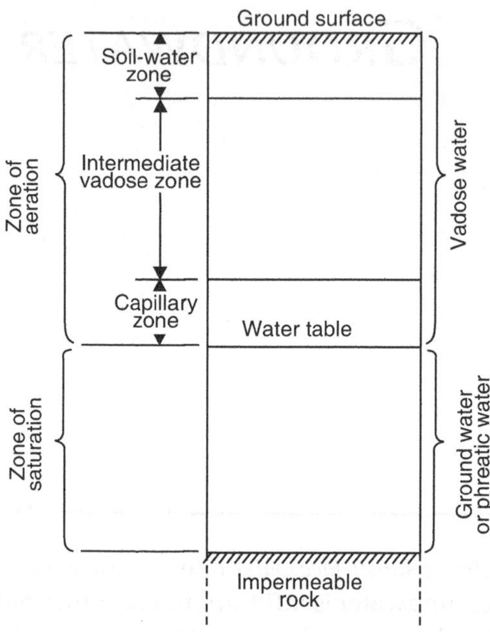

FIGURE 9.1 Definition sketch of zones of saturation and aeration.

9.2 OCCURRENCE OF GROUNDWATER

Groundwater occurs at various locations below the earth's surface depending on the physical properties of various formations that exist. Fig. 9.2 is a typical cross-section of the earth's crust showing the occurrence of groundwater.

Aquifer. Formations which contain groundwater and at the same time which are sufficiently permeable to transmit and yield water in usable quantities are called the aquifers. They may get recharged directly from above through the processes of precipitation and infiltration, or they may have the recharge area somewhere else on the earth's surface. The amount of water contained by the aquifer depends on the *porosity* of the aquifer formation while the amount of water that it can yield depends on its *permeability*.

Unconfined Aquifer. An aquifer having water table in it is called an unconfined aquifer. For this aquifer water table serves as the upper surface of zone of saturation while a less permeable or essentially impermeable layer defines its lower boundary. The impermeable layer underlying an unconfined aquifer may be made of clay, shale, solid limestone, igneous rock or bed rock. Unconfined aquifer is also known as a *free aquifer, phreatic aquifer, water table aquifer*, or *nonartesian aquifer*. The water table in an unconfined aquifer varies in undulating form and in slope, depending upon the areas of recharge and discharge, pumpage from wells and its permeability. A well driven into an unconfined aquifer will indicate a static water level corresponding to the water table level at that location.

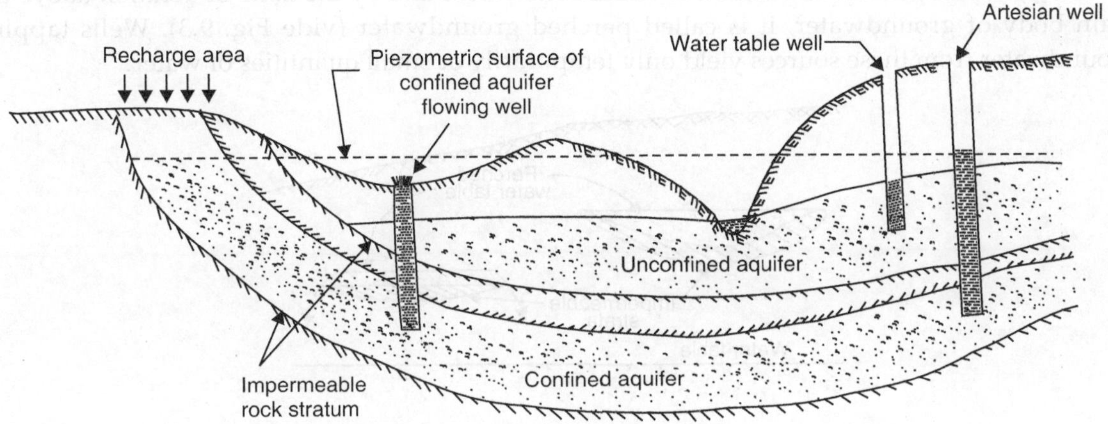

FIGURE 9.2 Groundwater in confined and unconfined aquifers.

Confined Aquifer. When an aquifer is sandwiched between two layers of much less permeable material then it is called a confined aquifer. For example, a sandy layer between two clay layers, sandstone layer between two layers of shale etc. It is also known as a *pressure aquifer*, or an *artesian aquifer*. Confined aquifers are completely filled with water and they do not have a free water table. They will have recharge area somewhere on the earth surface. The water in confined aquifers will be under pressure and the pressure condition in the aquifer is characterised by the *piezometric surface,* which is the surface obtained by connecting equilibrium water levels in the tubes or *piezometers* penetrating the confined aquifer (Fig. 9.2). A horizontal surface in level with the water table at the recharging area may be taken as the approximate piezometric surface of a confined aquifer. If a well penetrates into the confined aquifer, water level in the well will rise to piezometric level. If the piezometric level at the place of the well is above the upper confining layer the static water level in the well will be above the aquifer. Such a well is called an *artesian well*. If the piezometric surface at the place of the well is above the ground level, the confined aquifer will yield a *free-flowing well* also known simply as a *flowing well*. A well penetrating into an unconfined aquifer, on the other hand, is called a *water table well*. The groundwater can also move to ground surface through natural passages like *faults* or *sinkholes*, where it then produces a *spring*.

Aquiclude. A geological formation which is saturated and which may contain large amount of water because of its high porosity but cannot transmit water as it is relatively impermeable is called an aquiclude. A clay layer is an example of an aquiclude.

Aquitard. It is a saturated geological formation which is poorly permeable and hence it does not yield water freely to wells. It may transmit vertically appreciable quantities of water to or from adjacent aquifers. Sandy clay is an example of aquitard.

Aquifuge. An impervious geological formation which neither contains nor transmits water is known as an aquifuge. Solid granite rock is an example of aquifuge.

Leaky Aquifer. An aquifer bound by one or two aquitards is called a leaky aquifer or a *semi-confined* acquifer.

Perched groundwater. When a groundwater body. is separated from the main groundwater by a relatively impermeable stratum of small areal extent and by the zone of aeration above the main body of groundwater, it is called perched groundwater (vide Fig. 9.3). Wells tapping groundwater from these sources yield only temporarily or small quantities of water.

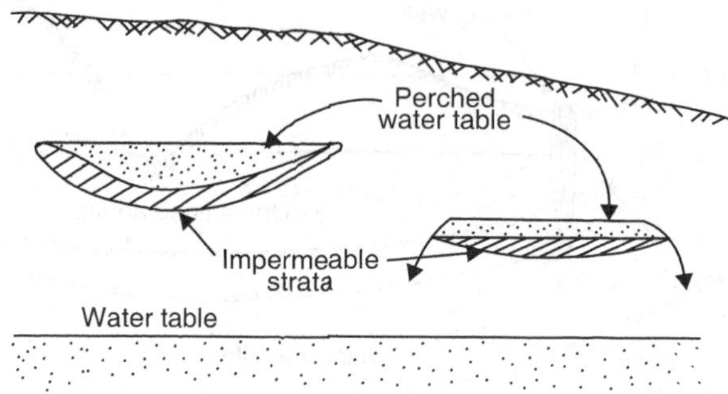

FIGURE 9.3 Perched water table.

Influent and Effluent Streams. If the water table of an unconfined aquifer intersects a surface stream, then aquifer contributes water to the stream. Such a stream is called an effluent stream. If a surface stream runs well above the water table, water flows from the stream to the aquifer. Such a stream is called an influent stream (vide Fig. 9.4). Most of the flow of *perennial streams* in non-rainy periods originate from groundwater when they behave as effluent streams. Likewise a large part of the flow of *ephemeral streams* (which go dry during prolonged rainless period) may percolate and become groundwater.

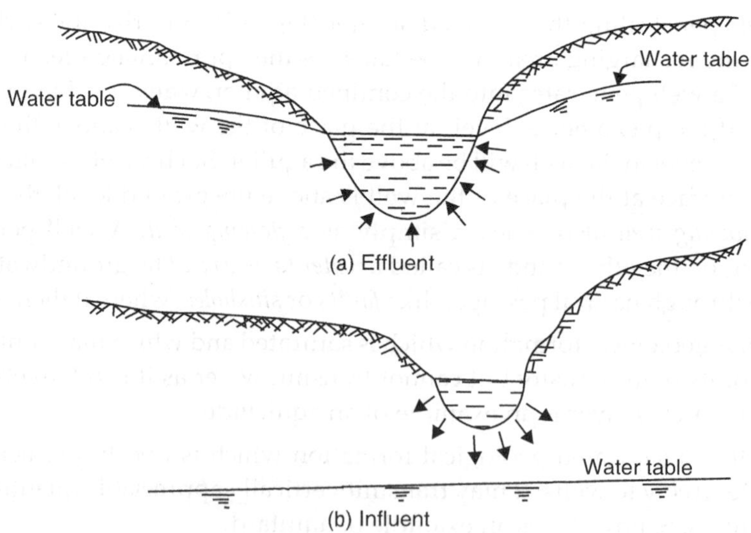

FIGURE 9.4 Effluent and influent streams.

9.3 AQUIFER PARAMETERS

An aquifer has the ability to store and transmit water. The quantity of water stored by an aquifer and the quantity of water released by it depend on the nature and composition of the aquifer which are quantified through certain parameters like *porosity, specific yield, storage coefficient, permeability* and *transmissivity*. These aquifer parameters are now briefly discussed.

Porosity. The ratio of volume of pore space, V_v to the volume of the formation V is called the porosity, usually denoted by n. Thus we have

$$n = \frac{V_v}{V} \quad ...(9.1)$$

In unconsolidated materials, the porosity depends on the shape and size distribution of the particles and their packing arrangement. The *original porosity* of a material is that which existed at the time the material was formed. *Secondary porosity* results from fractures and solution channels.

Original porosity can be measured in the laboratory using an undisturbed sample. The sample is ovendried and its weight w_1 measured. It is then saturated with some liquid (the liquid may have to be forced into the sample under pressure to ensure complete saturation) and its weight w_2 is again measured. Finally the saturated sample is immersed in a vessel containing the same liquid and the weight of the liquid displaced by the sample, w_3 is noted. The weight of liquid required to saturate the sample, that is $(w_2 - w_1)$, divided by the weight of the liquid displaced by the saturated sample, that is w_3, is the porosity of the sample. Multiplication of this ratio with 100 gives the porosity by percentage. Secondary porosity cannot be measured without an impossibly large sample.

Specific Yield. The specific yield of an aquifer is the ratio of the volume of water which will drain freely from the material to the volume of the formation. Therefore,

$$S_y = \frac{w_y}{V} \quad ...(9.2)$$

where w_y is the volume of water drained and S_y is the specific yield. Values of specific yield depend on grain size, shape and distribution of pores and compaction of the formation. Specific yield of fine-grained aquifers will be small, while coarse-grained material will have high specific yield. When the material is drained, some water will always be retained in the small pore spaces under capillary tension and therefore the volume of water drained will never be equal to the volume of pore spaces. That is $w_y < V_v$. Hence the specific yield is always less than porosity.

Specific Retention. The specific retention S_r is defined as the ratio of the volume of water retained w_r in the material to the total volume of the material, when a saturated material is dewatered. Thus

$$S_r = \frac{w_r}{V} \quad ...(9.3)$$

Since w_r and w_y together constitute the total volume of water which is the total pore space, it is obvious that the specific retention and the specific yield are complimentary quantities whose sum represents the porosity of the aquifer. That means,

$$n = S_r + S_y \quad ...(9.4)$$

Storage coefficient. The storage coefficient S is defined as the volume of water that an aquifer releases from or takes into storage per unit surface area of the aquifer per unit drop of water table in the case of an unconfined aquifer and per unit drop of piezometric surface in the case of a confined aquifer. It is also known as *storativity*. The concept of storage coefficient is better illustrated through Fig. 9.5.

As depicted in Fig. 9.5 (*b*), when there is a change in water table water physically enters or leaves the unconfined aquifer. Therefore for this case the storage coefficient equals the specific yield. Fig. 9.5 (*a*) shows the situation in a confined aquifer. When there is a change in the piezometric surface it does not represent the water released from the aquifer because a confined aquifer remains completely saturated at all times. The volume of water released from or taken into storage in this case is because of the elastic response of the aquifer, confining beds and water to the pressure changes.

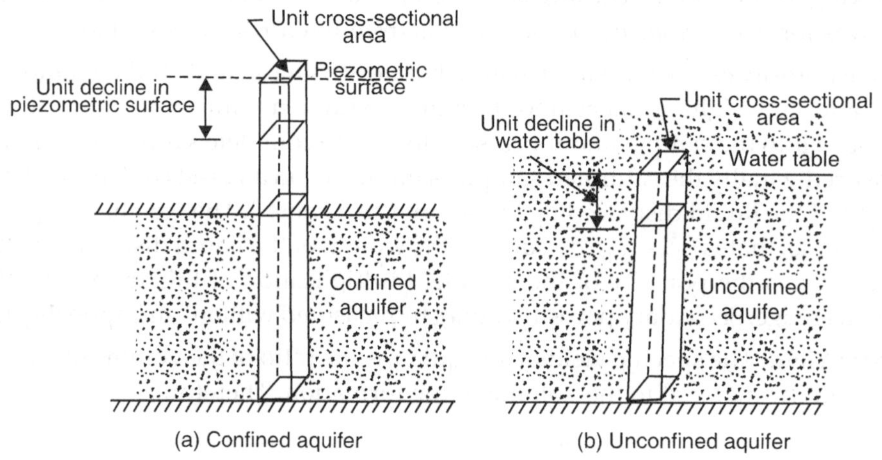

FIGURE 9.5 Definition sketch of storage coefficient.

The storage coefficient of a confined aquifer will be very small in the range of 0.00005 to 0.005, indicating that large pressure changes over extensive areas are required to produce substantial water yields. The storage coefficient of unconfined aquifer on the other hand is usually large and of the order 0.01 to 0.3.

Permeability. The permeability of a material is a measure of its capacity to transmit water or any other fluid through its interstices. Groundwater is transmitted through aquifers at very small velocities ranging from 1 to 500 m/year. The laminar flow principles are shown to be applicable for the flow of groundwater in aquifers by Darcy. The *Darcy's law* states that the rate of flow per unit area of an aquifer is proportional to the gradient of the potential head measured in the direction of flow. Since the velocities are so small that the velocity heads are negligible and the gradient of the potential head equals the hydraulic gradient. And since the flow rate per unit area denotes the velocity of flow, the Darcy's law may be written as

$$v = Ki = -K\frac{dh}{dl} \qquad \text{...(9.5)}$$

where the constant of proportionality K is known as the *hydraulic conductivity* or the *coefficient of permeability*, v is the velocity of water through porus medium, and i or dh/dl is the hydraulic gradient. The negative sign indicates that the flow takes place in the direction of falling head.

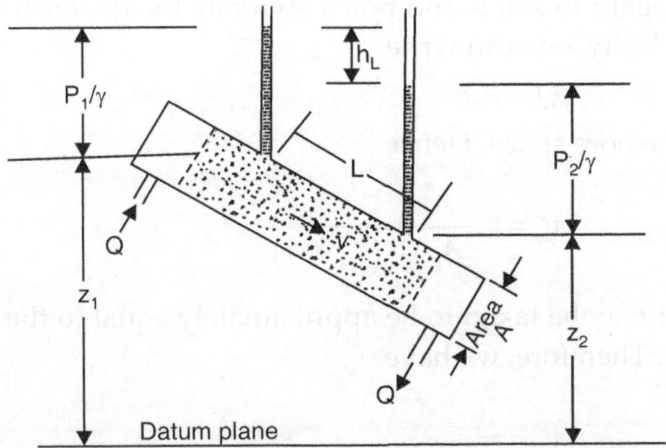

FIGURE 9.6 Head loss in flow through porus media.

The verification of Darcy's law follows from a simple experiment shown in Fig. 9.6. Applying Bernoulli energy equation for the two sections where piezometers are inserted

$$\frac{p_1}{\gamma} + z_1 + \frac{v_1^2}{2g} = \frac{p_2}{\gamma} + z_2 + \frac{v_2^2}{2g} + h_L \quad \text{...(9.6)}$$

where p is pressure, γ is the specific weight of water, v is the velocity of flow, g is the acceleration due to gravity, z is the elevation and h_L is the head loss. Subscripts 1 and 2 refer to the two sections separated by a distance L. Neglecting the velocity heads as they will be very very small compared to other terms and on rearrangement we obtain

$$h_L = \left(\frac{p_1}{\gamma} + z_1\right) - \left(\frac{p_2}{\gamma} + z_2\right) \quad \text{...(9.7)}$$

which shows that the head loss is independent of the inclination of the cylinder. Darcy's experiments showed that the discharge Q is directly proportional to head loss h_L and area of flow, and inversely proportional to the length L. That is

$$Q \propto \frac{A}{L} \cdot h_L; \quad \text{or} \quad \frac{Q}{A} \propto \frac{h_L}{L}; \quad \text{or} \quad v \propto \frac{h_L}{L}.$$

In the limiting case as the two sections become closer and closer, $\frac{h_L}{L}$ tends to $\frac{dh}{dl}$. Introducing a constant of proportionality, $v = -K\frac{dh}{dl}$; which is the Darcy's law. In Eq. (9.5), if $\frac{dh}{dl}$ equals unity v equals K. Hence the hydraulic conductivity may also be defined as the rate of flow per unit area of an aquifer under a unit hydraulic gradient. Since hydraulic gradient is dimensionless, it follows from Eq. (9.5) that the hydraulic conductivity has the same units as velocity and can be expressed in cm/s or m/s. But when hydraulic conductivity of aquifers is expressed in these units the value will be very small and therefore it is usually expressed in m/day.

The velocity v in Eq. (9.5) is variously known as the *Darcy's velocity, discharge velocity*, or *apparent velocity*. It is based on the assumption that the flow occurs through the entire cross-section of the material without regard to solids and pores. Actually the flow is limited to pore space only. Thus if v_a is the actual velocity, one can write

$$Q = A.v = A_p.\, v_a$$

where A_p is the area of the pore space. Hence

$$v_a = v.\, \frac{A}{A_p} = \frac{v}{(A_p / A)}.$$

But the ratio A_p/A may be taken to be approximately equal to the porosity of the material and can be replaced by n. Therefore, we have

$$v_a = \frac{v}{n} \qquad \text{...(9.8)}$$

Since the pore space vary continuously with location within the medium, the actual velocity is non-uniform and varies from point to point.

The Darcy's law is found to be valid for the flows with *Reynold's number* less than unity. The Reynold's number R_e is given by

$$R_e = \frac{\rho\, v d_{10}}{\mu} \qquad \text{...(9.9)}$$

where ρ and μ are the density and viscosity of the fluid, v is the Darcy's velocity and d_{10} is an effective grain size. The Darcy's law does not depart seriously upto $R_e < 10$ which may be taken as the upper limit. Most groundwater flows occur with $R_e < 1$ and so Darcy's law is applicable. Deviations from Darcy's law can occur where steep hydraulic gradients exist, such as near pumped wells and in the formations with large underground openings.

The hydraulic conductivity is dependent on both the properties of the fluid and the properties of the medium. It can be proved that K is given by

$$K = \frac{cd^2 \rho g}{\mu} \qquad \text{...(9.10)}$$

where ρ and μ are the density and viscosity of the fluid, g is the acceleration due to gravity, d is the average pore size of the medium and c is a factor accounting for the shape, packing and other characteristics of the material in the medium. The *intrinsic permeability* which is the property of the medium only and which does not depend on the fluid properties is given by

$$k = \frac{K\mu}{\rho g} = cd^2 \qquad \text{...(9.11)}$$

where k is the intrinsic permeability, and K is the hydraulic conductivity for the given fluid with properties μ and ρ. It can be seen that k has the units of area. The customary unit of k is the *darcy* which is equal to 0.987×10^{-12} m^2. The intrinsic permeability is not used in groundwater studies. It is mainly applied in the petroleum and natural gas industries which deal with underground flow of fluids of different viscosity and density including liquids and gases.

The hydraulic conductivity can be measured in a laboratory using *permeameters*. The values obtained through laboratory tests are not truly representative of the actual hydraulic conductivity in the field. This is because undisturbed samples of unconsolidated materials are difficult to obtain, while disturbed samples experience changes in porosity, packing and grain orientation. Also, flow in solution cavities or rock fractures, and the effect of large boulders in gravel aquifers cannot be duplicated in a laboratory sample. The average hydraulic conductivity of an aquifer for a large area around the test well can be determined from pumping tests using the principles of well hydraulics that are discussed later in the chapter.

Table 9.1 gives the approximate average values of porosity, specific yield and permeability of some typical materials.

Table 9.1 Approximate average porosity, specific yield and permeability of various materials

Materials	*Porosity* %	*Specific yield* %	*Hydraulic conductivity m/day*	*Intrinsic permeability darcys*
Clay	45	3	0.0004	0.0005
Sand	35	25	41.0	50
Gravel	25	22	4100.0	5000
Gravel and Sand	20	16	410.0	500
Sandstone	15	8	4.10	5
Dense limestone and shale	5	2	0.041	0.05
Quartzite, granite	1	0.5	0.0004	0.0005

Sometimes the porus medium may be stratified and each strata may be possessing different horizontal and vertical hydraulic conductivities. Suppose a porus medium consists of n layers with thicknesses $B_1, B_2, \ldots\ldots, B_n$ with horizontal hydraulic conductivities $K_{x_1}, K_{x_2}, K_{x_3}, \ldots\ldots, K_{x_n}$ and with vertical hydraulic conductivities $K_{y_1}, K_{y_2}, K_{y_3}, \ldots\ldots, K_{y_n}$. Then the equivalent horizontal hydraulic conductivities K_x of the medium is given by

$$K_x = \frac{\Sigma\left(K_{x_i} \cdot B_i\right)}{\Sigma B_i} \qquad \text{...(9.12)}$$

And the equivalent vertical hydraulic conductivity K_y is given by

$$K_y = \frac{\Sigma B_i}{\Sigma\left(B_i / K_{y_i}\right)} \qquad \text{...(9.13)}$$

where all the summations are taken from $i = 1$ to $i = n$.

Transmissivity. The transmissivity of an aquifer is also known as *transmissibility*. It is the product of the hydraulic conductivity K, and the thickness of the aquifer b and is expressed in units of m^2/day. If we consider a unit width of an aquifer of thickness b, the discharge through it is given by

$$Q = \text{Area} \times \text{Velocity} = (b \times 1) \times K\left(-\frac{dh}{dl}\right)$$

$$Q = Kb\left(-\frac{dh}{dl}\right) = T\left(-\frac{dh}{dl}\right) \quad ...(9.14)$$

where $T = Kb$ is the transmissivity. When the hydraulic gradient equals unity, the discharge is equal to T itself. In other words, the transmissivity of an aquifer may be defined as the discharge through the aquifer of one metre width under a unit hydraulic gradient. The value of T ranges from about 12 m²/day to about 1200 m²/day depending on the type and thickness of the aquifer. It is usually evaluated from the pumping tests.

Example 9.1. *An undisturbed rock sample has an oven dry weight of 0.655 kg. After saturation with kerosene its weight is 0.732 kg. It is then immersed in kerosene and found to displace 0.301 kg. What is the porosity of the sample?*

Solution. Oven-dry weight $w_1 = 0.655$ kg

Weight of saturated sample $w_2 = 0.732$ kg

Weight of kerosene required to saturate the sample

$$= (w_2 - w_1)$$
$$= 0.732 - 0.655 = 0.077 \text{ kg}$$

Weight of kerosene displaced by the saturated sample

$$w_3 = 0.301 \text{ kg}$$

Porosity of the sample $\quad n = \dfrac{(w_2 - w_1)}{w_3}$

$$= \frac{0.077}{0.301} = 0.2558$$
$$= 25.58\%$$

Example 9.2. *When 3.68 million m³ of water was pumped out from an unconfined aquifer of 6.2 km² areal extent, the water table was observed to go down by 2.6 m. What is the specific yield of the aquifer? During a monsoon season if the water table of the same aquifer goes up by 10.8 m what is the volume of recharge?*

Solution. Water released from the aquifer = $(6.2 \times 10^6) \times 2.6 \times s_y$

Water pumped out $= 3.68 \times 10^6$ m³

Equating these two quantities $s_y = \dfrac{3.68}{6.2 \times 2.6} = 0.2283$

∴ Specific yield of the acquifer = 0.2283 or 22.83%

Volume of recharge $= 6.2 \times 10^6 \times 10.8 \times s_y$

$= 6.2 \times 10^6 \times 10.8 \times 0.2283$

$= 15.287$ million m³.

Example 9.3. *The water table levels in two observation wells 350 m apart are + 210.5 and + 206.25 m respectively. If the hydraulic conductivity and porosity of the aquifer are 12.5 m/day and 15 per cent, what is the actual velocity of flow in the aquifer?*

Solution.
$$v = K\left(-\frac{dh}{dl}\right)$$

$$v = 12.5 \times \left[-\frac{(206.25 - 210.5)}{350}\right]$$

$$= 0.1518 \text{ m/day}$$

$$v_a = \frac{v}{n} = \frac{0.1518}{0.15} = 1.012 \text{ m/day}$$

$\therefore$ The actual velocity in the aquifer = 1.012 m/day.

Example 9.4. *A sample has a hydraulic conductivity of 10 m/day. What would be its intrinsic permeability? What is its hydraulic conductivity in cm/s? What would be its hydraulic conductivity at 30°C?*

Solution.
$$k = \frac{K.\mu}{\rho g}$$

$$K = 10 \text{ m/day} = 0.011574 \text{ cm/s}$$

$$\mu = 0.01 \text{ gm-cm/s}$$

$$\rho = 1 \text{ gm/c.c.}$$

$$g = 981 \text{ cm/s}^2$$

$$\therefore \quad k = \frac{0.011574 \times 0.01}{981} = 1.17982 \times 10^{-7} \text{ cm}^2$$

But one darcy = 0.987×10^{-12} m^2 = 0.987×10^{-8} cm^2

$$\therefore \quad k = \frac{1.17982 \times 10^{-7}}{0.987 \times 10^{-8}} = 11.954 \text{ darcys}$$

Intrinsic permeability of the sample = 11.954 darcys

Hydraulic conductivity in cm/s = 0.011574.

From Eq. (9.10),
$$K = \frac{cd^2 \rho g}{\mu}$$

$$= \frac{cd^2 g}{(\mu/\rho)} = \frac{cd^2 g}{v}$$

where v is the kinematic viscosity. Thus for a given porus medium the hydraulic conductivity is inversely proportional to the kinematic viscosity of the fluid. Therefore, we have

$$\frac{K}{K_t} = \frac{v_t}{v} \quad \text{or} \quad K_t = K \cdot \frac{v}{v_t}$$

where K_t and v_t are the hydraulic conductivity and kinematic viscosity at some temperature t and K and v are the hydraulic conductivity and kinematic viscosity at standard temperature of 20°C.

$$K_{30} = K \cdot \frac{v}{v_{30}} = 10 \times \frac{0.01}{0.008} = 12.5 \text{ m/day}$$

(Since kinematic viscosity for water at 20°C is 0.01 cm²/s and at 30°C, it is 0.008 cm²/s)

∴ Hydraulic conductivity of the sample at 30°C = 12.5 m/day.

9.4 GROUNDWATER MOVEMENT

Groundwater moves very slowly in the aquifers under differential head. When a groundwater basin is filled the excess water is discharged by several ways until a quasi-equilibrium is reached. All the perennial streams are generally effluent through at least a portion of their length gathering contributions from groundwater. When the water table is close to the surface, groundwater may be discharged by direct evaporation and transpiration. The groundwater movement in aquifers is governed by the established hydraulic principles. The theory of groundwater movement is based on Darcy's law. The study of groundwater movement helps in predicting the yields from a well and the effects of future pumping, evaluating aquifer parameters, defining the aquifer boundaries, analysing the salt water intrusion problems etc. The three dimensional continuity equation (principle of conservation of mass) for the flow of fluid through a porus medium is given by

$$\frac{\partial (\rho u)}{\partial x} + \frac{\partial (\rho v)}{\partial y} + \frac{\partial (\rho w)}{\partial z} = \frac{\partial}{\partial t}(\rho n) \quad \text{...(9.15)}$$

where ρ is the density of the fluid flowing through the medium, u, v and w are the components of velocity along the three mutually perpendicular coordinate axes x, y and z respectively, n is the porosity of the medium and t is the time variable. Many groundwater flow situations can be treated as near steady state flow cases. When the fluid flowing is water which is incompressible its density does not vary with either time of space. Hence for steady state incompressible fluid flow, the continuity equation reduces to

$$\frac{\partial u}{\partial x} + \frac{\partial v}{\partial y} + \frac{\partial w}{\partial z} = 0 \quad \text{...(9.16)}$$

If we assume that the porus medium is isotropic and homogeneous the hydraulic conductivity will be same in all directions at a point and also at all points. From Darcy's law $u = -K\frac{\partial h}{\partial x}$, $v = -K\frac{\partial h}{\partial y}$ and $w = -K\frac{\partial h}{\partial z}$. Substituting these for u, v and w in Eq. (9.16), we get

$$-K\left(\frac{\partial^2 h}{\partial x^2}+\frac{\partial^2 h}{\partial y^2}+\frac{\partial^2 h}{\partial z^2}\right)=0.$$

Since $K \neq 0$ we have

$$\frac{\partial^2 h}{\partial x^2}+\frac{\partial^2 h}{\partial y^2}+\frac{\partial^2 h}{\partial z^2}=0 \qquad ...(9.17)$$

or

$$\nabla^2 h = 0 \qquad ...(9.18)$$

Equation (9.17) or (9.18) is called Laplace equation.

In sections (9.5) to (9.7) we present some of the unidirectional groundwater flow problems. The theory of groundwater movement around wells is commonly termed the *well hydraulics*. There are two cases to be considered; *steady state* and *unsteady state* flows The steady state case rarely occurs in practice but the conditions after a prolonged pumping from a well may be approximated as steady state. The unsteady state case, or the *transient flow* case is of more practical significance. The sections 9.8 and 9.9 deal with steady and unsteady radial groundwater flow towards wells.

9.5 STEADY UNIDIRECTIONAL FLOW IN UNCONFINED AQUIFERS

When an unconfined aquifer exists between two streams which are nearly parallel with a difference in the elevations of their water surfaces we realise this kind of situation which is depicted in Fig. 9.7. Let us take the origin as shown.

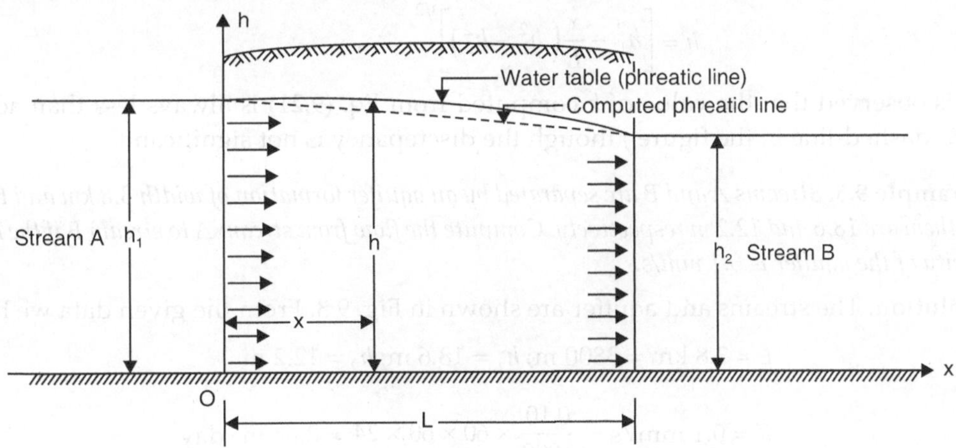

FIGURE 9.7 Unidirectional groundwater flow in an unconfined aquifer.

Consider a vertical section of height h at a distance x from the origin. If we take unit thickness in a direction normal to the plane of flow, the area of flow will be $(h \times 1) = h$. Strictly speaking the velocity of flow has to be taken as $-K\frac{dh}{ds}$, where s is the direction of flow which is slightly slanted. In other words the velocity has to be taken proportional to sine of the angle of the tangent drawn to the phreatic line. But we assume that the velocity can be taken as $-K.\frac{dh}{dx}$, that is we take that the velocity is proportional to the *tan* of the angle of the tangent drawn to the phreatic line. Therefore the discharge q through the area is given by

$$q = \text{area} \times \text{velocity} = h \times -K\frac{dh}{dx} = -Kh\frac{dh}{dx}$$

$q \,.\, dx = -K h .\, dh$ Integrating this equation on both sides

$$\int q.dx = -K\int h.\,dh \;;\; q\,.\,x = -K\,.\frac{h^2}{2} + C$$

where C is the constant of integration which can be evaluated using the boundary condition that when $x = 0$, $h = h_1$. Hence $C = K\frac{h_1^2}{2}$. Therefore we have

$$q = \frac{K}{2x}\left(h_1^2 - h^2\right) \qquad ...(9.19)$$

This equation is called the Dupuit equation and indicates that the phreatic line is a parabola. Substituting the other boundary condition that when $x = L$, $h = h_2$, we get an expression for the discharge

$$q = \frac{K}{2L}\left(h_1^2 - h_2^2\right) \qquad ...(9.20)$$

From Eq. (9.19) and Eq. (9.20), we get an expression for h in the form

$$h = \left[h_1^2 - \frac{x}{L}\left(h_1^2 - h_2^2\right)\right]^{1/2} \qquad ...(9.21)$$

It is observed that the value of h computed from Eq. (9.21) is always less than actual (as shown by dashed-line in the figure) though the discrepancy is not significant.

Example 9.5. *Streams A and B are separated by an aquifer formation of width 3.8 km and the depths of flow in them are 18.6 and 12.2 m respectively. Compute the flow from stream A to stream B if the hydraulic conductivity of the aquifer is 0.1 mm/s.*

Solution. The streams and aquifer are shown in Fig. 9.8. From the given data we have

$$L = 3.8 \text{ km} = 3800 \text{ m};\ h_1 = 18.6 \text{ m};\ h_2 = 12.2 \text{ m}$$

$$K = 0.1 \text{ mm/s} = \frac{0.10}{1000} \times 60 \times 60 \times 24 = 8.64 \text{ m/day}$$

$$q = \frac{K}{2L}\left(h_1^2 - h_2^2\right) = \frac{8.64}{2 \times 3800} \times \left(18.6^2 - 12.2^2\right)$$

$$= 0.224 \text{ m}^3/\text{day}.$$

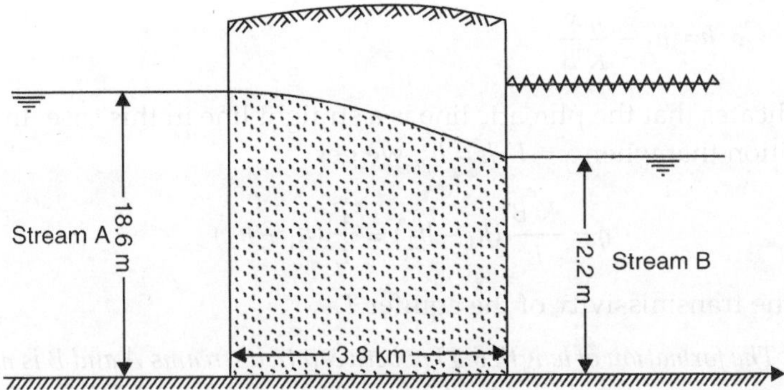

FIGURE 9.8 Unconfined aquifer of Example 9.5.

9.6 STEADY UNIDIRECTIONAL FLOW IN CONFINED AQUIFER

The case of unidirectional flow through a confined aquifer is shown in Fig. 9.9. Consider a section at a distance x from the origin where the height of the phreatic line is h. The area of flow at this section, (if we consider unit thickness in a direction perpendicular to the flow), is $B \times 1 = B$. Again the velocity is taken as $-K\dfrac{dh}{dx}$. Therefore we have

$$q = B \times \left(-K\frac{dh}{dx}\right) = -K\,B\frac{dh}{dx};\ \text{or } q.\,dx = -K\,B\,.\,dh.$$

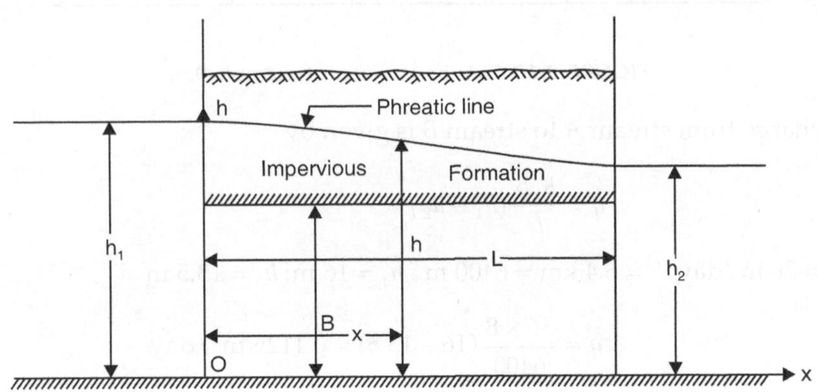

FIGURE 9.9 Unidirectional groundwater flow through a confined aquifer.

Integrating this on both sides

$$q\,.\,x = -K\,B.\,h + C_1,\ \text{where } C_1 \text{ is constant of integration}$$

C_1 can be evaluated using the boundary condition that when $x = 0$, $h = h_1$.

$\therefore$ $$0 = -K\,.\,B\,h_1 + C_1 \text{ or } C_1 = K\,B\,.\,h_1$$

Now $$q\,.\,x = -K\,B\,h + K\,B\,h_1 = K\,B\,(h_1 - h) \qquad \text{...(9.22)}$$

or
$$h = h_1 - \frac{q\,x}{K\,B} \qquad \text{...(9.23)}$$

Eq. (9.23) indicates that the phreatic line is a straight line in this case. In Eq. (9.22) if we use the boundary condition that when $x = L$, $h = h_2$, we get

$$q = \frac{K\,B}{L}(h_1 - h_2) = \frac{T}{L}(h_1 - h_2) \qquad \text{...(9.24)}$$

Since $B\,K$ is the transmissivity of the aquifer T.

Example 9.6. *The formation of length 6.4 km between two streams A and B is made up of two layers. The bottom layer is pervious with a thickness of 8 m and hydraulic conductivity of 36 m/day. The upper layer is impervious. Find the discharge from stream A to stream B if the depths of water in them are 16 m and 13.5 m respectively.*

Solution. The given flow situation is shown in Fig. 9.10.

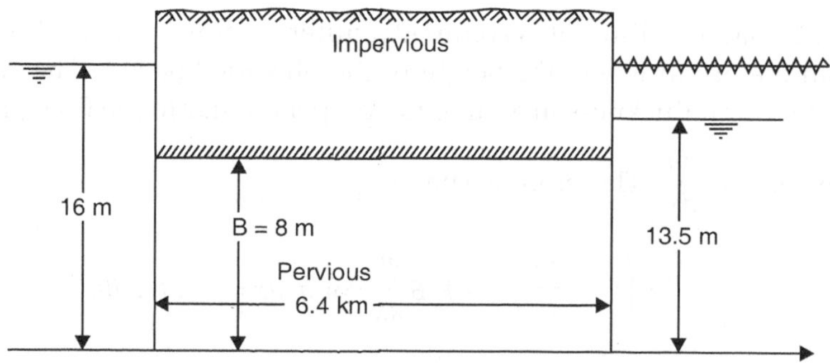

FIGURE 9.10 Confined aquifer of example 9.6.

The discharge from stream A to stream B is given by

$$q = \frac{K\,B}{L}(h_1 - h_2)$$

Now $K = 36$ m/day; $L = 6.4$ km $= 6400$ m, $h_1 = 16$ m; $h_2 = 13.5$ m

$$\therefore \qquad q = \frac{36 \times 8}{6400}(16 - 13.5) = 0.1125 \text{ m}^3\text{/day.}$$

9.7 STEADY UNIDIRECTIONAL FLOW IN UNCONFINED AQUIFER WITH RECHARGE FROM RAINFALL

This case deals with groundwater flow through an unconfined aquifer including the precipitation which recharges the aquifer. The details are as shown in Fig. 9.10. Let P be the rainfall intensity and we assume that P is always more than infiltration capacity rate ($P \geq f$) so that the recharge takes place at infiltration capacity rate. The discharge through a vertical section of height h at a distance x from the origin is given by

$$q = (h \times 1)\left(-K\frac{dh}{dx}\right) = -K\,h\,\frac{dh}{dx}$$

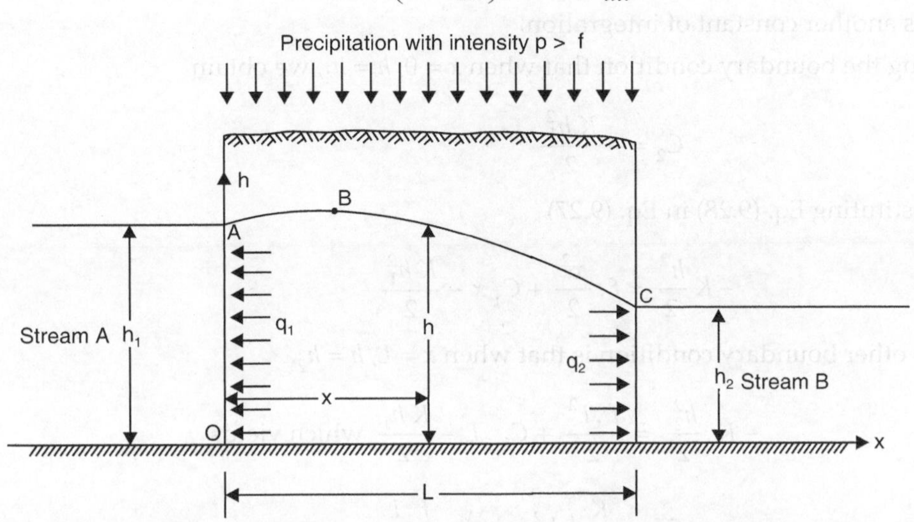

FIGURE 9.11 Unidirectional groundwater flow in an unconfined aquifer with precipitation as recharge.

Water joining the flow due to recharge can be taken as

$$dq = f \times (1 \times dx) = f \,.\, dx$$

$$\frac{dq}{dx} = f \qquad \text{...(9.25)}$$

By differentiating q w. r. t. x, we get

$$\frac{dq}{dx} = \frac{d}{dx}\left(-K\,h\,.\frac{dh}{dx}\right) = -K\,\frac{d}{dx}\left(h\,.\frac{dh}{dx}\right) \qquad \text{...(9.26)}$$

From Eq. (9.25) and Eq. (9.26)

$$-K\,\frac{d}{dx}\left(h\,.\frac{dh}{dx}\right) = f$$

$$-K\,d\left(h\,.\frac{dh}{dx}\right) = f\,.\,dx$$

Integrating the above equation once

$$-K\,h\,.\frac{dh}{dx} = f\,.\,x + C_1, \qquad \text{where } C_1 \text{ is constant of integration}$$

$$-K\,.\,h\,.\,dh = f.\,x\,.\,dx + C_1\,dx$$

Integrating again

$$-K\frac{h^2}{2} = f.\frac{x^2}{2} + C_1.x + C_2 \qquad ...(9.27)$$

where C_2 is another constant of integration.

Using the boundary condition that when $x = 0$, $h = h_1$, we obtain

$$C_2 = -\frac{K\,h_1^2}{2} \qquad ...(9.28)$$

Substituting Eq. (9.28) in Eq. (9.27)

$$-K\frac{h^2}{2} = f.\frac{x^2}{2} + C_1\,x - \frac{K\,h_1^2}{2} \qquad ...(9.29)$$

The other boundary condition is that when $x = L$, $h = h_2$.

$$-K.\frac{h_2^2}{2} = \frac{f.L^2}{2} + C_1.L - \frac{K\,h_1^2}{2} \text{ which yields}$$

$$C_1 = \frac{K}{2L}\left(h_1^2 - h_2^2\right) - \frac{f.L}{2} \qquad ...(9.30)$$

Substituting Eq. (9.30) in Eq. (9.29), we get

$$h^2 = h_1^2 - \frac{f}{K}(L - x)\,x - \frac{x}{L}\left(h_1^2 - h_2^2\right) \qquad ...(9.31)$$

or

$$h = \sqrt{h_1^2 - \frac{f}{K}(L - x)\,x - \frac{x}{L}\left(h_1^2 - h_2^2\right)} \qquad ...(9.32)$$

The phreatic line represented by Eq. (9.32) is shown as curve *ABC* in Fig. 9.11 which has a peak at *B*. The value of x where this peak occurs can be obtained by equating $\frac{dh}{dx}$ to zero.

To get an expression for the discharge, we get $h\,.\,\frac{dh}{dx}$ by differentiating Eq. (9.31) and substitute the same in the expression $q = -K\,h.\,dh/dx$. This gives the expression for q as

$$q = \frac{K}{2L}\left(h_1^2 - h_2^2\right) + f\left(x - \frac{L}{2}\right) \qquad ...(9.33)$$

The discharge q to the stream on the left is computed by substituting $x = 0$ in Eq. (9.33). Likewise when we make $x = L$, in Eq. (9.33), we get the discharge q_2 flowing in to the stream on the right.

Example 9.7. *Find the discharges into streams A and B of Example 9.5 when the rain falls with an intensity which is always more that the infiltration capacity of the aquifer and which may be taken as 0.3 cm/h.*

Solution. For this case with recharge by rainfall

$$q = \frac{K}{2L}\left(h_1^2 - h_2^2\right) + f.\left(x - \frac{L}{2}\right)$$

From example 9.5 $K = 8.64$ m/day, $L = 3800$ m, $h_1 = 18.6$ m, $h_2 = 12.2$ m

$$\frac{K}{2L}\left(h_1^2 - h_2^2\right) = 0.224 \text{ m}^3/\text{day}$$

$$f = 0.3 \text{ cm/h} = \frac{0.3}{100} \times 24 = 0.072 \text{ m/day}$$

$$q = \frac{K}{2L}\left(h_1^2 - h_2^2\right) + f\left(x - \frac{L}{2}\right)$$

$$= 0.224 + 0.072\,(x - 1900)$$

The discharge into the stream on left

$$q_1 = 0.224 + 0.072\,(0 - 1900) = 0.224 - 136.8$$

$$= -\,136.576 \text{ m}^3/\text{day}$$

Since the positive direction is towards the stream on right, the negative answer indicates that the flow of 136.576 m^3/day is taking place into the stream on left.

Similarly the discharge into the stream on right is obtained as

$$q_2 = 0.224 + 0.072\,(3800 - 1900) = 0.224 + 136.8$$

$$= 137.024 \text{ m}^3/\text{day}$$

The reader may easily verify that $q_1 + q_2$ is exactly equal to the total recharge into the formation between the stream. That is $q_1 + q_2 = 136.576 + 137.024 = 273.6$ m^3/day

Total recharge $= (3800 \times 1) \times 0.072 = 273.6$ m^3/s

9.8 STEADY RADIAL FLOW TO WELLS

The analysis of groundwater movement will be greatly simplified if the vertical flow components are neglected and the groundwater is assumed to move primarily in the lateral direction. Accordingly the following assumption are made in the theory of groundwater movement.

1. The flow is horizontal and uniformly distributed in a vertical section.
2. The velocity of flow is proportional to the *tan* of the hydraulic gradient instead of *sine* of the hydraulic gradient.

   The above two assumptions are called *Dupuit-Forchheimer* assumptions. In addition, the derivation of the steady flow discharge equation generally assumes the following.

3. The well is pumped at a constant rate.
4. The well fully penetrates the aquifer. (Equations can be derived for partially penetrating wells also. They are not discussed in this book).
5. The aquifer is homogeneous, isotropic, horizontal and of infinite horizontal extent.
6. Water is released from storage in the aquifer in immediate response to a drop in water table or piezometric surface.

Though the above assumptions are not strictly valid in all cases, they are required for making the analysis simpler and easier.

Steady Radial Flow to a Well in an Unconfined Aquifer. The flow system around a pumped well in an unconfined aquifer with an initially horizontal water table is depicted in Fig. 9.12. The thickness of aquifer is H and the well fully penetrates the aquifer.

If water is pumped at a constant rate from the well, a gradient in the water table towards the well is created which results in a depressing form of the water table. This is called the *cone of depression*. After a sufficiently long time when the contribution from the aquifer to the well is equal to the rate of pumping, the shape of the cone of depression remains fixed in space and will not change any further. When this condition is attained the flow is said to be steady. In practice this condition is rarely fulfilled and for the case of analysis a steady state flow is assumed if the changes in water table elevation have become negligible with time. The decrease in the water level at the well is called the *drawdown*. If the drawdown is small compared to the thickness of the aquifer, the streamlines of the flow to the well may be assumed to be horizontal.

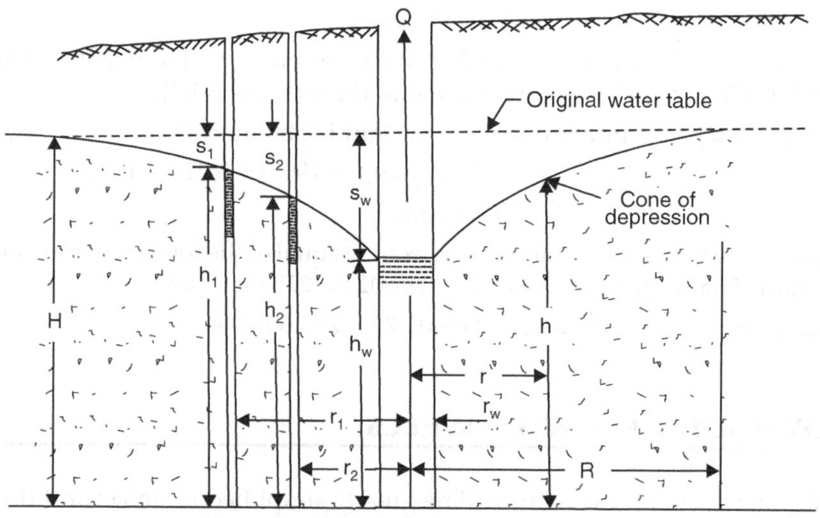

FIGURE 9.12 Steady flow to a well in an unconfined aquifer.

In a steady state radial flow, as shown in Fig. 9.12, the depth of flow at any radial distance r from the well is h, and the area of flow at this section is $2\pi rh$. According to Darcy's law the velocity of flow v is given by $v = K \,.\, i$, where i is the hydraulic gradient. Since the tangent of the water table is taken as hydraulic gradient, the velocity at this section becomes $K \,.\, \dfrac{dh}{dr}$, and the discharge is given by

$$Q = \text{Area} \times \text{Velocity} = 2\pi\, rh.\, K.\, \frac{dh}{dr}$$

or

$$h \,.\, dh = \frac{Q}{2\pi K} \cdot \frac{dr}{r}$$

Integration of the above equation with the known boundary conditions at the two observation bores yields.

$$\int_{h_1}^{h_2} h \, . \, dh = \frac{Q}{2\pi K}\int_{r_1}^{r_2}\frac{dr}{r}$$

$$\frac{\left(h_1^2 - h_2^2\right)}{2} = \frac{Q}{2\pi K} \, . \, \ln\frac{(r_1)}{(r_2)}$$

or
$$Q = \frac{\pi K\left(h_1^2 - h_2^2\right)}{\ln\left(\dfrac{r_1}{r_2}\right)} \qquad ...(9.34)$$

where h_1 and h_2 are the water levels in the two observation wells located at the radial distances r_1 and r_2 respectively from the main well. The radius at which the cone of depression just commences is known as the *radius of influence* denoted by R. If the radius of influence is known, the discharge can be computed, even in the absence of observation wells using the boundary condition at the well. That is

$$Q = \frac{\pi K\left(H^2 - h_w^2\right)}{\ln\left(\dfrac{R}{r_w}\right)} \qquad ...(9.35)$$

where h_w and r_w are the water level in the well and radius of the well, respectively. But $K\left(H^2 - h_w^2\right)$ $= \frac{2K(H + h_w)}{2}.(H - h_w)$. And $(H - h_w)$ is nothing but the drawdown in the well whereas $\frac{(H + h_w)}{2}$ may be taken as the average thickness of the aquifer. Therefore $K\left(H^2 - h_w^2\right) = 2T. \, s_w$, where T is the average transmissivity of the aquifer and s_w is the drawdown in the well. Now the discharge can be also expressed as

$$Q = \frac{2\pi T s_w}{\ln\left(\dfrac{R}{r_w}\right)} \qquad ...(9.36)$$

The following empirical equation may sometimes be useful to predict the radius of influence R.

$$R = 3000 \, s_w \sqrt{K} \qquad ...(9.37)$$

where s_w is in m and K is in m/s.

Equation (9.34) or (9.35) can be used (for a given drawdown) to estimate the yield from a well when the hydraulic conductivity is known, or to estimate the hydraulic conductivity of the aquifer if the discharge from the well is known.

This equation was derived by Dupuit in 1863 and modified by Thiem in 1906. It fails to describe accurately the drawdown curve near the well because the streamlines there are markedly curved, with large vertical flow components contradicting the assumption. Another serious drawback of this equation is that the true steady state equilibrium condition required for its applicability occurs only after a very long pumping period because of the low velocities of groundwater flow.

Example 9.8. *An unconfined aquifer has a thickness of 30 m. A fully penetrating 20 cm diameter well in this aquifer is pumped at a rate of 35 lit/s. The drawdown measured in two observation wells located at distances of 10 m and 100 m from the well are 7.5 m and 0.5 m respectively. Determine the average hydraulic conductivity of the aquifer. At what distance from the well the drawdown is insignificant?*

Solution.

$$Q = \frac{\pi K\left(h_1^2 - h_2^2\right)}{\ln\left(\frac{r_1}{r_2}\right)}$$

or

$$K = \frac{Q \,.\ln\left(\frac{r_1}{r_2}\right)}{\pi\left(h_1^2 - h_2^2\right)}$$

Here

$$Q = 35 \text{ lit/s} = 0.035 \text{ m}^3/\text{s}$$

$$r_1 = 100 \text{ m};\ r_2 = 10 \text{ m};\ \ln\left(\frac{r_1}{r_2}\right) = 2.302585$$

$$H = 30 \text{ m};\ s_1 = 0.5 \text{ m and } s_2 = 7.5 \text{ m}$$

$$h_1 = H - s_1 = 30 - 0.5 = 29.5 \text{ m}$$

$$h_2 = H - s_2 = 30 - 7.5 = 22.5 \text{ m}$$

$\therefore$

$$K = \frac{0.035 \times 2.302585}{\pi(29.5 + 22.5)(29.5 - 22.5)}$$

$$= 7.04745 \times 10^{-5} \text{ m/s}$$

$$= 6.089 \text{ m/day}$$

Let R be the radius of influence where the drawdown is zero.

Then

$$Q = \frac{\pi K\left(H^2 - h_1^2\right)}{\ln\left(\frac{R}{r_1}\right)}$$

$$\ln\left(\frac{R}{r_1}\right) = \frac{\pi K\left(H^2 - h_1^2\right)}{Q}$$

$$= \frac{\pi \times 7.04745 \times 10^{-5}\left(30^2 - 29.5^2\right)}{0.035} = 0.1882$$

$$\frac{R}{r_1} = 1.207$$

$$\therefore \qquad R = 1.207 \times 100 = 120.7 \text{ m.}$$

Example 9.9. *A well with a radius of 0.5 m, completely penetrates an unconfined aquifer of thickness 50 m and K = 30 m/day. The well is pumped so that the water level in the well remains at 40 m above the bottom. Assuming that pumping has essentially no effect on water table at r = 500 m, what is the steady-state discharge?*

Solution.
$$Q = \frac{\pi K\left(H^2 - h_w^{\,2}\right)}{\ln\left(\dfrac{R}{r_w}\right)}$$

Here $H = 50$ m; $h_w = 40$ m; $R = 500$ m; $r_w = 0.50$ m; $K = 30$ m/day.

$$\therefore \qquad Q = \frac{\pi \times 30 \times \left(50^2 - 40^2\right)}{\ln\left(\dfrac{500}{0.5}\right)} = 12279.387 \text{ m}^3/\text{day}$$

$$= 0.1421225 \text{ m}^3/\text{s} = 142.12 \text{ lit/s.}$$

Steady Radial Flow to a Well in a Confined Aquifer. The flow around a well penetrating fully into a confined aquifer of thickness 'b' under steady-state conditions is shown in Fig. 9.13. In this case the depth of flow remains constant everywhere. The discharge flowing into the well through a section which is situated at a radial distance r from the well becomes

$$Q = 2\pi rbK.\frac{dh}{dr}$$

$$dh = \frac{Q}{2\pi bK}\cdot\frac{dr}{r}.$$

Integration the above equation for the known boundary conditions at the two observation wells.

$$h_1 - h_2 = \frac{Q}{2\pi kb}\ln\left(\frac{r_1}{r_2}\right)$$

or
$$Q = \frac{2\pi T(h_1 - h_2)}{\ln\left(\dfrac{r_1}{r_2}\right)} \qquad \text{...(9.38)}$$

where $T = Kb$ is the transmissivity of the aquifer.

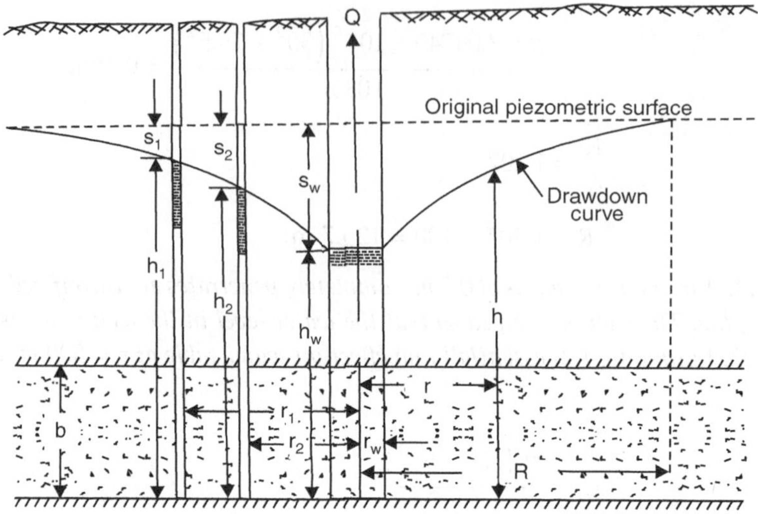

FIGURE 9.13 Steady flow to a well in a confined aquifer.

In terms of drawdowns in the wells

$$Q = \frac{2\pi T(s_2 - s_1)}{\ln\left(\frac{r_1}{r_2}\right)} \qquad \text{...(9.39)}$$

In the absence of observation wells, utilising the boundary conditions at the well,

$$Q = \frac{2\pi T(H - h_w)}{\ln\left(\frac{R}{r_w}\right)} \qquad \text{...(9.40)}$$

where h_w and r_w are the height of water in the well and radius of the well respectively and R is the radius of influence. Since $(H - h_w)$ represents the drawdown in the well s_w, the discharge may also be written as

$$Q = \frac{2\pi T s_w}{\ln\left(\frac{R}{r_w}\right)} \qquad \text{...(9.41)}$$

Example 9.10. *In an artesian aquifer of 8 m thick, a 10 cm diameter well is pumped at a constant rate of 100 lit/minute. The steady state drawdown observed in two wells located at 10 m and 50 m distances from the centre of the well are 3 m and 0.05 m respectively, compute the transmissivity and the hydraulic conductivity of the aquifer.*

Solution.

$$Q = \frac{2\pi T(s_2 - s_1)}{\ln\left(\frac{r_1}{r_2}\right)}$$

$$Q = 100 \text{ lit/min} = 0.00167 \text{ m}^3/\text{s}$$

$$r_2 = 10 \text{ m; and } r_1 = 50 \text{ m}$$

$$\therefore \quad \ln\left(\frac{r_1}{r_2}\right) = 1.60944$$

$$(s_2 - s_1) = 3 - 0.05 = 2.95 \text{ m}$$

$$\therefore \quad T = \frac{0.00167 \times 1.60944}{2\pi \times 2.95} = 1.4472 \times 10^{-4} \text{ m}^2/\text{s}$$

$$= 12.5 \text{ m}^2/\text{day}$$

The permeability of the aquifer

$$K = \frac{T}{b} = \frac{1.4472 \times 10^{-4}}{8}$$

$$= 1.809 \times 10^{-5} \text{ m/s}$$

$$= 1.563 \text{ m/day.}$$

Example 9.11. *A well of 0.5 m diameter penetrates fully into a confined aquifer of thickness 20 m and hydraulic conductivity 8.2 × 10^{-4} m/s. What is the maximum yield expected from this well if the drawdown in the well is not to exceed 3 m. The radius of influence may be taken as 260 m.*

Solution.

$$Q = \frac{2\pi T s_w}{\ln\left(\frac{R}{r_w}\right)}$$

$$T = K \cdot b$$

$$= 8.2 \times 10^{-4} \times 20 = 0.0164 \text{ m}^2/\text{s}$$

$$s_w = 3\text{m};$$

$$\frac{R}{r_w} = \frac{260}{0.25} = 520$$

$$\ln\left(\frac{R}{r_w}\right) = 6.25383$$

$$\therefore \quad Q = \frac{2\pi \times 0.0164 \times 3}{6.25383}$$

$$= 0.04943 \text{ m}^3/\text{s} = 49.43 \text{ lit/s.}$$

9.9 UNSTEADY RADIAL FLOW TO A WELL IN A CONFINED AQUIFER

When pumping is started with an initially horizontal piezometric surface, the effect of discharge is to create a cone of depression in the piezometric surface near the well which extends outward with the continued pumping. In other words the position of the piezometric surface near the well changes continuously with time. This is clearly a case of unsteady flow. Theis, in 1935, developed the non-steady-state analysis of flow in a confined aquifer. This equation is derived from the analogy between the flow of ground water and conduction of heat and it is based on the following assumptions.

(*i*) the aquifer is confined,

(*ii*) the well diameter is very small so that the storage within the well can be neglected, and

(*iii*) the aquifer is dewatered instantaneously as the piezometric head drops. The equation is given as

$$s = \frac{Q}{4\pi T}\int_u^{\infty}\frac{e^{-u}}{u}.du \qquad ...(9.42)$$

where s = The drawdown in m in an observation well at a distance r from the pumping well after 't' seconds since the pumping began

Q = The constant pumping rate from the well in m^3/s

T = The transmissivity of the aquifer in m^2/s

and
$$u = \frac{r^2 S}{4Tt} \qquad ...(9.43)$$

where S is the storage coefficient of the aquifer. If a time unit of day is preferred t, T and Q should be expressed in days, m^2/day and m^3/day, respectively.

The exponential integral in Eq. (9.42) is symbolically written as $W(u)$ and is generally called *well function* or *Theis well function*. In terms of well function Eq. (9.42) becomes

$$s = \frac{Q}{4\pi T}W(u) \qquad ...(9.44)$$

The well function can be evaluated from the series expansion given by

$$W(u) = \left[-0.577216 - \ln u + u - \frac{u^2}{2.2!} + \frac{u^3}{3.3!} - \frac{u^4}{4.4!} + ...\right] \qquad ...(9.45)$$

The well function is tabulated in Table 9.2. For example, for $u = 0.005$ the value of $W(u)$ may be read as 4.73 from this table. In the Theis equation it is assumed that the confining beds are water tight and do not transmit any water. But in the case of leaky aquifers there may be considerable flow through the confining beds. Modified equations have been developed which take into account such leakage. Equations are also available for unsteady radial flow to partially penetrating wells in confined aquifers. However, they are not discussed here.

Table 9.2 Well Function: Values of W (u) for values of u

u	1.0	2.0	3.0	4.0	5.0	6.0	7.0	8.0	9.0
× 1	0.219	0.049	0.013	0.0038	0.0011	0.00036	0.00012	0.000038	0.000012
$\times 10^{-1}$	1.82	1.22	0.91	0.70	0.56	0.45	0.37	0.31	0.26
$\times 10^{-2}$	4.04	3.35	2.96	2.68	2.47	2.30	2.15	2.03	1.92
$\times 10^{-3}$	6.33	5.64	5.23	4.95	4.73	4.54	4.39	4.26	4.14
$\times 10^{-4}$	8.63	7.94	7.53	7.25	7.02	6.84	6.69	6.55	6.44
$\times 10^{-5}$	10.94	10.24	9.84	9.55	9.33	9.14	8.99	8.86	8.74
$\times 10^{-6}$	13.24	12.55	12.14	11.85	11.63	11.45	11.29	11.16	11.04
$\times 10^{-7}$	15.54	14.85	14.44	14.15	13.93	13.75	13.60	13.46	13.34
$\times 10^{-8}$	17.84	17.15	16.74	16.46	16.23	16.05	15.90	15.76	15.65
$\times 10^{-9}$	20.15	19.45	19.05	18.76	18.54	18.35	18.20	18.07	17.95
$\times 10^{-10}$	22.45	21.76	21.35	21.06	20.84	20.66	20.50	20.37	20.25
$\times 10^{-11}$	24.75	24.06	23.65	23.36	23.14	22.96	22.81	22.67	22.55
$\times 10^{-12}$	27.05	26.36	25.96	25.67	25.44	25.26	25.11	24.97	24.86
$\times 10^{-13}$	29.36	28.66	28.26	27.97	27.75	27.56	27.41	27.28	27.16
$\times 10^{-14}$	31.66	30.97	30.56	30.27	30.05	29.87	29.71	29.58	29.46
$\times 10^{-15}$	33.96	33.27	32.86	32.58	32.35	32.17	32.02	31.88	31.76

The Theis equation can be used to evaluate the average aquifer parameters S and T in the viscinity of the well by measuring the changes in the drawdown with time in one or more observation wells near the pumping well under the influence of constant pumping rate. Pumping need not be continued for long duration till the so-called steady state conditions are attained.

Because of the mathematical difficulties encountered in applying Eq. (9.42) or its equivalent Eq. (9.45), several investigators have developed simple approximate solutions that can be readily applied for field purposes. Three methods by Theis, Cooper and Jacob, and Chow and the Theis recovery test method are only discussed here.

Theis Method. Eq. (9.43) can be rearranged to take the form

$$\frac{r^2}{t} = \left(\frac{4T}{S}\right) u \qquad \text{...(9.46)}$$

Similarly Eq. (9.44) can be written as

$$s = \left(\frac{Q}{4\pi T}\right) W(u)$$

Comparing these two equations it can be seen that the relation between $W(u)$ and u must be similar to that between s and $\frac{r^2}{t}$, because the terms in the parenthesis in both the equations are constants. Based on this similarity Theis suggested a graphical procedure to obtain an approximate solution for S and T.

The procedure is as follows. During a pumping test the drawdown in an observation well is continuously measured with time. A curve is plotted on a transparent logarithmic paper between drawdown s and the values of $\frac{r^2}{t}$. This curve is known as the *data curve*. Another curve is prepared on a logarithmic paper between the values of u and $W(u)$ obtained from Table 9.2, which is called the *type curve*. Of course the scale for one log cycle adopted for both these curves must be same. The data curve is now superimposed on the type curve such that the axes of both curves are parallel and it is then adjusted until some portion of the data curve coincides with the type curve. This procedure is called the *matching*. The coordinates of any convenient point on the matched portion namely u, $W(u)$, $\frac{r^2}{t}$ and s are determined. Then the values of T and S can be obtained by substituting these coordinate values in Eqs. (9.44) and (9.46).

Example 9.12. *A 20 cm diameter well penetrating a confined aquifer is pumped at a uniform rate of 30 lit/s and the drawdowns measured in an observation well 60 m away are given below.*

t, hours	0.0167	0.0333	0.0666	0.1333	0.5	1.0	2.0	3.0	4.0
s, metres	0.20	0.30	0.41	0.53	0.76	0.90	1.00	1.07	1.12

Determine the aquifer parameters T and S using the Theis method and estimate the drawdown in the well at the end of 7 days of pumping.

Solution. Values of r^2/t in m^2/min are computed using r = 60 m and they are shown in Table 9.3.

Table 9.3 Pumping Test Data of Example 9.9 (r = 60 m)

t in hours	s in m	r^2/t in m^2/min
0.0167	0.20	3593
0.0333	0.30	1802
0.0666	0.41	901
0.1333	0.53	450
0.5	0.76	120
1.0	0.90	60
2.0	1.00	30
3.0	1.07	20
4.0	1.12	15

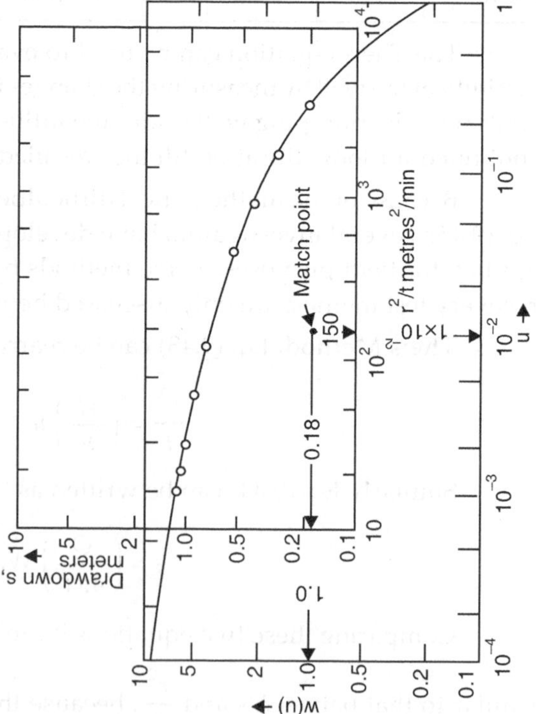

FIGURE 9.14 Theis method of solution for S and T.

Values of s and r^2/t are plotted on a logarithmic paper. Values of $W(u)$ and u from Table 9.2 are plotted on another sheet of logarithmic paper and a curve is drawn through the points. The two curves are superposed and shifted with coordinate axes parallel until the curve of the observational points (data curve) coincides with the type curve, as shown in Fig. 9.14. For convenience a match point is selected with $W(u) = 1.00$ and $u = 1 \times 10^{-2}$ for which s and r^2/t are read as 0.18 m and 150 m²/min respectively. Now from Eq. (9.44)

$$T = \left(\frac{Q}{4\pi S}\right). W(u)$$

$$Q = 30\ l/s = 0.03\ m^3/s = 2592\ m^3/day$$

$$\therefore \quad T = \frac{2592 \times 1}{4\pi \times 0.18} = 1145.92\ m^2/day.$$

and from Eq. (9.46),

$$S = \frac{4Tu}{(r^2/t)}$$

$$r^2/t = 150\ m^2/min = 216000\ m^2/day$$

$$\therefore \quad S = \frac{4 \times 1145.92 \times 1 \times 10^{-2}}{216000} = 0.0002122.$$

From Eq. (9.46),

$$u = \frac{S}{4T} \cdot \frac{r^2}{t}$$

At the end of seven days, $\frac{r^2}{t} = \frac{60 \times 60}{7} = 514.286\ m^2/day$

and $$u = \frac{0.0002122}{4 \times 1145.92} \times 514.286 = 2.3809 \times 10^{-5}$$

From Table 9.2, when $u = 2.3809 \times 10^{-5}$ $W(u) = 10.088$

$$s = \frac{Q}{4\pi \times T} W(u) = \frac{2592}{4\pi \times 1145.92} \times 10.088 = 1.816\ m.$$

Cooper and Jacob Method. It was noted by Cooper and Jacob that for small values of r and large values of t, u is small, so that the series terms in Eq. (9.45) become negligible after the first two terms. Let s_1 and s_2 be the drawdowns in an observation well near the pumping well at two different times t_1 and t_2 where both of them are large. Then, we have

$$s_1 = \frac{Q}{4\pi T} W(u_1)$$

where $u_1 = \frac{r^2 S}{4Tt_1}$

and $$s_2 = \frac{Q}{4\pi T} W(u_2)$$

where $$u_2 = \frac{r^2 S}{4Tt_2}$$

Therefore, $$\Delta s = s_2 - s_1$$

$$= \frac{Q}{4\pi T}\left[W(u_2) - W(u_1)\right]$$

Since u_1 and u_2 are small, from the approximation mentioned above

$$\Delta s = \frac{Q}{4\pi T}\left[-0.5772 - \ln u_2 - (-0.5772 - \ln u_1)\right]$$

$$= \frac{Q}{4\pi T}(\ln u_1 - \ln u_2)$$

$$= \frac{Q}{4\pi T} \ln\left(\frac{u_1}{u_2}\right)$$

$$= \frac{Q}{4\pi T} \ln\left(\frac{t_2}{t_1}\right)$$

or $$T = \frac{Q}{4\pi \Delta s} \ln\left(\frac{t_2}{t_1}\right) \quad ...(9.47)$$

Changing this to decimal logarithm

$$T = \frac{2.3026\, Q}{4\pi \Delta s} \log\left(\frac{t_2}{t_1}\right) \quad ...(9.48)$$

The drawdown s is plotted on an arithmetic scale against time t on a logarithmic scale as shown in Fig. 9.15. This time-drawdown curve will tend to be a straight line for large values of t (or small value of u). If Δs is taken as the drawdown difference for one log cycle on time, that is $t_2 = 10\, t_1$, T is easily evaluated from

$$T = \frac{2.3026\, Q}{4\pi \Delta s} \quad ...(9.49)$$

since $$\log\left(\frac{t_2}{t_1}\right) = 1.$$

To evaluate S, the following procedure is adopted. Assuming the present approximation holds good for the entire range,

$$s = \frac{Q}{4\pi T} W(u)$$

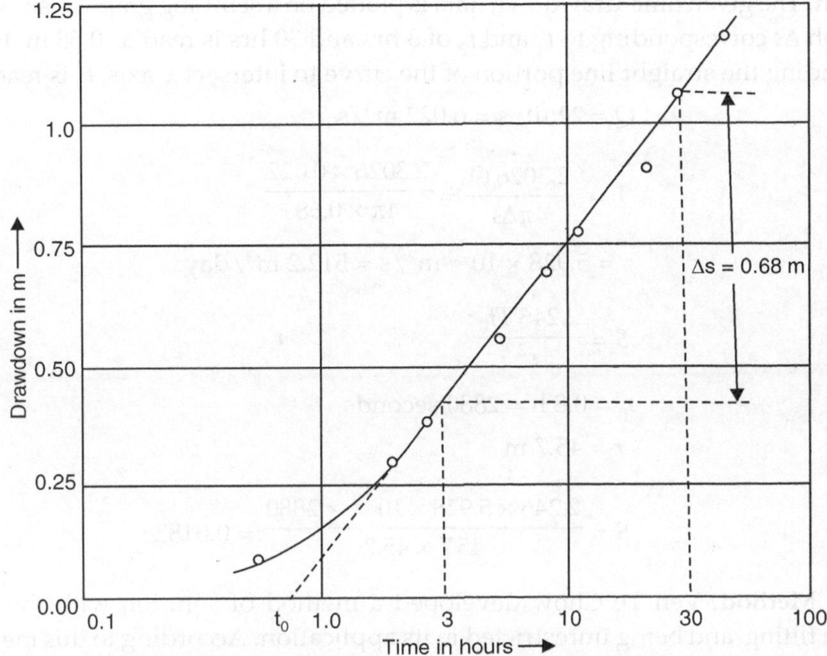

FIGURE 9.15 Time-drawdown plot for Cooper-Jacob method.

When s is zero, $W(u)$ must be equal to zero. That is

$$0 = W(u)$$
$$= -0.5772 - \ln u.$$

which yields $u = 0.56146$. The straight line portion of the curve is extended back to intersect the x-axis to give t_0, the time corresponding to $s = 0$. Hence at $s = 0$,

$$u = 0.56146 = \frac{Sr^2}{4Tt_0}$$

or

$$S = \frac{2.246\, Tt_0}{r^2} \qquad \text{...(9.50)}$$

Equations (9.49) and (9.50) are used to evaluate T and S after obtaining Δs and t_0 from the time-drawdown curve.

Example 9.13. *A pumping test was carried out on a new irrigation bore well penetrating fully into a confined aquifer at a rate of 22 lit/s. The drawdown measured in an observation well located at 45.7 m from the pumping well during the test is as given below. Determine T and S of the aquifer, using Cooper-Jacob method.*

Time t in hours	0.5	1.8	2.7	5.4	9.0	12.0	18	30	54
Drawdown s in m	0.091	0.294	0.382	0.55	0.701	0.785	0.911	1.06	1.24

Solution. The given time-drawdown data is plotted on a semi-log graph as shown in Fig. 9.15. From this graph Δs corresponding to t_1 and t_2 of 3 hrs and 30 hrs is read as 0.68 m. From the same graph by extending the straight line portion of the curve to intersect x-axis, t_0 is read as 0.8 hours.

$$Q = 22 \text{ lit/s} = 0.022 \text{ m}^3/\text{s}$$

$$T = \frac{2.3026\,Q}{4\pi\Delta s} = \frac{2.3026 \times 0.022}{4\pi \times 0.68}$$

$$= 5.928 \times 10^{-3} \text{ m}^2/\text{s} = 512.2 \text{ m}^2/\text{day}$$

$$S = \frac{2.246\,Tt_0}{r^2}$$

$$t_0 = 0.8 \text{ h} = 2880 \text{ seconds}$$

$$r = 45.7 \text{ m}$$

$$\therefore \qquad S = \frac{2.246 \times 5.928 \times 10^{-3} \times 2880}{45.7 \times 45.7} = 0.01836$$

Chow's Method. Ven Te Chow developed a method of solution with the advantage of avoiding curve fitting and being unrestricted in its application. According to this method, the data on time-drawdown in an observation well is plotted on a semilogarithic paper in the same manner as for Cooper-Jacob method. On the plotted curve an arbitrary point with co-ordinates t and s is chosen. At this chosen point a tangent is drawn to the drawdown curve and the drawdown difference Δs in m per log cycle is determined. Then the value of the new function $F(u)$ is computed from

$$F(u) = \frac{s}{\Delta s} \qquad \text{...(9.51)}$$

And the values of u, and $W(u)$ corresponding to $F(u)$ are found from Fig. 9.16 given by Chow.

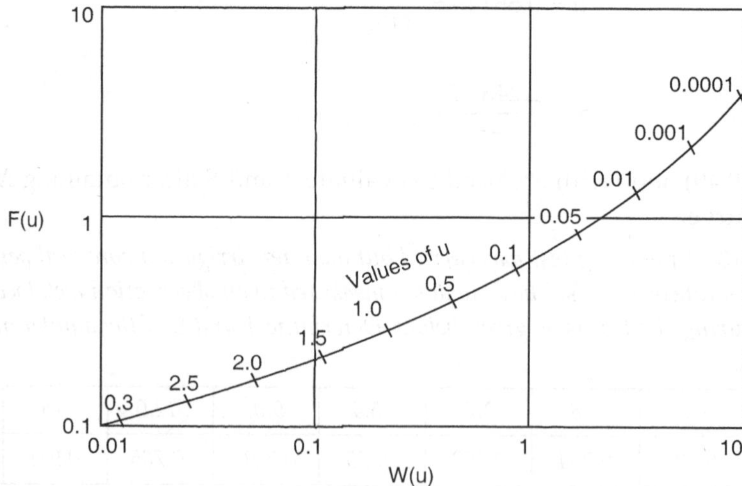

FIGURE 9.16 Relation between F(u), W(u) and u for Chow's method.

Since the relation between $F(u)$, $W(u)$ and u is given by

$$F(u) = \frac{W(u)\, e^u}{2.3} \qquad \text{...(9.52)}$$

the value of u may also be found out by trial and error without resorting to Fig. 9.14 such that along with the corresponding value of $W(u)$ obtained from well function table, it satisfies Eq. (9.52). If $F(u) > 2.0$; even the trial and error procedure is avoided since in that case $W(u) = 2.3\, F(u)$ and u is then obtained from well function table.

After knowing u and $W(u)$, the aquifer parameters T and S are finally computed using Eqs. (9.44) and (9.46).

Example 9.14. *The time-drawdown measurements in an observation well at a radial distance of 200 m from the pumping well are as given under. What are the values of the parameters T and S if the constant rate of pumping is 1000 m³/day? Use Chow's method.*

t in days	*0.001*	*0.005*	*0.01*	*0.05*	*0.10*	*0.50*	*1.0*	*5.0*	*10*
s in m	*0.017*	*0.097*	*0.145*	*0.267*	*0.322*	*0.449*	*0.504*	*0.632*	*0.687*

Solution. The given *s vs t* data is plotted on a semilogarithmic paper as shown in Fig. 9.17. At point *A*, with co-ordinates of $s = 0.2$ m and $t = 0.02$ days, a tangent is drawn which gives $\Delta s = 0.18$ m for one log cycle.

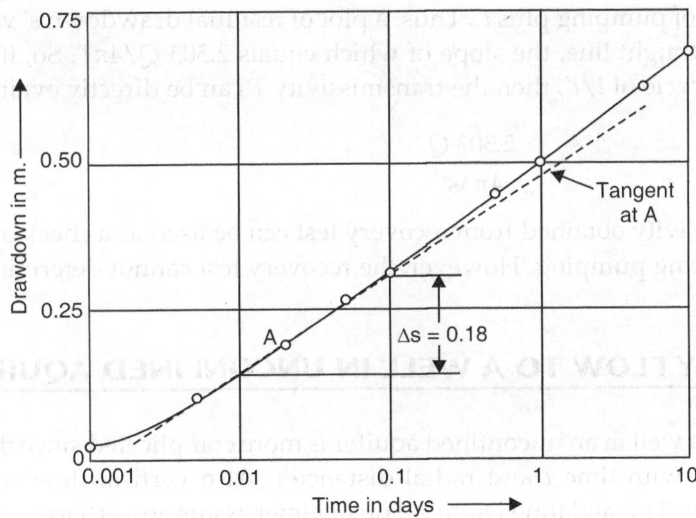

FIGURE 9.17 Determination of *T* and *S* by Chow's method.

$$\therefore \qquad F(u) = \frac{s}{\Delta s} = \frac{0.2}{0.18} = 1.11$$

From Fig. (9.16), when $F(u) = 1.11$

$$W(u) = 2.2 \text{ and } u = 0.065$$

From Eq. (9.44),
$$T = \frac{Q}{4\pi s} W(u)$$

$$= \frac{1000 \times 2.2}{4\pi \times 0.2} = 875.35 \text{ m}^2/\text{day} = 0.01 \text{ m}^2/\text{s}$$

From Eq. (9.46),
$$S = \frac{4\,Tu\,.\,t}{r^2}$$

$$= \frac{4 \times 875.35 \times 0.065 \times 0.02}{200 \times 200} = 1.14 \times 10^{-4}$$

Recovery test. At the end of a pumping test, when the pump is stopped, the water levels in pumping and observation wells will begin to rise. This is referred to as the *recovery* of groundwater. The measurements of drawdown below the original static water level (prior to pumping) during the recovery period are known as *residual drawdowns*, denoted by s'. Theis showed that for r small and t' large the residual drawdown s' can be given as,

$$s' = \frac{2.303\,Q}{4\pi T} \log\left(\frac{t}{t'}\right) \qquad \text{...(9.53)}$$

where t is the time since pumping was started and t' is the time since pumping was stopped. That is, t is the total time of pumping plus t'. Thus, a plot of residual drawdown s' versus the logarithm of (t/t') plots as a straight line, the slope of which equals $2.303\ Q/4\pi T$. So, if $\Delta\ s'$ is the residual drawdown per log cycle of t/t', then the transmissivity T can be directly evaluated as

$$T = \frac{2.303\,Q}{4\pi \Delta s'} \qquad \text{...(9.54)}$$

The transmissivity obtained from recovery test can be used as a check on T calculated from drawdown data during pumping. However, the recovery test cannot determine the value of S.

9.10 UNSTEADY FLOW TO A WELL IN UNCONFINED AQUIFER

Unsteady flow into a well in an unconfined aquifer is more complicated since the transmissivity of the aquifer changes with time t and radial distance r. Also vertical flow components may be significant near the well invalidating Dupuit- Forchheimer assumptions. However, if the drawdown s is small compared to the thickness of the aquifer H, the Theis Eq. (9.44) of confined aquifers can be used for unconfined aquifers. For larger drawdowns more exact solutions are available.

9.11 EFFECTS OF BOUNDARIES ON GROUNDWATER MOVEMENT

In the analysis of steady and unsteady flows towards a well in an aquifer, it has been assumed that the aquifer is of infinite extent having uniform thickness, and the water table or piezometric surface is horizontal before the pumping commenced. Such ideal conditions are rarely met with in the field although in many cases the conditions are fairly satisfactory to yield results of reasonable accuracy. In several other cases the boundary conditions are far from ideal and need some special attention. For example, when two wells are closely located and simultaneously pumped the effect of pumping in each will be felt by other and the cone of depression will not be symmetrical. The water table in the case of unconfined aquifers may have significant slope because of various topographic conditions. A recharge source such as a stream may exist within the range of well drawdown limits.

The *method of images* is a convenient way to treat many of the boundary problems encountered in the field. A detailed description of the method of images can be found in any text book exclusively dealing with groundwater hydrology.

9.12 GROUNDWATER EXPLORATION

The occurrence and movement of groundwater are closely related to the geological formations. Therefore a thorough understanding of geology of the area under investigation will be of immense help to the hydrologists in search of groundwater. Apart from drilling, which is the direct method, there are some geophysical methods which can be used to detect the presence of groundwater. These include electrical resistivity method and seismic refraction method.

In the electrical resistivity method, the resistivity of the formation is measured by sending electricity through it from the ground surface. In the seismic refraction method a seismic shock is produced by striking the ground with a heavy hammer and the velocity of travel of the shock wave is recorded. The resistivity measurements or the velocities of shock waves are then analysed and interpreted to know the characteristics of the formations and also the presence of water if any.

The technology of remote sensing has developed rapidly in the recent years. Photographs of earth taken from aircraft or satellite at various electromagnetic wavelength ranges can provide useful information regarding groundwater conditions.

For a detailed description of these methods a good textbook on geohydrology or groundwater hydrology may be consulted.

9.13 SAFE YIELD

The *safe yield* of a groundwater basin may be defined as the amount of water which can be withdrawn from it annually without producing any undesirable effect. Any withdrawal in excess of safe yield is called *overdraft*.

In many cases the safe yield may simply be taken to be the amount of water that enters the groundwater basin every year. When the withdrawals exceed the long-term mean annual recharge to the basin, the excess must come from the storage within the aquifer. Such permanent depletion is often referred to as *mining* of groundwater. In most basins the quantity of water in storage is many times the annual recharge. Therefore, in any one year the draft can exceed the recharge without causing permanent depletion. But on a long-term basis, if the draft exceeds the average annual recharge it becomes over draft.

The safe yield may be less than the average annual recharge to the basin due to many factors that include economics, water quality and water rights. Higher drafts may lower the water table resulting in increased pumping costs. Pumping in a coastal aquifer could induce sea water intrusion in to the basin. Higher draft rates may cause interference with prior water rights within a basin or in adjacent basins. If no economic, quality, or legal problems are created by pumpage from a groundwater basin, its safe yield is then governed by the average annual recharge to the basin.

9.14 YIELD OF AN OPEN WELL

Large diameter wells dug down to the water bearing strata are known as the *open wells*. They derive water from the formations close to the ground surface. Thus open wells are always shallow, with depths not exceeding 10 to 20 m. The large size of the open wells allows the storage of large quantities of water in the well. The side walls of the open well are usually lined with stone masonry and the water finds entry into the well only from the bottom. When the pumping is not taking place the water level in the well is same as the general water table level in the surroundings of the well. When the pumping is taking place, the water level in the well is depressed and the difference between the water table level and the water level in the well is known as the depression head. The rate of flow into the well is proportional to the depression head. Thus $Q \propto h$ or $Q = K \,.\, h$ where Q is the rate of flow into the well, h is the depression head and K is a constant reflecting the characteristics of the formation surrounding the well. When the depression head is large, it may cause the dislodging of soil particles due to high velocities accompanying it. The depression head at which this happens is known as the *critical depression head*. Normally the *safe depression head* also known as the *working head* is taken to be one-third of the critical depression head. The yield of the well corresponding to the critical depression head is known as the *maximum yield* or the *critical yield* and that corresponding to the working head is known as the *maximum safe yield*. Thus if the water is pumped out under constant depression head equal to the working head H, the safe yield is given by

$$Q = KH$$

or

$$Q = \frac{K}{A} \,.\, A \,.\, H$$

or

$$Q = C \,.\, A \,.\, H \qquad \text{...(9.55)}$$

where $C = \dfrac{K}{A}$ is known as the *specific yield* or the *specific capacity* of an open well. It has dimensions of T^{-1}. A is the cross-sectional area of flow into the well at its base which is taken to be equal to $\dfrac{4}{3}$ times the plan area of the well due to cavity formation at the bottom.

Eq. (9.55) can be used to estimate the safe yield from an open well provided the working head, the size and the specific yield of the well are known. The specific yield of the well can be determined by the *recuperation test*. The safe yield of an open well can also be determined by the *pumping test*.

Pumping Test. It is an accurate method of determining the safe yield of an open well. In this method the water level in the well is depressed through large withdrawals of water till the working head is reached. At this stage the discharge from the pump is regulated so that the rate of withdrawal equals the rate of inflow into the well; *i.e.* a constant water level is maintained in the well. Under this equilibrium condition, the volume of water pumped out in unit time gives the safe yield of the well. The safe yield is usually expressed in m^3/h.

Recuperation Test. This is not as accurate as the pumping test, but is resorted to when it is difficult to regulate the pump to maintain a constant level in the well. In this method, the water level in the well is depressed to some level below the normal water table level and the pumping is stopped. The water level in the well starts recuperating. The time taken by the water to rise to some other level is noted. From this data the specific yield of the well can be found out as explained below.

As shown in Fig. 9.18, let the water level be depressed by h_1 m through pumping. After a time of T hours let the depression head be h_2 m. Let h be the depression head in the well at a time t after the pumping stopped. Let dh be the change in depression head in a small time interval of dt.

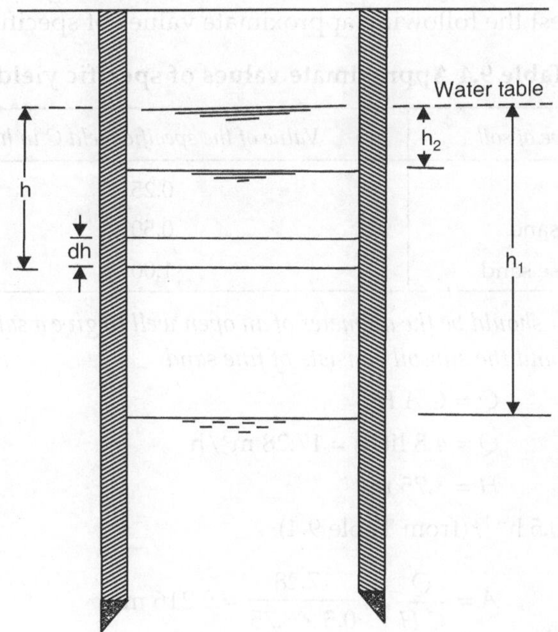

FIGURE 9.18 Recuperation test.

Water flown into well = $Q \,.\, dt = K \,.\, h \,.\, dt.$

Volume change in water content of the well = $-A \,.\, dh$

(The minus sign indicates that h decreases as t increases)

Equating these two,

$$K \,.\, h.dt. = -A \,.\, dh$$

Integrating the above equation between the limits;

$t = 0$ when $h = h_1$

$t = T$ when $h = h_2$

we obtain
$$\frac{K}{A}\int_0^T dt = \int_{h_1}^{h_2} -\frac{dh}{h}.$$

or
$$\frac{K}{A}[t]_0^T = -[\ln h]_{h_1}^{h_2} = [\ln h]_{h_2}^{h_1}$$

$$\frac{K}{A} = \frac{1}{T}\ln\frac{h_1}{h_2}$$

or
$$C = \frac{2.303}{T}\log\left(\frac{h_1}{h_2}\right) \quad ...(9.56)$$

There value of C obtained by substituting the recuperation test data in Eq. (9.56) may now be used to estimate the safe yield of the open well for the known working head from Eq. (9.55). In the absence of recuperation test the following approximate values of specific yield may be adopted.

Table 9.4 Approximate values of specific yield

Type of soil	*Value of the specific yield C in h^{-1}*
Clay	0.25
Fine sand	0.50
Coarse sand	1.00

Example 9.15. *What should be the diameter of an open well to give a safe yield of 4, 8 lit/s? Assume the working head as 3.75 m and the subsoil consists of fine sand.*

Solution. $Q = C\,A\,H$

$Q = 4.8$ lit/s $= 17.28$ m^3/h

$H = 3.75$ m

For fine sand, $C = 0.5$ h^{-1}, (from Table 9.4)

$$\therefore \quad A = \frac{Q}{C\,H} = \frac{17.28}{0.5 \times 3.75} = 9.216 \text{ m}^2$$

$$\text{Diameter of the well} = \sqrt{\frac{4A}{\pi}} = \sqrt{\frac{4 \times 9.216}{\pi}} = 3.43 \text{ m.}$$

Example 9.16. *The following data is obtained from a recuperation test on an open well of diameter 6.5 m.*

R.L. of water table = 237.8 m

R.L. of water level in the well when the pumping is just stopped = 231.2 m.

R.L. of water level in the well 2.5 hours after the pumping is stopped = 234.5 m.

Estimate the safe yield of the well, if the working head is 3 m.

Solution. $h_1 = 237.8 - 231.2 = 6.2$ m

$h_2 = 237.8 - 234.5 = 3.3$ m

$T = 2.5$ h

$$C = \frac{2.303}{T} \log\left(\frac{h_1}{h_2}\right) = \frac{2.303}{2.5} \log\left(\frac{6.2}{3.3}\right) = 0.2523 \text{ h}^{-1}$$

$$Q = CAH = 0.2523 \times \left(\frac{\pi}{4} \times 6.5 \times 6.5\right) \times 3$$

$$= 25.17 \text{ m}^3/\text{h} = 6.98 \text{ lit/s}.$$

Example 9.17. *During a recuperation test conducted on an open well in a region, the water level in the well was depressed by 3 m and it was observed to rise by 1.75 m in 75 minutes. (a) What is the specific yield of open wells in that region? (b) What could be the yield from a well of 5 m diameter under a depression head of 2.5 m? (c) What should be the diameter of the well to give a yield of 12 lit/s under a depression head of 2.0 m?*

Solution. $h_1 = 3$ m

$h_2 = (3 - 1.75) = 1.25$ m

$T = 75 \text{ minutes} = 1.25$ h

(*a*)
$$C = \frac{K}{A} = \frac{2.303}{T} \cdot \log\left(\frac{h_1}{h_2}\right)$$

$$= \frac{2.303}{1.25} \times \log\left(\frac{3}{1.25}\right) = 0.700 \text{ h}^{-1}$$

(*b*)
$$Q = CAH = 0.7 \times \left(\frac{\pi}{4} \times 5 \times 5\right) \times 2.5 = 34.36 \text{ m}^3/\text{h}$$

(*c*)
$$A = \frac{Q}{C\,H}; \quad Q = 12 \text{ lit/s} = 43.2 \text{ m}^3/\text{h}$$

$\therefore$
$$A = \frac{43.2}{0.7 \times 2} = 30.86 \text{ m}^2$$

$$d = \sqrt{\frac{4A}{\pi}} = 6.27 \text{ m}.$$

REVIEW EXERCISES

Descriptive Questions

1. Distinguish between
   (*i*) Aquifer and Aquifuge
   (*ii*) Confined aquifer and water table aquifer
   (*iii*) Aquiclude and Aquitard
   (*iv*) Groundwater and Perched groundwater.

2. Distinguish between
   (*i*) Water table and piezometric surface (*ii*) Vadose zone and phreatic zone
   (*iii*) Artesian well and flowing well (*iv*) Influent and effluent streams.
3. Distinguish between
   (*i*) Hydraulic conductivity and intrinsic permeability
   (*ii*) Specific yield and storage coefficient (*iii*) Darcy velocity and actual velocity
   (*iv*) Specific retention and specific yield (*v*) Transmissivity and storativity
   (*vi*) Pumping test and recovery test.
4. Explain the terms
   (*i*) Cone of depression (*ii*) Radius of influence
   (*iii*) Drawdown (*iv*) Secondary porosity.
5. List out the assumptions made in the analysis of steady radial flow into well.
6. Adopting the usual notation, derive an expression for the steady state discharge of a well in an unconfined aquifer.
7. Derive an expression for the steady state discharge of well fully penetrating into a confined aquifer.
8. Write down the Theis equation for the unsteady radial flow into a well in a confined aquifer. Explain each term in the equation.
9. Describe the Theis method of determining the aquifer parameters using the pumping test data.
10. What do you understand by data curve, type curve and matching?
11. Explain, with a neat sketch, the Cooper and Jacob method of determining the aquifer parameters.
12. Explain the Chow's method of determining the aquifer parameters using the pumping test data.
13. Describe the Recovery test method to determine the transmissivity of the aquifer.
14. How do you compare the results of aquifer parameters obtained from Theis, Cooper and Jacob, and Chow's methods?
15. What do you understand by *safe yield* of a groundwater basin? What are the factors influencing the safe yield? What is meant by *overdraft* and mining?
16. Explain how the yield of an open well can be determined using recuperation test.
17. Define the terms critical depression head, working head, and specific capacity in the case of an open well.
18. What are the advantages of groundwater compared to surface water.
19. Derive Eq. (9.15).
20. In the case of steady unidirectional flow through an unconfined aquifer with rainfall recharge derive and expression for x * where x^* is the distance from the origin at which the phreatic line has maximum height. Use standard notation.

Analytical Questions

1. An undisturbed rock sample has an oven-dry weight of 1305 gm. When it is completely saturated with kerosene it weighed 1463 gm. The saturated sample, when immersed in kerosene, displaced 605 gm of kerosene. What is the porosity of the sample.
2. An unconfined aquifer has an areal extent of 15 km^2. When 9.5 million m^3 of water was pumped out, the water table was observed to go down by 2.4 m. What is the specific yield of the aquifer? If the water table of the same aquifer rises by 12.5 m during a monsoon season, what is the volume of recharge?
3. The elevation of water table in an unconfined aquifer at two locations separated by a distance of 100 m is 1026.2 m and 1025.0 m respectively. If the permeability of the aquifer is 12 m/day and porosity is 15 per cent, what is the actual velocity of flow in the aquifer?
4. A soil sample has a hydraulic conductivity of 15 m/day. What would be its intrinsic permeability? What would be its hydraulic conductivity at 30°C?
5. In order to determine the permeability of an aquifer, a tracer is introduced in an observation well and it is traced in another downstream well 78 m away from the first after 46.5 h. If the elevations of water levels in the two wells differ by 2.9 m and porosity of the aquifer is 18%, calculate the coefficient of permeability of the aquifer.
6. An unconfined aquifer has an area of 325 km^2, thickness of 24.5 m and a porosity of 30%. What is its specific retention if it can yield 1890 million m^3 of free-draining water?
7. What is the hydraulic conductivity of a porus medium with fluid density 1 gm/c.c. fluid viscosity 1 centipoise and intrinsic permeability 7.5 darcys.
8. In a water table aquifer of 50 m thickness, a 20 cm diameter well is pumped at a uniform rate of 0.05 m^3/s. If the steady state drawdown measured in the observation wells located at 10 m and 100 m distances from the well are 6.5 m and 0.25 m respectively, determine the hydraulic conductivity of the aquifer.
9. Determine the radius of influence and the drawdown in the well of analytical question 8.
10. Determine the yield from a 30 cm diameter well under a drawdown of 10 m in the well, if the radius of influence and hydraulic conductivity are 150 m and 5.0 m/day respectively. The aquifer is unconfined with a thickness of 60 m.
11. In analytical question 10, if all other conditions remain unchanged, what is the percentage change in the yield under the following cases

    (*a*) The diameter of the well is increased to 45 cm

    (*b*) The drawdown is limited to 6 m

    (*c*) The hydraulic conductivity of the aquifer becomes 12.5 m/day.
12. A 20 cm diameter well penetrates fully a confined aquifer of thickness 25 m. When the well is pumped at a rate of 200 litres/minute the steady state drawdown in the two observations wells located at 10 m and 100 m distance from the pumping well are found to be 3.5 m and 0.05 m respectively. Calculate the permeability and the transmissivity of the aquifer.
13. Determine the radius of influence and the drawdown in the well for the well of analytical question 12.

14. A well with a radius of 0.5 m penetrates completely a confined aquifer of thickness 40 m and permeability 30 m/day. The well is pumped so that the water level in the well remains at 7.5 m below the original piezometric surface. Assuming that the radius of influence is 500 m, compute the steady state discharge from the well.
15. For the well in analytical question 14, what is the percentage decrease in drawdown for the same discharge if the well diameter is doubled? What is the percentage increase in discharge for the same drawdown if the well diameter is doubled? Assume the radius of influence to be same in all the cases.
16. A well is pumped at the constant rate of 0.004 m^3/s in a confined aquifer of transmissivity $T = 0.004\ m^2/s$ and storativity $S = 0.0005$. Calculate the drawdown 24 h after the start of pumping in an observation well located at a distance of 250 m from the pumped well.
17. A well is pumped at the constant rate of 0.008 m^3/s. A match of the well function [u versus $W(u)$] with drawdown versus time data from an observation well located at a distance of 430 m from the pumped well has produced the following matching values: $u = 1$, $W(u) = 1$, $s = 0.21$ m and $r^2/t = 2055\ m^2$/minute. Calculate the transmissivity and storativity of the aquifer.
18. The drawdown measured in an observation well located at a distance of 100 m from a pumped well is recorded below.

Time in days t	*Drawdown in m* s
0.001	0.083
0.005	0.196
0.01	0.249
0.05	0.376
0.10	0.431
0.50	0.559
1.0	0.614
5.0	0.742
10.0	0.797

The well is in a confined aquifer and the uniform pumping rate from the well is 1000 m^3/day. Determine the aquifer parameters using the Theis matching curve method.
19. Determine the parameter of the aquifer of analytical question 18 using the Cooper-Jacob method.
20. Determine the parameters of the aquifer of analytical question 18 using the Chow's method.
21. Design an open well to give a yield of 10 lit/s. Assume that the well is dug in Coarse sand soil and the working head is 4 m.
22. In a recuperation test on an open well, the water level was depressed by 4 m and it was observed to rise by 2.5 m in 90 minutes. What is the specific capacity of the well? What would be the yield from the well under a depression head of 3 m, if the diameter of the well is 7.5 m?

23. Two rivers A and B with water flow depths of 20 m and 12 m, respectively run almost parallel and are separated by an aquifer formation of 5.0 km width. If the hydraulic conductivity of the aquifer is 15 m/day compute the seepage flow from stream A to stream B.

24. In the analytical question 23 assume that the formation is pervious only for a thickness of 10 m at the bottom and therefore acts as a confined aquifer. What is the flow from stream A to stream B now?

25. Solve analytical question 23 when rain falling at an intensity of 2 mm/h is recharging the aquifer. Assume that the minimum infiltration capacity rate is equal to rainfall intensity.

Objective Type Questions

1. An aquifer is a geological formation which
   (*a*) does not contain water (*b*) contains water but does not transmit
   (*c*) contains and also transmits water (*d*) is a rock outcrop.
2. Which of the following formations does not contain any groundwater?
   (*a*) aquifer (*b*) aquifuge
   (*c*) aquitard (*d*) aquiclude.
3. A geological formation which may contain water but is essentially impermeable to the flow of water through it is known as
   (*a*) aquifer (*b*) aquifuge
   (*c*) aquitard (*d*) aquiclude.
4. Water existing in capillary zone is a part of
   (*a*) phreatic water (*b*) ground water
   (*c*) gravity water (*d*) vadose water.
5. The surface obtained by joining the water levels in several observation well's penetrating a confined aquifer represents
   (*a*) piezometric surface (*b*) water-table surface
   (*c*) capillary frinze (*d*) cone of depression.
6. In the case of a flowing well, the piezometric surface is
   (*a*) below the ground level (*b*) above the ground level
   (*c*) between the ground level and the water surface in the well
   (*d*) below the water surface in the well.
7. In the case of a water table well, the piezometric surface
   (*a*) is above the ground level (*b*) is below the water level in the well
   (*c*) coincides with the water level in the well
   (*d*) is between the water level in the well and ground level.

8. The ratio of the volume of water retained by the formation when it is freely drained to the volume of the formation is known as
(*a*) specific yield (*b*) specific retention
(*c*) specific storage (*d*) porosity.
9. Specific yield of an aquifer is defined as the ratio of the
(*a*) volume of pore space to the volume of soil
(*b*) volume of water freely drained from a saturated soil to the volume of soil
(*c*) volume of water retained, when a saturated soil is freely drained to the volume of soil
(*d*) volume of pore space to the volume of soil.
10. An aquifer which is underlain by an impermeable layer at the bottom and not confined at the top is known as
(*a*) confined aquifer (*b*) unconfined aquifer
(*c*) semiconfined aquifer (*d*) perched aquifer.
11. An unconfined aquifer is also known as
(*a*) an artesian aquifer (*b*) a leaky aquifer
(*c*) a perched aquifer (*d*) a water table aquifer.
12. The permeability of an aquifer
(*a*) increases with increase in temperature (*b*) increases with decrease in temperature
(*c*) is independent of temperature (*d*) decreases with decrease in temperature.
13. An influent stream is one which
(*a*) contributes runoff to the groundwater (*b*) derives runoff from groundwater
(*c*) neither contributes nor desines runoff from groundwater
(*d*) flows only below the ground.
14. A stream which contributes runoff to groundwater is known as
(*a*) influent stream (*b*) effluent stream
(*c*) perennial stream (*d*) ephemeral stream.
15. Darcy's law for groundwater movement states that the velocity is proportional to
(*a*) the hydraulic gradient (*b*) the square of the hydraulic gradient
(*c*) the logarithm of the hydraulic gradient
(*d*) the reciprocal of the hydraulic gradient.
16. A laboratory test on a sample from an aquifer revealed a porosity of 35%. The specific yield of the aquifer will be
(*a*) equal to 0.35 (*b*) less than 0.35
(*c*) more than 0.35 (*d*) difficult to predict.
17. The dimensions of the intrinsic permeability are
(*a*) $L\,T^{-1}$ (*b*) $L^2\,T^{-1}$
(*c*) $L^2\,T^0$ (*d*) $L^2\,T^2$.

18. The dimensions of the transmissivity of the aquifer T are

(*a*) $L\,T^{-1}$ (*b*) $L^2\,T^{-1}$

(*c*) $L^2\,T^2$ (*d*) $L^2\,T^{-2}$.

19. The unit of intrinsic permeability in C.G.S. system is

(*a*) cm^2 (*b*) cm^2/s

(*c*) cm^2/s^2 (*d*) cm/s.

20. The upper limit of Reynold's number for the Darcy's law to be valid for groundwater flow is

(*a*) 0.01 (*b*) 0.1

(*c*) 1 (*d*) 10.

21. Radius of influence is the horizontal distance between the centre of the pumped well and

(*a*) a point on the cone of depression of maximum drawdown

(*b*) a point on the cone of depression of zero drawdown

(*c*) the first observation well (*d*) the second observation well.

22. The equation for the steady state radial flow to a well was first developed by

(*a*) Darcy (*b*) Theis

(*c*) Dupuit (*d*) Jacob.

23. The dimensions of the specific capacity of an open well are

(*a*) $L\,T^{-1}$ (*b*) L^{-1}

(*c*) $T\,L^{-1}$ (*d*) T^{-1}.

24. The safe depression head for open wells is generally taken to be .... of the critical depression head

(*a*) $\frac{1}{3}$ (*b*) $\frac{1}{2}$

(*c*) $\frac{2}{3}$ (*d*) $\frac{3}{2}$.

25. The dimensions of the storativity S are

(*a*) $L\,T$ (*b*) $L^{-1}\,T^{-1}$

(*c*) $L\,T^{-1}$ (*d*) $L^0\,T^0$.

ANSWERS

Analytical Questions

1. $n = 0.2612$ or 26.12%

2. $s_y = 0.264$; Volume of recharge = 49.5×10^6 m^3

3. 0.96 m/day **4.** 17.93 darcys; 18.75 m/day

5. 8.123 m/h **6.** 0.0626

7. 6.274 m/day **8.** 5.433 m/day

9. 110.354 m; 23.044 m

10. 0.029 m^3/s

11. (*a*) + 6.236% (*b*) – 37.82% (*c*) + 150%

12. 1.2237 m/day; 30.59 m^2/day

13. 103.4 m; 10.4 m

14. 0.0984 m^3/s

15. 10.034%; 11.15%

16. 0.259 m

17. 261.92 m^2/day; 0.000354

18. T = 1000.7 m^2/day; S = 0.000107

19. T = 1011 m^2/day; S = 0.000091

20. T = 875 m^2/day; S = 0.00011

21. 6.77 m

22. 0.654 h^{-1}; 86.68 m^3/h

23. 0.384 m^3/day

24. 0.24 m^3/day

25. $q_1 = -119.616$ m^3/day; $q_2 = 120.384$ m^3/day

Objective Type Questions

1. (*c*)	2 (*b*)	3. (*d*)	4. (*d*)	5. (*a*)	6. (*b*)
7. (*c*)	8. (*b*)	9. (*b*)	10. (*b*)	11. (*d*)	12. (*b*)
13. (*a*)	14. (*a*)	15. (*a*)	16. (*b*)	17. (*c*)	18. (*b*)
19. (*a*)	20. (*c*)	21. (*b*)	22. (*c*)	23. (*d*)	24. (*a*)
25. (*d*).					

Chapter 10 RUNOFF

10.1 INTRODUCTION

The portion of precipitation which appears in the surface streams of either perennial of intermittent nature is called the *runoff*. This is the flow collected from a drainage basin and appearing at an outlet of the basin. In a general sense it is the precipitation excess after meeting the evapotranspiration demands. While the precipitation is sporadic and irregular in nature, the runoff is more uniform compared to precipitation. This contrast between precipitation and runoff is due to the storage effects of the surface layers of the earth.

Runoff is usually expressed as volume per unit time, the common unit being m^3/s or *cumec*. When runoff from different drainage basins is to be compared, a convenient way of expressing the runoff is, perhaps, cumec/sq. km. Runoff may also be expressed as depth equivalent over a basin, that is, mm/day, mm/month of mm/year. This is a particularly useful unit for comparing runoff rate or runoff volume with precipitation since precipitation is almost invariably expressed in this way.

10.2 COMPONENTS OF RUNOFF

According to the source from which the flow is derived, the total runoff is visualised to consist of *surface runoff, subsurface runoff,* and *groundwater runoff*. Fig. 10.1 shows the inter-relationship between various items involved in the process of transforming the total precipitation into the total runoff.

Before reaching the ground a small portion of precipitation evaporates and joins back the atmosphere while another small portion is intercepted which also eventually becomes atmospheric moisture through evaporation. This portion is usually small and insignificant in the runoff analysis.

The remaining precipitation reaching the ground is called the *throughfall*. It may fall either on land surface or directly on to the water surfaces of streams, lakes, and reservoirs. The precipitation falling on water surfaces is called *channel precipitation* and it immediately becomes the streamflow. A very small fraction of the precipitation falling on the land surface may be absorbed in the storage like small ponds, swamps etc., before it becomes evaporation and infiltration. This is called the *depression storage*. The rest of the precipitation falling on land surface, after satisfying the infiltration demand, is temporarily detained on the ground surface and when sufficient depth is built-up it travels over the ground surface towards the stream channels either as quasi-laminar sheet flow or in the form of minor rivulets. This is called the *overland flow*. The overland flow ceases shortly after the rainfall stops. Surface runoff may then be defined as that part of the total runoff which travels over the ground surface to reach a stream channel and then through the channel to reach the basin outlet. It is the combination of the overland flow and the channel precipitation.

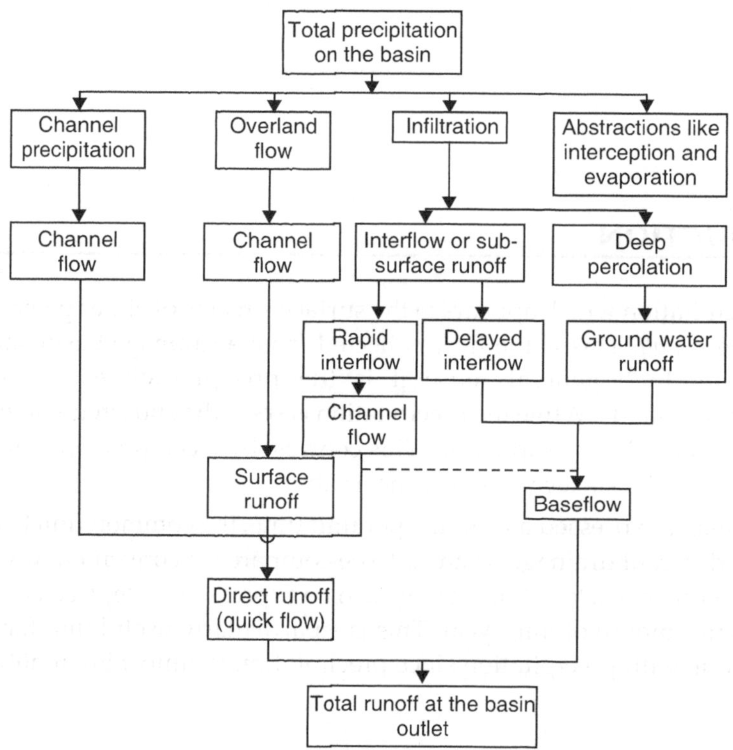

FIGURE 10.1 Schematic representation of runoff process.

Water which infiltrates the soil surface and then moves laterally through the upper soil horizons towards the stream channels above the main groundwater table is known as the *interflow*. It is also known as *subsurface runoff, subsurface storm flow, storm seepage,* and *secondary base flow*. If the lateral hydraulic conductivity of the surface layers is substantially greater than the overall vertical hydraulic conductivity, it is a favourable condition for the generation of interflow. Generally interflow moves more slowly than surface runoff and therefore it may take longer period to reach the stream. In certain formations a part of interflow may enter the stream rapidly.

The infiltrated water which percolates deeply becomes groundwater and when groundwater table rises and intersects the stream channels of the basin it discharges into streams as the groundwater runoff. Since the movement of groundwater is very slow the groundwater runoff to streams may lag behind the occurrence of precipitation by several days, or weeks. The groundwater flow also tends to be very regular and it represents the main long-term component of total runoff and is particularly important during dry spells when surface runoff is absent.

For the practical purpose of analysis total runoff in stream channels is generally classified as *direct runoff* and *base flow*. The direct runoff the *quick flow* is that part of runoff which enters the stream promptly and is equal to the sum of surface runoff and rapid interflow. This clearly represents the major runoff contribution during storm periods and is also the major contribution to most floods. The base flow is defined as the *sustained* or *fair-weathered* runoff and it is composed of groundwater runoff and delayed interflow. Some hydrologists prefer to include the entire interflow as a part of base flow as shown by the broken line in Fig. 10.1. The distinction between direct runoff and base flow is on the basis of time of arrival of the runoff in the stream rather than on the path followed by the runoff.

The *precipitation excess* is that part of total precipitation which contributes directly to the surface runoff. When the precipitation is in the form of rainfall only it is known as *rainfall excess.* That part of precipitation which contributes to the direct runoff may be called the *effective precipitation* or *effective rainfall* if only rainfall is involved.

10.3 FACTORS AFFECTING RUNOFF

The runoff from a drainage basin is influenced by various factors which may be put under two groups, namely the climatic factors and the physiographic factors.

The climate factors include

(*i*) Type of precipitation
(*ii*) Intensity of rainfall
(*iii*) Duration of rainfall
(*iv*) Areal distribution of rainfall
(*v*) Direction of storm movement
(*vi*) Antecedent precipitation
(*vii*) Other climatic factors that effect evaporation and transpiration.

The physiographic factors are

(*i*) Land use
(*ii*) Type of soil
(*iii*) Area of the basin
(*iv*) Shape of the basin
(*v*) Elevation
(*vi*) Slope

(*vii*) Orientation
(*viii*) Type of drainage network
(*ix*) Indirect drainage
(*x*) Artificial drainage.

Though it is easy to identify the various factors which affect runoff, it is rather difficult to quantify some of them and still more difficult to express either maximum runoff (flood), or minimum runoff or the average flow in terms of the above factors. The effect of these factors on the volume of runoff and on the rate of runoff (time distribution) will now be discussed.

10.3.1 Effect on Volume of Runoff

The runoff volume over a period of a year can generally be taken as the difference between the total precipitation received by the basin in the year and the evapotranspiration losses. So, when runoff volume on annual basis is considered it is the total annual precipitation which matters and the factors like intensities, duration, areal distribution etc. become insignificant or irrelevant. The climate factors like temperature, humidity, radiation etc., will set the broader upper limits to the total volume of runoff through their influence on evaporation and transpiration. In the case of physiographic parameters it is the area of the catchment which affects the volume of runoff most. Other factors being equal the total amount of precipitation is large on larger areas. In a region where annual precipitation exceeds the potential evapotranspiration larger areas produce larger volumes of runoff. However, if the annual precipitation is less than potential evapotranspiration a larger area is just as likely as a smaller one to have a zero or very low runoff volume. However, when volume of runoff per unit area of the basin is considered, perhaps area may not have any significant effect. Slope and type of soil may indirectly influence the total runoff from a basin through their effects in delaying water movement after precipitation thereby possibly affecting the amount of evapotranspiration. In general, the highest annual runoff would be expected from steeply sloping areas having thin soils and impermeable rocks. The elevation of the catchment, and its orientation may affect total runoff, again indirectly through its orographic influence on precipitation.

10.3.2 Effect on Time Distribution of Runoff

The effect on the distribution of runoff in time tends to be more variable and unpredictable since the factors change rapidly over a period of few days, hours, or even minutes. The following discussion is with respect to the runoff hydrograph produced by a storm at the basin outlet. The important features of the runoff hydrograph are the peak discharge, the time to peak discharge and the rate of increase of the discharge till the peak is attained. These features are shown in Fig. 10.2.

Type of precipitation. If the precipitation falls in the form of rain its effect on runoff is almost immediately felt provided its intensity is sufficiently large. The resulting runoff, of course, will also depend upon the interaction of the other factors. On the other hand if the precipitation during a given period is entirely in the form of snow it may not contribute anything to runoff at that time. But the melting of this snow in subsequent warm periods may contribute heavily to the runoff rate causing even floods sometimes.

Intensity of precipitation. For rainfall intensities exceeding the infiltration capacity, the runoff increases with increase in intensity. Owing to the storage effects of the basin, however, the increase in the runoff rate is not the same as the increase in rainfall intensity.

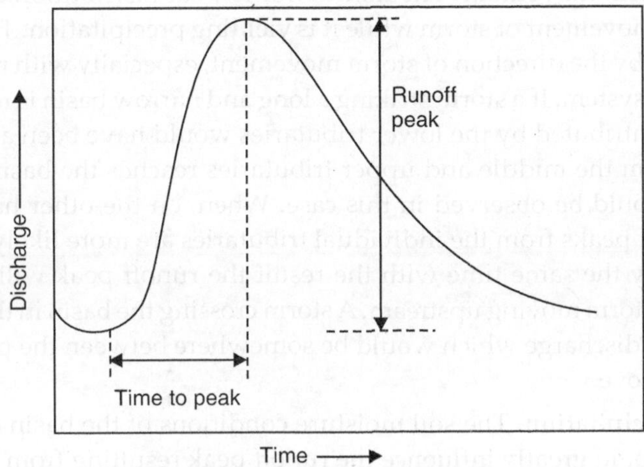

FIGURE 10.2 Features of hydrograph.

Duration of rainfall. If rainfall is of sufficiently prolonged duration, infiltration capacity rate is greatly reduced resulting in high runoff rates. Under favourable conditions the infiltrated water may even raise the water table to the ground surface reducing the infiltration to zero leading to serious flood. As a consequence, rains of long duration may produce high rates of runoff even though the intensity is relatively mild. It is pertinent here to consider the duration of rainfall in conjunction with *critical concentration time of the basin* t_c, which is defined as the time taken by a rain drop falling on the remotest point of the basin to reach the basin outlet. If the duration of rainfall is greater than or equal to t_c, the whole of the basin is likely to contribute to runoff during the later stages of the storm so that the potential runoff rate is at a maximum. On the other hand, if the rainfall duration is less than t_c the potential runoff will be lower than the maximum because only part of the basin will be contributing to runoff before rainfall ceases. It is also apparent that importance of the rainfall duration tends to vary with size and nature of the catchment. In a small catchment with steep slopes maximum potential runoff is likely to be caused by a rainfall of much shorter duration than the duration required in a large gently undulating catchment.

Rainfall distribution over the basin. Uniform areal distribution of rainfall over a basin is rarely observed in nature. Some portions of the basin will receive rainfall which is much less than the average rainfall over the basin while the remaining portion will receive rainfall which is more than the average of the basin. Heavy rainfalls in the lower portions of the basin will produce a runoff hydrograph with a rapid rise and early peak discharge, while heavy rainfall in the upper portions will reverse the hydrograph trend with a slow rise and late peak. On the other hand, rainfall distributed uniformly over the entire basin will tend to produce increased base flow and consequently a long and slow increase in the hydrograph. A means of quantifying the areal distribution is perhaps provided by the *distribution co-efficient* which is obtained by dividing the maximum rainfall observed at any point in the basin by the average rainfall over the basin. Hence for any given total rainfall, all other conditions being same, the greater the distribution co-efficient the greater will be the peak runoff. If, however, the storm is centered in the lower part of the basin near the basin outlet a higher peak runoff would result than from a storm having the same distribution co-efficient but which is centered in the upstream portion of the basin.

Direction of storm movement. The non-uniform areal distribution of precipitation over a region is the result of movement of storm while it is yielding precipitation. The runoff from a basin is therefore influenced by the direction of storm movement, especially with respect to the direction of flow of the drainage system. If a storm striking a long and narrow basin is moving in an upstream direction the runoff contributed by the lower tributaries would have been already drained out by the time the runoff from the middle and upper tributaries reaches the basin outlet and therefore less peak discharge would be observed in this case. When, on the other hand, the storm moves downstream the runoff peaks from the individual tributaries are more likely to arrive at the basin outlet at approximately the same time with the result the runoff peak will be many times more than that in the case of storm moving upstream. A storm crossing the basin in the transverse direction would produce a peak discharge which would be somewhere between the peaks produced by the two cases described above.

Antecedent precipitation. The soil moisture conditions of the basin existing at the time of occurrence of storm would greatly influence the runoff peak resulting from that storm. Even very intense rains falling in late summer, when the soil moisture is at its least, rarely produce high discharges because most of the water enters the soil moisture under the existing high infiltration capacity rates and is held there. In the winter and rainy seasons, since the precipitation occurred in the earlier periods would have raised the soil moisture to high level, one can expect very low infiltration capacity rates. The groundwater runoff also is considerable under these conditions and therefore rains falling in these periods may produce significantly high runoff peaks and even disastrous floods.

The *antecedent precipitation index* (API) as given below may be taken as a measure of the soil moisture conditions existing on the day of storm under consideration.

$$I_t = KI_{t-1} \qquad \text{...(10.1)}$$

where I_t is the API of any day t, K is a constant known as the *recession factor*, and I_{t-1} is the API of the day $(t-1)$. If rain occurs on any day, the amount of rain is added to the index and this value is used in the computation of API of the next day. If there is no rainfall in the previous t days, then

$$I_t = K^t I_0 \qquad \text{...(10.2)}$$

where I_0 is the initial value of API. The value of K ranges between 0.85 and 0.98. There are also other methods of defining API.

Example 10.1. *Rainfall of 12, 30, 40, 44 and 17 mm were recorded on 3rd, 9th, 10th, 16th and 17th days of particular month. Compute the antecedent precipitation index for the first 20 days of the month and sketch its variation with time. Assume that API of the last day in the previous month is 85 mm and the value of the recession factor K is 0.90.*

Solution. The API of any day t, denoted by I_t is obtained from the equation.

$I_t = KI_{t-1} + P_t$, where P_t is the precipitation on tth day. The computations to obtain the API for the first 20 days in the month are set out in Table 10.1, with $I_0 = 85$ and $K = 0.90$.

Table 10.1 Determination of Antecedent Precipitation Index

Day t	*Precipitation in mm* P_t	I_{t-1}	$K.I_{t-1}$	$I_t = KI_{t-1} + P_t$
1	0	85.00	76.50	76.50
2	0	76.50	68.85	68.85
3	12	68.85	61.97	73.97
4	0	73.97	66.57	66.57
5	0	66.57	59.91	59.91
6	0	59.91	53.92	53.92
7	0	53.92	48.53	48.53
8	0	48.53	43.68	43.68
9	30	43.68	39.31	69.31
10	40	69.31	62.38	102.38
11	0	102.38	92.14	92.14
12	0	92.14	82.93	82.93
13	0	82.93	74.64	74.64
14	0	74.64	67.18	67.18
15	0	67.18	60.46	60.46
16	44	60.46	54.41	98.41
17	17	98.41	88.57	105.57
18	0	105.57	95.01	95.01
19	0	95.01	85.51	85.51
20	0	85.51	76.96	76.96

The variation of the antecedent precipitation with time t in days is shown in Fig. 10.3 wherein the precipitation observed is indicated as a rectangular bar on the day of its occurrence.

Other climatic conditions. Apart from precipitation, the other climatic factors like temperature, wind velocity, atmospheric pressure, radiation, humidity etc. ; also affect runoff. Their influence, however, instead of being direct, is to determine how much of precipitation that falls on the basin shall be transpired and evaporated, the residual only being available for runoff. Their effect is more pronounced on the volume of runoff (over a long period of time such as a year) than on the time distribution of runoff from individual storm bursts.

Land use. Land use affects the runoff through its influence on interception, evapotranspiration and soil moisture movement. In urban areas there is little scope for infiltration and transpiration and all the rainfall immediately becomes direct runoff producing high discharges. Rain falling on a thickly forested area may not produce any runoff because in the forests a thick layer of *mulch* of leaves and grass exists and this can absorb a lot of rainfall leaving very little for runoff. Even this little runoff encounters a lot of resistance during the overland flow. In non-forested areas the infiltration, interception and evapotranspiration losses are less and therefore high runoff rates are expected. Thus, it is apparent that the land use is one of the most important factors that affects the runoff of any area. Further, the land use may not be uniform over the entire basin and the land use pattern may be altered rather rapidly.

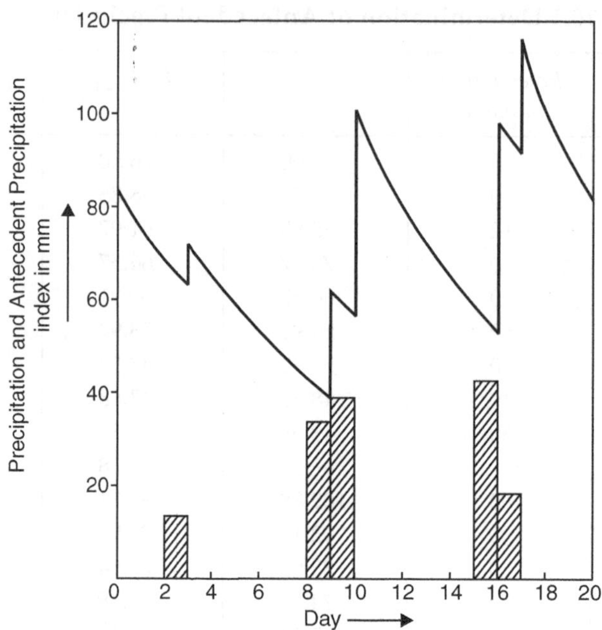

FIGURE 10.3 Variation of antecedent precipitation index.

Type of soil. The type of soil has direct influence on its infiltration capacity rate and consequently it also affects the runoff. Open textured sandy soils will tend to have higher infiltration rates and therefore tend to produce less peak discharges. Fine-grained and closely compacted clay soils will have the opposite effect on runoff.

Area of the basin. The method of fixing the area of drainage basin from topographic maps for any given outlet point (or the gauging station) on the stream was discussed in Chapter 4. It was noted that the area of the basin increases as the outlet point shifts downstream.

Area of the basin affects the peak flow and minimum flow in different ways and hence the effects on these flow characteristics will be considered separately.

If all other factors including depth and intensity of rainfall remain constant in all instances, the total runoff expressed as depth in cm over the basin will be same for all the basins irrespective of their size. The total runoff expressed as volume will be more in the case of larger basins associated with higher peak discharges. However, when peak flow is expressed in cumecs per sq km of the basin it will be less in the case of larger basins. This is because in larger basins the base period of the hydrograph broadens out as more time is required for the water to drain out. It was assumed that the depth and intensity of rainfall is the same in all instances. But it was seen in Chapter 5 that the maximum intensity of rainfall of given frequency decreases with increase in area. Therefore, larger basins are likely to experience less intense storms and peak discharge per unit area of the basin will be lower.

As far as the minimum flow in the stream is concerned, the larger basins will produce higher rates than smaller catchments. This is because after the surface runoff ceases, the flow is entirely drawn from ground water storage which is large for larger basins.

Shape of the catchment. In an approximately square or circular drainage basin the tributaries often tend to come together and join the main stream near the center of the area. Consequently the separate runoff peaks generated by a heavy rainfall in the individual tributaries are likely to reach the main stream in approximately the same locality and at approximately the same time, thereby resulting in a large and rapid increase in the runoff. On the other hand, if the basin is long and narrow the tributaries will tend to be relatively short and are more likely to join the main stream at intervals along its length. This means that after a heavy rainfall over the area the runoff peaks of the lower tributaries would have left the catchment before the peaks of the upstream tributaries have reached the basin outlet. Elongated catchments are thus less subjected to high runoff peaks. The numerical indices like form factor, compactness co-efficient, circularity ratio and elongation ratio will express the shape of the basin quantitatively. These indices were described in detail in Chapter 4.

Elevation. The variation in elevation and also the mean elevation of a basin may influence runoff in as much as they decide the proportion of precipitation falling in the form of snow and the evaporation and transpiration losses. For large basins the mean elevation is determined by the intersections method. A topographic map of the basin is subdivided into squares of equal size by enough number of lines so that at least 100 intersections fall within the area. The mean elevation of the basin is then computed as the average of the elevations at all the intersections.

Slope. The slope of the catchment area is an important factor as it decides the relative importance of infiltration, interflow and overland flow. Infiltration is more predominant in flat areas while the overland flow and interflow are more predominant in steeply sloping areas. Hence steeply sloped catchments will tend to produce higher runoff peaks.

The shape of the runoff hydrograph depends not only on the speed with which water enters the stream channel but also on the speed with which it moves down the channel to the basin outlet. Stream channel slope may therefore be as important as the catchment slope and has frequently yielded more significant correlations with runoff. The methods of determining the stream channel slope were described in Chapter 4.

Orientation. The orientation of the basin decides the amount of solar radiation received from the sun. Thus it may affect the runoff through its influence on evaporation, transpiration and snowmelt processes.

Drainage network. The characteristics of the runoff hydrograph are greatly influenced by the efficiency of the drainage system of the basin. If the basin is well drained with large number of tributaries shortening the length of overland flow, the surface runoff concentrates quickly resulting in high peaks. Correspondingly the minimum flow in such cases is low.

The characteristics of the drainage network may be fairly well described by (*i*) the order of the stream (*ii*) stream density and (*iii*) drainage density. These were discussed in Chapter 4.

Indirect drainage. There are two cases in which the basin is not drained through the surface drainage network that has been assumed to exist in all the preceding discussions. They are (*i*) *Karst* topography and (*ii*) very pervious surface with flat slope.

Regions underlain by soluble rock formations, especially lime stone, often have characteristics undulating surfaces with conical round hillocks and circular sinks. Such areas are said to have a karst topography. The runoff from these areas usually enters the ground through sink holes and fissures and pursues its course to an outlet through a system of underground passages or solution channels into an adjacent or far distant streams, lakes, or the sea. The resulting underground drainage network is often complex and is somewhat similar to a surface drainage network in its characteristics. In determining runoff from basins containing karst topography it is best to exclude all karst areas since they rarely contribute anything to either surface runoff or groundwater runoff. On the other hand, in estimating the runoff of a drainage basin containing highly permeable depression areas, these areas should usually be excluded only in estimating surfaces runoff but should be included in making studies of groundwater runoff.

Artificial drainage. Conditions of artificial drainage exist in areas provided with a system of open ditches or tile drains and also in urban areas provided with storm sewers and where considerable portions of the area are covered by buildings or paved surfaces. Runoff from such areas would naturally be accelerated, but the effect would be purely local and no significant increase in peak flow of large streams can result in general.

Artificial and natural reservoirs act favourably on the distribution of runoff and produce more regulated flow in the stream, even though they increase the evaporation losses somewhat from the basin.

10.4 BASIN YIELD

In general sense, yield refers to the quantity of any product resulting from the exploitation of natural resources. Specifically, basin yield refers to the quantity of water available from a stream at a given point over a specified duration of time. The duration of time in the definition of yield would normally be a month or longer. The emphasis is on water volumes rather than instantaneous discharges. Therefore yield from a basin is the summation of the continuous hydrograph of flow at its outlet over the specified time period. It is the consequence of all hydrologic events causing flow, including storms of all durations and intensities and the climatic, geologic and land use environment.

The hydrologic water balance equation for any basin under consideration may be written as

$$(S_1 + S_{s_1} + S_{g_1} + S_{sm_1} + P)$$

$$= E + \int_{t_1}^{t_2} Q \cdot dt + \left(S_2 + S_{s_2} + S_{g_2} + S_{sm_2}\right) \qquad ...(10.3)$$

where S = Volume of water in storage in channels and reservoirs of the basin

S_s = Volume of water in storage on the surface of the ground on leaves, pavements, etc.

S_g = Volume of water in storage as groundwater

S_{sm} = Volume of water in storage as soil moisture

P = Total average depth of precipitation over the basin between times t_1 and t_2

E = Total evaporation and evapotranspiration from the basin between times t_1 and t_2

Q = Instantaneous rate of flow from the basin

and the subscripts 1 and 2 indicate the beginning and end of the time period respectively. The term $\int_{t_1}^{t_2} Q \, . \, dt$ provides the estimate of basin yield which may be the most easily measured of all the terms in the water balance equation. When continuous flow measurements are not available the yield may be determined as residual in the water balance equation (10.3). For this purpose, measurements or estimates of the two parameters P and E, and several storage terms are required. The equation may be reduced to

$$\int Q \, . \, dt = P - E - \Delta S \qquad \text{...(10.4)}$$

where ΔS is the change in storage. With judicious selection of the time period $(t_2 - t_1)$ like a year, the magnitude of the term ΔS may be made significant. Substitution of potential evapotranspiration in place E may provide a faster estimate of basin yield. This may introduce considerable error in the estimates of yield depending on the descrepancy between potential and actual evapotranspirations. Here it may be mentioned that the potential evapotranspiration is governed by meteorological conditions and will be fairly stable from year to year and may be reliably determined. But the determination of the actual evapotranspiration is really a very difficult task. Some empirical approaches to estimate the yield from a basin will be discussed in the next section.

Dependable Yield. The annual yield (or the annual runoff volume) from a basin is the effect of various processes such as precipitation, infiltration, evapotranspiration, storage etc., which are random in nature. Therefore the annual yield is also a random variable whose magnitude varies from year to year. The annual yield in any year in future can be estimated only with some degree of dependability associated with it. Thus for example, 75% dependable yield of a basin is the annual runoff volume from the basin which can be expected to be equalled or exceeded, on the average, 3 years out of 4 years. Similarly 50% dependable yield can be expected, on the average, every alternate year.

If the annual yield of a basin is measured for a sufficiently long period of time, the yield corresponding to any stipulated level of dependability can be determined using the frequency analysis or from an appropriate probability distribution fitted to the observed data. Like annual rainfall, the annual yield also follows the normal distribution in majority of the cases. Both these methods have already been described in detail in Chapter 3. The reader is advised to go through Example 3.7 and Example 5.3, for the details of working. When the data is not available on annual yield, or when the available data is too small, long yield data may be obtained by using established and justifiable rainfall runoff correlations developed for neighbouring basins (discussed in the next section) and generally available long rainfall records. The frequency analysis may then be carried out on this extended (in a way *synthetic data, i.e.* not observed data) data to obtain the dependable yields.

Example 10.2. *The annual yields of a basin in million m^3 observed for a period of 19 years from 1971 to 1989 are 832, 672, 488, 1632, 608, 1128, 960, 560, 880, 984, 640, 832, 696, 632, 896, 464, 688, 432 and 1440 respectively. Using the frequency analysis determine the following.*

(a) 75% dependable yield

(b) 50% dependable yield

(c) *Dependability of yield of 1300 million m³.*

Solution. Following the same procedure as in Example 3.7, one will obtain the following answers.

75% dependable yield =536 million m^3

50% dependable yield = 720 million m^3

Dependability of yield of 1300 million m^3 = 12%.

Example 10.3. *Example 10.2 assuming that Normal distribution is a good fit for the data of annual yields.*

Solution. Following the same procedure as in Example 5.3, one will get the following answers.

Arithmetic mean of the annual yields = 813.9 million m^3

Standard deviation of the annual yields = 317.7 million m^3

75% dependable yield = 600.0 million m^3

50% dependable yield = 813.9 million m^3

Dependability of yield of 1300 million m^3 = 6.3%.

10.5 RAINFALL RUNOFF RELATIONSHIPS

In hydrological analysis and design, it is often necessary to develop relations between precipitation and runoff. Such relations are useful for extrapolation or interpolation of runoff records from the precipitaion records which are generally available for longer periods. Through these relations the estimates of the runoff of ungauged catchments may be obtained. The prediction of runoff including its time distribution generated by individual storm bursts is essential for efficient flood forecasting and for the operation of hydraulic projects. Estimates of monthly, seasonal or annual runoffs may be required for operational purposes or to provide a data base for evaluating reservoir storage requirements.

The relationships between precipitation and the time distribution of runoff are best studied through hydrograph analysis and unit hydrographs which will be discussed thoroughly in the next chapter. This subsection deals with relations between only runoff volume and precipitation.

10.5.1 Rainfall Runoff Relationships for Individual Storms

Since the factors affecting runoff vary considerably on a drainage basin during storms, a direct plotting of rainfall against runoff volume for individual storms does not usually produce a satisfactory correlation. The result is likely to be a wide scatter of points indicating no definite trend. However, by introducing some important factors influencing the runoff into the relation the results may be greatly improved.

Coaxial Graphical Method. In 1951 Kohler and Linsley developed a method known as graphical method of *coaxial correlation* in which the storm runoff is estimated using the known values of the storm rainfall, duration of the storm, antecedent precipitation index and the week of the year in which the storm occurred as the independent variables. The details of the method can be found in the book *Hydrology for Engineers* by Linsely, Kohler and Paulhus. Methods using the computer to develop a runoff relation based on the variables used in the coaxial method are also available.

Infiltration Method. The storm runoff volume can also be computed using the infiltration approach. If the rainfall intensity is always more than the infiltration capacity rate, the problem reduces to defining an infiltration rate curve, superimposing it on the rainfall hyetograph of the storm and then find the area of the hyetograph above the infiltration rate curve to give the runoff volume. Again the antecedent moisture conditions play an important role in selecting the infiltration rate curve. If rainfall intensity fluctuate above and below the infiltration rate curve, the use of infiltration indices such as ϕ-index or W-index may be resorted to. These methods are already described in Chapter 8.

Strange's Tables and Curves. W.L. Strange, one of the British Engineers of pre-independent days, developed curves and tables to compute the runoff volume from daily rainfall amounts. His results were based on the then available data on rainfall and runoff in the border areas of present-day Maharashtra and Karnataka. For the purpose of computing runoff, Strange classified the catchment as *good, normal* and *bad.* The initial wetness of the ground was considered under three categories namely *dry, damp* and *wet.* The runoff coefficients and yields given by Strange for a normal catchment are shown in Table 10.2.

Table 10.2 Strange's Tables for Daily Runoff

Daily Rainfall mm	*Runoff percentages and yields when the original state of ground is*					
	Dry		*Damp*		*Wet*	
	Percentage %	*Yield mm*	*Percentage %*	*Yield mm*	*Percentage %*	*Yield mm*
6.25	—	—	—	—	8	0.50
12.50	—	—	6	0.75	12	1.50
18.75	—	—	8	1.50	16	3.00
25.00	3	0.75	11	2.75	18	4.50
31.25	5	1.56	14	4.37	22	6.88
37.50	6	2.25	16	6.00	25	9.37
43.75	8	3.50	19	8.31	30	13.10
50.00	10	5.00	22	11.00	34	17.00
62.50	15	9.37	29	18.25	43	26.90
75.00	20	15.00	37	27.75	55	41.25
100.00	30	30.00	50	50.00	70	70.00

The values of Table 10.2 are to be increased by 25 percent for good catchments and reduced by 25 percent for bad catchments. Strange has also provided Tables for estimating the runoff from monsoon rainfall.

10.5.2 Relationship between Runoff and Seasonal or Annual Rainfall

The estimated runoff volumes of individual storms using the methods described above may be added to give the seasonal or the annual runoff. Such an estimate of annual runoff may be in error to the extent of groundwater runoff entering the stream in between the individual storms which may be quite considerable. Also, if the estimates of seasonal or annual runoff are required for longer periods, the computations would be extremely tedious. This need for rainfall runoff relationships on seasonal and annual basis was recognised by the engineers in the past and number of such relations have been developed. Some relationships take one or two additional independent variables to reflect the climatic or catchment characteristics, apart from the rainfall.

Linear Regression. In many cases a simple linear relationship, though approximate, between precipitation and runoff may be obtained when considered on annual basis. For this purpose it is better to use the annual runoff based on a *water year* than a calender year. The 12 month period starting and finishing at the time of seasonal minimum flow is defined as the water year. For most of the Indian rivers the water year may be taken as the period from June 1st to May 31st. Assuming that all other factors on annual basis do not change appreciably over the basin, the runoff may be linearly regressed with precipitation alone in the form

$$R = a P - b \qquad ...(10.5)$$

where a and b are constants to be obtained from the regression analysis of the observed data on annual precipitaion P and annual runoff R.

Inglis formula. The formula developed by Inglis using the data collected from catchments in the Western Ghats is of this type. It is given as

$$R = 0.85\,P - 30.5 \qquad ...(10.6)$$

where both R and P are in cm.

For large river basins there may be a considerable time lag between the precipitation and the subsequent discharge reaching the streams from the groundwater recharged by the same precipitation. Therefore, when groundwater is an appreciable part of total annual runoff, introduction of precipitation of the previous year as an additional parameter may improve the regression. That is

$$R_i = aP_i + bP_{i-1} + c \qquad ...(10.7)$$

where R_i and P_i are the runoff and precipitation in the year i, P_{i-1} is the precipitation in the year $(i - 1)$ and a, b, and c are the regression constants. Relations similar to Eq. (10.5) or (10.7) can be established for seasons shorter than a year, for example a month or south west monsoon season.

Sometimes nonlinear regression equation between precipitation and runoff may prove to be a better fit. For example, the Inglis formula for the catchments in the non-Ghat areas (Deccan plateau) is of this type. It is given by

$$R = \frac{1}{254}(P - 17.8)\,P \qquad ...(10.8)$$

where R and P are the annual runoff and annual precipitation.

Binnie's Method. Based on rainfall and runoff measurements of a small catchment near Nagpur, Sir Alexander Binnie has suggested the following percentages of runoff.

Table 10.3 Binnie's Runoff Percentages

Annual rainfall (cm)	*Runoff volume (percent)*
50	15
60	21
70	25
80	29
90	34
100	38
110	40

These percentages have found acceptance for the estimation of yield in Madhya Pradesh and Vidarbha region of Maharashtra.

Barlow's Tables. According to Barlow the runoff from a catchment can be expressed as

$$R = K_b\, P \qquad ...(10.9)$$

where P is the monsoon rainfall and K_b is a runoff coefficient which depends on the type of catchment and the nature of monsoon rainfall. Based on the study of small catchments (with areas less than 130 km^2) in Uttar Pardesh, Barlow gave values of K_b as percentage which are given in Table 10.4. For this purpose the catchment is categorised into 5 types and the rainfall pattern of monsoon rainfall into 3 seasons.

Table 10.4 Barlow's Table of Runoff Coefficient K_b in percentage

Class	*Description of catchment*	*K_b in percentage*		
		Season 1	*Season 2*	*Season 3*
A	Flat, cultivated and black cotton soils	7	10	15
B	Flat, partly cultivated various soils	12	15	18
C	Average catchment	16	20	32
D	Hills and plains with little cultivation	28	35	60
E	Very hilly, and steep with hardly any cultivation	36	45	81

Season 1: Light rain, no heavy downpour

Season 2: Average or varying rainfall, no continuous downpour

Season 3: Continuous downpour.

Strange's Tables. Strange also developed runoff coefficient as a function of the catchment character to compute the runoff from monsoon rainfall which are given in Table 10.5.

Table 10.5 Strange's Runoff Coefficients in percentage

Total monsoon rainfall (cm)	*Runoff coefficient in percentage*		
	Good catchment	*Average catchment*	*Bad catchment*
25	4.3	3.2	2.1
50	15.0	11.3	7.5
75	26.3	19.7	13.1
100	37.5	28.0	18.7
125	47.6	35.7	23.8
150	58.9	44.1	29.4

Khosla's Method. Since the runoff on seasonal or annual basis may be viewed as the precipitation less the evaporation and evapotranspiration losses from the basin and since these losses are influenced to a great extent by the temperature, often temperature is introduced as a salient parameter in the rainfall-runoff relations. The temperature again varies with time and in space. However, the mean temperature of any region does not vary much from year to year. Based on the assumption that mean temperature can be taken to be a complete measure of all the factors that are responsible for the losses, Khosla proposed the following formula for runoff. The loss in any month L_m is computed from

$$L_m = 5\,T_m, \text{ for } T_m > 4.5°\text{C} \qquad ...(10.10)$$

where T_m is the mean temperature in the month m in °C. When $T_m \leq 4.5°$, the loss may be assumed as given below. Then the runoff

T_m *in* °C	– 18°	– 12°	– 7°	–1°	4.5°
L_m *in mm*	10	12.5	15	18	21

for the month m, denoted by R_m is determined from the equation

$$R_m = P_m - L_m \text{ and } R_m \geq 0 \qquad ...(10.11)$$

where P_m is the precipitation in the month m in mm. The annual runoff R_A in mm is then given by

$$R_A = \Sigma R_m \qquad ...(10.12)$$

The Khosla's formula gives fairly good results for the annual yield and hence is recommended for use in preliminary studies. The formula can also be used to generate annual runoff data for ungauged catchments from the generally available rainfall and temperature data.

In the modified Khosla's formula, the annual runoff is expressed as a fraction of annual precipitation in the form

$$\frac{R}{P} = C^T \qquad ...(10.13)$$

where R and P are the annual runoff and precipitation expressed in the same units, T is the mean annual temperature in °C and C is a regression constant whose values ranges from 0.952 to 0.971.

Another version of the Khosla's formula is given by

$$R = P\, e^{T/a} \quad ...(10.14)$$

where a is a regression constant, and R, P and T have the same significance as in Eq. (10.13). The constants a and c in the above empirical equations are generally obtained using the limited data available on certain basins. When they are applied to other basins they may not be found satisfactory. Therefore while applying these equations to predict runoff from ungauged catchments, one should exercise caution in selecting the empirical constants.

Example 10.4. *The mean monthly temperatures over a basin for a 12 month period from June to May are 25.8, 24.4, 23.8, 23.5, 23.6, 20.2, 17.1, 16.6, 18.5, 23.3, 27.6 and 28.4 °C respectively. The observed rainfalls in mm in the corresponding months are 86, 229, 208, 115, 15, 182, 10, 24, 24, 12, 0 and 0. Determine the annual runoff from the basin using Khosla's formula.*

Solution. The computations shown in Table 10.6 are self explanatory. It may be noted that when L_m exceeds P_m and produce a negative value for R_m, then R_m is set equal to zero.

Table 10.6 Khosla's Method to Compute Annual Runoff

Month	*m*	T_m	P_m *(mm)*	$L_m = 5T_m$	$R_m = P_m - L_m$ *(mm)*
June	1	25.8	86	129	0
July	2	24.4	229	122	107
Aug.	3	23.8	208	119	89
Sept.	4	23.5	115	117.5	0
Oct.	5	23.6	15	118.0	0
Nov.	6	20.2	182	101.0	81
Dec.	7	17.1	10	85.5	0
Jan.	8	16.6	24	83	0
Feb.	9	18.5	24	92.5	0
Mar.	10	23.3	12	116.5	0
Apr.	11	27.6	0	138.0	0
May	12	28.4	0	142.0	0
					$R_A = 277$ mm

10.6 FLOW - DURATION CURVE

When the runoff rate observed at the catchment outlet is continuously measured, it can be plotted as the ordinate against time on abscissa to provide the runoff hydrograph. The variability in runoff is expressed is terms of the fluctuations of the runoff rate in the hydrograph. There are two other ways of evaluating the variability in the runoff. They are the *flow-duration* curve and the *flow-mass curve* which are described in this section and the next section respectively.

A graph representing the rate of runoff in the stream as the ordinate against the percentage of time such flow is equalled or exceeded as the abscissa is called the flow duration curve.

To determine the flow duration curve for a particular location, it is necessary to obtain daily flow data for a certain period of time, either 1 year or a number of years. The length of the records indicate the total number of days in the series. The daily flows are then arranged in descending order of magnitude, from the highest to the lowest flow value, with each flow value being assigned a rank. The highest flow would get a rank of 1, the next highest flow a rank of 2, and so on, and lowest flow would get a rank n where n is the total number of days in the record. For each flow value, the percent of time is computed as $\frac{m}{n} \times 100$, where m is the rank assigned to the flow. The flow duration curve is obtained by plotting percent of time as the abscissa and the flow value as the ordinate. The flow duration curve can also be constructed using weekly, ten daily and monthly flow values. Flow duration curve determined from monthly flows is given in Example 10.5 and shown in Fig. 10.4. The flow duration curve is also known as the *discharge-frequency curve* or the cumulative frequency curve. The amount of work involved in preparing a flow duration curve can be reduced by dividing the flow data into class intervals instead of handling each individual observation. Thus, for example, if in a record length of 365 days the daily flow is between 40 m^3/s and 50 m^3/s on 73 days, then the mid value of class interval, *i.e.*, 45 m^3/s, is plotted against percent of time given by $\frac{73}{365} \times 100 = 20\%$, on the flow duration curve. A flow duration curve constructed based on daily flows will be steeper than the flow duration curve prepared from the monthly flow data for the same period of record. This is because the larger interval data will smoothen out the variation in the shorter interval data.

Uses of Flow Duration Curve

1. A flow duration curve allows the evaluation of low level flows. For example, from Fig. 10.4, the flow expected to be exceeded 75 percent of the time can be readily determined to be 20 m^3/s, and for 90 per cent of the time it is 16 m^3/s, etc. If the flow is regulated by providing storage to yield a minimum flow of 25 m^3/s all the time, the flow duration curve gets modified as shown by the broken line curve in Fig. 10.4.
2. The flow duration curve is highly useful in the planning and design of water resources projects. In particular, for hydropower studies, the flow duration curve serves to determine the potential for *firm power* generation. In the case of a run- of-the-river plant, with no storage facilities, the firm power is usually computed on the basis of flow available 90 to 97 percent of the time. The firm power is also known as the *primary power*. *Secondary power* is the power generated at the plant utilising water other than that used for the generation of firm power.
3. If a sediment rating curve is available for the given stream, the flow duration curve can be converted into cumulative sediment transport curve by multiplying each flow rate by its rate of sediment transport. The area under this curve represents the total amount of sediment transported.
4. The flow duration curve also finds use in the design of drainage systems and in flood control studies.

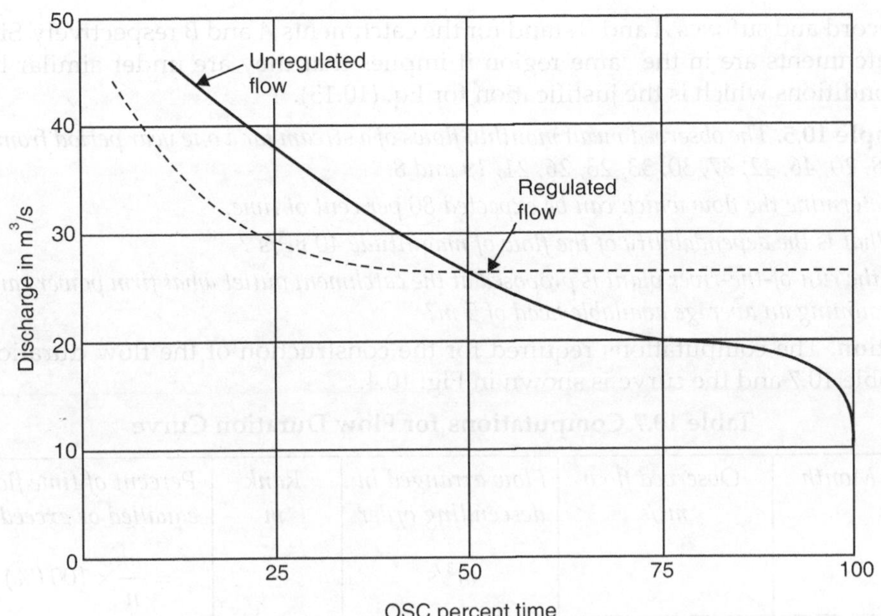

FIGURE 10.4 Flow duration curve.

5. A flow duration curve plotted on a log-log paper provides a qualitative description of the runoff variability in the stream. If the curve is having steep slope throughout, it indicates a stream with highly variable discharge. This is typical of the conditions where the flow is mainly from surface runoff. A flat slope indicates small variability which is a characteristic of the streams receiving both surface runoff and groundwater runoff. A flat portion at the lower end of the curve indicates substantial contribution from groundwater runoff, while the flat portion at the upper end of the curve is characteristic of streams with large flood plain storage, such as lakes and swamps, or where the high flows are mainly derived from snowmelt.
6. The shape of the flow duration curve may change with the length of record. This aspect of the flow duration curve can be utilised for extrapolation of short records. Suppose there is a catchment *A* with a long record (say 1941 to 1980) which also covers the period of the short record of catchment *B* (say 1971 to 1980) in the same region. Then it is assumed that relationship which existed during the concurrent period is also valid for the long period of catchment *A* for which catchment *B* has no record. First the flow duration curves are prepared for both the catchments for the concurrent period of record (1971 to 1980). Next the flow duration curve of *A* is constructed for the long period (1941 to 1980). Now the ordinates of flow duration curve of catchment *B* for the long period are obtained from the following equation.

$$Q_B = Q_A \frac{q_B}{q_A} \quad ...(10.15)$$

where, for any given percent of time, q denotes the ordinate of flow duration curve based on short record, Q denotes the ordinate of flow duration curve based on long record and suffixes A and B stand for the catchments A and B respectively. Since both the catchments are in the same region it implies that they are under similar hydrological conditions which is the justification for Eq. (10.15).

Example 10.5. *The observed mean monthly flows of a stream for a one year period from June to May in* m^3/s *are 18, 20, 46, 42, 37, 30, 33, 23, 26, 21, 19 and 8.*

(*a*) *Determine the flow which can be expected 80 per cent of time.*

(*b*) *What is the dependability of the flow of magnitude 40* m^3/s *?*

(*c*) *If the run-of-the-river plant is proposed at the catchment outlet what firm power can be produced assuming an average available head of 5 m?*

Solution. The computations required for the construction of the flow duration curve are shown in Table 10.7 and the curve is shown in Fig. 10.4.

Table 10.7 Computations for Flow Duration Curve

Month	*Observed flow* m^3/s	*Flow arranged in descending order* m^3/s	*Rank* m	*Percent of time flow equalled or exceeded* $= \frac{m}{n} \times 100$ (%)
June	18	46	1	8.33
July	20	42	2	16.67
Aug	46	37	3	25.00
Sep	42	33	4	33.33
Oct	37	30	5	41.67
Nov	30	26	6	50.00
Dec	33	23	7	58.33
Jan	23	21	8	66.67
Feb	26	20	9	75.00
Mar	21	19	10	83.33
Apr	19	18	11	91.67
May	8	8	12	100.00

From the flow duration curve shown by thick line in Fig. 10.4, the following answers are obtained.

(*a*) Flow expected 80 percent of the time = 18.5 m^3/s

(*b*) Dependability of flow of magnitude 40 m^3/s = 18.4%

(*c*) For estimating the firm power the discharge available 90 to 97% of the time is used. Let us adopt 95% dependable flow. From flow duration curve the magnitude of the flow available 95% of time is 14 m^3/s. Available head is 5 m. Assuming an overall efficiency of 80% for the turbine-generator system, we obtain

$$P = \eta \gamma Q H$$

$$= 0.80 \times 9810 \times 14 \times 5 = 5.494 \times 10^5 \text{ watts}$$

$$= 549.4 \text{ kW}$$

$$= 747.18 \text{ h.p.}$$

10.7 FLOW-MASS CURVE

For many water resources investigations it is important to know the total volumes of water that have to be dealt with over long periods of time. For such purposes the most convenient means of presenting the runoff data is the mass curve. When it is desired to use water at uniform or nearly uniform rate greater than the minimum discharge in the stream, it is necessary to provide storage where water is impounded during periods of high flow for use during periods of low flows. Mass curve serves as a very useful tool to determine the required storage capacity for any uniform rate of demand.

A graph of the cumulative value of any hydrologic quantity, generally as the ordinate, plotted against the time as abscissa is known as a *mass* curve. When cumulative volume of runoff is taken as the hydrologic quantity, the mass curve may be called a *flow mass* curve or simply a mass curve. It is the integration of the hydrograph and therefore represents the area under the hydrograph from one time to another. Mathematically, the mass curve is expressed by

$$V(t) = \int_0^t Q(\tau).\, d\tau \qquad ...(10.16)$$

where $V(t)$ = cumulative volume of flow up to time t from the start of the record, that is, the ordinate of the mass curve at any time t.

$Q(\tau)$ = the discharge as a function of time, *i.e.*, the ordinate of the hydrograph, and

τ = a dummy variable.

A typical mass curve is shown in Fig. 10.5. The mass curve concept is believed to have been first suggested by Rippl in the year 1883.

Hence, the mass curve is also known by the name *Rippl curve* or *Rippl diagram*. The difference in the ordinates of any two points on the curve is the summation of the flow during the intervening period of time. Thus, if two points such as P and Q in Fig. 10.5 are joined by a straight line, the

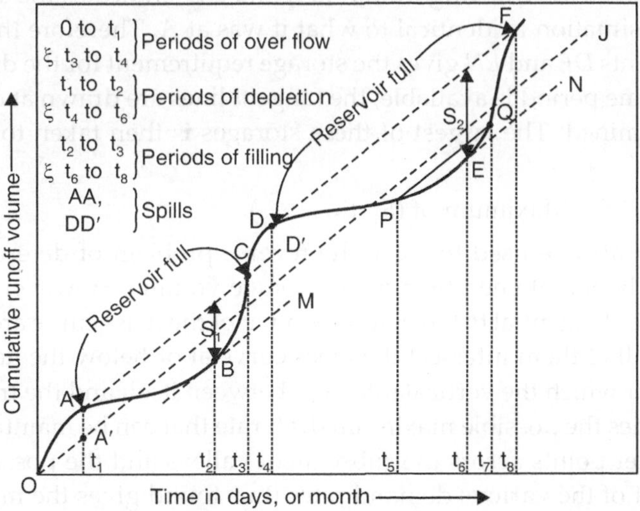

FIGURE 10.5 Flow mass curve.

slope of the line PQ would give the average rate of flow during the period t_5 to t_7. The rate of flow at any time is indicated by the slope of the tangent drawn to the mass curve at the corresponding time. If the mass curve is horizontal, *i.e.*, having zero slope, for a particular period of time it means that there is no discharge in the stream during that period.

The rate of draft or the rate of demand is the rate at which water is required for any use. A straight line having the slope equal to this rate is called a *draft line* or *use line*. Where the mass curve is steeper than the draft line, the flow in the stream is at a higher rate than the rate of draft and consequently some of the streamflow is available for storage, or it will spill if the reservoir is already full. Where the mass curve has flatter slope than the draft line, water will be drawn from storage in order to maintain the required rate of draft.

The storage required to meet any given draft rate may be found out from mass curve as follows. Referring to Fig. 10.5 again, let the line OJ represent the cumulative demand the slope of which is equal to the rate of draft. Let us assume that the reservoir is full at the beginning. From O to A the mass curve is steeper than OJ, indicating that the flow rate into the reservoir is more than the draft rate. Since the reservoir is already, full, no water can be stored though there is surplus water. Hence, the reservoir will be overflowing during the period from 0 to t_1. The ordinate AA' indicates the total volume of the spill in this period. From A onwards the mass curve is flatter than OJ . So withdrawls from the reservoir start from the time t_1. As the amount of water indicated by AA' is already wasted, it cannot be included further in the analysis for the purpose of cumulative inflow. So all the inflows into the reservoir and the withdrawls from it are to be reckoned from A. In other words the origin will be shifted from O to A. So a line drawn parallel to OJ and tangential to mass curve at A, that is AC, will now become the cumulative demand. From A to B the mass curve is flatter than OJ. The reservoir is continuously depleted from t_1 to t_2 and S_1, the vertical intercept between the line AC and B, would indicate the maximum withdrawl from it. Thus S_1 determines the storage requirement for the dry period t_1 to t_2. Form B to D the mass curve is again steeper than OJ and the reservoir starts filling from t_2 and it will be full at t_3. S_1 may be more easily measured as the vertical intercept between the two tangent lines AC and BM which are parallel to OJ and drawn to the mass curve at *peak* (*or the ridge*) point A and the *trough* (*or the valley*) point B. Again the reservoir will be overflowing from t_3 to t_4 and DD' indicates the amount of spill during this period. At D, the situation is identical to what it was at A. Therefore the vertical intercept S_2 between the two tangents DF and EN gives the storage requirement for the dry period t_4 to t_8. If the flow data for a large time period is available, the tangent lines are drawn at other ridge points and the storages are determined. The largest of these storages is then taken to be the actual storage requirement. That is,

$$S = \text{Maximum of } (S_1, S_2, \ldots\ldots\ldots).$$

Mass curve can also be used to solve the reverse problem of determining the maximum demand rate that can be maintained by a given storage volume. However, it is a trial and error procedure. Straight lines tangential to the first peak point of the mass curve are drawn with different trial slopes such that all of them intersect the mass curve at or below the next higher peak point. The slope of the line for which the vertical intercept between itself and the trough point is equal to the given storage defines the possible maximum draft rate that can be maintained. Similar straight lines are drawn at other points across the subsequent valleys and the possible demand rates are obtained. The smallest of the various demand rates thus found gives the maximum demand rate that can be maintained by a given storage.

In the above procedure the ordinate of mass curve is ever increasing and if the record length is large, it requires a graph sheet of large size. This problem can be eliminated by plotting the *residual mass curve* instead of the flow mass curve. In the residual mass curve, cumulative net inflow into the reservoir, that is the difference between the cumulative runoff and the cumulative demand, is plotted against time as shown in Fig. 10.6. In such a curve the inclined draft line *OJ* of Fig 10.5 is transformed into the horizontal line *OJ* of Fig. 10.6 and so are the tangent lines *AC*, *BM* and *DF* and *EN*. The storage required for the first dry period is now given as the difference between the ordinates of the residual mass curve at *A* and *B* and for the next dry period as that between the ordinates of *D* and *E*.

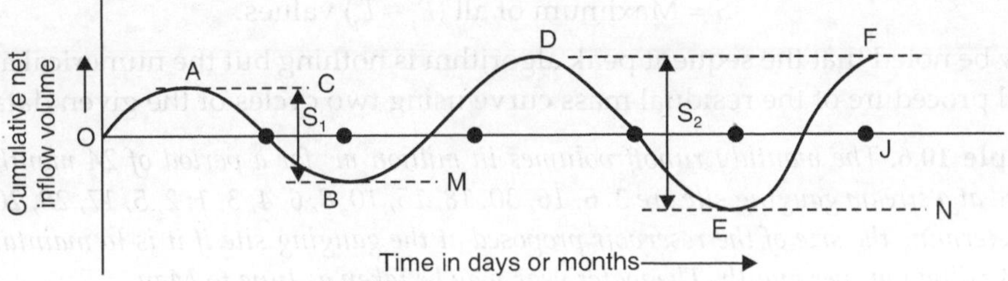

FIGURE 10.6 Residual mass curve.

Sequent Peak Algorithm

In the mass curve analysis, the reservoir is assumed to be full at the beginning of the dry period and the storage required to pass the dry period is estimated. If the mass curve contains only one ridge point and if there is no well defined subsequent trought point, it may be necessary to repeat the given data for one more cycle to arrive at the desired storage requirement. Also, the demand rate is invariably non-uniform due to seasonal variations in the demand. The sequent peak algorithm is a simple and straightforward analytical procedure leading to faster determination of the storage requirement. Unlike in the simple mass curve, it can easily take care of variable demands. Evaporation and seepage losses from the reservoir which are not considered so far can be easily taken care of by adding them to the demands when reasonable estimates of them are available. This procedure, which was first given by Thomas in 1963, is now described.

(*i*) Calculate the net inflow into the reservoir ($V_i - D_i$) for all the time periods $i = 1, 2, \ldots\ldots, 2N$, where N is the number of periods of the data series and V_i for $i = N + 1$, to $2N$ are same as V_i for $i = 1$ to N. That is, the given data is repeated once again to have a series of length $2\,N$. V_i and D_i denote the volume of runoff and volume of Demand in the ith period.

(*ii*) Calculate the cumulative net inflow volumes, namely

$$\sum_{i=1}^{t}(V_i - D_i) \quad \text{for } t = 1, 2, \ldots\ldots\ldots, 2N.$$

(*iii*) Locate the first peak (local maximum) P_1 in the cumulative net inflows.

(*iv*) Locate the sequent peak P_2 which is the next peak of greater magnitude than the first.

(v) Between this pair of peaks find the lowest trough (local minimum) T_1, and calculate $(P_1 - T_1)$.

(vi) Starting with P_2, find the next sequent peak P_3, that has magnitude greater than P_2.

(vii) Find the lowest trough T_2 between P_2 and P_3 and calculate $(P_2 - T_2)$.

(viii) Starting with P_3, find P_4 and T_3 ; calculate $(P_3 - T_3)$.

(ix) Repeat the procedure for all sequent peaks available in the 2N period and compute $(P_j - T_j)$ for all j values.

(x) The required reservoir storage is given by

$$S = \text{Maximum of all } (P_j - T_j) \text{ values.}$$

It may be noted that the sequent peak algorithm is nothing but the numerical adoptation of the graphical procedure of the residual mass curve using two cycles of the given data series.

Example 10.6. *The monthly runoff volumes in million m³ for a period of 24 months (two water years) recorded at a stream gauging site are 3, 6, 16, 30, 18, 15, 10, 8, 6, 4, 3, 1, 2, 5, 17, 28, 20, 15, 12, 7, 5, 4, 3 and 2. Determine the size of the reservoir proposed at the gauging site if it is to maintain an assured supply of 8.33 million m³ per month. The water year may be taken as June to May.*

Solution. From the given data the cumulative runoff volumes are calculated as shown in Table 10.8. The mass curve is plotted as shown in Fig. 10.7. The line MN in the inset indicates the draft rate of 50 million m³ per 6 months or 8.33 million m³/ month. Tangents AC and BD are drawn to the mass curve at the peak point A and the trough point B. These two tangents are parallel to MN. The vertical intercept between these two tangents, which is 30 million m³ in the present case, is the required storage capacity.

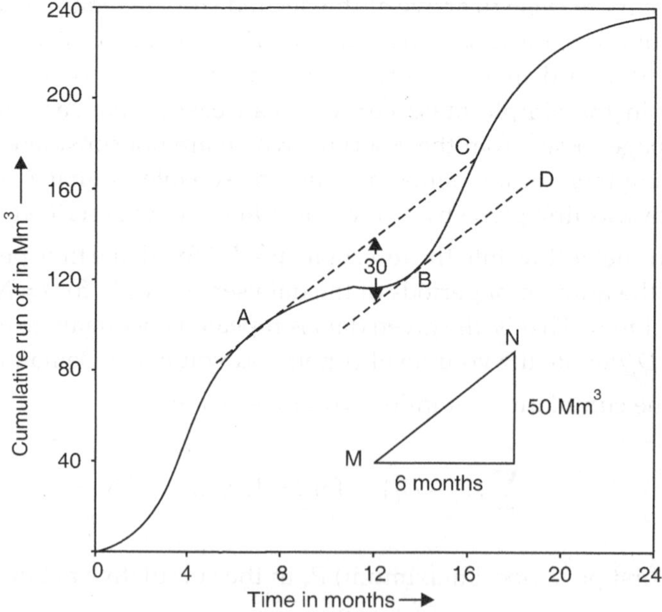

FIGURE 10.7 Mass curve of Example 10.6.

Table 10.8 Calculations for the mass curve of Example 10.6

Month	*Runoff volume million m^3*	*Cumulative runoff volume million m^3*
June	3	3
July	6	9
Aug	16	25
Sep	30	55
Oct	18	73
Nov	15	88
Dec	10	98
Jan	8	106
Feb	6	112
Mar	4	116
April	3	119
May	1	120
June	2	122
July	5	127
Aug	17	144
Sept	28	172
Oct	20	192
Nov	15	207
Dec	12	219
Jan	7	226
Feb	5	231
Mar	4	235
April	3	238
May	2	240

Example 10.7. *Obtain the storage requirement of Example 10.6 without the use of the mass curve.*

Solution. When the runoff volume in any time period is more than the demand in the same time period, there is *surplus* water. This surplus water either goes to build up the storage or it will be spilled if the reservoir is already full. Therefore nothing is drawn from the reservoir during the surplus period. When the runoff in the stream is less than the demand, then there is a *deficit*. Water is drawn from the storage to meet the deficit. Starting from the beginning of the dry period, if the deficit in each time period is found out and if these deficits are added, the cumulative deficit at the end of dry period denotes the maximum amount of water drawn from the storage and the capacity of the reservoir must at least be equal to this.

Similarly, if the cumulative surplus is found out for the following surplus period it must exceed the cumulative deficit to ensure that the reservoir will be full again and we can also find out when this happens. Table 10.9 shows the calculations to determine the storage capacity by this method.

Table 10.9. Storage Determination by Arithmetic Calculations

Month	*i*	*Runoff volume* V_i	*Demand* D_i	$(V_i - D_i)$	*Cumulative deficit*	*Cumulative surplus*
June	1	3	8.33	– 5.33	– 5.33	
July	2	6	8.33	– 2.33	– 766	
Aug	3	16	8.33	+ 7.67		+ 7.67
Sept	4	30	8.33	+ 21.67		+ 29.34
Oct	5	18	8.33	+ 9.67		+ 39.01
Nov	6	15	8.33	+ 6.67		+ 45.68
Dec	7	10	8.33	+ 1.67		+ 47.35
Jan	8	8	8.33	– 0.33	– 0.33	
Feb	9	6	8.33	– 2.33	– 2.66	
Mar	10	4	8.33	– 4.33	– 6.99	
April	11	3	8.33	– 5.33	– 12.32	
May	12	1	8.33	– 7.33	– 19.65	
June	13	2	8.33	– 6.33	– 25.98	
July	14	5	8.33	– 3.33	– 29.31	
Aug	15	17	8.33	+ 8.67		+ 8.67
Sept	16	28	8.33	+ 19.67		+ 28.34
Oct	17	20	8.33	+ 11.67		+ 40.01
Nov	18	15	8.33	+ 6.67		+ 46.68
Dec	19	12	8.33	+ 3.67		50.35
Jan	20	7	8.33	– 1.33	– 1.33	
Feb	21	5	8.33	– 3.33	– 4.66	
Mar	22	4	8.33	– 4.33	– 8.99	
April	23	3	8.33	– 5.33	– 14.32	
May	24	2	8.33	– 6.33	– 20.65	

Table 10.9 shows that the maximum cumulative deficit is 29.31 million m^3. Therefore the reservoir should have a capacity of 29.31 million m^3. This answer is differing slightly from what was obtained in the previous example using mass curve procedure. This is because of the smooth curve drawn through the points of mass curve.

If the storage capacity of the reservoir is fixed as 29.31 million m^3 and if it is full at the beginning of the record, it will be depleted by 7.66 units by the end of July, full again by end of August, overflowing during September to December, empty by end of July, full again sometime in October and so on.

Example 10.8. *The following table gives the average monthly runoff rates at a proposed reservoir site, the pan evaporation and the rainfall recorded at a nearby meteorological observatory, and the estimated monthly demands. The downstream riparian rights require the release of natural flow or 18.5 million m^3 in each month whichever is less. The pan coefficient may be taken as 0.75. The average area of submergence at the reservoir site is 7.5 km^2 for which the runoff coefficient may be taken as 0.35. Determine the storage of the reservoir to meet the demands.*

Month	Average monthly runoff rate m³/s	Evaporation mm	Precipitation mm	Monthly demand M m³
Jan	18.8	120	18	19.6
Feb	15.0	125	22	21.5
Mar	11.0	130	9	23.4
April	7.5	190	11	27.0
May	6.0	200	5	24.0
June	19.0	160	135	26.0
July	75.0	120	225	20.0
Aug	81.2	120	185	20.0
Sept	60.0	110	205	18.5
Oct	30.0	100	10	18.5
Nov	26.4	90	45	16.0
Dec	22.0	110	30	19.0

Solution. The runoff volume in any month in M m³ is given by

$$V = \frac{Q \times N \times 24 \times 60 \times 60}{10^6} = 0.0864\,QN,$$ where Q is the mean monthly runoff rate in m³/s and N is the number of days in the month. Therefore the runoff volume in May is equal to $0.0864 \times 6 \times 31 = 16.07$ M m³ and in February, it is equal to $0.0864 \times 15 \times 28 = 36.29$ M m³.The runoff volumes in M m³ thus computed are shown in col. (2) of Table 10.10.

The volume of water evaporated from the reservoir is obtained from the expression $\left(A \times 10^6 \times \frac{E}{1000} \times C\right)$ m³, where A is the waterspread area in km², E is the evaporation in mm and C is the pan coefficient. Taking A as 7.5 and C as 0.75, the volume in M m³, is given by 0.005625 E. These evaporation losses are given in col. (4).

Except in the month of May, the natural flow is more than 18.5 M m³ in all the months. There fore the riparian releases will be equal to 16.07 Mm³ in May, (which is the natural flow) and 18.5 Mm³ in all other months. These are shown in col. (5).

The runoff coefficient is 0.35. In the absence of the reservoir only 35% of rainfall on the water spread area becomes runoff which is already included in the monthly runoff volumes. But, when the reservoir is built 100% rainfall on the water spread area becomes runoff. Thus the additional volume of water added to the storage when the reservoir is built has to be estimated as $(1 - 0.35) \times A \times P$, and taking A in km² and P in mm, it is equal to 0.004875 P million m³. These volumes due to rainfall are given in col. (6). The monthly demands are shown in col. (3). Thus [col. (3) +col. (4) +col. (5) – col. (6)], which is entered in col. (7) indicates the net demand in each month. The difference between runoff volume V and the net demand D is computed for each month and entered in col. (8). When $(V - D)$ is positive, it is surplus and when it is negative, it is deficit. The cumulative deficit and the cumulative surplus are given in col. (9) and col. (10) respectively.

The storage required is equal to the maximum cumulative deficit, which in the present case is 72.309 M m³.

Table 10.10 Storage Determination of Example 10.8

Month	*Monthly flow V M m^3*	*Demand M m^3*	*Evaporation M m^3*	*Riparian release M m^3*	*Rainfall M m^3*	*Net Demand D M m^3*	*(V – D) M m^3*	*Cumulative deficit M m^3*	*Cumulative surplus M m^3*
(1)	(2)	(3)	(4)	(5)	(6)	(7) = (3)+(4) +(5)– (6)	(8) =(2) – (7)	(9)	(10)
Jan	50.35	19.6	0.675	18.5	0.089	36.686	+ 13.664	—	+ 13.664
Feb	36.29	21.5	0.703	18.5	0.107	40.596	– 4.306	– 4.306	—
Mar	29.46	23.4	0.731	18.5	0.044	45.287	– 15.827	– 20.133	—
April	19.44	27.0	1.069	18.5	0.054	46.515	– 27.075	– 47.208	—
May	16.07	24.0	1.125	16.07	0.024	41.171	– 25.101	– 72.309	—
June	49.25	26.0	0.900	18.5	0.658	44.742	+ 4.508	—	+ 4.508
July	200.88	20.0	0.675	18.5	1.097	38.078	+ 162.802	—	+167.310
Aug	217.49	20.0	0.675	18.5	0.902	38.273	+ 179. 217	—	+346.527
Sep	155.52	18.5	0.619	18.5	0.999	36.620	+ 118.900	—	+ 465.427
Oct	80.35	18.5	0.563	18.5	0.049	37.514	+ 42.836	—	+ 508.263
Nov	68.43	16.0	0.506	18.5	0.219	34.787	+ 33.643	—	+ 541.906
Dec	58.93	19.0	0.619	18.5	0.146	37.973	+ 20.957	—	+562.863

Example 10.9. *The monthly runoff volume into a reservoir in M m^3 for a 12 month period from January to December are 22.8, 18.3, 13.7, 11.4, 9.1, 13.9, 91.4, 102.8, 68.4, 41.1, 32.1 and 27.4 respectively. Determine the storage capacity of the reservoir to meet a demand which is equal the average flow in the river.*

Solution. The average flow in the river is given by

$$\frac{22.8 + 18.3 + 13.7 + 11.4 + ... + 32.1 + 27.4}{12}$$

$$= \frac{452.4}{12} = 37.7 \text{ M m}^3 \text{ per month.}$$

If a mass curve is constructed with the given flow and if the cumulative demand line is superimposed on it, it will be observed that there is no trough following the first ridge point. Hence we would adopt the sequent peak algorithm using two cycles of the given data. The working of the method is given in Table 10.11.

Table 10.11 Determination of storage by the Sequent Peak Algorithm

Month	*Runoff, V_i M m^3*	*Demand, D_i M m^3*	*$(V_i – D_i)$*	*$\Sigma (V_i – D_i)$*	*Remarks*
Jan	22.8	37.7	–14.9	–14.9	
Feb	18.3	37.7	– 19.4	– 34.3	
Mar	13.7	37.7	– 24.0	– 58.3	
April	11.4	37.7	– 26.3	– 84.6	

May	9.1	37.7	– 28.6	– 113.2	
June	13.9	37.7	– 23.8	– 137.0	
July	91.4	37.7	53.7	– 83.3	
Aug	102.8	37.7	65.1	– 18.2	
Sept	68.4	37.7	30.7	12.5	
Oct	41.1	37.7	3.4	15.9	P_1
Nov	32.1	37.7	– 5.6	10.3	
Dec	27.4	37.7	– 10.3	0	
Jan	22.8	37.7	– 14.9	– 14.9	
Feb	18.3	37.7	– 19.4	– 34.3	
Mar	13.7	37.7	– 24.0	– 58.3	
April	11.4	37.7	– 26.3	– 84.6	
May	9.1	37.7	– 28.6	– 113.2	
June	13.9	37.7	– 23.8	– 137.0	T_1
July	91.4	37.7	53.7	– 83.3	
Aug	102.8	37.7	65.1	– 18.2	
Sept	68.4	37.7	30.7	12.5	
Oct	41.1	37.7	3.4	15.9	P_2
Nov	32.1	37.7	– 5.6	10.3	
Dec	27.4	37.7	– 10.3	0	

The required storage capacity $= P_1 - T_1$

$= 15.9 - (-137.0)$

$= 152.9 \text{ M m}^3$.

10.8 ASSESSMENT OF SURFACE WATER RESOURCES OF INDIA

The success of any water resources project depends, primarily, on the reliability of data used in its planning. It is unfortunate that reliable discharge data of river basins in India are not available. Based on Khosla's formula, the total mean annual runoff in all the river basins of India has been estimated to be 1672.5×10^9 m^3. According to K.L. Rao (1975), it is 1645×10^9 m^3. The correct figure can be obtained only when continuous streamflow measurements are made available for a long period at least over a period of three to four decades—from a large number of gauging sites covering all the river basins in the country.

The two major North-Indian rivers Bramhaputra and Ganga with mean annual runoffs of 510×10^9 m^3 and 493×10^9 m^3 respectively occupy the 8th and 10th places in world's largest rivers. The two major rivers of South India namely Godavari and Krishna with mean annual runoff of 105×10^9 m^3 and 68×10^9 m^3 respectively occupy 22nd and 31st places.

REVIEW EXERCISES

Descriptive Questions

1. Describe briefly how the total precipitation is transformed into the total runoff.
2. What are the various components of runoff ? Describe how each component is derived in the runoff process.
3. Distinguish between

   (*i*) Surface runoff and subsurface runoff (*ii*) Direct runoff and base flow

   (*iii*) Rainfall excess and effective rainfall (*iv*) Overland flow and interflow

   (*v*) Depression storage and detention storage.
4. List out the various climate factors affecting runoff. Specifically discuss the effect of the direction of storm movement and the antecedent precipitaion on the peak discharge.
5. List out the various physiographic factors which affect runoff. Discuss their influence on the volume of runoff and on the time distribution of runoff.
6. How is antecedent precipitation index computed from daily rainfall data ? What is the range of the recession factor?
7. What do you understand by *Karst* topography.
8. Define basin yield. How is it obtained from the water balance equation?
9. What is *dependable yield* ? How is it estimated from a given record ?
10. Explain why rainfall-runoff relationships are needed.
11. Write down the water balance equation of a drainage basin and explain the significance of each term.
12. How is runoff estimated using Strange's tables and Barlow's tables?
13. Explain how runoff is estimated using Khosla's method.
14. What is a flow duration curve? How is it constructed?
15. What are the uses of a flow duration curve?
16. What is a flow mass curve ? How is it constructed?
17. Explain with a neat sketch how the storage requirement to meet a uniform demand rate can be determined using a mass curve.
18. What is a residual mass curve? How is storage determined by using it?
19. Describe the use of sequent peak algorithm in determining storage requirements.
20. Explain how the storage requirement can be computed by arithmetic calculations without using mass curve.

Analytical Questions

1. The rainfalls recorded on 3rd, 5th and 9th day of a month are 35, 17 and 25 mm respectively. Compute the antecedent precipitation index for the first 12 days of the month using a recession factor of 0.92 and assuming that the antecedent precipitation index on the last day of the previous month is 70 mm.
2. The annual yields from a basin for a period of 19 years in M m^3 are 1220, 1780, 1060, 750, 618, 1480, 1340, 516, 508, 1380, 1450, 1090, 944, 397, 919, 1137, 857, 412 and 523. Determine the 75% dependable yield, 25% dependable yield and the dependability of the yield of magnitude 900 M m^3 using the frequency analysis.
3. Obtain the answers for analytical question 2 assuming that the annual yields follow the Normal distribution.
4. The average depth of rainfall over a basin for a period of 12 months from June to May are 95, 255, 230, 132, 50, 200,12, 26, 30, 14, 0 and 2 respectively. The mean monthly temperatures in the corresponding months are 26, 25, 24, 24, 22, 21,18, 17, 19, 23, 28 and 29°C respectively. Determine the annual runoff from the basin using Khosla's formula.
5. For the data of analytical question 4, compute the annual runoff using the modified Khosla's formula. Take C = 0.955 in the modified equation.
6. Draw a flow duration curve from the following data on mean monthly discharge in m^3/s.

Month	*Year 1*	*Year 2*
January	62.5	102.0
February	58.0	67.0
March	55.0	50.0
April	48.0	45.0
May	40.0	32.0
June	35.0	29.5
July	25.5	26.7
August	38.0	20.0
September	46.5	34.0
October	76.0	42.6
November	116.5	55.7
December	80.5	72.0

(*a*) Determine the flow which can be expected 80 per cent of the time.

(*b*) What is the percent of time a flow of magnitude 80 m^3/s can be expected ?

(*c*) If a hydro-power plant is proposed at the site to which the above data refers and where the head available is 15 m, what would be the capacity of the power plant to produce firm power? What is the corresponding annual energy production? Assume an overall efficiency of 75% for the turbine-generator system.

7. The two-year monthly runoff of a river into a reservoir are given below.

Month	*Runoff in M m³*	
	Year 1	*Year 2*
Jan	58	10
Feb	66	31
Mar	29	50
Apr	33	53
May	37	39
June	25	39
July	10	36
Aug	2	12
Sept	2	12
Oct	3	14
Nov	4	18
Dec	8	20

Draw a mass curve and find the storage capacity of the reservoir to meet a uniform demand of 20 M m^3 per month.

8. Determine the storage requirement of the reservoir of analytical question 7 by arithmetic calculations without using mass curve.

9. The mean monthly runoffs (in m^3/s) of a river are as given below.

Jan	Feb	Mar	April	May	June	Jul	Aug	Sept	Oct	Nov	Dec
5.5	8.5	9.0	9.0	6.5	7.0	10.0	14.0	19.5	17.0	11.0	8.0

Allowing a riparian release of 2.0 m^3/s and reservoir loss of 0.5 m^3/s, what storage capacity is required to meet a uniform demand of 15 M m^3 per month? Assume 30-day months.

10. The following table gives the average monthly runoff rates at a proposed reservoir site, the pan evaporation and the rainfall recorded at a nearby meteorological observatory, and the estimated monthly demands. The riparian rights require the release of natural flow or 20 M m^3 in each month whichever is less. The pan coefficient may be taken as 0.72. The average area of submergence at the reservoir site is 18 km^2 for which the runoff coefficient may be taken as 0.55. Determine the storage capacity of the reservoir to meet the demands.

Month	*Average monthly runoff rate m³/s*	*Evaporation mm*	*Rainfall mm*	*Monthly demand M m³*
Jan	22.6	130	20	28.5
Feb	18.5	135	25	32.0
Mar	13.2	120	10	34.0
April	9.0	205	7	39.0
May	7.2	220	4	35.0
June	22.8	160	130	37.7

July	90.0	130	225	29.0
Aug	97.5	120	190	32.0
Sept	72.0	110	205	27.0
Oct	36.0	100	10	25.0
Nov	31.7	95	40	23.2
Dec	26.4	95	35	27.0

11. Estimate the reservoir capacity of Exercise 10.30 ignoring the information on, evaporation and rainfall. By what percent the reservoir capacity would be underestimated in this case?
12. The average monthly inflow into a reservoir during a dry year is given below.

Month	June	July	Aug	Sept	Oct	Nov	Dec	Jan	Feb	Mar	April	May
Flow in M m^3	52	160	536	778	535	388	267	214	150	107	78	67

If a uniform demand of 230 M m^3 per month is to be met by this reservoir, what storage capacity is required? (Use sequent peak algorithm).

13. The ordinate of the flow duration curve of a certain stream can be expressed as follows:

$$Q = 10 + \frac{950}{T}$$

in which Q is the discharge in m^3/s and T is the percent of time Q is equalled or exceeded (restricted to the range 1 to 100%). What flow can be expected 95% of the time, 75% of the time and 50% of the time?

14. Collect data on annual rainfall and annual runoff of a basin and establish a relationship between them of the form given by Eq. (10.5) or Eq. (10.7).
15. A hydro-electric plant is proposed at a site on a stream which has the following monthly runoff rates in a dry year.

Month	Jan	Feb	Mar	April	May	June	July	Aug	Sept	Oct	Nov	Dec
Flow in M m^3	120	105	72	32	35	68	146	133	115	92	75	108

The average net head available for power generation in 30 m. The combined efficiency of the turbine-generation system is 80%. Estimate the storage required to produce a firm power of 20 MW.

Objective Type Questions

1. Direct runoff is the sum of
   (*a*) the surface runoff and the base flow
   (*b*) the base flow and the groundwater runoff

(c) the delayed subsurface runoff and the deep percolation

(d) the surface runoff and the prompt subsurface runoff.

2. The base flow is the difference between

(a) the total runoff and the direct runoff

(b) the total runoff and the delayed subsurface runoff

(c) the prompt subsurface runoff and the delayed subsurface runoff

(d) the total runoff and the prompt subsurface runoff.

3. The base flow of a stream represents

(a) the groundwater runoff and the prompt subsurface runoff

(b) the groundwater runoff and the subsurface runoff

(c) the groundwater runoff and the delayed subsurface runoff

(d) the runoff due to only snow melt.

4. The observed annual runoff from a basin of area 500 km^2 is 150 M m^3 and the corresponding annual rainfall over the basin during the same year in 750 mm. What is the runoff coefficient?

(a) 0.2 (b) 0.67

(c) 0.4 (d) 0.5.

5. *Virgin flow* means

(a) the flow in the stream corresponding to no infiltration and evaporation losses

(b) the flow in the stream which is not affected by the works of man

(c) the flow in the stream which does not contain the flow from the watershed leakage of the neighbouring basin

(d) none of the above.

6. The water year in India may be taken as

(a) from June to May (b) from January to December

(c) from April to March (d) from October to September.

7. The anticident precipitation index on a certain day is I_0. If the recession factor is k, and if there is no rainfall for the next n days, the API on nth day is given by

(a) $I_n = \dfrac{I_0}{k^n}$ (b) $I_n = I_0 k^n$

(c) $I_n = \dfrac{k^n}{I_0}$ (d) $I_n = \dfrac{1}{I_0 k^n}$.

8. The subsurface runoff is also known as

(a) Interflow (b) Storm seepage

(c) Secondary base flow (d) All the above.

9. The monthly runoff is estimated, using Khosla's formula, as $R_m = P_m - L_m$. The loss term L_m here

(a) depends on the mean monthly temperature of the basin

(*b*) depends on the antecedent precipitation index

(*c*) is expressed as a function of land use and the slope of the basin

(*d*) is a constant.

10. Flow mass curve is the graph drawn between

(*a*) the flow rate and time (*b*) cumulative volume of flow and time

(*c*) cumulative volume of flow and cumulative time

(*d*) cumulative discharge and time.

11. Flow mass curve is used

(*a*) to determine the storage capacity of the reservoir to meet a given uniform demand

(*b*) to check the consistency of the flow record at a given site

(*c*) to develop the unit hydrograph

(*d*) to develop the synthetic unit hydrograph.

12. The flow mass curve is an integral curve of

(*a*) rainfall hyetograph (*b*) infiltration curve

(*c*) discharge hydrograph (*d*) rating curve of the gauging site.

13. If a tangent drawn parallel to the demand line from a ridge point of a flow mass curve does not intersect the mass curve again, it represents that

(*a*) the runoff inflow into the reservoir cannot meet the demand as the reservoir will not be full again

(*b*) the reservoir is overflowing resulting in wastage.

(*c*) there is enough streamflow to meet a higher demand rate

(*d*) none of the above.

14. Flow duration curve is the graph drawn between

(*a*) the discharge in the stream and time

(*b*) the accumulated discharge and time

(*c*) the discharge and the per cent of time such discharge is equalled or exceeded

(*d*) cumulative volume of flow and time.

15. The peculiarity of areas with Karst topography is that

(*a*) they do not produce any runoff

(*b*) they produce only direct runoff but no base flow

(*c*) they produce only base flow but no direct runoff

(*d*) none of the above.

ANSWERS

Analytical Questions

1. 64.4, 59.25, 89.51, 82.35, 92.76, 85.34, 78.51, 72.23, 91.45, 84.13, 73.4, 67.53

2. 523 M m^3, 1340 M m^3 , 56.5%

3. 691.3 M m^3, 1243.53 M m^3, 43.44%
4. 347 mm
5. 362.75 mm
6. (*a*) 33.5 m^3/s (*b*) 12%
(*c*) 95% dependable flow = 25.8 m^3/s, 2847.35 kW and 2.3696×10^6 kWh
7. 104 M m^3
8. 101 M m^3
9. 7.968 M m^3
10. 105.02 M m^3
11. 96.57 M m^3 ; 8.05%
12. 782 M m^3
13. 20 m^3/s , 22.67 m^3/s, 29.0 m^3/s
15. 349.65 M m^3

Objective Type Questions

1. (*d*)	2. (*a*)	3. (*c*)
4. (*c*)	5. (*b*)	6. (*a*)
7. (*b*)	8. (*d*)	9. (*a*)
10. (*b*)	11. (*a*)	12. (*c*)
13. (*a*)	14. (*c*)	15. (*a*)

Chapter 11

HYDROGRAPH ANALYSIS

11.1 INTRODUCTION

In the previous chapter, the relations between rainfall and runoff volume were discussed. In this chapter the time distribution of runoff produced by a given precipitation on a basin is analysed. The runoff measured at the basin outlet, when plotted against time gives the hydrograph. As the runoff includes the contributions from surface runoff, interflow and groundwater runoff, the hydrograph can be regarded as an integral expression of the physiographic and climatic characteristics that govern the relations between rainfall and runoff. It shows the time distribution of runoff at the basin outlet defining the complexities of the basin characteristics by a single empirical curve. Thus it forms a basis to relate rainfall and the time distribution of runoff produced by it. Detailed analysis of flood hydrographs is usually important in flood control and flood forecasting, or in establishing design flows for hydraulic structures which must pass flood waters.

11.2 FEATURES OF A HYDROGRAPH

The shape of a hydrograph resulting from a single short duration storm occurring over the drainage area follows a general pattern. If the subsequent storm burst does not occur before the direct runoff of the present storm ceases, a typical single peaked hydrograph as shown in Fig. 11.1 results. Such a storm is known as an *isolated storm*. On the other hand if the next rainfall occurs before the direct runoff of the previous rainfall ceases, there will be a multi-peaked hydrograph such as the one shown in Fig. 11.3 (*a*). Such a storm may be called a *complex storm*. A complex hydrograph may be resolved into the corresponding simple hydrographs. However, single peaked hydrographs resulting from isolated storms are generally preferred for hydrograph analysis.

As shown in Fig. 11.1, a typical single peaked hydrograph produced by an isolated storm consists of a *rising limb*, a *crest segment* and a *recession limb*. The rising limb *BC* has a well defined point of rise *B* followed by increasing discharge. The crest segment *CE* contains the peak discharge within it, and from *E* it is the recession limb with decreasing discharge. The recession limb is also known as the *falling limb* or the *depletion curve*. Points *C* and *E* are the inflection points on the rising and recession limbs respectively. The segment *AB* is called the approach segment which indicates the base flow in the river prior to the storm.

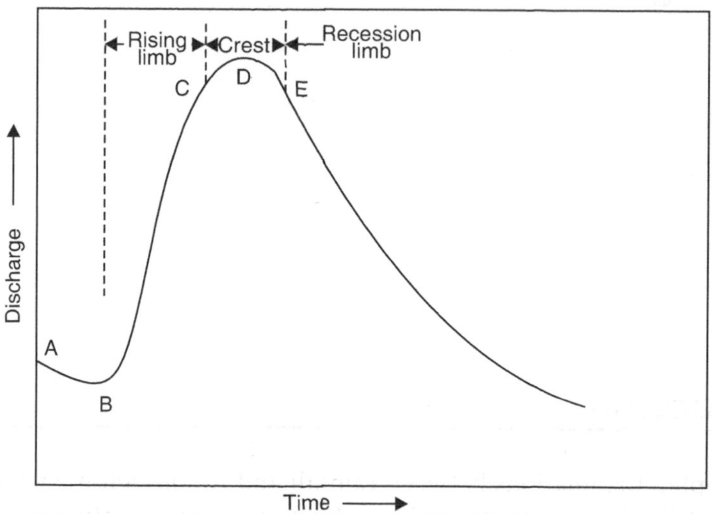

FIGURE 11.1 Typical single peaked hydrograph.

The shape of the rising limb depends mainly on the duration and the intensity distribution of rainfall, and to some extent on the antecedent condition and the shape of the *time area diagram* of the basin (See Chapter 12). The peak discharge, included in the crest segment, represents the highest concentration of runoff from the basin. It occurs usually at a certain time after the rainfall has ended and this time depends on the aerial distribution of rainfall. The point of inflection on the recession limb is commonly assumed to mark the time at which surface inflow to the channel system or the overland flow ceases. Thereafter the recession curve represents withdrawal of water from storage within the channel system. Therefore, it is more or less independent of variations in rainfall and infiltration. On large basins where the storms produce rainfall only locally the shape of the recession curve may vary from storm to storm. The recession curve is also called a depletion curve since it represents the depletion from the channel storage.

Though there are number of equations to describe the recession curve, the following is most generally used.

$$Q_t = Q_0\ K_r^{(t-t_0)} \qquad ...(11.1)$$

where Q_0 and Q_t are the flows at times t_0 and t and K_r is a *recession constant* with a value of less than unity. Eq. (11.1) can be expressed in a different form as

$$Q_t = Q_0 e^{-a(t-t_0)} \qquad ...(11.2)$$

where $a = -\ln\ K_r$. The value of the recession constant K_r depends on the time unit selected.

The recession curve starting from E represents the depletion from the channel storage. When the direct runoff ceases after some time, it represents the depletion from groundwater storage since the discharge in the stream then is essentially due to base flow. When a time unit of one day is adopted the value of K_r for the direct runoff recession curve will be between 0.05 to 0.20 and for the base flow recession curve it will be between 0.85 to 0.9.

Equation (11.2) will plot as a straight line on a semilogarithmic paper with Q plotted on logarithmic scale. The slope of this line should represent the recession constant K_r. Determination of recession constants of direct runoff and base flow using the semilogarithmic plot is explained in Example 11.1. The storages S_t remaining in the basin at time t is obtained as

$$S_t = \int_t^{\infty} Q_t . dt = \int_t^{\infty} Q_0 . e^{-a(t-t_0)} . dt$$

or

$$S_t = \frac{Q_t}{a} = \frac{Q_t}{(-\ln K_r)} \qquad \text{...(11.3)}$$

Example 11.1. *The ordinates of the recession limb of a hydrograph are listed below. Assuming that the total runoff consists of only direct runoff and base flow, find the recession constants for the base flow and direct runoff.*

Time (days)	*Flow (m^3/s)*	*Time (days)*	*Flow (m^3/s)*
14	411	20	100
15	358	21	83
16	250	22	70
17	192	23	58
18	151	24	49
19	122	25	41

Solution. The given hydrograph is plotted on a semilogarithmic paper with discharge taken on logarithmic scale as shown in Fig. 11.2.

The recession limb is represented by *DEFG*. The last portion of the curve *FG* is a straight line which represents the base flow recession. This line is projected back as *GFH*. The point *F* denotes the end of direct runoff. The difference between the total runoff hydrograph ordinate and the ordinate of the line *FH* at any time therefore denotes the direct runoff at that time. The direct runoff thus computed at different times is listed below.

Time (days)	*Total runoff*	*Baseflow measured as the ordinate of the line GFH in Fig. 11.2*	*Direct runoff*
15	358	229	129
16	250	193	57
17	192	189	30
18	151	136.5	14.5
19	122	115	7

20	100	97	3
21	83	81.3	1.7
22	70	70	0
23	58	58	0

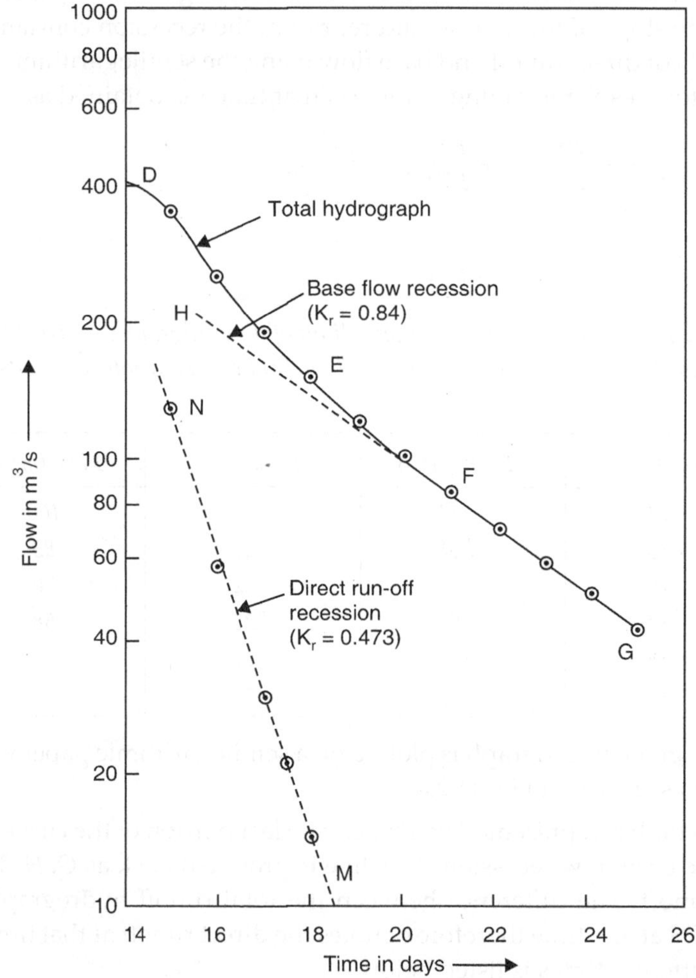

FIGURE 11.2 Recession analysis on semilogarithmic plot.

These direct runoffs are plotted on the same paper taking the direct runoff on logarithmic scale again. The line *MN* which represents this plot is the recession curve of the direct runoff. From Eq. (11.1),

$$\frac{Q_t}{Q_0} = K_r^{(t-t_0)}$$

Taking t_0 =16 days and t = 18 days, Q_0 and Q_t are read as 193 m^3/s and 136 m^3/s respectively on the recession curve *FH* of Fig. 11.2.

$$\therefore \quad \frac{136}{193} = K_r^{(18-16)} = K_r^2$$

$$K_r^2 = 0.7046 \text{ or } K_r = 0.84$$

Therefore the recession constant for base flow is 0.84.

Similarly for $t_0 = 16$ days and $t = 18$ days, Q_0 and Q_t are read as 61.7 and 13.8 respectively on the recession curve *MN* of Fig. 11.2.

$$K_r^2 = \frac{13.8}{61.7} = 0.2237 \text{ or } K_r = 0.473$$

Therefore the recession constant for direct runoff is 0.473.

Example 11.2. *In a stream the base flow is observed to be 30 m³/s on Feb. 1 and 23 m³/s on Feb. 10. If there is no rain during February, estimate the base flow on Feb. 28 and the volume of water in groundwater storage on Feb. 1 and Feb. 28. Assume base flow recession curve of the stream can be described by Eq. (11.1).*

Solution. $Q_0 = 30 \text{ m}^3/\text{s} \qquad t_0 = 1$

$Q_t = 23 \text{ m}^3/\text{s} \qquad t = 10$

$$\frac{Q_t}{Q_0} = K_r^{(t-t_0)}$$

$$\frac{23}{30} = K_r^{(10-1)} = K_r^9$$

$$\therefore \quad K_r = 0.971$$

when $t = 28$, $Q_t = Q_0\,(0.971)^{(28-1)}$

$Q_t = 30 \times 0.971^{27} = 13.52 \text{ m}^3/\text{s}$.

$\therefore$ The flow in the stream on Feb. 28 = 13.52 m³/s.

From Eq. (11.3),

$$S_t = \frac{Q_t}{(-\ln K_r)} = \frac{Q_t}{-\ln 0.971} = 33.98\,Q_t.$$

In the estimation of K_r a time unit of day has been used. Hence the storage computed from the above expression will be in cumec-days.

Storage on Feb. 1 = 33.98 × 30 = 1019.4 cumec-days

= 88.08 × 10⁶ m³

= 88.08 Mm³

Storage on Feb. 28 = 33.98 × 13.52 = 459.41 cumec-days

= 39.69 × 10⁶ m³

= 39.69 Mm³.

11.3 HYDROGRAPH SEPARATION

It is necessary to separate the observed hydrograph into its component parts namely the surface runoff, interflow and groundwater runoff for subsequent analysis. However, in many engineering applications it is the standard practice to separate the hydrograph into two parts only, *i.e.*, direct runoff and base flow.

If a continuous record of discharge in the stream over a period of few years is available a *master depletion curve* also called a *type curve* or a *composite recession curve* is constructed and used for separating the baseflow. The construction of the master depletion curve is as follows. From the continuous hydrograph as shown in Fig. 11.3 (*a*) recession segments which represent only base flow contribution at their respective tails are selected. These segments should be selected at as many different stages in the river as possible. They are then plotted on transparent papers with coincident time axis and are adjusted such that all of them will have a common overlapping portion as shown in Fig. 11.3 (*b*). The common curve tangential to all the independent recession curves is nothing but the master depletion curve.

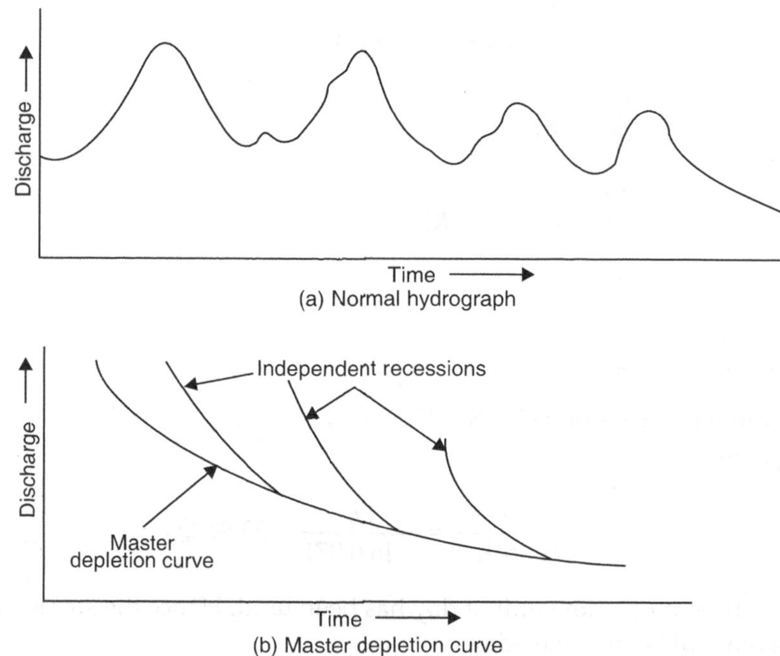

FIGURE 11.3 Construction of master depletion curve.

It can now be applied to the hydrograph of a particular storm as shown in Fig. 11.4 where the recession curve of the hydrograph and the master depletion curve are fitted together such that they are coincident below point C, and are divergent above C. Then the point C represents the end of direct runoff and a straight line joining it to the point of rise represents the base flow line.

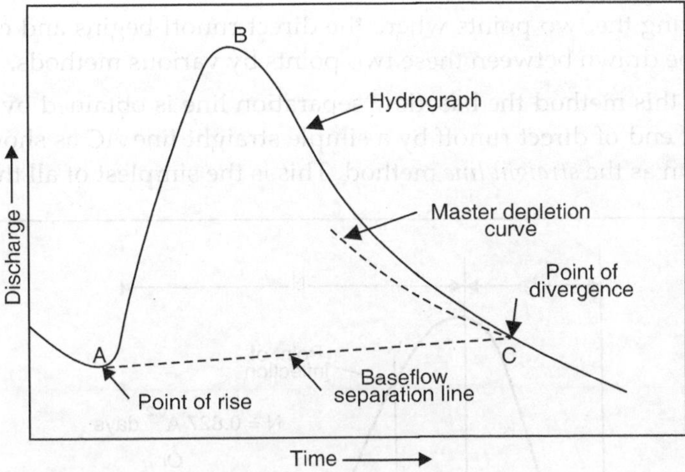

FIGURE 11.4 Baseflow separation using master depletion curve.

While the procedure described above is perhaps the best available, it requires previously observed data on streamflow for a long period which may not be available in all cases. In such situations a point on the recession limb of the hydrograph to mark the end to direct runoff may be fixed *N* days after the peak of the hydrograph where *N* is calculated from the empirical equation.

$$N = 0.827\, A^{0.2} \qquad \text{...(11.4)}$$

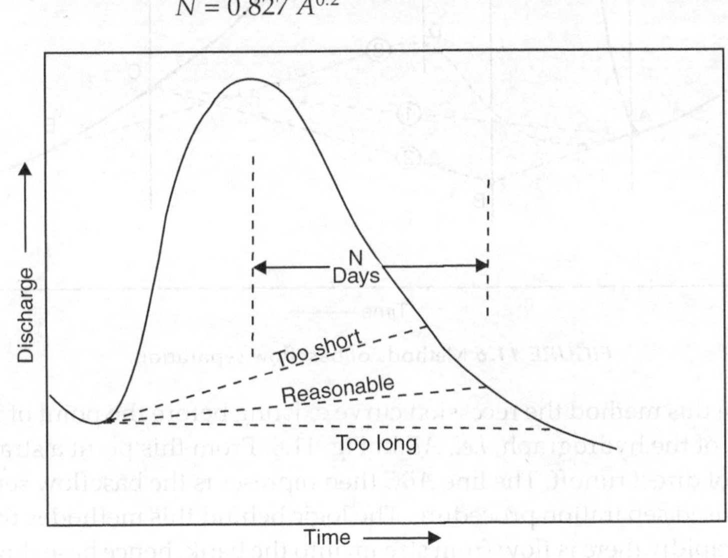

FIGURE 11.5 Selection of baseflow separation line.

in which, *A* is the area of the drainage basin in km^2. However, *N* as given by Eq. (11.4) may be taken as a rule of thumb and need not be strictly adhered to, if it is giving either too long or too short base flow separation line. Another empirical equation to determine *N* is given by $N = a\, t_p$, where t_p is time to peak and *a* is a constant whose value may be taken between 2 and 4 for small basins and more than 4 for large basins. *N* is probably better determined by inspection of number of hydrographs of the basin. Fig. 11.5 illustrates some reasonable and unreasonable estimates of *N*.

After identifying the two points where the direct runoff begins and ceases, the base flow separation line can be drawn between these two points by various methods.

Method 1. In this method the baseflow separation line is obtained by joining the point of rise and the point of end of direct runoff by a simple straight line AC as shown in Fig. 11.6. This method is also known as the *straight line* method. This is the simplest of all the methods.

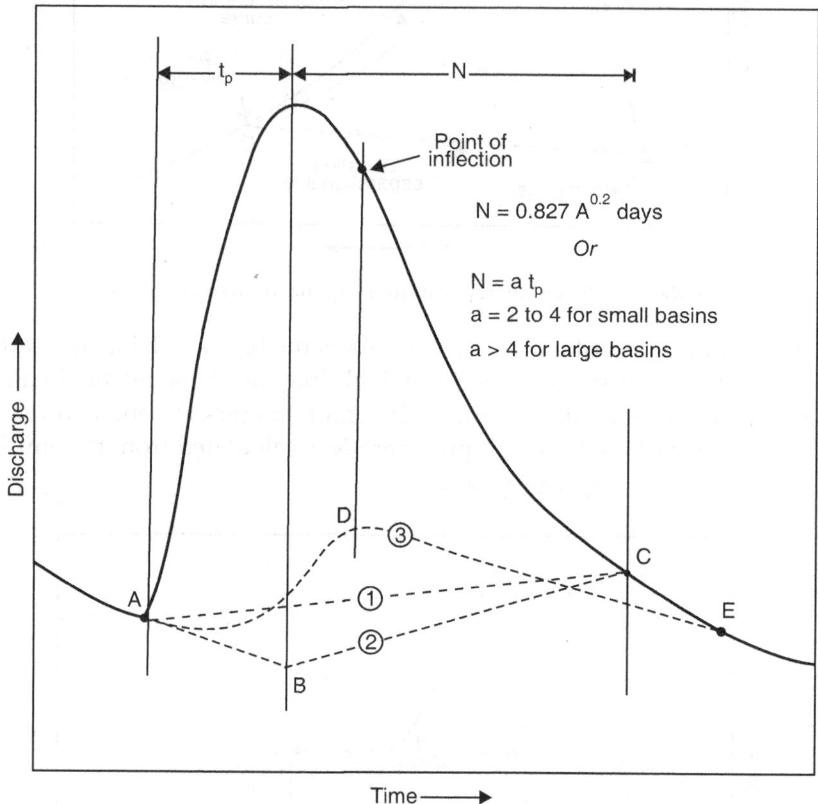

FIGURE 11.6 Methods of baseflow separation.

Method 2. In this method the recession curve existing before the point of rise is extended to a point under peak of the hydrograph, *i.e.*, AB in Fig. 11.6. From this point a straight line is drawn to the point of end of direct runoff. The line ABC then represents the baseflow separation line. This is the most widely used separation procedure. The logic behind this method is that when the stage in the stream rises rapidly there is flow from stream into the bank, hence base flow should decrease until the stages in the stream begin to fall. When the stage in the stream recedes the bank storage returns to the stream making the baseflow contribution large.

Method 3. In this method the baseflow curve on the recession side of the hydrograph is extrapolated backwards to the time of point of inflection on the recession limb and then it is connected to the point of rise by a smooth curve. ADE in Fig. 11.6 represents the baseflow separation line of this method. This method is preferred where the stream and groundwater table are hydraulically connected and flow from groundwater storage reaches the stream fairly rapidly.

Differences in baseflow due to various separation methods are likely to be small when compared to the direct runoff hydrograph volume. These differences are probably unimportant as long as one method is used consistently. The hydrograph obtained as the difference between total runoff hydrograph and the baseflow separation line is called the *direct runoff hydrograph* often abbreviated as *DRH*.

11.4 THE UNIT HYDROGRAPH THEORY

The next step in the hydrograph analysis after obtaining the direct runoff by separating the baseflow from the total runoff would be to relate it to the effective rainfall which produced it. This is accomplished by the use of the concept of *unit hydrograph* introduced by Sherman in 1932. The unit hydrograph theory is essentially based on the results reported in the study of heavy floods in New England in 1927 by a committee of Boston Society of Civil Engineers.

Definition. The unit hydrograph of a drainage basin is defined as a hydrograph of direct runoff resulting from 1 cm of effective rainfall applied uniformly over the basin area at a uniform rate during a specified period of time.

Assumptions. The above definition along with the following basic assumptions constitute the so-called unit hydrograph theory. The unit hydrograph is originally named the *unit graph*.

(*a*) The effective rainfall is uniformly distributed within its duration.

(*b*) The effective rainfall is uniformly distributed throughout the whole area of the basin.

(*c*) The base periods of the direct runoff hydrographs produced by effective rainfall of same duration (though the intensities may be different) are also same.

(*d*) The ordinates of the direct runoff hydrographs of a common base period (or the direct runoff hydrographs produced by effective rainfalls of same duration) are directly proportional to the total volume of direct runoff represented by the respective hydrographs.

(*e*) For a given drainage basin the hydrograph of runoff due to a given period of rainfall reflects the unchanging characteristics of the basin.

Assumptions (*a*) and (*b*) are incorporated in the definition of unit hydrograph itself. Assumption (*d*) is known as the *principle of linearity* or *principle of superposition*. Assumptions (*c*) and (*e*) imply that the shape of the runoff hydrograph remains same irrespective of at whatever time it may occur as long as the duration of rainfall producing it is same. This is known as the *principle of time invariance*. Sherman originally used the word *unit* to denote a specified period of time or a unit duration which is later misunderstood as though to refer to *unit depth*.

Limitations. Since uniform intensity over long durations is less likely the storms selected for unit hydrograph analysis should be of short duration. Short duration storms would most likely produce an intense and nearly uniform effective rainfall yielding a well-defined single peaked hydrograph.

Since the uniform aerial distribution of rainfall over large areas is less likely, the unit hydrographs can be applied only to drainage basins with small areas. In the case of large basins, the basin has to be divided into small sub-basins and each sub-basin is subjected to analysis for storms covering the whole sub-basin. The unit hydrographs are best suited to areas not more than 5000 km^2 although they have been applied to fairly large areas with varying degree of success. The unit hydrograph is also not suitable for areas less than 200 hectares. Basins with odd shapes, particularly those which are long and narrow, will commonly have very uneven rainfall distribution and hence unit hydrographs are not well adopted to such basins.

The base period of the direct runoff hydrograph should depend on the intensity of rainfall also, since channel storage is more in intense storms elongating the base period. However the assumption of constant base period for storms of same duration may not produce excessive error since the recession curve approaches zero asymptotically.

The principle of linearity is not valid strictly. Unit hydrographs derived from very small rainfall events will generally have lower peak than those derived from large storms. In other words the unit hydrographs derived from two storms of same duration but with different runoff volumes will not be same. The peak ordinate in the unit hydrograph derived from the storm of larger runoff volume will be more than the peak ordinate of the unit hydrograph derived from the storm of smaller runoff volume. If such non-linearity is significant, unit hydrographs should be used with caution. In the absence of any simple non-linear approach, the principle of linearity may be assumed. The peak discharge in the unit hydrograph derived from ordinary floods is often increased by 5 to 20 per cent before using it for predicting the extreme flood peaks with the belief that channel-flow time shortens as flood magnitude increases. However, if extreme floods overflow into flood plains, the opposite effect might result.

According to principle of time invariance, the direct runoff hydrograph from a drainage basin due to a given pattern of effective rainfall, at whatever time it may occur, is invariable. It is known that basin characteristics change with seasons, man-made adjustments etc. These changes may not produce significant effect on the direct runoff in many cases.

The unit hydrograph method cannot be applied when a major portion of storm precipitation is in the form of snow.

11.5 DERIVATION OF UNIT HYDROGRAPH

The procedure adopted in deriving a unit hydrograph is simple and straight forward. It follows from the assumptions of unit hydrograph theory. The various steps involved in the derivation are described below.

Step 1. The rainfall records of the drainage basin are scanned through to find a storm of desired duration that gave a fairly uniform distribution over the basin and also within its duration and having a runoff volume of nearly 1 cm or greater. The rainfall hyetograph of this storm is constructed using a convenient uniform time interval.

Step 2. The corresponding runoff hydrograph is plotted on a graph and the baseflow is separated using any one of the methods described earlier. The ordinate of the direct runoff hydrograph at any time is then obtained as the difference between the total runoff and the baseflow at that time. The volume of direct runoff is computed by first finding the area of the direct runoff hydrograph and then multiplying it by an appropriate scale factor. For example, if a time scale of

1 cm = 2 hr and a discharge scale of 1 cm = 10 m^3/s are adopted then 1 cm^2 area of the direct runoff hydrograph represents 10 × 2 × 60 × 60 m^3, that is, 72000 m^3. The depth of runoff is then computed by dividing the volume of direct runoff by the area of the basin.

Alternatively the depth of direct runoff can be directly obtained from the equation given below which is nothing but the numerical integration of the direct runoff hydrograph using some uniform time interval.

$$d = \frac{0.36\ \Delta t \sum Q}{A} \quad \text{...(11.5)}$$

where d = Depth of direct runoff in cm.

Δt = The uniform time interval in hours at which the ordinates of direct runoff hydrograph are measured.

$\sum Q$ = Sum of all the ordinates of direct runoff hydrograph in m^3/s.

A = The area of the drainage basin in km^2.

Step 3. The ordinates of direct runoff hydrographs are now divided by the depth of direct runoff to yield the ordinates of unit hydrograph.

Step 4. The duration of the unit hydrograph is determined by drawing a horizontal line on the rainfall hyetograph in such a way that the excess rainfall above this line is exactly equal to the observed depth of runoff. The area below this line represents the abstractions. It may be recalled here that this line is nothing but the ϕ-index line. The duration for the unit hydrograph obtained in step 3 is the time elapsed between the beginning and the end of the effective rainfall. Example 11.3 demonstrates the use of these steps in the derivation of a unit hydrograph.

Averaging of Unit Hydrographs

A unit hydrograph derived from a single storm may be in error since the storms of same duration and intensity may produce different unit hydrographs due to different storm orientations and different sequences of rainfall increments. Therefore, it is desirable to average the unit hydrographs from several storms of the same duration to get the representative unit hydrograph. Straight numerical averaging of the discharges at various times is not appropriate, as such averaging may result in flattened crest segment as shown in Fig. 11.7. The proper procedure is to compute the average of the peaks and the average of times of peak. Therefore the co-ordinate position of the peak point in the average unit hydrograph would be $\left(\frac{t_1 + t_2 + t_3}{3}, \frac{Q_{p1} + Q_{p2} + Q_{p3}}{3}\right)$, where t_1, t_2 and t_3 are the times at which the peak discharges Q_{p1}, Q_{p2} and Q_{p3} occur in the three individual unit hydrographs, respectively, used in the averaging. The average unit hydrograph is then sketched to conform to the general shape of the other unit hydrographs, passing through the computed average peak point, and having a volume of 1 cm of runoff.

Example 11.3. *Derive the unit hydrograph for a drainage basin of area 104 km^2 from the observed total runoff hydrograph given below.*

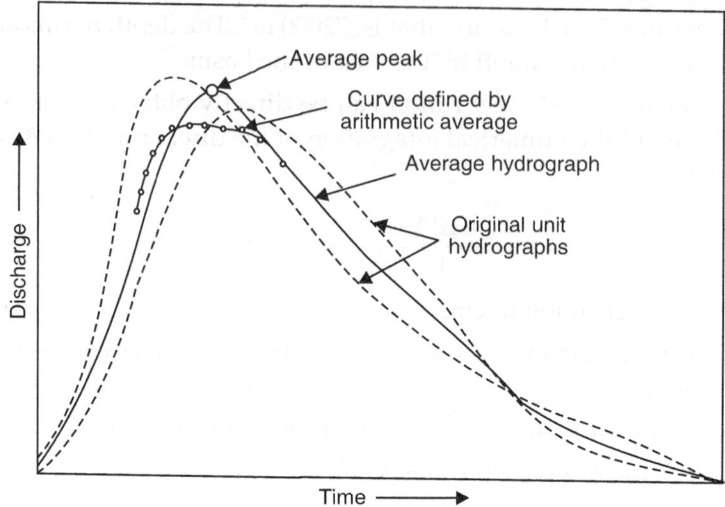

FIGURE 11.7 Construction of an average unit hydrograph.

Date	*Time hr*	*Runoff m^3/s*	*Date*	*Time hr*	*Runoff m^3/s*
June 10	2	16.0	*June 11*	2	76.4
	4	15.0		4	65.0
	6	14.2		6	55.2
	8	158.5		8	46.7
	10	260.0		10	39.6
	12	286.0		12	34.0
	14	221.0		14	28.3
	16	186.5		16.	22.7
	18	157.0		18	21.5
	20	133.0		20	20.6
	22	113.0		22	19.9
	24	93.4		24	19.2

The ordinates of the rainfall mass curve which produced the above runoff are as given below.

Date	*Time hr*	*Cumulative rainfall cm*
June 10	2	0
	4	1.5
	6	3.0
	8	10.5
	10	19.0
	12	20.0
	14	20.0

Solution. From the given storm data, the rainfall hyetograph is constructed adopting a time interval of 2 hr. The rainfall intensities of this hyetograph are 0.75, 0.75, 3.75, 4.25, 0.50 and 0.0 cm/h respectively. The hyetograph is shown above the hydrograph in Fig. 11.8. The point of rise *A* is easily located at the end of 6th hour on June 10. The time distance *N* in days between the peak of the hydrograph and the end of direct runoff from Eq. (11.4) is

$$N = 0.827\ (104)^{0.2}$$
$$= 2.0937 \text{ days}$$
$$= 50.25 \text{ hr.}$$

But from the plotted hydrograph this appears to be too long and hence the end of direct runoff C is arbitrarily fixed at 16th hour on June 11. The base flow line *ABC* is then constructed by joining *A* and *C* by a straight line.

The ordinates of the direct runoff hydrograph obtained as the difference between the total runoff and the base flow are entered in Table 11.1. The depth of direct runoff obtained from Eq. (11.5) with Δt = 2 hr is

Table 11.1 Derivation of Unit Hydrograph

Date	*Time h*	*Total runoff m³/s*	*Base flow*	*Direct runoff*	*Ordinates of unit hydrograph m³/s*	*Time after start of direct runoff h*
1	2	3	4	5	6	7
June 10	6	14.2	14.2	0	0	0
	8	158.5	14.7	143.8	12.52	2
	10	260.0	15.2	244.8	21.32	4
	12	286.0	15.7	270.3	23.54	6
	14	221.0	16.2	204.8	17.84	8
	16	186.5	16.7	169.8	14.79	10
	18	157.0	17.2	139.8	12.18	12
	20	133.0	17.7	115.3	10.04	14
	22	113.0	18.2	94.8	8.26	16
	24	93.4	18.7	74.7	6.51	18
June 11	2	76.4	19.2	57.2	4.98	20
	4	65.0	19.7	45.3	3.95	22
	6	55.2	20.2	35.0	3.05	24
	8	46.7	20.7	26.0	2.26	26
	10	39.6	21.2	18.4	1.60	28
	12	34.0	21.7	12.3	1.07	30
	14	28.3	22.2	6.1	0.53	32
	16	22.7	22.7	0	0	34

ΣQ= 1658.4

$$d = \frac{0.36\,\Delta t \sum Q}{A}$$

$$= \frac{0.36 \times 2 \times 1658.4}{104}$$

$$= 11.48 \text{ cm.}$$

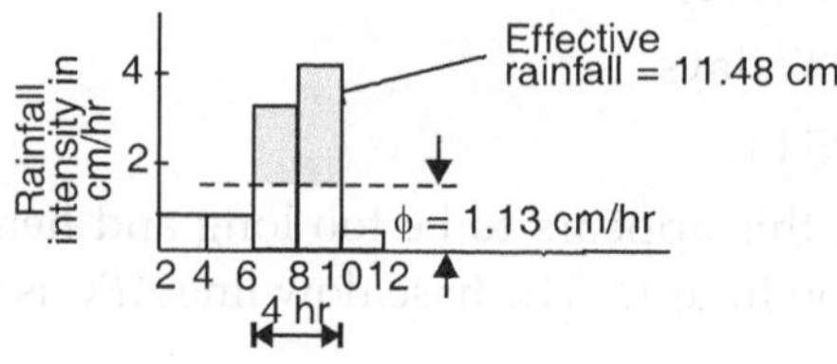

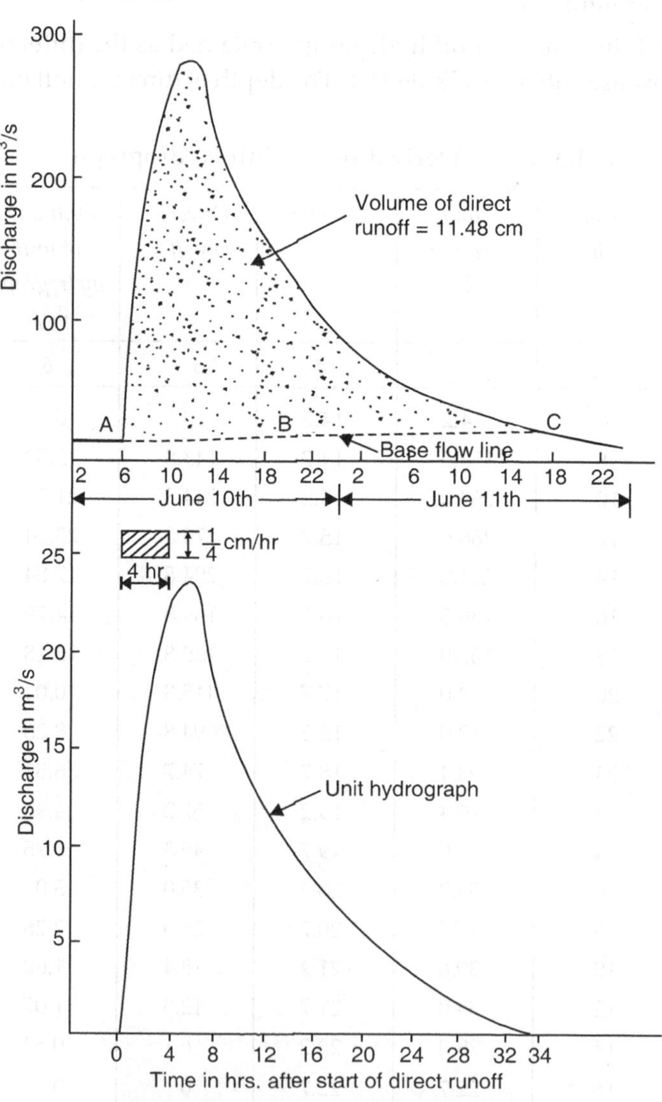

FIGURE 11.8 Derivation of unit hydrograph.

The ordinates of direct runoff in Col. (5) are divided by the depth of runoff in cm that is 11.48 and the resulting unit hydrograph ordinates are entered in Col. (6). To fix the duration of unit hydrograph, by trial and error ϕ-index line of 1.13 cm/h is drawn which would give an effective rainfall exactly equal to 11.48 cm. From the figure the duration of effective rainfall is found to be 4 h. That is, the unit hydrograph derived for the basin is 4 h unit hydrograph, usually abbreviated as 4 h. U.H. As seen from the hyetograph the effective rainfall is not uniformly distributed within its duration. However, the small variation of 0.5 cm/h in the rainfall intensity is considered to be insignificant.

11.6 THE CONVOLUTION EQUATION

Once the unit hydrograph has been derived, it may be used to find the direct runoff hydrographs. For this, of course, the hyetograph of the effective rainfall has to be obtained after the estimated abstractions are deducted from the observed rainfall. The time interval adopted in obtaining the effective rainfall hyetograph must be same as the duration of the unit hydrograph. This is because the unit hydrograph of a particular duration cannot be used to determine the direct runoff resulting from storms of other duration. In general, the base period of the unit hydrograph is adjusted to be an integral multiple of its duration and the ordinates are specified at a uniform time interval equal to its duration. Thus if U_1, U_2, U_3, ..., U_N are the non-zero ordinates of the unit hydrograph, its base period will be $(N + 1)$. D hours where D is the duration of the unit hydrograph in hours.

If the effective rainfall hyetrograph of a storm consists of only one time interval of D hours with an intensity of I cm/h, the effective rainfall in the storm equals $(I.D)$ cm. Since the unit hydrograph always represents only 1 cm of runoff, the ordinates of the unit hydrograph are to be multiplied by $(I.D)$ to give the required direct runoff hydrograph. This follows from the principle of linearity of the unit hydrograph theory.

If the effective rainfall hyetograph, on the other hand, consists of more than one storm bursts each with a duration of D hours and with intensities of I_1, I_2, ..., I_M, the resulting direct runoff hydrograph may be obtained by superimposing the direct runoff from individual storm bursts with appropriate time lags as shown in Fig. 11.9. The effective rainfall in the first time interval is given by $P_1 = I_1 . D$ cm. The runoff due to this rainfall P_1 is obtained by multiplying the ordinates of the unit hydrograph by P_1 and it starts from the commencement of the storm. Similarly, the runoff due to rainfall P_2 in the second time interval (equal to $I_2 . D$) is obtained by multiplying the ordinates of the unit hydrograph by P_2. However this runoff commences D hours later and therefore has to be lagged by D hours while superimposing. The runoff due to the rainfall P_3 in the third time interval (equal to $I_3 . D$), obtained in a similar manner, has to be lagged by 2 D hours, and so on and so forth. The total runoff at any time is then computed as the sum of the runoffs contributed by the individual storms at that time. If N is the number of ordinates of the unit hydrograph and M is the number of storm bursts in the effective rainfall hyetograph, from Fig. 11.9 we observe that there will be $(N + M - 1)$ ordinates in the direct runoff hydrograph. Therefore the base period of the direct runoff hydrograph will be $(N + M) . D$ hours, which is $(M - 1) . D$ hours longer than the base period of the unit hydrograph. Referring to Fig. 11.9 again, the direct runoff at different times may be written as

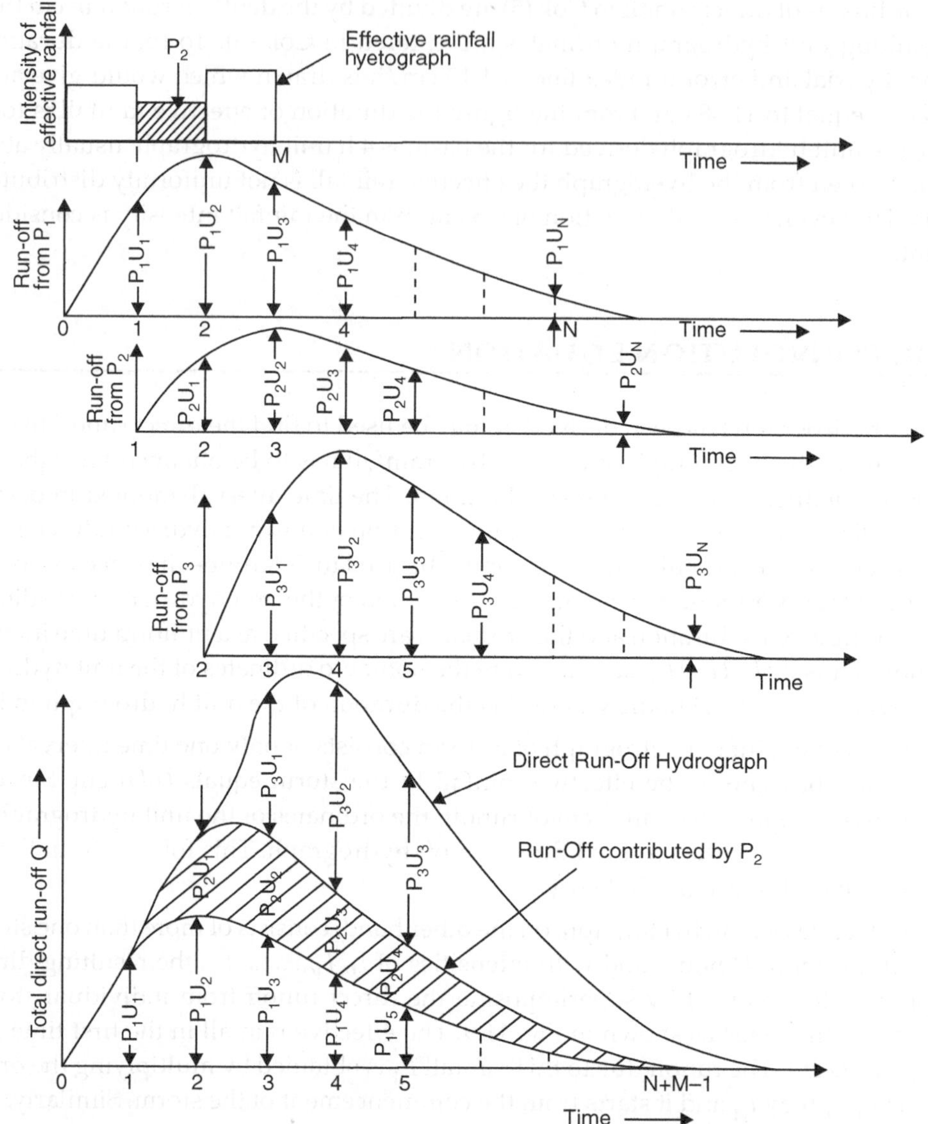

FIGURE 11.9 Graphical representation of the convolution equation.

$$\left.\begin{aligned} Q_1 &= P_1\,U_1 \\ Q_2 &= P_1\,U_2 + P_2\,U_1 \\ Q_3 &= P_1\,U_3 + P_2\,U_2 + P_3\,U_1 \\ &\vdots \\ &\text{etc.} \end{aligned}\right\} \quad \text{...(11.6)}$$

The equation of (11.6) can be written in a more general form as given below.

$$Q_n = \sum_{m=1}^{k} P_m\,U_{n-m+1} \quad \text{...(11.7)}$$

with $n = 1, 2, 3, \ldots\ldots\ldots, (N + M - 1)$

and $k = n$, if $n \le M$

$k = M$, if $n > M$

This is because the rain occurring after the time step n, cannot contribute to runoff at time step n. For instance if $N = 7$ and $M = 4$, the direct runoff hydrograph will have $(7 + 4 - 1) = 10$ ordinates, and specifically

$$Q_3 = \sum_{m=1}^{k} P_m U_{3-m+1} = \sum_{m=1}^{3} P_m U_{4-m} \qquad (\because n < 4)$$

$$= P_1 U_3 + P_2 U_2 + P_3 U_1$$

$$Q_6 = \sum_{m=1}^{k} P_m U_{6-m+1} = \sum_{m=1}^{4} P_m U_{7-m} \qquad (\because n > 4)$$

$$= P_1 U_6 + P_2 U_5 + P_3 U_4 + P_4 U_3$$

$$Q_9 = \sum_{m=1}^{k} P_m U_{9-m+1} = \sum_{m=1}^{4} P_m U_{10-m} \qquad (\because n > 4)$$

$$= P_1 U_9 + P_2 U_8 + P_3 U_7 + P_4 U_6$$

$$= P_3 U_7 + P_4 U_6 \;(\because U_8 = U_9 = 0\text{, as the unit hydrograph has only 7 ordinates})$$

Equation (11.7) is called the *discrete convolution equation*, which becomes handy when the direct runoff is determined using the arithmetical procedure especially when implemented on a computer.

11.7 UNIT HYDROGRAPH FROM COMPLEX STORMS

The simple method of deriving unit hydrograph outlined in section (11.5) requires that an isolated storms with fairly uniform intensity can be selected from the available record. However, it may not be possible in practice to find such isolated storms always. Then it becomes necessary to derive the unit hydrograph from complex storms involving varying intensities of rainfall with the corresponding multipeaked hydrographs. Two methods of deriving unit hydrograph from complex storms are now described.

Deconvolution Method. The selected complex storm hydrograph is analysed and the base flow is separated from it to give the direct runoff hydrograph. After determining the total volume of direct runoff, a ϕ-index line is established on the hyetograph to define the pattern of effective rainfall. Let the effective rainfall hyetograph consist of M consecutive storm bursts of intensities I_1, I_2, ..., I_M each having a duration of D hours. Then the runoff volume contributions of the individual storms are $P_1, P_2, \ldots\ldots\ldots, P_M$, where $P_1 = I_1 D$, $P_2 = I_2 D, \ldots\ldots\ldots\ldots, P_M = I_M \cdot D$. Obviously $(P_1 + P_2 + \ldots\ldots\ldots + P_M)$ should be equal to the depth of runoff represented by the direct runoff hydrograph. Let $Q_1, Q_2, \ldots\ldots\ldots, Q_R$ be the ordinates of the direct runoff hydrograph. Now if $U_1, U_2, \ldots\ldots\ldots, U_N$ are the ordinates of the unit hydrograph to be determined, the relations between $P_1, P_2, \ldots\ldots\ldots, P_M, U_1, U_2, \ldots\ldots, U_N$ and $Q_1, Q_2, \ldots\ldots, Q_R$ are given by Eq. (11.6). In the convolution procedure, P and U are known and Q are determined. Now Q and P are known and U are to be determined. This problem is thus the inverse of the convolution and hence the method of finding the ordinates of the unit hydrograph using Eq. (11.6) may be termed as the *deconvolution method.*

$$Q_1 = P_1 U_1$$

Since Q_1 and P_1 are known, U_1 can be determined.

$$Q_2 = P_1 U_2 + P_2 U_1$$

Since Q_2, P_1, P_2 and U_1 (from the earlier step) are known, U_2 can be determined. Thus the set of equations given by (11.6) can be solved progressively for $U_1, U_2, \ldots\ldots\ldots, U_N$, where $N = (R + 1 - M)$.

In the above procedure all the rain periods, although of different intensities, are assumed to be of same duration of D hours. This is a condition for the use of this method. No doubt the procedure described appears to be very simple and straightforward. But there are numerous difficulties. Each computation depends on all preceding computations for values of U. Errors in observed streamflow, and variations in intensity and aerial distribution of rainfall during several storm periods can lead to cumulative errors which may become large in the later portion of the unit hydrograph. Sometimes large negative ordinates, which cannot be physically justified, are also obtained.

Solution of Equation (11.6) may be obtained using optimisation methods such as *linear programming*, linear regression, etc. ; which will provide the best fit solution to the observed data.

Example 11.4. *The effective rainfall hyetograph of a complex storm has a duration of 12 h, with rainfall intensity of 2.0, 0.75 and 4.0 cm/h respectively in successive 4 h periods. The ordinates of the corresponding direct runoff hydrograph read at 4 h intervals are 160, 300, 570, 636, 404, 234, 105 and 48 m³/s respectively. Determine the ordinates of the 4 h unit hydrograph using the deconvolution method.*

Solution. From the given data, we have

$$D = 4 \text{ h} ; I_1 = 2.0 \text{ cm/h}, I_2 = 0.75 \text{ cm/h and } I_3 = 4 \text{ cm/h}$$

$$P_1 = I_1 . D = 2 \times 4 = 8 \text{ cm}$$

$$P_2 = I_2 . D = 0.75 \times 4 = 3 \text{ cm}$$

$$P_3 = I_3 . D = 4 \times 4 = 16 \text{ cm}$$

$$Q_1 = 160, Q_2 = 300, \ldots\ldots\ldots\ldots \text{ and } Q_8 = 48$$

Also $M = 3$, and $R = 8$. Therefore the 4 h U.H. will have ($N = R + 1 - M = 8 + 1 - 3 = 6$) 6 ordinates which are computed as given below.

$$Q = P_1 U_1 \text{ or } U_1 = \frac{Q_1}{P_1} = \frac{160}{8} = 20 \text{ m}^3/\text{s}$$

$$Q_2 = P_1 U_2 + P_2 U_1 = 8 U_2 + 3 \times 20 = 8 U_2 + 60$$

$$U_2 = \frac{Q_2 - 60}{8} = \frac{300 - 60}{8} = 30 \text{ m}^3/\text{s}$$

$$Q_3 = P_1 U_3 + P_2 U_2 + P_3 U_1 = 8 U_3 + 3 \times 30 + 16 \times 20$$

$$= 8 U_3 + 410$$

$$U_3 = \frac{Q_3 - 410}{8} = \frac{570 - 410}{8} = 20 \text{ m}^3/\text{s}$$

$$Q_4 = P_1U_4 + P_2U_3 + P_3U_2 = 8\ U_4 + 3 \times 20 + 16 \times 30$$
$$= 8\ U_4 + 540$$

$$U_4 = \frac{Q_4 - 540}{8} = \frac{636 - 540}{8} = 12 \text{ m}^3/\text{s}$$

$$Q_5 = P_1U_5 + P_2U_4 + P_3U_3 = 8\ U_5 + 3 \times 12 + 16 \times 20$$
$$= 8\ U_5 + 356$$

$$U_5 = \frac{Q_5 - 356}{8} = \frac{404 - 356}{8} = 6 \text{ m}^3/\text{s}$$

$$Q_6 = P_1U_6 + P_2U_5 + P_3U_4 = 8\ U_6 + 3 \times 6 + 16 \times 12$$
$$= 8\ U_6 + 210$$

$$U_6 = \frac{Q_6 - 210}{8} = \frac{234 - 210}{8} = 3 \text{ m}^3/\text{s}$$

It is fortuitous that no negative values are obtained in this particular example. Thus the ordinates of the 4 h unit hydrograph at 4 h intervals are 20, 30, 20, 12, 6 and 3 m^3/s respectively.

Collin's Method. In the Collin's method of determining the unit hydrograph from a complex storm, the following steps are involved.

Step 1. An approximate unit hydrograph is assumed. Applying this unit hydrograph the direct runoff hydrograph that would result from all the effective rainfall blocks of the hyetograph except the largest is estimated using the convolution equation.

Step 2. The runoff hydrograph obtained in step (1) is subtracted from the actual direct runoff hydrograph, and the residuals are reduced to the unit hydrograph ordinates by dividing the residuals by the effective rain in the largest block.

Step 3. The unit hydrograph ordinates obtained in step (2) should represent the unit hydrograph assumed in step (1). If the correspondence between these two is poor, the weighted average of the assumed unit hydrograph of step (1) and the residual unit hydrograph of step (2) is used as the revised approximation for the next trial.

Step 4. The above three steps are repeated until the residual unit hydrograph does not differ from the assumed unit hydrograph very much.

This method of successive approximations is particularly useful when the largest rainfall block in the effective rainfall hyetograph is very great compared with the other blocks since rapid convergence of the solution takes place then. The unit hydrograph is assumed to be a *distribution graph*, which is discussed in section (11.10).

11.8 UNIT HYDROGRAPHS FOR VARIOUS DURATIONS

A unit hydrograph whose duration is *D* h can be applied only to storms of duration *D* h to obtain direct runoff. However a tolerance of ± 20 to 25% is often allowed. Thus a 8 h U.H. may be used to estimate the runoff from a storm of 6 to 10 h duration without appreciable errors. Nevertheless, in runoff prediction studies, it is preferable to use unit hydrograph of same duration as that of the storm. The unit hydrographs of various durations are best obtained by analysis of data of storms covering a required range of durations. If this is not possible for want of data, they may be developed by conversion of the available U.H. For example if a *D* h U.H. is available, then a 2*D* h, 3*D* h etc. unit hydrographs can be easily developed using the following procedure.

Suppose a *D* h U.H. is given and a 2 *D* h U.H. is required. The given *D* h U.H. is plotted along with its effective rainfall. If another rainfall of *D* h duration is assumed to follow the first rainfall immediately, the direct runoff of the second rainfall is the *D* h U.H. itself, but shifted to the right by *D* h. If the two *D* h graphs are now added graphically, the total hydrograph thus obtained represents the direct runoff from 2 *D* h rainfall at an intensity of 1/*D* cm/h, or a total runoff volume of 2 cm. Since the U.H. should always represent one cm of runoff, the 2 *D* h U.H. is obtained by dividing the ordinates of total hydrograph by 2. This process of conversion is shown schematically in Fig. 11.10. The same procedure can be extended to obtain U.H. of any duration which is an integral multiple of the duration of given U.H. by adding sufficient number of unit graphs with appropriate lags. It can be easily seen that as the duration of U.H. increases, its base period also increases or the hydrograph broadens. Since it has to give always 1 cm of runoff, its peak gets diminished and the time to peak increases. The conversion of U.H. from shorter duration to longer duration can be conveniently carried out in a tabular statement which is demonstrated in Example (11.5).

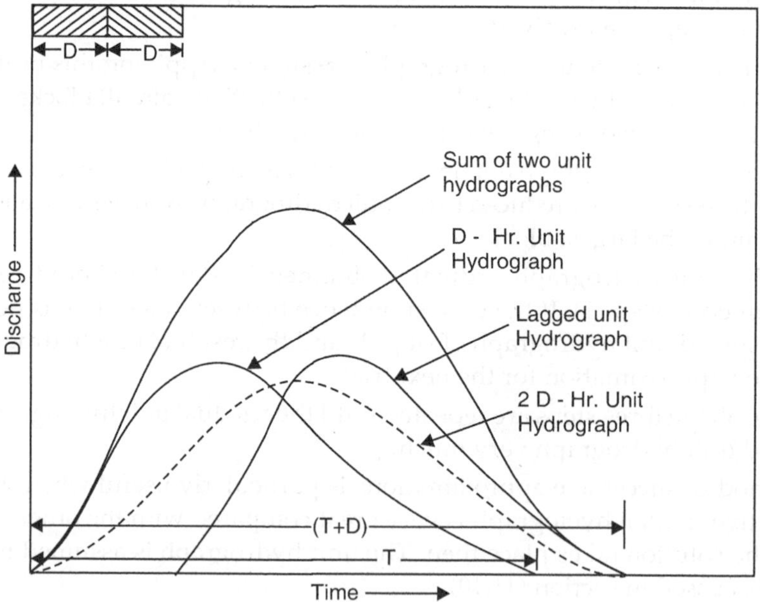

FIGURE 11.10 Conversion of a D h U.H. into a 2D h U.H.

Example 11.5. *From the 4 h unit hydrograph of Example (11.3), derive the 8 h and 12 h unit hydrograph.*

Solution. The necessary computations required to obtain the ordinates of 8 h unit hydrograph are given in Table 11.2.

Table 11.2 Calculations of 8 h U.H. from 4 h U.H.

Time h	*Ordinates of 4 h U.H. in m³/s* without lag	lagged by 4 h	*Combined hydrograph m³/s*	*Ordinates of 8 h U.H. m³/s*
(1)	(2)	(3)	(4) = (2) + (3)	(5) = (4)/2
0	0		0	0
2	12.52		12.52	6.26
4	21.32	0	21.32	10.66
6	23.54	12.52	36.06	18.03
8	17.84	21.32	39.16	19.58
10	14.79	23.54	38.33	19.17
12	12.18	17.84	30.02	15.01
14	10.04	14.79	24.83	12.42
16	8.26	12.18	20.44	10.22
18	6.51	10.04	16.59	8.30
20	4.98	8.26	13.24	6.62
22	3.95	6.51	10.46	5.23
24	3.05	4.98	8.03	4.02
26	2.26	3.95	6.21	3.11
28	1.60	3.05	4.65	2.33
30	1.07	2.26	3.33	1.67
32	0.53	1.60	2.13	1.07
34	0	1.07	1.07	0.54
36		0.53	0.53	0.27
38		0	0	0

In Table 11.2,

Column (1) = time with uniform interval of 2 h

Column (2) = ordinates of 4 h unit hydrograph taken from Example 11.3

Column (3) = ordinates of 4 h U.H. lagged by 4 h

Column (4) = Col. (2) + Col. (3)

= represents the ordinates of direct runoff hydrograph of 8 h storm with 2 cm of runoff

Column (5) $= \dfrac{\text{Col. (4)}}{2}$ = ordinates of 8 h U.H.

The computations to obtain the ordinates of 12 h U.H. are shown in Table (11.3), which is similar to Table (11.2) except that now 3 unit hydrograph are added and the ordinates of the combined hydrograph are divided by 3.

Table 11.3 Calculation of 12 h U.H. from 4 h U.H.

Time h	Ordinates of 4 h U.H. in m³/s			Combined hydrograph m³/s	Ordinates of 12 h U.H.
	without lag	lagged by 4 h	lagged by 8 h		
(1)	(2)	(3)	(4)	(5) = (2) + (3) + (4)	(6) = (5)/3
0	0			0	0
2	12.52			12.52	4.17
4	21.32	0		21.32	7.11
6	23.54	12.52		36.06	12.02
8	17.84	21.32	0	39.16	13.05
10	14.79	23.54	12.52	50.85	16.95
12	12.18	17.84	21.32	51.34	17.11
14	10.04	14.79	23.54	48.37	16.12
16	8.26	12.18	17.84	38.28	12.76
18	6.51	10.04	14.79	31.34	10.45
20	4.98	8.26	12.18	25.42	8.47
22	3.95	6.51	10.04	20.50	6.83
24	3.05	4.98	8.26	16.29	5.43
26	2.26	3.95	6.51	12.72	4.24
28	1.60	3.05	4.98	9.63	3.21
30	1.07	2.26	3.95	7.28	2.43
32	0.53	1.60	3.05	5.18	1.73
34	0	1.07	2.26	3.33	1.11
36		0.53	1.60	2.13	0.71
38		0	1.07	1.07	0.36
40			0.53	0.53	0.18
42			0	0	0

The features of the three unit hydrographs are compared below.

	4 h U.H.	8 h U.H.	12 h U.H.
Peak ordinate	23.54 m³/s	19.58 m³/s	17.11 m³/s
Time to peak	6 h	8 h	12 h
Base period	34 h	38 h	42 h

11.9 S-CURVE HYDROGRAPH

The simple procedure of converting the unit hydrograph of short duration into unit hydrograph of longer duration, explained in the previous section, cannot be adopted if the duration of the required unit hydrograph is either less than, or not an integral multiple of the duration of the given unit hydrograph. In such situations the U.H. of any duration can be obtained from the U.H. of given duration using the *S*-curve technique. A *S*-curve hydrograph may be defined as the hydrograph of direct runoff resulting from a continuous effective rainfall of uniform intensity $\frac{1}{D}$ cm/h. As shown in Fig. 11.11, the *S*-curve is constructed by adding together a series of *D* h unit hydrographs, each lagged by *D* h with respect to the previous one. The *S*-curve hydrograph attains a constant ordinate, called the *equilibrium discharge* denoted by Q_e, approximately at the end of the base period *T* of the unit hydrograph. Thus the number of unit hydrographs needed to produce the *S*-curve is $\frac{T}{D}$. The *S*-curve ordinates are sometimes found to oscillate in the top portion at and around the equilibrium discharge. This is called the *hunting* of *S*-curve. It may be the result of the errors in the unit hydrograph such as that the U.H. used in the *S*-curve derivation might not be representing the runoff from a storm of uniform time distribution. Nevertheless, an average *S*-curve usually be drawn to attain the equilibrium discharge smoothly. The name of the *S*-curve is derived from its deformed shape of the upper case letter *S*. Since the rainfall rate is equal to the runoff rate at the equilibrium state, it follows that

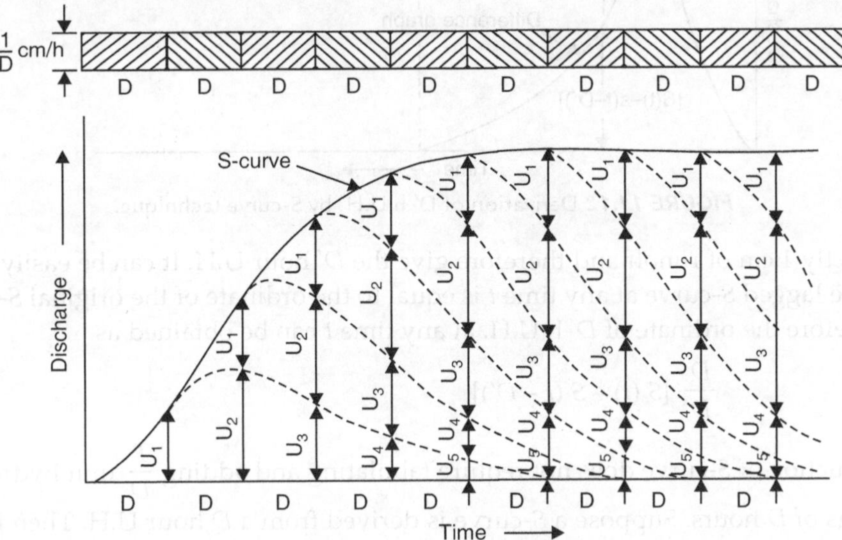

FIGURE 11.11 S-curve hydrograph.

$$Q_e = A \cdot \frac{1}{D} \text{ km}^2\text{-cm/h} = 2.778 \frac{A}{D} \text{ m}^3/\text{s} \qquad ...(11.8)$$

where *A* is the area of the basin in km² and *D* is the duration of unit hydrograph in hours which is used in the construction of the *S*-curve.

If the *D*′ hour U.H. is required (where $D' \lessgtr D$ such that $\frac{D'}{D}$ is a rational number) it can be obtained from *S*-curve as follows. Two *S*-curves are drawn on the same time axis lagged by *D*′ hours as shown in Fig. 11.12. The rainfall hyetographs which produced these *S*-curves are also shown in the same figure at the top. The difference between the two hyetographs represents a storm of duration *D*′ with an intensity of $\frac{1}{D}$ cm/h and hence a rainfall of $\frac{D'}{D}$ cm. The difference graph between the two *S*-curves has to then represent the direct runoff hydrograph of this storm. Now if the ordinates of the difference graph are divided by $\frac{D'}{D}$ (or multiplied by $\frac{D}{D'}$), they

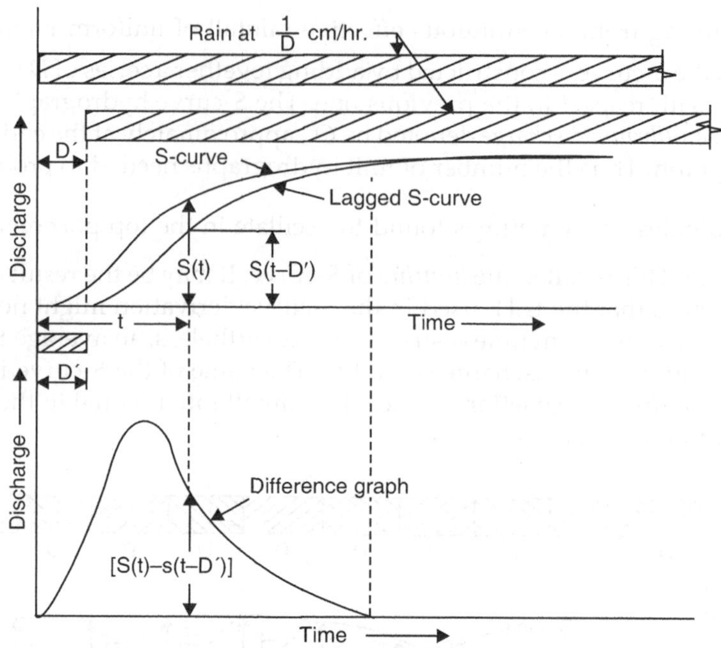

FIGURE 11.12 Derivation of D′ h U.H. by S-curve technique.

represent exactly 1 cm of runoff and therefore give the D' hour U.H. It can be easily seen that the ordinate of the lagged *S*-curve at any time *t* is equal to the ordinate of the original *S*-curve at time $(t - D')$. Therefore the ordinate of D' h U.H. at any time *t* can be obtained as

$$\frac{D}{D'}\ [S\ (t) - S\ (t - D')]$$

Construction of *S*-curve does not require tabulating and adding $\frac{T}{D}$ unit hydrographs with successive lags of *D* hours. Suppose a *S*-curve is derived from a *D* hour U.H. Then if we take the difference of two *S*-curves lagged by *D* h itself, the difference graph is nothing but the unit hydrograph itself. Therefore, the ordinate of *D* h unit hydrograph at any time *t* is given by

$$U\ (t) = S\ (t) - S\ (t - D)$$

or

$$S\ (t) = U\ (t) + S\ (t - D) \qquad ...(11.9)$$

The second term on R.H.S of Eq. (11.9) is called the *S-curve addition*. Thus the ordinate of *S*-curve at any time *t* is obtained as the sum of the unit hydrograph ordinate at *t* and the *S*-curve addition, where *S*-curve addition is also an ordinate of *S*-curve itself but at time $(t - D)$. It may be noted that for $t \le D$, the ordinates of *S*-curve and *D* h U.H. are identical. For $t > D$, Eq. (11.9) provides an easy way of constructing the *S*-curve. The construction of *S*-curve hydrograph from a U.H. of a given duration *D* and the derivation of U.H. of any other duration D' using the *S*-curve are demonstrated in the Examples 11.6 and 11.7.

Example 11.6. *The ordinates of a 4 h U.H. of a basin of area 300 km^2 measured at 1 h intervals are 6, 36, 66, 91, 106, 93, 79, 68, 58, 49, 41, 34, 27, 23, 17, 13, 9, 6, 3 and 1.5 m^3/s respectively. Obtain the ordinates of a 3 h U.H. for the basin using the S - curve technique.*

Solution. The computations of the solution are shown in Table 11.4.

Column (1) = The time axis at a uniform interval of 1 h.
Column (2) = The given ordinates of 4 h U.H.
Column (3) = The S-curve additions.
Column (4) = Col. (2) + Col. (3) = The *S* - curve ordinates.
Column (5) = *S* - curve lagged by 3 h (since $D' = 3$ h).
Column (6) = Col. (4) – Col. (5) = The difference graph.

$$\text{Column (7)} = \text{Col. (6)} \times \frac{4}{3}\left(\text{since } \frac{D}{D'} = \frac{4}{3}\right)$$

= The ordinates of 3 h U.H.

Table 11.4 Derivation of 3 h U.H. from 4 h U.H. by S-curve Technique

Time t hours	*Ordinates of 4 h U.H. U (t) m³/s*	*S-curve additions S (t – 4) m³/s*	*S-curve ordinates S (t) m³/s*	*S-curve lagged by 3 h, S (t – 3) m³/s*	*Difference graph m³/s*	*Ordinates of 3h U.H.*
(1)	(2)	(3)	(4) = (2) + (3)	(5)	(6) = (4) – (5)	(7) = (6) × 4/3
0	0		0		0	0
1	6		6		6	8.0
2	36		36		36	48.0
3	66		66	0	66	88.0
4	91	0	91	6	85	113.3
5	106	6	112	36	76	101.3
6	93	36	129	66	63	84.0
7	79	66	145	91	54	72.0
8	68	91	159	112	47	62.7
9	58	112	170	129	41	54.7
10	49	129	178	145	33	44.0
11	41	145	186	159	27	36.0
12	34	159	193	170	23	30.7
13	27	170	197	178	19	25.3
14	23	178	201	186	15	20.0
15	17	186	203	193	10	13.3
16	13	193	206	197	9	12.0
17	9	197	206.5*	201	5.5	7.3
18	6	201	207.0	203	4.0	5.3
19	3	203	207.5*	206	1.5	2.0
20	1.5	206	208.0*	206.5	0**	0
21	0	206.5	208.5*	207.0	0**	

* Adjustments required to see that *S*-curve attains the equilibrium discharge smoothly at the end of the base period of the given U.H.

** Adjustments required to see that the 3 h U.H. will have a smooth tail and a base period of 20 h.

Since the *S*-curve ordinates and U.H. ordinates are same for $t \le D$, the values in Col. (4) are first filled up for $t = 0, 1, 2, 3, 4$. Again since the *S*-curve addition at time *t* is same as the *S*-curve ordinate at $(t - D)$, the values in Col. (4) for $t = 0, 1, \ldots\ldots\ldots, 4$ are placed in Col. (3) for $t = 4, 5, 6, 7, 8$ and then the values of Col. (4) are obtained for $t = 4, 5, \ldots\ldots\ldots\ldots, 8$. The procedure is repeated in alternative steps till all the *S*-curve ordinates are obtained.

The equilibrium discharge in the present case is given by

$$Q_e = \frac{2.778\,A}{D} = \frac{2.778 \times 300}{4} = 208.4 \text{ m}^3/\text{s}.$$

The ordinates of the *S*-curve towards the end are slightly adjusted such that the *S*-curve reaches the equilibrium discharge smoothly.

The base period of the given 4 h U.H. is 21 hours. Therefore the 3 h U.H. is expected to have a base period of 20 hours. Accordingly the ordinates of the difference graph are also adjusted such that the tail is a smooth curve and the base period is 20 hours.

Example 11.7. *The ordinates of a 4 h U.H. of a basin of area 295 km^2 measured at 4 h intervals are 54.0, 68.6, 43.2, 22.0, 10.7, 4.8, and 1.4 m^3/s respectively. Obtain the ordinates of 3 h U.H. and 5 h U.H. of the basin using the S-curve method.*

Solution. This is similar to the previous example except that the U.H. ordinates are specified at 4 h intervals instead of 1 h intervals. Since the durations of the required unit hydrographs are 3 h and 5 h, the *S*-curve needs to be constructed with a time interval of 1 h or less. (If the durations of the required unit hydrographs were to be 2 h and 6 h, it is sufficient if the *S*-curve is derived with a time interval of 2 h). Therefore the given U.H. ordinates are plotted on a graph to a large scale (which is not shown here) and a smooth curve is drawn passing through these points. Then the ordinates of U.H. at 1 h interval are interpolated from this curve. These ordinates are shown in column (2) of Table 11.5, wherein the figures with bold face are the given values and the rest are the interpolated values. The remaining computations of Table 11.5 are similar to those presented in Table 11.4 of example 11.6.

11.10 DISTRIBUTION GRAPH

The distribution graph is also a unit hydrograph but in a different form. The base period of the unit hydrograph is divided into number of uniform time periods each equal to the duration of the unit hydrograph. The volume of runoff in each time period of the U.H. is expressed as a percentage of the total volume of runoff represented by the entire U.H. Thus, for example, the area under U.H. in any time period divided by the total area under U.H. gives the distribution percentage for that time period. These percentages drawn as rectangular bars or steps against the respective time periods define the distribution graph. Evidently the total of all the ordinates of the distribution graph equals 100. The concept of the distribution graph was first introduced by Bernard in 1935. Fig. 11.13 shows a typical unit hydrograph and the corresponding distribution graph.

Table 11.5 Derivation of 3 h U.H. and 5 h U.H. from 4 h U.H. by S-curve Technique

(All values except those in column (1) are in units of m^3/s)

Time hours	4 h U.H.	S-curve additions	S-curve ordinates	S-curve lagged by 3 h	Difference graph	3 h U.H.	S-curve lagged by 5 h	Difference graph	5 h U.H.
(1)	(2)	(3)	(4) = (2) + (3)	(5)	(6) = (4) – (5)	(7) = (6) × $\frac{4}{3}$	(8)	(9) = (4) – (8)	(10) = (9) × $\frac{4}{5}$
0	0		0		0	0		0	0
1	5.0		5.0		5.0	6.7		5.0	4.0
2	18.0		18.0		18.0	24.0		18.0	14.4
3	35.2		35.2	0	35.2	46.9		35.2	28.2
4	**54.0**	0	54.0	5.0	49.0	65.3		54.0	43.2
5	67.8	5.0	72.8	18.0	54.8	73.1	0	72.8	58.2
6	72.9	18.0	90.9	35.2	55.7	74.3	5.0	85.9	68.7
7	72.4	35.2	107.6	54.0	53.6	71.5	18.0	89.6	71.7
8	**68.6**	54.0	122.6	72.8	49.8	66.4	35.2	87.4	69.9
9	62.9	72.8	135.7	90.9	44.8	59.7	54.0	81.7	65.4
10	56.4	90.9	147.3	107.6	39.7	52.9	72.8	74.5	59.6
11	49.7	107.6	157.3	122.6	34.7	46.3	90.9	66.4	53.1
12	**43.2**	122.6	165.8	135.7	30.1	40.1	107.6	58.2	46.6
13	37.1	135.7	172.8	147.3	25.5	34.0	122.6	50.2	40.2
14	31.4	147.3	178.7	157.3	21.4	28.5	135.7	43.0	34.4
15	26.3	157.3	183.6	165.8	17.8	23.7	147.3	36.3	29.0
16	**22.0**	165.8	187.8	172.8	15.0	20.0	157.3	30.5	24.4
17	18.3	172.8	191.1	178.7	12.4	16.5	165.8	25.3	20.2
18	15.4	178.7	194.1	183.6	10.5	14.0	172.8	21.3	17.0
19	13.0	183.6	196.6	187.8	8.8	11.7	178.7	17.9	14.3
20	**10.7**	187.8	198.5	191.1	7.4	9.9	183.6	14.9	11.9
21	8.8	191.1	199.9	194.1	5.8	7.7	187.8	12.1	9.7
22	7.2	194.1	201.3	196.6	4.7	6.3	191.1	10.2	8.2
23	5.9	196.6	202.5	198.5	4.0	5.3	194.1	8.4	6.7
24	**4.8**	198.5	203.3	199.9	3.4	4.5	196.6	6.7	5.4
25	3.8	199.9	203.7	201.3	2.4	3.2	198.5	5.2	4.2
26	2.9	201.3	204.2	202.5	1.7	2.3	199.9	4.3	3.4
27	2.1	202.5	204.6	203.3	1.3	1.7	201.3	3.3	2.6
28	**1.4**	203.3	204.7	203.7	1.0	1.3	202.5	2.2	1.8
29	0.8	203.7	*204.8	204.2	0.6	0.8	203.3	1.5	1.2
30	0.4	204.2	*204.9	204.6	0.3	0.4	203.7	1.2	1.0

31	0.2	204.6	*205.0	204.7	**0	0	204.2	***0.7	0.6
32	0	204.7	*205.0	204.8	**0		204.6	***0.3	0.2
33		204.8	*205.0	204.9	**0		204.7	***0	0

* Adjustments for S-curve ;

** Adjustments for the difference graph so that 3 h U.H. has a base period of 31 h.

*** Adjustments for the difference graph so that 5 h U.H. has a base period of 33 h.

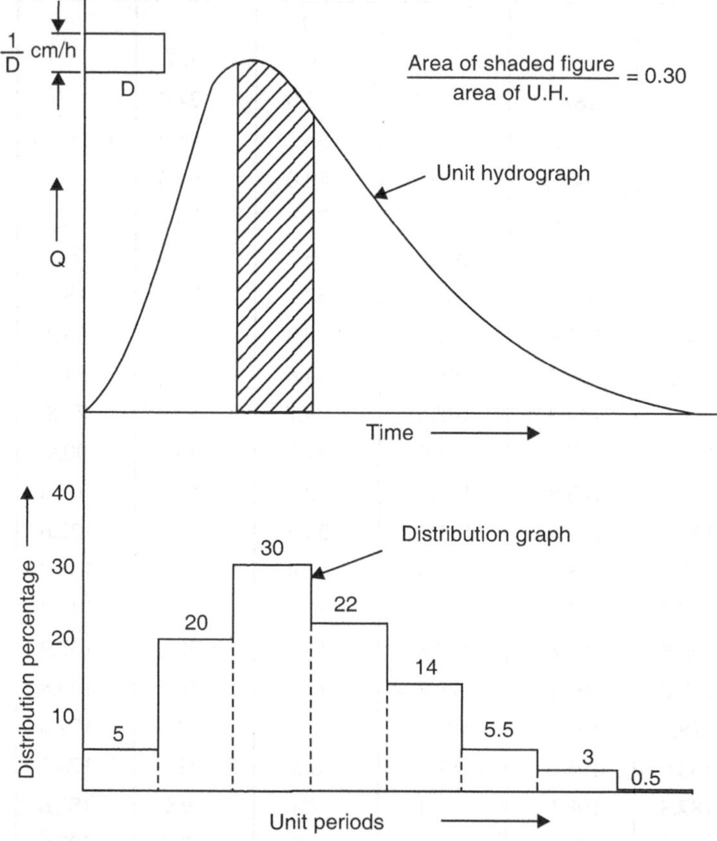

FIGURE 11.13 The distribution graph.

The use of the distribution graph to determine direct runoff from a known effective rainfall hyetograph is exactly the same as that of a unit hydrograph. See example 11.11. Once the runoff ordinates are determined from the distribution graph, a smooth curve has to be drawn through the steps to obtain the direct runoff hydrograph. The distribution graph is therefore less precise than the unit hydrograph. But it is much better suited for the Collin's iterative process of deriving U.H. from complex storms which was explained in section 11.7.

11.11 USE AND APPLICATION OF UNIT HYDROGRAPH

The unit hydrographs are used in many hydrological problems such as

(*i*) in the development of flood hydrograph corresponding to design storms which are required for the design of hydraulic structures,

(*ii*) in the watershed simulation models,

(*iii*) in the studies of flood forecasting and flood warning systems, and

(*iv*) in extending the flood flow records based on rainfall records. The limitations of the unit hydrograph theory which were discussed in section 11.4 must be kept in mind to justify the use of unit hydrograph in any hydrological studies.

The application of unit hydrograph to obtain the runoff hydrographs from the rainfall data is essentially based on the convolution equation (11.7). The procedure consists of the following steps.

(*i*) The rainfall hyetograph is constructed for the storm under consideration and the effective rainfall is determined by drawing an appropriate ϕ-index line.

(*ii*) The effective rainfall hyetograph is split into periods of approximately uniform intensity and durations that meet the requirements of the unit hydrograph available.

(*iii*) The direct runoff hydrograph resulting from each rainfall period is obtained by multiplying the ordinates of the appropriate U.H. by the depth of effective rainfall in that period.

(*iv*) These direct runoff hydrographs are plotted and graphically added with appropriate time lag to yield the composite direct runoff hydrograph.

(*v*) The estimated baseflow is then added to the composite direct runoff hydrograph to give the total runoff hydrograph.

The computations to obtain the runoff hydrographs can be conveniently carried out in tabular statements as indicated in the following examples.

Example 11.8. *A storm produced rainfall intensities of 0.75, 2.25 and 1.25 cm/h on the drainage basin of example 11.3 in 3 successive time periods of 4 h each. Assuming a base flow of 10 m*^{3}*/s and a* ϕ*-index of 2.5 mm/h compute total runoff hydrograph resulting from this storm.*

Solution. Since each rainfall is of 4 h period, we can use 4 h U.H. of example 11.3 to compute the direct runoff

$$\phi\text{-index} = 2.5 \text{ mm/h} = 0.25 \text{ cm/h}$$

Depth of effective rainfall in the first 4 h period

$$= (0.75 - 0.25) \times 4 = 2 \text{ cm}$$

Depth of effective rainfall in the second 4 h period

$$= (2.25 - 0.25) \times 4 = 8 \text{ cm}$$

Depth of effective rainfall in the third 4 h period

$$= (1.25 - 0.25) \times 4 = 4 \text{ cm.}$$

Table 11.6 Computation of Flood Hydrograph due to known Storm Rainfalls

Time h	*4 h U.H. m³/s*	*Direct runoff in m³/s from effective rainfalls during three successive 4 h periods*				*Baseflow m³/s*	*Total runoff m³/s*
		2 cm	*8 cm*	*4 cm*	*Composite direct runoff*		
(1)	(2)	(3)	(4)	(5)	(6) = (3) + (4) + (5)	(7)	(8) = (6) + (7)
0	0	0			0	10	10
2	12.52	25.04	—	—	25.04	10	35.04
4	21.32	42.64	0	—	42.64	10	52.64
6	23.54	47.08	100.16	—	147.24	10	157.24
8	17.84	35.68	170.56	0	206.24	10	216.24
10	14.79	29.58	188.32	50.08	267.98	10	277.98
12	12.18	24.36	142.72	85.28	252.36	10	262.36
14	10.04	20.08	118.32	94.16	232.56	10	242.56
16	8.26	16.52	97.44	71.36	185.32	10	195.32
18	6.51	13.02	80.32	59.16	152.50	10	162.50
20	4.98	9.96	66.08	48.72	124.76	10	134.76
22	3.95	7.90	52.08	40.16	100.14	10	110.14
24	3.05	6.10	39.84	33.04	78.98	10	88.98
26	2.26	4.52	31.60	26.04	62.16	10	72.16
28	1.60	3.20	24.40	19.92	47.52	10	57.52
30	1.07	2.14	18.08	15.80	36.02	10	46.02
32	0.53	1.06	12.80	12.20	26.06	10	36.06
34	0	0	8.56	9.04	17.60	10	27.60
36			4.24	6.40	10.64	10	20.64
38			0	4.28	4.28	10	14.28
40				2.12	2.12	10	12.12
42				0	0	10	10

Column (2) : ordinates of 4 h U.H. taken from example (11.3)

Column (3) : ordinates of 4 h U.H. multiplied by the first rainfall of 2 cm

Column (4) : ordinates of 4 h U.H. multiplied by the second rainfall of 8 cm and lagged by 4 hours since the runoff from the second rainfall commences from 4th hour.

Column (5) : ordinates of 4 h U.H. multiplied by the third rainfall of 4 cm and lagged by 8 hours since the runoff from the third rainfall commences from 8th hour.

Column (6) : Cols. (3) + (4) + (5) = ordinates of composite direct runoff hydrograph

Column (7) : baseflow which is assumed to be constant at 10 m^3/s

Column (8) : ordinates of the total runoff hydrograph, or the flood hydrograph.

Example 11.9. *Determine the peak discharge and time to peak in the flood hydrograph from the drainage basin of example 11.7 resulting from a 12 h storm which has intensities of 1 cm/h for the first 3 h, 1.75 cm/h for the next 4 h and 0.9 cm/h for the last 5 h. Assume a ϕ-index of 0.5 cm/h and a constant baseflow of 15 m^3/s.*

Solution. The rainfall hyetograph consists of intensities over non-uniform time periods of 3, 4 and 5 h. We therefore need 3 h, 4 h and 5 h unit hydrographs to compute the corresponding direct runoffs. These unit hydrographs are taken from example 11.7 and the computations are shown in Table 11.7. The calculations are terminated after the recession of the flood hydrograph is noticed as we are interested only in the peak flow.

Effective rainfall in the first 3 h

$$= (1.0 - 0.5) \times 3 = 1.5 \text{ cm}$$

Effective rainfall in the next 4 h

$$= (1.75 - 0.5) \times 4 = 5.0 \text{ cm}$$

Effective rainfall in the last 5 h

$$= (0.9 - 0.5) \times 5 = 2.0 \text{ cm}$$

Column (1) : time axis with uniform interval of 1 h

Column (2) : ordinates of 3 h U.H. taken from example 11.7

Column (3) : ordinates of 3 h U.H. multiplied by the first rainfall of 1.5 cm

Column (4) : ordinates of 4 h U.H. lagged by 3 h, since the runoff from the next rain commences from third hour.

Column (5) : ordinates of 4 h U.H. multiplied by the second rainfall of 5.0 cm

Column (6) : ordinates of 5 h U.H. lagged by 7 h, since the runoff from the last rain commences from 7th hour.

Column (7) : ordinates of 5 h U.H. multiplied by the third rainfall of 2.0 cm

Column (8) : Cols. (3) + (5) + (7) = composite direct runoff

Column (9) : baseflow assumed to be constant at 15 m^3/s

Column (10) : Cols. (8) + (9) = flood hydrograph.

The peak flow in the flood hydrograph has a magnitude of 513.85 m^3/s and it occurs at the end of 12th hour.

Example 11.10. *In example 11.9 if 4 h U.H. only is used to compute the direct runoff, what would be the percentage error in peak flow ?*

Solution. If the effective rainfall hyetograph is assumed to have 3 equal time periods of 4 h each with rainfalls of 1.5, 5.0 and 2.0 cm respectively, the direct runoff can be computed using the 4 h U.H. only for all the three periods. Now the U.H. need not be tabulated at 1 h interval and it is sufficient if it is tabulated at 4 h intervals as the successive lags for the second and third rains are 4 h and 8 h. The calculations are shown in Table 11.8.

Table 11.7 Computations of Flood Hydrograph of Example 11.9

(All values except those in column 1 are in units of m³/s)

Time h	Ordinates of 3 *h U.H.*	*Runoff from first rain of 1.5 cm in first 3 h*	*Ordinates of 4 h U.H. lagged by 3 h*	*Runoff from second rain of 5 cm in the next 4 h*	*ordinates of 5 h U.H. lagged by 7 h*	*Runoff from third rain of 2 cm in the last 5 h*	*Composite direct runoff hydrograph*	*Baseflow*	*Flood hydrograph*
(1)	(2)	(3) = (2) × 1.5	(4)	(5) = (4) × 5	(6)	(7) = (6) × 2	(8) = (3) + (5) + (7)	(9)	(10) = (8) + (9)
0	0	0					0	15	15.00
1	6.7	10.05					10.05	15	25.05
2	24.0	36.00					36.00	15	51.00
3	46.9	70.35	0	0			70.35	15	85.35
4	65.3	97.95	5.0	25.0			122.95	15	137.95
5	73.1	109.65	18.0	90.0			199.65	15	214.65
6	74.3	111.45	35.2	176.0			287.45	15	302.45
7	71.5	107.25	54.0	270.0	0	0	377.25	15	392.25
8	66.4	99.60	67.8	339.0	4.0	8.0	446.60	15	461.60
9	59.7	89.55	72.9	364.5	14.4	28.8	482.85	15	497.85
10	52.9	79.35	72.4	362.0	28.2	56.4	497.75	15	512.75
11	46.3	69.45	68.6	343.0	43.2	86.4	498.85	15	513.85
12	40.1	60.15	62.9	314.5	58.2	116.4	491.05	15	506.05
13	34.0	51.00	56.4	282.0	68.7	137.4	470.40	15	485.40
14	28.5	42.75	49.7	248.5	71.7	143.4	434.65	15	449.65
15	23.7	35.55	43.2	216.0	69.9	139.8	391.35	15	406.35

*Peak flow.

Table 11.8 Computation of Flood Hydrograph of Example 11.10

Time h	*4 h U.H. ordinates* m^3/s	*Runoff from first rain* m^3/s	*Runoff from second rain* m^3/s	*Runoff from third rain* m^3/s	*Composite direct runoff* m^3/s
(1)	(2)	(3)	(4)	(5)	(6)
0	0	0			0
4	54.0	81.0	0		81.0
8	68.6	102.9	270	0	372.9
12	43.2	64.8	343	108	515.8
16	22.0	33.0	216	137.2	386.2
20	10.7	16.05	110	86.4	212.45

The peak ordinate = 515.8 m^3/s

The corresponding peak ordinate of the composite direct runoff in example (11.9) is 498.85 m^3/s.

$$\text{Percentage error in peak flow} = \frac{515.8 - 498.85}{498.85} \times 100 = 3.4\%$$

∴ The peak flow is overestimated by 3.4%.

Example 11.11. *A drainage basin of area 580 km² has experienced a 12 h storm with rainfall intensities of 2.5, 1.8 and 2.3 cm/h respectively in successive periods of 4 h each. The percentages of the distribution graph for storms of 4 h duration are 5, 15, 30, 20, 13, 9, 5.5 and 2.5. Determine the ordinates of discharge hydrograph. Assume an average φ-index of 0.8 cm/h and neglect baseflow.*

Solution. Effective rainfall in first 4 h period

$$= (2.5 - 0.8) \times 4 = 6.8 \text{ cm}$$

Effective rainfall in second 4 h period

$$= (1.8 - 0.8) \times 4 = 4 \text{ cm}$$

Effective rainfall in third 4 h period

$$= (2.3 - 0.8) \times 4 = 6 \text{ cm}$$

1 cm of runoff over an area of 580 km² drained in a period of 4 h is equivalent to a runoff rate of $580 \times 10^6 \times \frac{1}{100} \times \frac{1}{4} \times \left(\frac{1}{3600}\right) = 402.8 \text{ m}^3/\text{s}$.

Column (1) = time periods of 4 h each

Column (2) = distribution percentages

Column (3) = distribution ratios

Column (4) = Col. (3) × 6.8

Column (5) = Col. (3) × 4 and lagged by one time period

Column (6) = Col. (3) × 6 and lagged by two time periods

Column (7) = Col. (4) + (5) + (6) = runoff in cm drained in a period of 4 h

Column (8) = Col. (7) × 402.8 = runoff in m^3/s.

Table 11.9 Computing Direct Runoff Using Distribution Graph

Time interval h	*Distribution percentage*	*Distribution ratio*	*Distributed runoff in cm due to effective rainfalls of*			*Runoff*	
			6.8 cm in first 4h	*4 cm in second 4 h*	*6 cm in third 4 h*	*cm*	*m³/s*
(1)	(2)	(3)	(4)	(5)	(6)	(7) = (4) + (5)+ (6)	(8) = (7) × 402.8
0–4	5	0.05	0.34	—		0.34	136.95
4–8	15	0.15	1.02	0.2	—	1.22	491.42
8–12	30	0.30	2.04	0.6	0.3	2.94	1184.23
12–16	20	0.20	1.36	1.2	0.9	3.46	1393.69
16–20	13	0.13	0.884	0.8	1.8	3.484	1403.36
20–24	9	0.09	0.612	0.52	1.2	2.332	939.33
24–28	5.5	0.055	0.374	0.36	0.78	1.514	609.84
28–32	2.5	0.025	0.170	0.22	0.54	0.93	374.60
32–36				0.10	0.33	0.43	173.20
36–40					0.15	0.15	60.42

1 cm of runoff from an area of 580 km² drained in a period of 4 h is equivalent to a runoff rate of

$$580 \times 10^6 \times \left(\frac{1}{100}\right) \times \frac{1}{4} \times \frac{1}{3600} = 402.8 \text{ m}^3/\text{s}.$$

11.12 SYNTHETIC UNIT HYDROGRAPH

Unit hydrographs can be derived using the methods described earlier provided rainfall and runoff records are available for the basin under consideration. But there are many basins, especially in the developing countries, which are not gauged and for which unit-graphs may be required. Hence some method of deriving unit hydrographs for ungauged basins is necessary. This is usually done by relating the selected basin characteristics to the unit hydrograph shape. Once such relations are established between the basin parameters and unit hydrograph parameters for the basins having sufficient data, the same relations are applied to get the unit hydrograph of ungauged basins in the same hydrometeorologically homogeneous area from the known basin parameters. The unit hydrograph thus obtained is known as *Synthetic unit hydrograph*.

There are several approaches to obtain the synthetic unit hydrograph and the most widely used one is due to Snyder. He selected the peak discharge Q_p, the basin lag t_p, and the base width T, to define the shape of unit hydrograph. The basin lag is defined as the time from the centroid of rainfall to the peak of the hydrograph. These parameters, as shown in Fig. 11.14, and the fact that the volume of runoff represented by unit hydrograph must be 1 cm would permit the complete sketching of the unit hydrograph.

The basin characteristics considered by Snyder are the area and shape of the basin. The equations relating the unit hydrograph parameters to the basin parameters as proposed by Snyder are

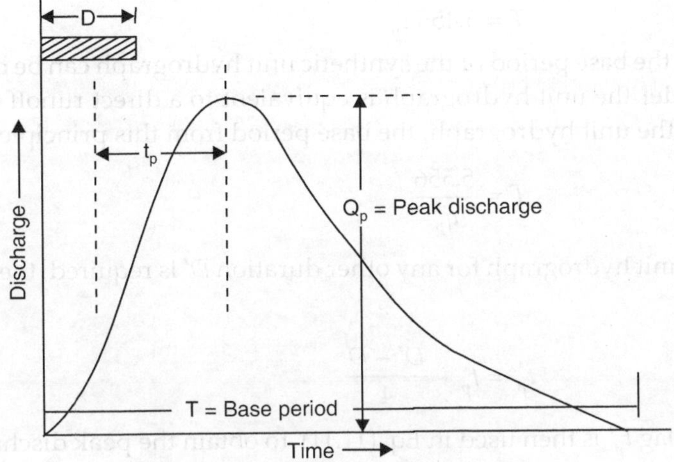

FIGURE 11.14 Synthetic unit hydrograph.

$$t_p = C_t (L L_c)^{0.3} \quad \text{...(11.10)}$$

$$q_p = \frac{Q_p}{A} = \frac{2.778\, C_p}{t_p} \quad \text{...(11.11)}$$

$$T = 72 + 3t_p \quad \text{...(11.12)}$$

where t_p = the basin lag in hours

L = the length of main stream in km

L_c = the distance along the main stream from the basin outlet to a point on the stream which is nearest to the centroid of the basin in km

A = the catchment area in km^2

T = the base width of the unit hydrograph in hours

Q_p = the peak discharge in the unit hydrograph in m^3/s

q_p = the peak discharge per unit area of the catchment in m^3/s/km^2

C_t = a coefficient accounting for catchment slope and the associated catchment storage, which is found to vary between 1.35 and 1.65 for the basins in Appalachian Mountain region studied by Snyder (the equivalent values in British system are 1.8 and 2.2 when L and L_c are in miles)

C_p = a dimensionless coefficient ranging from 0.56 to 0.69.

The standard duration for the above unit hydrograph was fixed by Snyder as

$$D = \frac{2}{11}\, t_p \quad \text{...(11.13)}$$

Eq. (11.12) gives unreasonably long base period for small catchments. It is recommended by some investigators that a base period which is 5 times the time to peak is about reasonable. That is

$$T = 5\left(\frac{D}{2} + t_p\right) = 5\left(\frac{t_p}{11} + t_p\right)$$

or
$$T = 5.455\, t_p \qquad ...(11.14)$$

Alternatively, the base period of the synthetic unit hydrograph can be determined using the fact that the area under the unit hydrograph is equivalent to a direct runoff of 1 cm. Assuming a triangular shape for the unit hydrograph, the base period from this principle is given by

$$T = \frac{5.556}{q_p} \qquad ...(11.15)$$

If a synthetic unit hydrograph for any other duration D' is required, then the modified basin lag is given by

$$t_p' = t_p + \frac{D' - D}{4} \qquad ...(11.16)$$

The modified lag t_p' is then used in Eq. (11.11) to obtain the peak discharge of D' hours unit hydrograph.

$$q_p' = \frac{Q_p'}{A} = \frac{2.778\, C_p}{t_p'}$$

or
$$Q_p' = \frac{2.778\, A\, C_p}{t_p'} \qquad ...(11.17)$$

Similarly the base period of the D' hour unit hydrograph is obtained from Eq. (11.12) or Eq. (11.14) replacing t_p by t_p' or from Eq. (11.15) replacing q_p by q_p' and the reasonable of these three values is adopted.

The sketching of the synthetic unit hydrograph is made easier by knowing the width of the hydrograph at a discharge equal to certain per cent of peak discharge. They are given by

$$W_{50} = 2.14\, (q_p')^{-1.08} \qquad ...(11.18)$$

$$W_{75} = 1.22\, (q_p')^{-1.08} \qquad ...(11.19)$$

where W_{50} and W_{75} are the widths of the hydrograph in hours at 50 and 75 percent of the peak discharge respectively. Usually one-third of this width is distributed before the unit hydrograph peak and two-thirds after the peak.

Snyder cautioned that the basin lag may tend to vary with flood magnitude and the synthetic unit hydrographs are likely to give better results for fan shaped catchments than those with irregular shapes. He recommended that the coefficients C_t and C_p be better determined on a regional basis as they are likely to vary from region to region. It is therefore advisable that the value of these coefficients are determined from the known parameters of the unit hydrograph of a meteorologically homogeneous catchment and then are used in the ungauged catchment under consideration.

Though Snyder observed a range of 1.35 to 1.65 for C_t, some investigators have reported a range of 0.3 to 6.0. In fact the value of C_t should depend on the slope of the basin, and it has been observed that it is inversely proportional to the square root of the slope. Linsley has suggested that the following modified equation may be used to find the basin lag.

$$t_p = C_L \left(\frac{L\, L_c}{\sqrt{S}} \right)^n \qquad ...(11.20)$$

where S is the basin slope expressed as a dimensionless ratio and C_L and n are the regression constants. For the basins in the United States studied by Linsley, n was found to be 0.38 and C_L had a value of 1.715 for the mountainous drainage areas, 1.03 for the foot hill drainage areas and 0.50 for the valley drainage areas. Likewise C_p was also found to vary quite considerably and a range of 0.31 to 0.93 has been reported by some investigators, though in the studies conducted by Snyder it was between 0.56 and 0.69.

Example 11.12. *From the topographical map of a drainage basin the following quantities are measured : $A = 3480$ km², $L = 148$ km and $L_c = 74$ km. The 12 h unit hydrograph derived for the basin has a peak ordinate of 155 m³/s occurring at 40 h. Determine the coefficients C_t and C_p for the synthetic unit hydrograph of the basin.*

Solution. $D' = 12$ h

$$t_p' = 40 - \frac{D'}{2} = 40 - 6 = 34 \text{ h}$$

From Eq. (11.16), $$t_p' = t_p + \frac{D' - D}{4}$$

and from Eq. (11.13), $$D = \frac{2}{11} t_p$$

$$\therefore \quad 34 = t_p + \frac{12}{4} - \frac{1}{4}\left(\frac{2}{11} t_p\right) = 3 + \frac{21}{22} t_p$$

or $t_p = 32.48$ h

From Eq. (11.10), $t_p = C_t (L\, L_c)^{0.3}$

$$C_t = \frac{t_p}{(L\, L_c)^{0.3}} = \frac{32.48}{(148 \times 74)^{0.3}} = 1.994$$

$$q_p' = \frac{Q_p'}{A} = \frac{155}{3480} = 0.04454$$

From Eq. (11.17), $$C_p = \frac{q_p'\, t_p'}{2.778}$$

$$C_p = \frac{0.04454 \times 34}{2.778} = 0.545.$$

Example 11.13. *A basin having a drainage area of 2500 km² with $L = 100$ km and $L_c = 50$ km is a sub-basin of the catchment of example (11.12). Compute a 4 h synthetic unit hydrograph for this sub-basin.*

Solution. The values of $C_t = 1.994$ and $C_p = 0.545$ in the previous example can be safely applied to this sub basin.

$$t_p = C_t (L\, L_c)^{0.3} = 1.994 (100 \times 50)^{0.3} = 25.67 \text{ h}$$

$$D = \frac{2}{11} t_p = \frac{2}{11} \times 25.67 = 4.67 \text{ h}$$

$$D' = 4\ h$$

$$t_p' = t_p + \frac{D' - D}{4} = 25.67 + \frac{4 - 4.67}{4}$$

$$= 25.5\ \text{h}$$

$$q_p' = \frac{2.778\ C_p}{t_p'} = \frac{2.778 \times 0.545}{25.5} = 0.0594\ \text{m}^3/\text{s}/\text{km}^2$$

$$Q_p = A \cdot q_p' = 2500 \times 0.0594 = 148.5\ \text{m}^3/\text{s}$$

From Eq. (11.15), $T' = \dfrac{5.556}{q_p'}$

$$= \frac{5.556}{0.0594} \approx 93.5\ \text{h}$$

From Eq. (11.18), $W_{50} = 2.14\ (q_p')^{-1.08}$

$$= \frac{2.14}{(0.0594)^{1.08}} = 45.2\ \text{h}$$

From Eq. (11.19), $W_{75} = 1.22\ (q_p')^{-1.08}$

$$= \frac{1.22}{(0.0594)^{1.08}} = 25.7\ \text{h}$$

The synthetic unit hydrograph is drawn in Fig. 11.15.

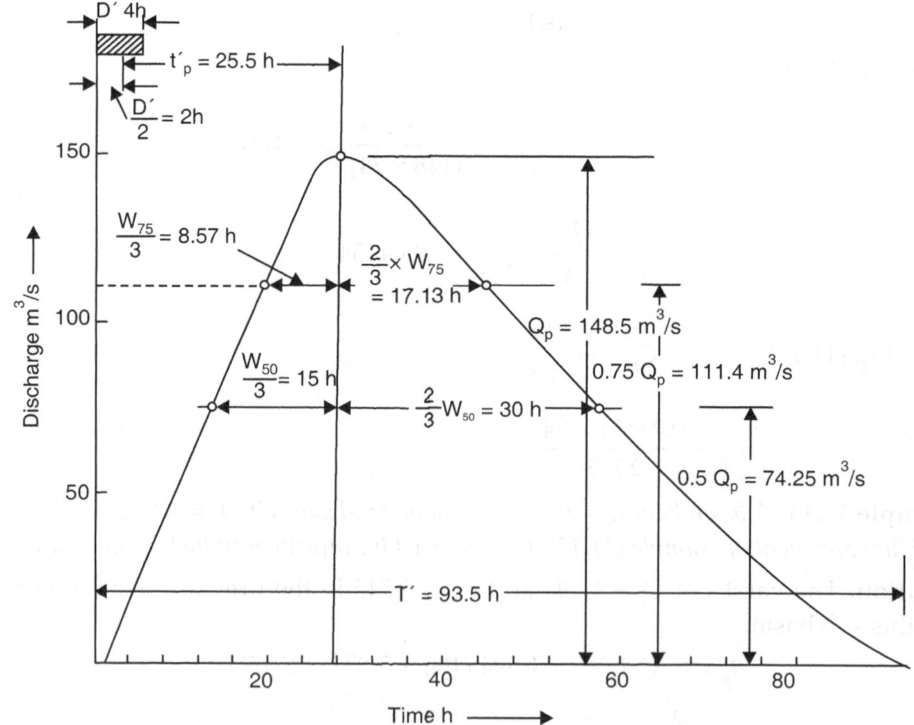

FIGURE 11.15 Synthetic unit hydrograph of example 11.13.

Example 11.14. *Derive a 3 h synthetic unit hydrograph of a basin with the following data : Basin area = 3000 km², Length of the main stream = 120 km ; Distance from centroid of the basin to the outlet =63 km. The Snyder's coefficients C_t and C_p may be assumed to be 1.60 and 0.64 respectively.*

Solution.
$$t_p = C_t\,(L\,L_c)^{0.3}$$
$$= 1.6\,(120 \times 6.3)^{0.3} = 23.32 \text{ h}$$

$$D = t_p \times \frac{2}{11} = 4.24 \text{ h}$$

$$D' = 3 \text{ h}$$

$$t'_p = t_p + \frac{D' - D}{4} = 23.32 - 0.31 = 23.01 \text{ h}$$

$$q_p' = \frac{2.778\,C_p}{t_p'} = \frac{2.778 \times 0.64}{23.01} = 0.0773 \text{ m}^3/\text{s/km}^2$$

$$Q_p' = A \,.\, q_p' = 231.9 \text{ m}^3/\text{s}$$

$$T' = \frac{5.556}{q_p'} = \frac{5.556}{0.0773} = 71.87 \text{ h} \approx 72 \text{ h}$$

$$W_{50} = \frac{2.14}{0.0773^{1.08}} = 33.98 \approx 34 \text{ h}$$

$$W_{75} = \frac{1.22}{0.0773^{1.08}} = 19.37 \text{ h}$$

with these values of Q_p', t'_p, T', W_{50} and W_{75} a smooth graph to give 1 cm of runoff can be sketched.

11.13 DIMENSIONLESS UNIT HYDROGRAPH

The dimensionless unit hydrograph developed by United States Soil Conservation Service (SCS) provides, a shape to the unit hydrograph and therefore leads to more reproducible results than the Snyder method. The plotting positions of the SCS dimensionless unit hydrograph are expressed as the ratios and are given in Table (11.10). Q_p is the peak discharge of the unit hydrograph and t_{pk} is the time to peak. The base period is taken to be five times the time to peak. In terms of Snyder's notation, if D' is the duration of the required unit hydrograph, then

$$t_{pk} = t_p' + \frac{D'}{2} \qquad \text{...(11.21)}$$

once Q_p and t_{pk} have been determined the SCS dimensionless unit hydrograph, shown in Fig. 11.16, is used to calculate the unit hydrograph ordinates. The shape of the dimensionless unit hydrograph is more in agreement with unit hydrographs that are likely to occur in nature.

Table 11.10 Coordinates of SCS Dimensionless Unit Hydrograph

$\frac{t}{t_{pk}}$	$\frac{Q}{Q_p}$	$\frac{t}{t_{pk}}$	$\frac{Q}{Q_p}$	$\frac{t}{t_{pk}}$	$\frac{Q}{Q_p}$
0	0	1.1	0.990	2.4	0.147
0.1	0.030	1.2	0.930	2.6	0.107
0.2	0.100	1.3	0.860	2.8	0.077
0.3	0.190	1.4	0.780	3.0	0.055
0.4	0.310	1.5	0.680	3.2	0.040
0.5	0.470	1.6	0.560	3.4	0.029
0.6	0.660	1.7	0.460	3.6	0.021
0.7	0.820	1.8	0.390	3.8	0.015
0.8	0.930	1.9	0.330	4.0	0.011
0.9	0.990	2.0	0.250	4.5	0.005
1.0	1.000	2.2	0.207	5.0	0.000

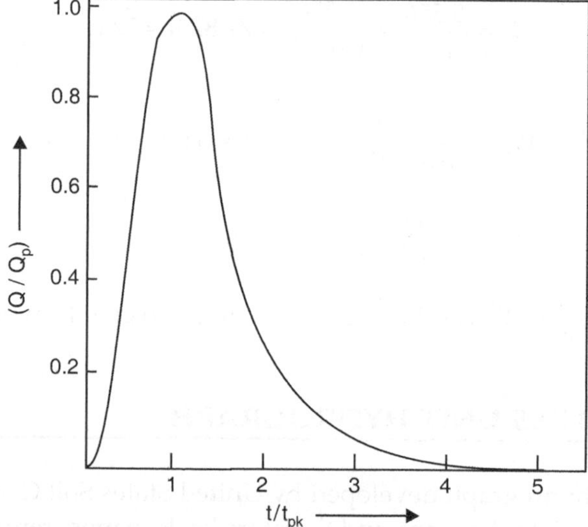

FIGURE 11.16 SCS dimensionless unit hydrograph.

11.14 CWC METHOD FOR INDIAN CATCHMENTS

Based on study of large number of catchments in India with varying shapes and areas (ranging from 25 to 500 km^2) the Central Water Commission of India developed the following method for constructing synthetic unit hydrographs.

The weighted mean slope S_m is computed from the equation

$$S_m = \left[\frac{L_c}{\frac{L_1}{\sqrt{S_1}} + \frac{L_2}{\sqrt{S_2}} + \dots\dots\dots + \frac{L_n}{\sqrt{S_n}}} \right]^2 \qquad \text{...(11.22)}$$

where L_c is the length of main stream in km from the basin outlet to a point on the main stream near the centre of gravity of the area, and S_1, S_2, ..., S_n are the slopes (expressed as ratio) of the main stream in reaches of lengths L_1, L_2, ..., L_n in km. Then the peak discharge of the D hour unit hydrograph is given by

$$Q_{pd} = 4.44\, A^{3/4} \text{ for } S_m > 0.0028 \qquad \text{...(11.23)}$$

and

$$Q_{pd} = 222\, A^{3/4}\, S_m^{\;2/3} \text{ for } S_m < 0.0028 \qquad \text{...(11.24)}$$

where A is the area in km^2. The lag time t_p in hours for the 1 h unit hydrograph (the time interval from the centroid of the effective rainfall to the peak) is given by

$$t_p = \frac{3.95}{\left(\frac{Q_{pd}}{A}\right)^{0.9}} \qquad \text{...(11.25)}$$

The duration of effective rainfall for the purpose of design flood peaks is taken as

$$D = 1.1\, t_p \qquad \text{...(11.26)}$$

Equations (11.22) through (11.26) provide the basis to determine the design unit hydrograph.

REVIEW EXERCISES

Descriptive Questions

1. Sketch a typical hydrograph resulting from an isolated storm and identify the features of the same.
2. Describe how the recession constants of direct runoff and baseflow curves are obtained from a semilogarithmic plot.
3. Why is baseflow separated from total runoff ?
4. What is a master depletion curve ? How is it used to separate the baseflow from total runoff ?
5. Describe any two methods of separating the baseflow from the total runoff.
6. What does the word *unit* refer to in the unit hydrograph ? Explain with sketches what do you understand by the principle of linearity and principle of time invariance in the unit hydrograph theory ?
7. Describe the step by step procedure of the derivation of a unit hydrograph from an isolated storm.

8. Define unit hydrograph. What are the assumptions underlying the unit hydrograph theory ? How do they limit the applicability of unit hydrograph ?
9. How is an average unit hydrograph determined from various unit hydrographs derived from different storms of identical duration ?
10. Write down the convolution equation and explain its physical interpretation with a neat sketch.
11. Describe two methods of deriving unit hydrograph from complex storms.
12. Explain how a 2*D* h unit hydrograph is derived from a *D* h unit hydrograph ?
13. What is a *S*-curve hydrograph ? How is it constructed ? What are its uses ?
14. Distinguish between unit hydrograph and distribution graph. What are the uses of distribution graph ?
15. What are the uses of unit hydrograph ?
16. Describe how unit hydrograph can be used to predict the runoff from storms.
17. What do you understand by synthetic unit hydrograph ? Explain how it is derived.
18. What is a dimensionless unit hydrograph ? How is it useful in constructing a synthetic unit hydrograph of the basin ?
19. Explain the CWC method of deriving the synthetic unit hydrograph for Indian catchments.
20. Define the lag time for a basin with a neat sketch.

Analytical Questions

1. Prove equation (11.5)
2. The ordinates of the recession limb of a hydrograph are listed below. Assuming that the interflow is negligible determine the direct runoff and baseflow recession constants.

Time from peak hours	*Discharge m^3/s*	*Time from peak hours*	*Discharge m^3/s*
0	507	72	124
12	442	84	103
24	308	96	86
36	237	108	72
48	186	120	60
60	151	132	50

3. If $Q_t = Q_0\, e^{-a(t-t_0)}$ describes the baseflow recession in a stream prove that the storage at any time S_t supplying the baseflow is given by $S_t = Q_t/a$.
4. In a stream the baseflow is observed to be 38.2 m^3/s on March 8 and 20.6 m^3/s on March 20. If there is no rain in March what would be the baseflow on March 31 ? What is the volume of water in groundwater storage on March 1 and March 31 ?
5. Given below are the ordinates of observed runoff produced by a storm whose duration is believed to be 3 hours on a drainage basin of area 665.3 sq. km. Assuming a constant baseflow of 60 cubic metre per second derive the ordinates of 3 hour U.H.

	Time	Flow in cubic metre per second
1st day	3 A.M.	60
	6 A.M.	60
	9 A.M.	600
	12 Noon	950
	3 P.M.	800
	6 P.M.	700
	9 P.M.	610
	12 mid-night	530
2nd day	3 A.M.	460
	6 A.M.	400
	9 A.M.	350
	12 Noon	310
	3 P.M.	270
	6 P.M.	240
	9 P.M.	210
	12 mid-night	190
3rd day	3 A.M.	170
	6 A.M.	150
	9 A.M.	130
	12 Noon	110
	3 P.M.	90
	6 P.M.	80
	9 P.M.	70
	12 mid-night	60

6. The observed hydrograph produced by a storm lasting for 3 h with a uniform intensity on a drainage basin of 835 km^2 area has the following ordinates.

Time hours	Flow m^3/s	Time hours	Flow m^3/s	Time hours	Flow m^3/s
0	12.7	24	99.0	48	30.3
3	156.0	27	85.0	51	26.9
6	255.0	30	73.6	54	23.8
9	212.4	33	62.6	57	21.2
12	184.0	36	53.5	60	18.7
15	158.6	39	46.0	63	16.9
18	134.0	42	39.6	66	16.0
21	116.0	45	34.5		

Separate the baseflow and compute the volume of direct runoff assuming that baseflow varies linearly from 12.7 m^3/s at the beginning to 16.0 m^3/s at the end. Compute the ordinates of the 3 h unit hydrograph of the basin.

7. Derive the 9 h unit hydrograph for the drainage basin of analytical question 5, using the 3 h U.H. derived there.

8. Derive the 6 h unit hydrograph for the drainage basin of analytical question 6, using the 3 h U.H. derived there.

9. The direct runoff hydrograph resulting from a 5.0 cm of effective rainfall of 6 h duration is given below. Determine the area of the catchment and the ordinates of the 6 h unit hydrograph.

Time hours	0	6	12	18	24	30	36	42	48	54	60	66	72
Direct runoff m^3/s	0	25	175	320	360	310	230	165	105	60	30	10	0

10. The peak ordinate of a flood hydrograph produced by a 4 h storm yielding 6.7 cm of rainfall is observed to be 832 m^3/s. If the baseflow and ϕ-index are 15 m^3/s and 0.5 cm/h. What is the peak ordinate of the 4 h unit hydrograph ?

11. A 6 h unit hydrograph of a basin has a peak ordinate of 96 m^3/s. When the baseflow in the stream is 25 m^3/s, and when the basin has reached its minimum infiltration capacity of 2.5 mm/h, a 6 h storm with 18.3 cm of total rainfall had occurred on the basin. What is the magnitude of the peak discharge in the flood hydrograph produced by this storm ?

12. Given below are the ordinates of 3 unit hydrograph in m^3/s derived from three separate storms on a small basin. All are believed to have resulted from 4 hour rains. Find the average unit hydrograph for the basin. Drainage area = 90 sq. km.

Time in hours	*Ordinates of 4 hour unit hydrograph from*		
	Storm 1	*Storm 2*	*Storm 3*
0	0	0	0
1	11	3	2
2	37	13	6
3	50	36	17
4	39	47	34
5	31	41	44
6	24	31	40
7	17	22	29
8	13	17	21
9	10	13	17
10	6	9	12
11	4	6	9
12	2	3.5	6
13	1	2	3.5
14	0.5	1.0	1.5
15	0	0	0

13. The ordinates of a runoff hydrograph resulting from a complex storm (after base flow separation) in m^3/s are given below. The storm is estimated to consist of three successive 6 h periods having runoff estimated as 15, 30 and 22.5 mm respectively. The area of the basin is 146 km^2. Using the method of Eq. (11.16) determine the 6 h unit hydrograph for the basin.

Time hours	*Direct runoff m^3/s*
0	0
3	34.5
6	114.0
9	182.6
12	183.0
15	147.1
18	100.9
21	64.6
24	42.3
27	23.7
30	9.3
33	2.2
36	0

14. The 4 h unit hydrograph of a drainage basin with an area of 764 km^2 can be approximated as a triangle with a base period of 80 h and time to peak of 16 h. Determine the ordinates of a 8 h unit hydrograph of this basin at 4 h intervals.

15. The 3 h unit hydrograph of a basin can be approximated as a triangle with a base period of 75 h and a peak discharge of 55.5 m^3/s. What is the area of the basin ? At what time the peak discharge occurs ?

16. Given below are the ordinates of a 4 h unit hydrograph of a basin in m^3/s at one hour intervals.

4, 25, 44, 60, 70, 61, 52, 45, 38, 32
27, 22, 18, 14, 11, 8, 6, 4, 2, 1

What is the area of the basin ?

17. Construct the S-curve hydrograph using the 4 h U.H. of analytical question 16. Hence derive the 2 hour and 6 hour unit hydrographs.

18. When the baseflow in the river is 150 cubic metre per sec, a 9 h storm occurred over the drainage basin of analytical question 5 producing rainfall intensities of 1.5 cm per hour in the first 3 h, 3 cm per hour in the second 3 h and 2.25 cm per hour in the last 3 hours. Assuming the average infiltration index to be 0.5 cm per hour, determine the magnitude and time of occurrence of peak runoff resulting from this storm.

19. The 3 h unit hydrograph of a basin with an area of 20 km^2 at one hour intervals are as given below.

0, 0.41, 1.38, 4, 7.72, 10.06, 9.24, 6.62,
4.57, 3.86, 2.76, 2.07, 1.38, 0.83, 0.41, 0.

If rainfall excess with intensity of 2.0 cm/h for a period of 4 h followed immediately by another 3 h storm with an intensity of 1 cm/h occurs on the basin, what is the peak flow produced by this rainfall and at what time after the commencement of rainfall would this peak flow occur ? Assume baseflow is negligible.

20. In analytical question 19, if 3 h unit hydrograph only is used to compute the direct runoff, what would be the percentage error in the peak flow ?

21. The percentage distributions (in successive time periods of 4 h) of a 4 distribution graph are 5, 20, 35, 20, 15 and 5. The basin has an area of 425 km^2 and experiences a 12 h storm with rainfalls intensities of 2.25, 1.75 and 2.5 cm/h respectively in successive periods of 4 h each. Assuming an average ϕ-index of 0.75 cm/h and neglecting the baseflow determine the ordinates of the runoff hydrograph.

22. The 2 h unit hydrograph derived for watershed *A* from the observed rainfall and runoff of an isolated storm on it has the following features : time to peak = 7 h, peak discharge = 108 m^3/s. The information on this watershed is as follows :

$$\text{area} = 250 \text{ km}^2 ; L = 38 \text{ km}, L_c = 16 \text{ km}.$$

Watershed *B*, which is hydrologically similar to watershed *A*, has the following characteristics ; area =180 km^2, L = 25 km, L_c = 15 km. Determine the one hour synthetic unit hydrograph for watershed *B* using the Snyder's method.

23. The peak discharge and time to peak in a 3 h unit hydrograph derived for a basin of area 250 km^2 with L = 30 km and L_c = 14 km are 5 m^3/s and 9 h respectively. Assuming that Snyder's synthetic unit hydrograph applies determine the coefficient C_t and C_p.
Determine the 2 h unit hydrograph for the upper 180 km^2 of the same watershed which has L = 20 km and L_c = 11.8 km.

24. Calculate the Snyder's synthetic unit hydrograph parameters for the following data: Catchment area = 950 km^2; L = 48 km; L_c = 21 km; C_t = 1.65; C_p = 0.57. Adopt standard duration.

25. A drainage basin has the following characteristics.

Basin area = 2500 sq. km

Length of the main stream L = 110 km

Distance from the centroid of the basin to outlet = 70 km

Construct the 4 hour synthetic unit hydrograph for the basin if C_t = 1.50 and C_p = 0.6.

26. Develop and plot the full synthetic unit hydrograph of exercise 11.45 by using the peak discharge and time to peak obtained there and the SCS dimensionless unit hydrograph.

Objective Type Questions

1. If Q_0 and Q_t are the discharges at times t_0 and t on the recession limb of the hydrograph and if a is a coefficient with positive value, then

(*a*) $\dfrac{Q_t}{Q_0} = e^{a(t-t_0)}$ (*b*) $\dfrac{Q_t}{Q_0} = e^{-a(t-t_0)}$.

(*c*) $\dfrac{Q_t}{Q_0} = e^{a(t-t_0)^2}$ (*d*) $\dfrac{Q_t}{Q_0} = e^{-a(t-t_0)^2}$.

2. The inflection point on the recession side of the hydrograph indicates the end of
(*a*) the baseflow (*b*) the direct rainfall
(*c*) the overland flow (*d*) rainfall.

3. The concept of unit hydrograph was first introduced by
(*a*) Dalton (*b*) Sherman
(*c*) Horton (*d*) Thiessen.

4. The unit hydrograph is the graphical relation between the time distributions of
(*a*) total rainfall and total runoff (*b*) total rainfall and direct runoff
(*c*) effective rainfall and total runoff (*d*) effective rainfall and direct runoff.

5. The word *unit* in the unit hydrograph refers to the
(*a*) unit depth of runoff (*b*) unit duration of the storm
(*c*) unit base period of the hydrograph (*d*) unit area of the basin.

6. If K_{rs}, K_{ri} and K_{rb} denote the recession constants of surface runoff, interflow and baseflow respectively, which of the following is true
(*a*) $K_{rs} > K_{ri} > K_{rb}$ (*b*) $K_{rs} < K_{ri} < K_{rb}$
(*c*) $K_{rs} > K_{ri} < K_{rb}$ (*d*) $K_{rs} < K_{ri} > K_{rb}$.

7. The peak ordinate of a 4 unit hydrograph of a basin is 270 m^3/s. Then, the peak ordinate of 8 h unit hydrograph of the same basin will be
(*a*) 270 m^3/s (*b*) less than 270 m^3/s
(*c*) more than 270 m^3/s (*d*) $(270 \times e^{-4/8})$ m^3/s.

8. The peak discharges in 4 and 8 hours unit hydrographs of a basin occur at t_1 and t_2. Then
(*a*) $t_1 = t_2$ (*b*) $t_1 > t_2$
(*c*) $t_1 < t_2$ (*d*) difficult to guess.

9. The base period of a 6 h unit hydrograph of a basin is 84 h. Then a 12 h hydrograph derived from this 6 h unit hydrograph will have a base period of
(*a*) 90 h (*b*) 78 h
(*c*) 42 h (*d*) 168 h.

10. The direct runoff hydrograph was found to have a peak ordinate at 20 h after the commencement of the effective rainfall and a base period of 100 h. The duration of the storm was
(*a*) less than 20 h (*b*) between 20 and 100 h
(*c*) more than 100 h (*d*) 100 h.

11. The basic principles of unit hydrograph theory are
(*a*) linearity and time invariance (*b*) nonlinearity and time invariance
(*c*) nonlinear time variance and linearity (*d*) nonlinear time variance and nonlinearity.

12. A storm occurring over a basin for a period of 6 h with uniform intensity produced an effective rainfall of 15 cm and the peak flow of 930 m^3/s. Over the same basin, if another storm of same duration but with an effective rainfall of 7.5 cm occurs, what is the corresponding peak flow ?

(*a*) 930 m^3/s (*b*) 1860 m^3/s
(*c*) 2790 m^3/s (*d*) 465 m^3/s.

13. The 4 unit hydrograph of a basin can be approximated as a triangle with a base period of 48 h and a peak ordinate of 200 m^3/s. Then the area of the basin is

(*a*) 1728 km^2 (*b*) 3456 km^2
(*c*) 864 km^2 (*d*) 5184 km^2.

14. The upper limit on the area of the basin for the applicability of unit hydrograph is generally taken to be

(*a*) 100 km^2 (*b*) 2500 km^2
(*c*) 5000 km^2 (*d*) 10000 km^2.

15. The statement that "the ordinates of the direct runoff hydrographs of a common base period are directly proportional to the volumes of runoff represented by the respective hydrographs" is known as

(*a*) principle of linearity (*b*) principle of time invariance
(*c*) principle of uniformity (*d*) none of the above.

16. The direct runoff hydrograph of a basin can be approximated as a triangle with a base period of 80 h and a peak flow of 200 m^3/s occurring at 16th hour. If the area of the basin is 1440 km^2, what is the depth of runoff indicated by the hydrograph ?

(*a*) 1 cm (*b*) 10 cm
(*c*) 2 cm (*d*) 20 cm.

17. The 12 h unit hydrograph of a basin of area 4320 km^2 can be approximated as a triangle with base period of 5 days. The peak flow in the direct runoff hydrograph of this basin produced by an effective rainfall of 5 cm in 12 h is equal to

(*a*) 500 m^3/s (*b*) 1000 m^3/s
(*c*) 1500 m^3/s (*d*) 2000 m^3/s.

18. An *S*-curve hydrograph is derived for a basin of 540 km^2 from a 6 h unit hydrograph. The equilibrium discharge in the *S*-curve is

(*a*) 277.8 m^3/s (*b*) 250 m^3/s
(*c*) 540 × 10^6 m^3/s (*d*) 3240 m^3/s.

19. The *S*-curve hydrograph is the summation of the

(*a*) unit hydrograph (*b*) total runoff hydrograph
(*c*) effective rainfall hyetograph (*d*) baseflow recession curve.

20. The S-curve hydrograph is used

(*a*) to estimate the peak flood flow of a basin resulting from a given storm
(*b*) to develop synthetic unit hydrograph

(*c*) to convert the unit hydrograph of any given duration into a unit hydrograph of any other desired duration

(*d*) to derive the unit hydrograph from complex storms.

21. The concept of distribution graph was first suggested by

(*a*) Barnes (*b*) Bernard

(*c*) Beard (*d*) Collins.

22. The lag time of the basin is the time interval between

(*a*) the centroid of the rainfall diagram and the peak of the hydrograph

(*b*) the beginning and end of direct runoff

(*c*) the beginning and end of effective rainfall

(*d*) the inflection points on the rising and recession limbs of the hydrograph.

23. The basin lag of the Snyder's synthetic unit hydrograph is given by

(*a*) $t_p = C_t\,(L\,L_c)^{0.3}$ (*b*) $t_p = \dfrac{C_t}{(L\,L_c)^{0.3}}$

(*c*) $t_p = C_t\left(\dfrac{L}{L_c}\right)^{0.3}$ (*d*) $t_p = \dfrac{C_t}{\left(\dfrac{L}{L_c}\right)^{0.3}}$

24. The peak discharge of the Snyder's synthetic unit hydrograph is given by

(*a*) $Q_p = \dfrac{2.778\,A}{C_p\,.\,t_p}$ (*b*) $Q_p = \dfrac{2.778\,A\,C_p}{t_p}$

(*c*) $Q_p = \dfrac{2.778\,C_p\,.\,t_p}{A}$ (*d*) $Q_p = \dfrac{2.778\,t_p}{A\,C_p}$.

ANSWERS

Analytical Questions

2. Direct runoff recession constant = 0.47

Baseflow recession constant = 0.85 (*from graphical solution*)

4. 11.696 m^3/s, 91.948 Mm^3 and 19.637 Mm^3.

5. The ordinates of 3 h U.H. at 3 hour intervals are

54, 89, 74, 64, 55, 47, 40, 34, 29, 25

21, 18, 15, 13, 11, 9, 7, 5, 3, 2, and 1 m^3/s.

6. Depth of direct runoff = 2 cm ; The ordinates of 3 h U.H. at 3 hours intervals are

71.6, 121.0, 99.6, 85.4, 72.6, 60.2, 51.1, 42.6, 35.5, 29.7, 24.1,

19.5, 15.7, 12.4, 9.8, 7.6, 5.8, 4.2, 2.8, 1.5, and 0.5 m^3/s.

7. The ordinates of 9 h U.H. at 3 h intervals are

18, 47.7, 72.3, 75.7, 64.3, 55.3, 47.3, 40.3, 34.3, 29.3, 25,

21.3, 18, 15.3, 13, 11, 9, 7, 5, 3.3, 2, 1, and 0.3 m^3/s.

8. The ordinates of 6 h U.H. at 3 h intervals are
35.8, 96.3, 110.3, 92.5, 79.0, 66.4, 55.7, 46.9, 39.1, 32.6, 26.9,
21.8, 17.6, 14.1, 11.1, 8.7, 6.7, 5, 3.5, 2.3, 1, and 0.3 m^3/s.
9. Area of the catchment = 773.28 km^2
The ordinates of 6 h U.H. at 6 h intervals are
5, 35, 64, 72, 62, 46, 33, 21, 12, 6 and 2 m^3/s.
10. 173.89 m^3/s
11. 1637.8 m^3/s
12. Peak of the average U.H. = 47 m^3/s
Time of peak in the average U.H. = 4 h
The ordinates at one hour interval to give 1 cm of runoff are
12.5, 32.5, 43.8, 47, 43, 22, 15, 11.3, 8.8, 6.2, 4, 2, 1 and 0.5 m^3/s.
13. The ordinates of 3 h U.H. at 3 h intervals are
23, 30, 28, 22, 13, 9, 6, 3 and 1 m^3/s.
14. The ordinates of 8 h U.H. at 4 h intervals are
7.5, 22.5, 37.5, 52.5, 58.1, 54.4, 50.6, 46.9, 43.1, 39.4, 35.6,
31.9, 28.1, 24.4, 20.6, 16.9, 13.1, 9.4, 5.6, and 1.9 m^3/s.
15. Area of the basin = 749.25 km^3

$$\text{Time to peak} = \frac{\text{Base period}}{5} = 15 \text{ h.}$$

16. 195.84 km^2
17. The ordinates of S-curve hydrograph at 1 intervals are :
0, 4, 25, 44, 60, 74, 86, 96, 105, 112, 118, 123, 127, 130, 132, 134, 135, 136, 136, ..., 136 m^3/s
The ordinate of 2 h U.H. at 1 h interval are :
8, 50, 80, 70, 60, 52, 44, 38, 32, 26, 22, 18, 14, 10, 8, 6, 4, and 2 m^3/s
The ordinates of 6 h U.H. at 1 h interval are :
2.7, 16.7, 29.3, 40, 49.3, 57.3, 61.3, 53.3, 45..3, 38.7, 32.7, 27.3, 18, 14.7, 11.3, 8.7, 6, 4, 2.7, 1.3 and 0.7 m^3/s.
18. 1364.25 m^3/s at the end of 12th hour.
19. 76.48 m^3/s at the end of 7th hour.
20. Now the peak flow would be 85.92 m^3/s and occurs at the end of 6th hour. The peakflow is overestimated by 12.34%.
21. 88.54, 413.19, 959.20, 1180.56, 1224.83, 678.82, 368.92 and 103.30 m^3/s.
22. $C_t = 0.842$ and $C_p = 0.933$
Peak flow = 93.3 m^3/s ; time of peak = 5.5 h ; lag = 5 h
Base width = 10.7 h, $W_{50} = 4.35$ h ; $W_{75} = 2.48$ h.

23. $C_t = 1.155$ and $C_p = 0.54$

Peak flow = 43.74 m^3/s ; time of peak = 7.18 h ; lag = 6.18 h

Base width = 22.86 h, W_{50} = 9.86 h ; W_{75} = 5.62 h.

24. $Q_p = 114.5\ m^3/s$; $t_p = 13.14$ h ; $D = 2.39$ h, $t_{pk} = 14.34$ h, $T = 45.9$ h , $W_{50} = 20.9$ h, $W_{75} = 11.94$ h.

25. Peak discharge = 190 m^3/s ; Base width = 73.27 h

lag = 22 h $\quad$ W_{50} = 34.69 h

Time to peak = 24 h $\quad$ W_{75} = 19.78 h.

26.

Time h	0	12	24	36	48	72	96	120
Discharge m^3/s	0	89.3	190	129.2	53.2	10.45	2.09	0

Objective Type Questions

1. (*b*)	**2.** (*c*)	**3.** (*b*)	**4.** (*d*)	**5.** (*b*)
6. (*a*)	**7.** (*b*)	**8.** (*c*)	**9.** (*a*)	**10.** (*a*)
11. (*a*)	**12.** (*d*)	**13.** (*a*)	**14.** (*c*)	**15.** (*a*)
16. (*c*)	**17.** (*b*)	**18.** (*b*)	**19.** (*a*)	**20.** (*c*)
21. (*b*)	**22.** (*a*)	**23.** (*a*)	**24.** (*b*).	

Chapter 12

INSTANTANEOUS UNIT HYDROGRAPH

12.1 INTRODUCTION

When a D h unit hydrograph is available other unit hydrographs whose duration is an integral multiple such as 2 D, 3 D, etc. can be easily derived by the application of principle of superposition. On the other hand, if the duration of the required unit hydrograph is not an integral multiple of D, or if it is less than D, it becomes necessary to derive S - curve hydrograph and then derive the unit hydrograph of required duration. To obtain the runoff hydrograph resulting from a storm of varying duration and varying intensities, it is preferable to have a unit graph with very short duration. Theoretically the shortest duration is zero. Accordingly if the duration of effective rainfall approaches zero, the unit hydrograph may be called as an *instantaneous units hydrograph,* generally abbreviated as IUH. Similarly the D h unit hydrograph may now be abbreviated as DUH. In the case of IUH the unit rainfall of 1 cm is visualised as though it is applied instantaneously over the basin. Of course, this is only a fictitious situation and a concept to be used in the hydrograph analysis. The advantage of IUH over DUH is that it eliminates the problem of unit duration and the restriction of uniform distribution of rainfall in time. Furthermore the use of IUH is better suited for the needs of theoretical investigations on the rainfall - runoff relationships. Clearly the IUH is a unique demonstration of a particular catchment response to rain, independent of duration and is thus a graphical expression of the integration of the catchment parameters like length, shape, slope etc., which control such a response. When the drainage basin is considered as a system with rainfall as the input and runoff as the output the IUH may be called as the *impulse response* of the system. As IUH is only an extension of unit hydrograph concept it is also based on the principle of linearity and principle of time invariance. Before dealing with various methods of deriving IUH of a drainage basin, the convolution process (that is, how an IUH can be applied to obtain the runoff hydrograph from a given storm) is briefly described.

12.2 CONVOLUTION INTEGRAL

Referring to Fig. 12.1, let $Q(t)$ be the direct runoff, $u(t)$ be the ordinate of IUH at any time t and let $I(\tau)$ be the ordinate of effective rainfall hyetograph at time τ. D_0 is the total duration of effective rainfall.

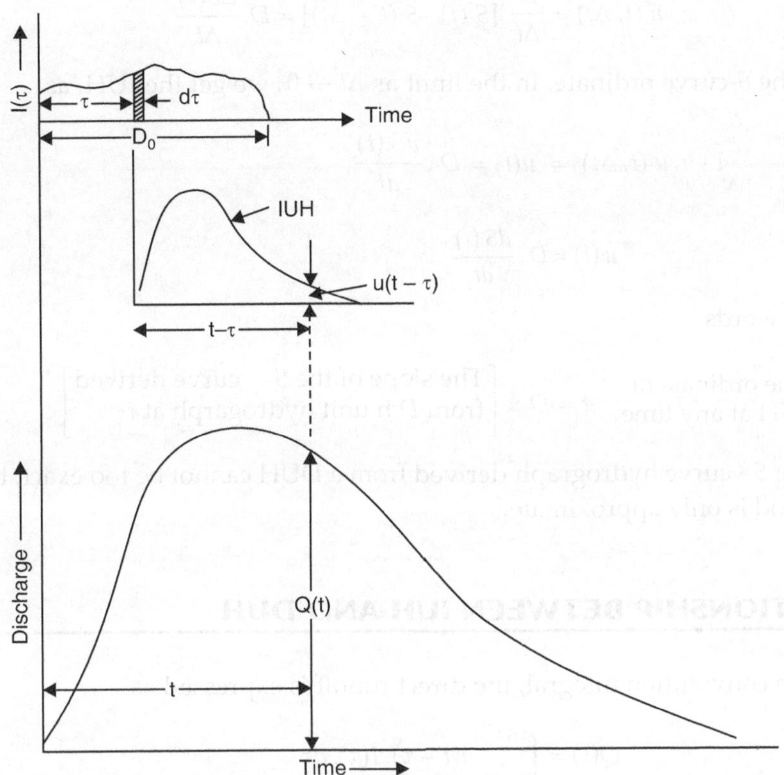

FIGURE 12.1 Convolution process.

The rainfall input to the basin can be considered as a succession of infinitesimal instantaneous inputs of depth $I(\tau) \,.\, d\tau$, the contribution to direct runoff of which can be obtained by placing the IUH at time τ and multiplying it by $I(\tau) . d\tau$ and this contribution would be $u(t-\tau).\ I(\tau).\ d\tau$. Since all such infinitesimal rainfalls upto 't' where $t \le D_0$ would contribute to the runoff at time t, the total direct runoff may be written as

$$Q(t) = \int_0^{t \le D_0} u(t-\tau).\ I(\tau).\ d\tau \qquad ...(12.1)$$

The above integral is called the *convolution integral* or the *Duhamel integral* and $u(t)$ is called the *Kernel* of the system. If IUH of the basin is available, the application of Eq. (12.1) yields the direct runoff due to any storm.

12.3 DERIVATION OF IUH FROM S-CURVE

Suppose a *S*-curve hydrograph is derived from an available unit hydrograph of duration *D* h. Let $u\ (t, \Delta t)$ denote the ordinates of Δt hour unit hydrograph at any time '*t*' . Then from *S*-curve technique of section (11.9), we have

$$u\ (t, \Delta t) = \frac{D}{\Delta t}\ [S\ (t) - S\ (t - \Delta t)] = D.\frac{\Delta S(t)}{\Delta t}$$

where $S\ (t)$ is the *S*-curve ordinate. In the limit as $\Delta t \rightarrow 0$, we get the IUH, as

$$\underset{\Delta t \rightarrow 0}{\text{Lt}}\ u\ (t, \Delta t) = u(t) = D\,.\frac{dS\ (t)}{dt}$$

$$\therefore \qquad u\,(t) = D\,.\frac{dS\,(t)}{dt} \qquad \text{...(12.2)}$$

In other words,

$$\text{The ordinate of IUH at any time } t = D \times \left[\begin{array}{l}\text{The slope of the } S \text{ - curve derived} \\ \text{from } D \text{ h unit hydrogarph at } t\end{array}\right]$$

Since the *S* - curve hydrograph derived from a DUH cannot be too exact, the IUH obtained from this method is only approximate.

12.4 RELATIONSHIP BETWEEN IUH AND DUH

As given by the convolution integral, the direct runoff is expressed as

$$Q(t) = \int_0^{t \le D_0} u(t - \tau).\ I(\tau).\ d\tau$$

If we consider the input rainfall of uniform intensity of $(1/D)$ cm/h for an infinite duration, the corresponding output runoff is nothing but the *S*-curve hydrograph. So replacing $Q\ (t)$, by $S\ (t)$ and $I(\tau)$ by $(1/D)$ in the above expression, we get

$$S\ (t) = \frac{1}{D}\int_0^t u\ (t - \tau).d\tau \qquad \text{...(12.3)}$$

Let the transformation that $z = (t - \tau)$ be introduced in the above integral. Then we have, $d\tau = -\,dz$, and the limits are, when $\tau = 0$, $z = t$ and when $\tau = t$, $z = 0$. Eq. (12.3) now becomes

$$S\ (t) = \frac{1}{D}\int_t^0 u\ (z)(-\,dz) = \frac{1}{D}\int_0^t u\ (z).dz$$

$$S\ (t) = \frac{1}{D}\int_0^t u\ (z).dz \qquad \text{...(12.4)}$$

So the ordinate of S-curve at t, resulting from an infinite rainfall of constant intensity $(1/D)$ cm/h can be obtained as $(1/D) \times$ (Integration of IUH from zero to t). Conversely, the ordinate of IUH at t, may be obtained as $D \times$(the differential of the S-curve at t). That is $u(t) = D\dfrac{dS(t)}{dt}$, which result is already proved in Eq. (12.2).

Let $u(t, D)$ be the ordinate of DUH at time t and let $S(t)$ be the S-curve obtained from this DUH. From Eq. (11.9)

$$u(t, D) = S(t) - S(t - D)$$

$$= \frac{1}{D}\int_0^t u(z)dz - \frac{1}{D}\int_0^{t-D} u(z).dz$$

$$u(t, D) = \frac{1}{D}\int_{t-D}^{t} u(z).dz \qquad \text{...(12.5)}$$

That means,

The ordinate of DUH at $t = (1/D) \times$ [the area of IUH in the limits between $(t - D)$, and t]

If the IUH is assumed to be linear between $(t - D)$ and t, Eq. (12.5) becomes

$$u(t, D) = u\left(t - \frac{D}{2}\right) \qquad \text{...(12.6)}$$

However, it is to be noted that Eq. (12.6) works well only when D is small and when the peak is not contained within the interval $(t - D)$ to t. Eq. (12.5) can be used to derive DUH of any duration D, provided the IUH is available thus eliminating the construction of S-curve hydrograph.

12.5 DERIVATION OF IUH BY THE OTHER METHODS

Though IUH may easily be derived from a S-curve hydrograph as outlined earlier, it is very approximate. In recent times various investigators have proposed various methods for the derivation of IUH. The approaches by Nash and Clark are only discussed here as they are more practical and the other methods proposed by Dooge, Diskin etc. are not considered here.

There is also considerable work reported in the literature on non-linear unit hydrograph theory which is beyond the scope of treatment here.

12.5.1 Nash's Model

Nash developed an equation for IUH by assuming that the operation performed by the catchment on an instantaneous rainfall is equivalent to a succession of routing through a number of linear reservoirs. The catchment is replaced by a series of 'n' reservoirs, each having the storage characteristic $S = KQ$, where S is the storage in the reservoir at any time 't' and Q is the outflow from the reservoir at the same time, and K is a constant (with units of time) called the *delay time* of the reservoir. The outflow from one reservoir becomes inflow to the next reservoir with the arrangement as shown in Fig. 12.2. For any reservoir the continuity equation may be written as

$$I - Q = \frac{dS}{dt} \qquad \text{...(12.7)}$$

where I is the rate of inflow into the reservoir at time 't'

Q is the rate of outflow from the reservoir at time 't'

and $\frac{dS}{dt}$ is the rate of change of storage in the reservoir.

Since for a linear reservoir $S = KQ$, $\frac{dS}{dt} = K.\frac{dQ}{dt}$.

So we have
$$I - Q = K.\frac{dQ}{dt}$$

$$Q + K.\frac{dQ}{dt} = I$$

$$\left(1 + K\frac{d}{dt}\right) Q = I \qquad \text{...(12.8)}$$

The solution of Eq. (12.8) is given by

$$Q = \frac{1}{K} e^{-t/k} \int e^{t/k} .I.dt \qquad \text{...(12.9)}$$

For the first reservoir, when the input is applied instantaneously, the inflow has the following characteristics.

$$I = 0, \text{ when } t \neq 0$$

$$\int_{-\infty}^{\infty} I.dt = V, \text{ where } V \text{ is the instantaneous input}$$

With these conditions the outflow from the first reservoir for an instantaneous input of V cm, becomes

$$Q_1 = \frac{V}{K} e^{-t/k} \qquad \text{...(12.10)}$$

The outflow from the second reservoir is given by

$$Q_2 = \frac{1}{K} e^{-t/k} \int e^{t/k} .I.dt$$

where I is to be now replaced by Q_1, as Q_1 becomes inflow to the second reservoir. Therefore,

$$Q_2 = \frac{1}{K} e^{-t/k} \int e^{t/k} .\frac{V}{K} e^{-t/k} .dt$$

$$= \frac{V}{K^2} . e^{-t/k} .t \qquad \text{...(12.11)}$$

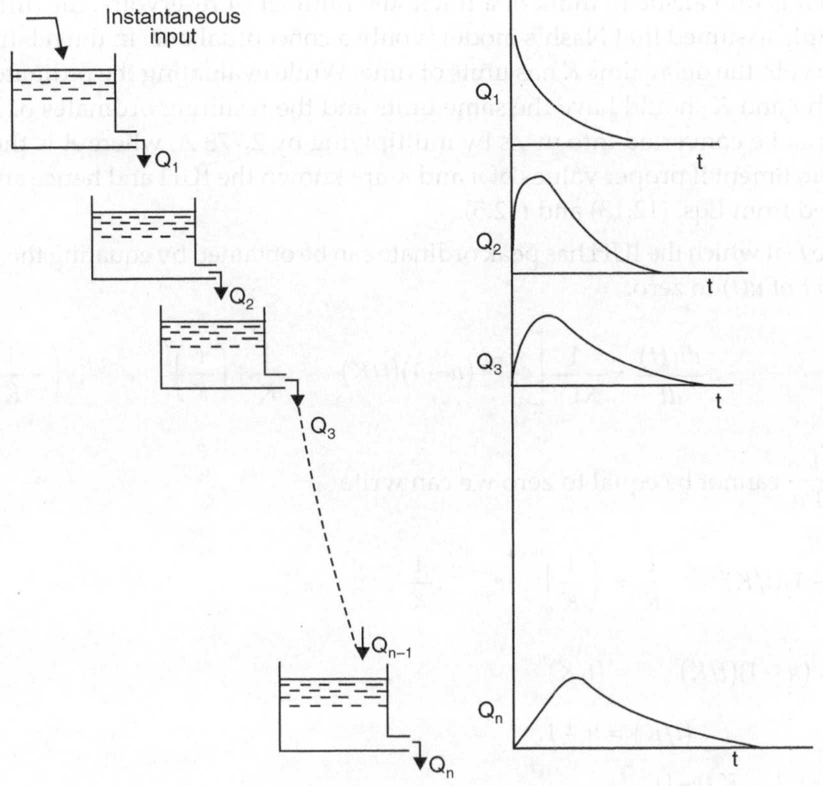

FIGURE 12.2 Arrangement of linear reservoirs in Nash Model.

Successive routing of the outflow through n reservoirs yields the outflow from nth reservoir as

$$Q_n = \frac{V}{K} e^{-t/k} . \left(\frac{t}{K}\right)^{n-1} . \frac{1}{(n-1)!} \qquad ...(12.12)$$

Instead of V cm of rainfall, if one cm of rainfall, is applied as the instantaneous input to the first reservoir, the outflow from the nth reservoir is nothing but IUH. That is,

$$u(t) = \frac{1}{K(n-1)!} e^{-t/K} \left(\frac{t}{K}\right)^{n-1}$$

The value of 'n' the number of reservoirs, as is shown later, is invariably a fraction. To allow for fractional values of n, the factorial in the above expression is replaced by Γn, since

$$\Gamma n = (n-1)!$$

Then the equation for IUH becomes

$$u(t) = \frac{1}{K\Gamma n} e^{-t/K} \left(\frac{t}{K}\right)^{n-1} \qquad ...(12.13)$$

Though it is unrealistic to think of a fractional number of reservoirs, the difficulty will not arise if it is simply assumed that Nash's model is only a conceptual one. In the relationship $S = KQ$ of a linear reservoir, the delay time K has units of time. While evaluating the ordinates of IUH from Eq. (12.13) both t and K should have the same units and the resulting ordinates of IUH will be in cm\hr. These can be converted into m^3/s by multiplying by 2.778 A, where A is the basin area in km^2. For any catchment if proper values of n and K are known the IUH and hence any DUH can be easily computed from Eqs. (12.13) and (12.5).

The time t_* at which the IUH has peak ordinate can be obtained by equating the first derivative with respect to t of $u(t)$ to zero.

$$\frac{du(t)}{dt} = \frac{1}{K\Gamma n}\left[e^{-t/K}(n-1)(t/K)^{n-2} \cdot \frac{1}{K} + \left(\frac{t}{K}\right)^{n-1} e^{-t/K} \cdot \left(-\frac{1}{K}\right)\right] = 0$$

Since $\frac{1}{K\Gamma n}$ cannot be equal to zero we can write

$$e^{-t/K}(n-1)(t/K)^{n-2} \cdot \frac{1}{K} = \left(\frac{t}{K}\right)^{n-1} e^{-t/K} \cdot \frac{1}{K}$$

$$(n-1)(t/K)^{n-2} = (t/K)^{n-1}$$

or

$$(t/K) = n-1.$$

Therefore $t_* = K\,(n-1)$ ...(12.13 *a*)

By substituting $(t/K) = (n-1)$ in Eq. (12.13), we get the magnitude of the peak ordinate.

$$u\,(t_*) = \frac{1}{K\Gamma n} e^{1-n}\,(n-1)^{n-1} \qquad \text{...(12.13 } b\text{)}$$

Replacing n by $(n + 1)$ in Eq. (12.13 *b*), we get the peak ordinate for the case of $(n + 1)$ reservoirs

$$u(t_*)_{n+1} = \frac{1}{K\Gamma n+1} . e^{-n}\, n^n.$$

Now

$$\frac{u(t_*)_{n+1}}{u(t_*)_n} = \frac{\frac{1}{K\Gamma n+1} . e^{-n}\, n^n}{\frac{1}{K\Gamma n} . e^{1-n}\,(n-1)^{n-1}} = \left(\frac{n}{n-1}\right)^{n-1} . \frac{1}{e} \qquad \text{...(12.13}c\text{)}$$

But $\underset{n\to\infty}{\text{Lt}}\left(\frac{n}{n-1}\right)^{n-1} \to e$. Hence for small n it will be less than e.

Therefore $$\frac{u(t_*)_{n+1}}{u(t_*)_n} < 1$$

or $$u(t_*)_{n+1} < u(t_*)_n \qquad ...(12.13\ d)$$

We can infer from Eq. (12.13 *a*) that the effect of each reservoir is to delay the occurrence of peak flow in IUH by *K* time units. Similarly Eq. (12.13 *c*) shows that each reservoir will attenuate the peak flow in IUH by a factor equal to $e^{-1}(n/n-1)^{n-1}$. In other word the peak discharge from any reservoir will be less than the peak flow from the pervious reservoir. This means that for a given *K* the IUH of a catchment with larger *n* will have lesser peak and larger time to peak compared to catchment with smaller *n*.

The gamma function is given by the integral

$$\Gamma n = \int_0^\infty e^{-x}\,.\,x^{n-1}.dx \qquad ...(12.14)$$

The values of Γn are obtained from standard gamma tables. It is usually tabulated for values of *n* ranging only form 1.0 to 2.0 as in Table 12.1. The relation that $\Gamma(n+1) = n\Gamma n$ can be utilised to evaluate Γn for any *n* either less than 1 or more than 2.

Suppose we want to find Γ0.8. Since 0.8 < 1, its gamma value will not be found in the Table 12.1. But we can write

$$\Gamma 1.8 = 0.8 \times \Gamma 0.8$$

or $$\Gamma 0.8 = \frac{\Gamma 1.8}{0.8}$$

Now Γ1.8 can be obtained from Table 12.1 as 0.931384

$$\therefore \qquad \Gamma 0.8 = \frac{0.931384}{0.8} = 1.16423$$

Similarly suppose we want to find Γ3.6. Again since 3.6 > 2, its gamma value will not be found in Table 12.1. But we can write

$$\Gamma 3.6 = 2.6 \times \Gamma 2.6$$
$$= 2.6 \times 1.6 \times \Gamma 1.6$$

Now Γ1.6 can be obtained from Table 12.1 as 0.893515

$$\therefore \qquad \Gamma 3.6 = 2.6 \times 1.6 \times 0.893515 = 3.7170224$$

Sometimes the following result may be found to be useful.

$$\Gamma 0.5 = (\pi)^{1/2} = 1.7725 \qquad ...(12.15)$$

Evaluation of n and K. The method of obtaining the values of *n* and *K* from the recorded rainfall and runoff of a storm event is now explained. The *m*th moment of any area about the origin is defined as

$$M_m = \frac{\int_A f(x)\,.\,dx\,.\,x^m}{\int_A f(x)\,.\,dx} \qquad ...(12.16)$$

From the above definition the *m*th moment of IUH about the origin is given by

$$M_m = \frac{\int_0^\infty u(t)\,.\,dt\,.\,t^m}{\int_0^\infty u(t)\,.\,dt} \qquad \text{...(12.17)}$$

Since the area of IUH is always unity

$$M_m = \frac{1}{K\Gamma n}\int_0^\infty e^{-t/K}.\left(\frac{t}{K}\right)^{n-1} t^m.\,dt$$

$$= \frac{K^m}{\Gamma n}\int_0^\infty e^{-t/K}.\left(\frac{t}{K}\right)^{n-1}\left(\frac{t}{K}\right)^m.\,d\left(\frac{t}{K}\right)$$

$$M_m = \frac{K^m}{\Gamma n}\int_0^\infty e^{-t/K}\left(\frac{t}{K}\right)^{n+m-1} d(t/K) \qquad \text{...(12.17 a)}$$

and from the definition of gamma function given by Eq. (12.14)

$$M_m = \frac{K^m\Gamma\,(m+n)}{\Gamma n}$$

Table 12.1 Gamma Function

n	Γn	n	Γn	n	Γn
1.00	1.000000	1.34	0.892216	1.68	0.905001
1.02	0.988844	1.36	0.890185	1.70	0.908639
1.04	0.978438	1.38	0.888537	1.72	0.912581
1.06	0.968744	1.40	0.887264	1.74	0.916826
1.08	0.959725	1.42	0.886356	1.76	0.921375
1.10	0.951351	1.44	0.885805	1.78	0.926227
1.12	0.943590	1.46	0.885604	1.80	0.931384
1.14	0.936416	1.48	0.885747	1.82	0.936845
1.16	0.929803	1.50	0.886227	1.84	0.942612
1.18	0.923728	1.52	0.887039	1.86	0.948687
1.20	0.918169	1.54	0.888178	1.88	0.955071
1.22	0.913106	1.56	0.889639	1.90	0.961766
1.24	0.908521	1.58	0.891420	1.92	0.968774
1.26	0.904397	1.60	0.893515	1.94	0.976099
1.28	0.900718	1.62	0.895924	1.96	0.983743
1.30	0.897471	1.64	0.898642	1.98	0.991708
1.32	0.894640	1.66	0.901668	2.00	1.000000

As there are two parameters to be evaluated, it is sufficient if we calculate the first two moments. Therefore,

for $m = 1$, the first moment of IUH,

$$M_1 = nK$$

and for $m = 2$, the second moment of IUH

$$M_2 = n(n+1)K^2$$

The effect of routing the input through a series of linear reservoirs is to transform each elementary block of rainfall $I(\tau).d\tau$ into an elementary outflow

$$\left[\frac{I(\tau).d\tau}{K\Gamma n} e^{-t/K}.\left(\frac{t}{K}\right)^{n-1}\right]$$

which is shown in Fig. 12.3. Consequently the centre of area of each elementary block is moved to the right by an amount nK, and the centre of area of the total outflow is also moved to the right by nK. So, we have

$$MQ_1 - MI_1 = nK \qquad ...(12.18)$$

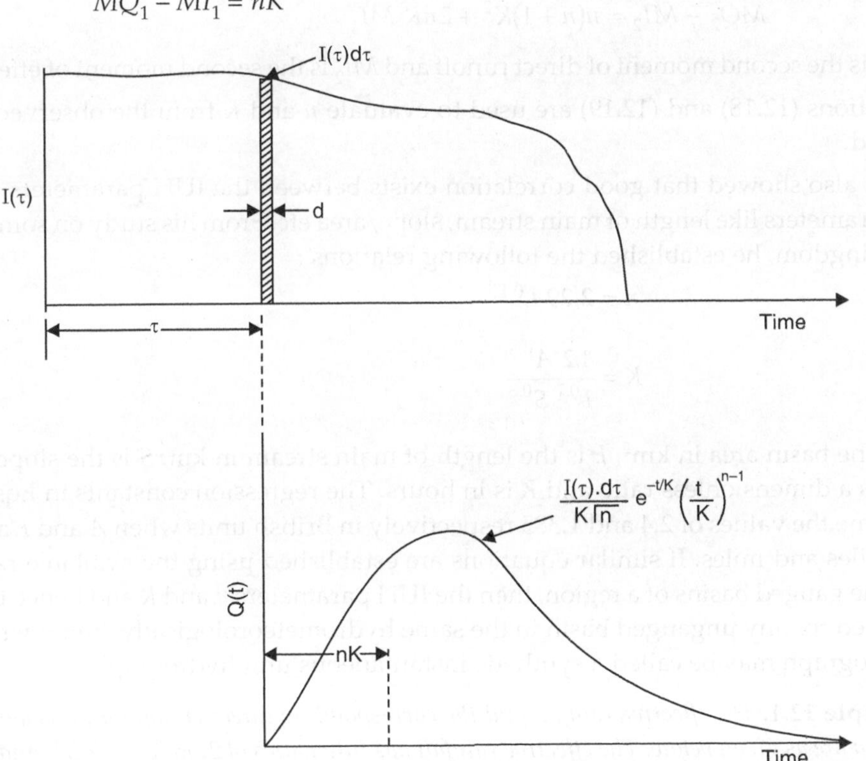

FIGURE 12.3 Transformation of rainfall into runoff.

where MQ_1 = The first moment of direct runoff, and

MI_1 = The first moment of effective rainfall.

Each elementary strip of Fig. 12.3 will give rise to an elementary outflow whose second moment about its own origin is $n(n+1)\,K^2$. Weighing the second moment of each routed strip with the area of each strip and transferring moments to the beginning of effective rainfall.

$$MQ_2 = \frac{1}{\int I(\tau)\,.\,d\tau}[n(n+1)K^2\int I(\tau)\,.\,d\tau$$

$$- n^2K^2\int I(\tau)\,d\tau + \int I(\tau)\,d\tau(t+nK)^2]$$

$$= n(n+1)K^2 - n^2K^2 + \frac{1}{\int I(\tau).\,d\tau}\int\left[t^2 + n^2K^2 + 2tnK\right]I(\tau).\,d\tau$$

$$= n(n+1)K^2 - n^2K^2 + \frac{\int I(\tau).\,d\tau\,.\,t^2}{\int I(\tau).\,d\tau} + 2nK\frac{\int I(\tau).\,d\tau\,.\,t}{\int I(\tau).\,d\tau} + n^2K^2$$

$$= n(n+1)\,K^2 + MI_2 + 2nK\,.\,MI_1$$

$$\therefore \qquad MQ_2 - MI_2 = n(n+1)K^2 + 2nK\,MI_1 \qquad \text{...(12.19)}$$

where MQ_2 is the second moment of direct runoff and MI_2 is the second moment of effective rainfall.

Equations (12.18) and (12.19) are used to evaluate n and K from the observed rainfall and runoff record.

Nash also showed that good correlation exists between the IUH parameters n and K, and the basin parameters like length of main stream, slope, area etc. From his study on some catchments in United Kingdom, he established the following relations :

$$n = 2.29\,L^{0.1} \qquad \text{...(12.20)}$$

$$K = \frac{1.2\,A^{0.3}}{L^{0.1}\,S^{0.3}} \qquad \text{...(12.21)}$$

where A is the basin area in km^2, L is the length of main stream in km, S is the slope of the basin expressed as a dimensionless ratio and K is in hours. The regression constants in Eqs. (12.20) and (12.21) assume the values of 2.4 and 1.385 respectively in British units when A and L are expressed in square miles and miles. If similar equations are established using the available rainfall runoff records of the gauged basins of a region, then the IUH parameters n and K and hence the IUH itself can be derived for any ungauged basin in the same hydrometeorologically homogeneous region. Such a hydrograph may be called a synthetic instantaneous unit hydrograph.

Example 12.1. *The effective rainfall and the corresponding direct runoff for an isolated storm on a drainage basin are as given below. The effective rainfall has intensities of 2.56, 2.52, 2.53, and 2.55 cm/h in four successive periods of one hour each. The runoff ordinates measured at one hour intervals are 2.83, 14.15, 40.50, 52.60,45.80, 34.50, 24.90, 19.20,14.70, 10.20, 6.80, 3.96, 2.26 and 0.90 m³/s. Determine the values of n and K.*

Solution.

$$MI_1 = \frac{(2.56 \times 1)0.5 + (2.52 \times 1)1.5 + (2.53 \times 1)2.5 + (2.55 \times 1)3.5}{2.56 \times 1 + 2.52 \times 1 + 2.53 \times 1 + 2.55 \times 1}$$

$$= \frac{20.30}{10.16} = 2 \text{ h}$$

$$MI_2 = \frac{(2.56 \times 1)0.5^2 + (2.52 \times 1)1.5^2 + (2.53 \times 1)2.5^2 + (2.55 \times 1)3.5^2}{10.16}$$

$$= \frac{53.35}{10.16} = 5.25 \text{ h}^2$$

The computations for calculating the first and second moments of direct runoff are shown in Table 12.2.

Table 12.2. Determination of n and K of Nash's IUH

Time h	*Direct runoff m^3/s*	*ΔQ = Vol. of direct runoff m^3/s-h*	*'t' distance of ΔQ from the origin h*	*t^2 h^2*	*$\Delta Q.t$*	*$\Delta Q.t^2$*
0	0					
1	2.83	1.42	0.5	0.25	0.71	0.355
2	14.15	8.49	1.5	2.25	12.75	19.100
3	40.50	27.33	2.5	6.25	68.30	171.000
4	52.60	46.50	3.5	12.25	163.00	570.0
5	45.80	49.20	4.5	20.25	221.00	998.0
6	34.50	40.15	5.5	30.25	221.00	1210.0
7	24.90	29.70	6.5	42.25	193.00	1250.0
8	19.20	22.05	7.5	56.25	165.00	1240.0
9	14.70	16.95	8.5	72.25	144.00	1225.0
10	10.20	12.45	9.5	90.25	118.00	1123.0
11	6.80	8.50	10.5	110.25	89.00	938.0
12	3.96	5.38	11.5	132.25	62.00	710.0
13	2.26	3.11	12.5	156.25	39.00	486.0
14	0.90	1.58	13.5	182.25	21.30	288.0
15	0.0	0.45	14.5	210.25	6.50	94.5
		Σ273.26			1524.56	10323

$$MQ_1 = \frac{\Sigma \Delta Q \,.\, t}{\Sigma \Delta Q} = \frac{1524.56}{273.26} = 5.6 \text{ h}$$

$$MQ_2 = \frac{\Sigma \Delta Q \,.\, t^2}{\Sigma \Delta Q} = \frac{10323}{273.26} = 37.8 \text{ h}^2$$

$$nK = MQ_1 - MI_1 = 5.6 - 2.0 = 3.6 \text{ h}$$

$$MQ_2 - MI_2 = n\,(n+1)\,K^2 + 2nK\,MI_1$$

$$37.8 - 5.25 = nK^2\,(n+1) + 2 \times 3.6 \times 2$$

$$32.65 = n^2K^2 + nK^2 + 14.4$$

$$nK^2 = 32.65 - 14.4 - (nK)^2$$

$$= 32.65 - 14.4 - (3.6)^2 = 5.29$$

$$K = \frac{nK^2}{nK} = \frac{5.29}{3.6} = 1.47 \text{ h}$$

$$n = \frac{nK}{K} = \frac{3.6}{1.47} = 2.45.$$

Example 12.2. *Derive the ordinates of IUH and hence the ordinates of 2h UH at 2 h interval for a drainage basin of area 200 km² having n = 3 and ; K = 5 h.*

Solution.
$$u(t) = \frac{1}{K\Gamma n}\, e^{-t/K} \cdot \left(\frac{t}{K}\right)^{n-1}$$

$$K = 5 \text{ h}\,;\ n = 3 \text{ and } \Gamma n = (n-1)!$$

Since n is an integer.

∴
$$\Gamma 3 = 2\,! = 2 \text{ and } K\Gamma n = 10\,;\ (n-1) = 2$$

∴
$$u(t) = \frac{1}{10} \cdot e^{-t/K} \cdot \left(\frac{t}{K}\right)^2$$

The computations are shown in a Table 12.3. In the first column the time is entered with increments of 2 h at each step. In Col. (2) t/k is evaluated for each t. Column (3) and (4) give $e^{-t/K}$ and $(t/K)^{n-1}$, respectively. Col. (5) is the ordinate of IUH in cm/h. By multiplying the values of Col. (5) by 2.778 A that is 556, the ordinates of IUH in m³/s are obtained and entered in Col. (6). Column (7) gives the ordinates of 2 h UH which are evaluated using Eq. (12.5).

Table 12.3. Computation of Nash's IUH

Time *t h* (1)	(t/K) (2)	$e^{-t/K}$ (3)	$(t/K)^2$ (4)	$u(t)$ *cm/h* (5)	$u(t)$ *m³/s* (6)	*Ordinates 2 h U.H m³/s* (7)
0	0	1.0000	0	0	0	0
2	0.4	0.6700	0.16	0.01072	5.96	2.98
4	0.8	0.4500	0.64	0.0288	16.00	10.98
6	1.2	0.3000	1.44	0.0432	24.00	20.00
8	1.6	0.2000	2.56	0.0512	28.40	26.20
10	2.0	0.1350	4.00	0.0542	30.10	29.30
12	2.4	0.0910	5.76	0.0524	29.10	29.60

14	2.8	0.0610	7.84	0.0480	26.70	27.90
16	3.2	0.0410	10.24	0.0420	23.40	25.10
18	3.6	0.0270	12.96	0.0350	19.40	21.40
20	4.0	0.0183	16.00	0.02928	16.30	17.80
22	4.4	0.0122	19.36	0.0236	13.10	14.70
24	4.8	0.0082	23.04	0.01896	10.50	11.80
26	5.2	0.0055	27.04	0.01488	8.30	9.40
28	5.6	0.0037	31.36	0.0116	6.45	7.38
30	6.0	0.0025	36.00	0.0090	5.00	5.73
32	6.4	0.0017	40.96	0.00696	3.83	4.42
34	6.8	0.0011	46.24	0.00508	2.82	3.32

For example,

The ordinate of 2 h UH at 16th h

$$= \frac{1}{2} \text{[Area of IUH between (16 – 2) h and 16 h]}$$

$$= \frac{1}{2} \text{[Area of IUH between 14 h and 16 h]}$$

Assuming the variation of IUH is approximately linear between 14 and 16 h.

The ordinate of 2 hUH at 16th h

$$= \frac{1}{2}\left[\frac{(\text{Ordinate of IUH at 14 h} + \text{Ordinate of IUH at 16 h}) \times 2}{2}\right]$$

$$= \frac{26.70 + 23.40}{2} = 25.05 \approx 25.10 \text{ m}^3/\text{s}$$

Similarly all other ordinates of 2 h UH are computed.

12.5.2 Clark's Model

In 1945, Clark showed that the IUH may be obtained by routing the rainfall excess, that is the effective rainfall. In fact, it is Clark who first used the concept of IUH. Clark has introduced the following two aspects of drainage basin in his approach.

(*i*) Every water particle has to travel certain distance before it reaches the gauging station and so the time of travel from various parts of the basin has to be taken into consideration.

(*ii*) As these quantities of water from different parts of the basin pass through the channel, the channel storage rises, to provide for the greater discharge and this in turn introduces a change in the time of travel of the water particles.

The first may be called the translation effect, and the second the storage effect.

The basin may be divided into zones by drawing what are known as the *isochrones* as shown in Fig. 12.4. An isochrone may be defined as the line joining points of equal time of travel. A water particle from any point on an isochrone labelled 6 h, for example, will take 6 h to reach the basin outlet. The area between successive isochrones is measured and a *time area concentration diagram* (also called simply the *time area diagram*) is prepared. This time area diagram can be conceived as the effective rainfall diagram with translation effect incorporated. Therefore, the runoff from the sub-area between 5th and 6th h isochrones will be distributed over a time period of one hour in the sixth hour. If the channel storage effects are neglected the time-area diagram itself will represent the IUH.

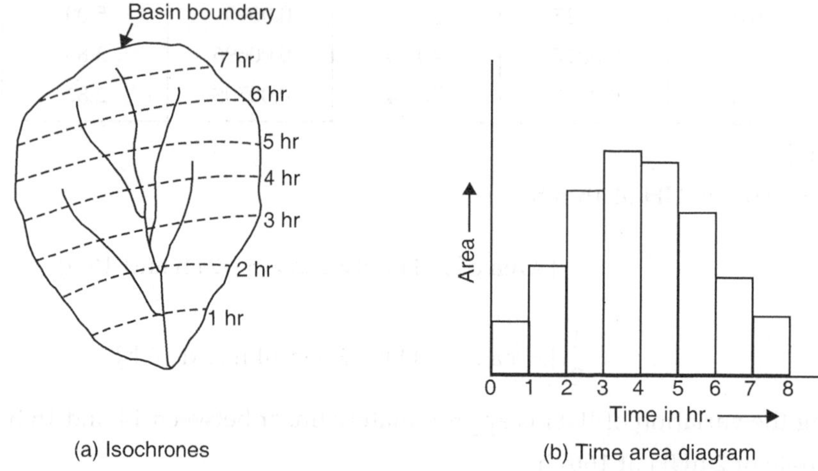

FIGURE 12.4 Construction of time area diagram.

The time area diagram is now routed through a linear reservoir whose storage is given by $S = KQ$ to incorporate the effects of channel storage. The ordinates of time area diagram can readily be converted into m^3/s by multiplying them with $(2.778/\Delta t)$, where Δt is the time interval between successive isochrones. The time area diagram may be routed through the linear reservoir by writing the continuity equation in the form

$$\left(\frac{I_1 + I_2}{2}\right)\Delta t - \left(\frac{Q_1 + Q_2}{2}\right)\Delta t = S_2 - S_1$$

where I, Q and S are the inflows, outflows, and storage respectively with sub-scripts 1 and 2 denoting the beginning and end of the routing period Δt.

Using the relation $S = KQ$, we get

$$Q_2 = \left(\frac{I_1 + I_2}{2}\right)C_0 + Q_1C_1 \qquad \text{...(12.22)}$$

where $C_0 = \dfrac{\Delta t}{K + \dfrac{\Delta t}{2}}$ and $C_1 = \dfrac{K - \dfrac{\Delta t}{2}}{K + \dfrac{\Delta t}{2}}$

Eq. (12.22) can be solved iteratively provided the outflow at the start is known, which in the present case is zero.

Evaluation of K. An estimate of K may be obtained from the recession curve of an observed runoff hydrograph. The continuity equation for the reservoir is

$$I - Q = \frac{dS}{dt} = K.\frac{dQ}{dt}$$

when the inflow into channel storage ceases, I becomes zero and the above equation yields an expression for K given by

$$K = -\frac{Q}{\left(\dfrac{dQ}{dt}\right)} \qquad \text{...(12.23)}$$

If Q is expressed in m^3/s and $\dfrac{dQ}{dt}$ is cumec/h, K will be obtained in hours and it will be positive since $\dfrac{dQ}{dt}$ is always negative on the recession limb. The value of K is usually obtained by using the derivative of the recession curve at the inflection point as shown in Fig. 12.5, since the overland flow (the inflow into channel storage) is supposed to cease at this point of time.

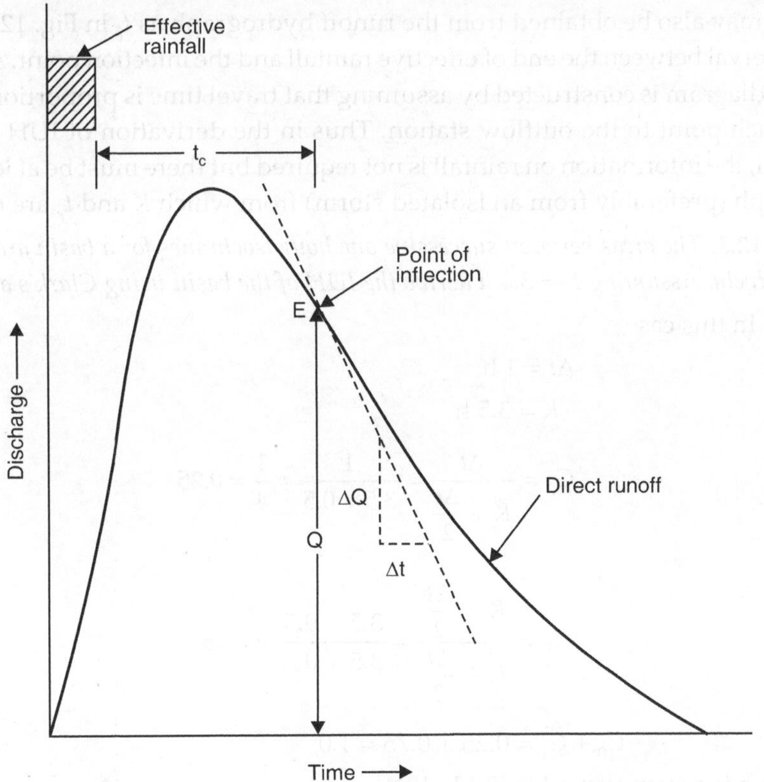

FIGURE 12.5 Determination of K from runoff hydrograph.

Alternatively, referring to Eq. (11.3), the value of K may be obtained as, $\left(\frac{1}{-\ln K_r}\right)$ where K_r is the recession constant of the direct runoff, provided of course the value of K_r is already determined from the analysis of the recession curve of the hydrograph on a semilogarithmic plot adopting a time unit of one hour.

Clark has also suggested an empirical relationship given by Eq. (12.24) to find K and Linsley suggested further modification to this as given by Eq. (12.25).

$$K = \frac{cL}{\sqrt{S}} \qquad \text{...(12.24)}$$

$$K = \frac{bL\sqrt{A}}{\sqrt{S}} \qquad \text{...(12.25)}$$

where K is the storage constant of the reservoir in hours, L is the length of main stream in km, S is the mean Channel slope expressed as a ratio, A is the drainage basin area in km^2, c is a constant varying from about 0.5 to 1.4 (0.8 to 2.2 with L in miles), and b is another constant varying from 0.015 to 0.03 (0.04 to 0.08 with L in miles and A in square miles).

The base period of the time area diagram denotes the maximum travel time through the catchment. This may also be obtained from the runoff hydrograph as t_c in Fig. 12.5, which is taken to be the time interval between the end of effective rainfall and the infection point. After determining t_c, the time area diagram is constructed by assuming that travel time is proportional to the channel distance from each point to the outflow station. Thus in the derivation of IUH for a basin using Clark's approach, the information on rainfall is not required but there must be at least one observed runoff hydrograph (preferably from an isolated storm) from which K and t_c are determined.

Example 12.3. *The areas between successive one hour isochrones for a basin are 15, 36, 81, 49, 28 and 9 km² respectively. Assuming K = 3.5 h derive the IUH of the basin using Clark's approach.*

Solution. In this case,

$$\Delta t = 1 \text{ h}$$

$$K = 3.5 \text{ h}$$

$$C_0 = \frac{\Delta t}{K + \frac{\Delta t}{2}} = \frac{1}{3.5 + 0.5} = \frac{1}{4} = 0.25$$

$$C_1 = \frac{K - \frac{\Delta t}{2}}{K + \frac{\Delta t}{2}} = \frac{3.5 - 0.5}{3.5 + 0.5} = 0.75$$

Check. $C_0 + C_1 = 0.25 + 0.75 = 1.0$

The routing is accomplished in Table 12.4.

Table 12.4 Computation of Clark's IUH

Time *h*	*Time-Area diagram ordinate* km^2	*Ordinate of inflow* m^3/s	$\left(\frac{I_1+I_2}{2}\right)$	$C_0\left(\frac{I_1+I_2}{2}\right)$	C_1Q_1	*Ordinate of IUH* m^3/s
0						0
1	15	41.6	41.6	10.4	0	10.4
2	36	100.0	100.0	25.0	7.8	32.8
3	81	225.0	225.0	56.25	24.6	80.85
4	49	136.0	136.00	34.00	60.8	94.80
5	28	78.0	78.0	19.50	71.0	90.5
6	9	25.0	25.0	6.25	68.0	74.25
7	—	—	—	—	55.6	55.6
8	—	—	—	—	41.8	41.8
9	—	—	—	—	31.4	31.4
10	—	—	—	—	23.6	23.6
11	—	—	—	—	17.7	17.7
12	—	—	—	—	13.3	13.3
13	—	—	—	—	10.0	10.0
14	—	—	—	—	7.5	7.5
15	—	—	—	—	5.63	5.63
16	—	—	—	—	4.22	4.22
17	—	—	—	—	3.16	3.16

REVIEW EXERCISES

Descriptive Questions

1. Define instantaneous unit hydrograph. How does it differ from unit hydrograph of finite duration ?
2. What are the advantages of instantaneous unit hydrograph over a unit hydrograph of finite duration ?
3. How can IUH be derived from *S*-curve ?
4. Establish the relationship between IUH ordinates and DUH ordinates.
5. Explain the Nash's conceptual model for IUH and derive the equation for IUH.
6. How are values of *n* and *K* in the Nash's IUH evaluated from an observed rainfall-runoff record ?
7. What are catchment isochrones ? What is a time-area diagram ?
8. Explain the concept of Clark's IUH.

9. How can the linear reservoir storage coefficient K in the Clark's IUH be determined from runoff data ?
10. Explain the convolution integral with a neat sketch.

Analytical Questions

1. Show that the peak discharge in the Nash's IUH is given by

$$u(t)_{\max} = \frac{1}{K\Gamma n} e^{1-n} (n-1)^{n-1}.$$

2. Determine the values of n and K of IUH of the drainage basin of example (11.3) and hence derive the ordinates of IUH.
3. From the IUH obtained in analytical question 2 derive the ordinates of 4 h unit hydrograph and compare them with those of example 11.3.
4. Determine the IUH of a catchment of area 36 km^2 using the following data. Assume a constant base flow of 5 m^3/s

6 h periods	0	1	2	3	4	5	6	7	8	9	10
Rainfall cm/h	1.5	3.5	2.5	1.5							
Stream flow m^3/s	5	15	75	170	185	147	84	43	18	8	5

5. The area of a drainage basin can be approximated to be a square with each diagonal of length 40 km. The main channel is along one of the diagonals. Assume that the isochrones are perpendicular to the main channel and the time of travel along the main channel is 5 km/h. Determine the IUH of the basin using Clark's approach taking $K = 4.5$ h.
6. The IUH of a basin can be approximated as a triangle with a base period of 20 h and a peak of 16 m^3/s occurring at 4th hour. Determine the ordinates of the 2h U.H. of the basin.
7. Use the Clark's method to derive a 2 h unit hydrograph for a catchment with the following time-area diagram.

Time (h)	0		1		2		3		4		5		6
Area (km^2)		10		20		30		20		12		8	

Use $\Delta t = 1$ h and $K = 3.5$ h.

Objective Type Questions

1. If the duration of the unit hydrograph approaches zero, the resulting unit hydrograph is known as

(*a*) the limiting unit hydrograph (*b*) the threshold unit hydrograph

(*c*) the instantaneous unit hydrograph (*d*) the saturated unit hydrograph.

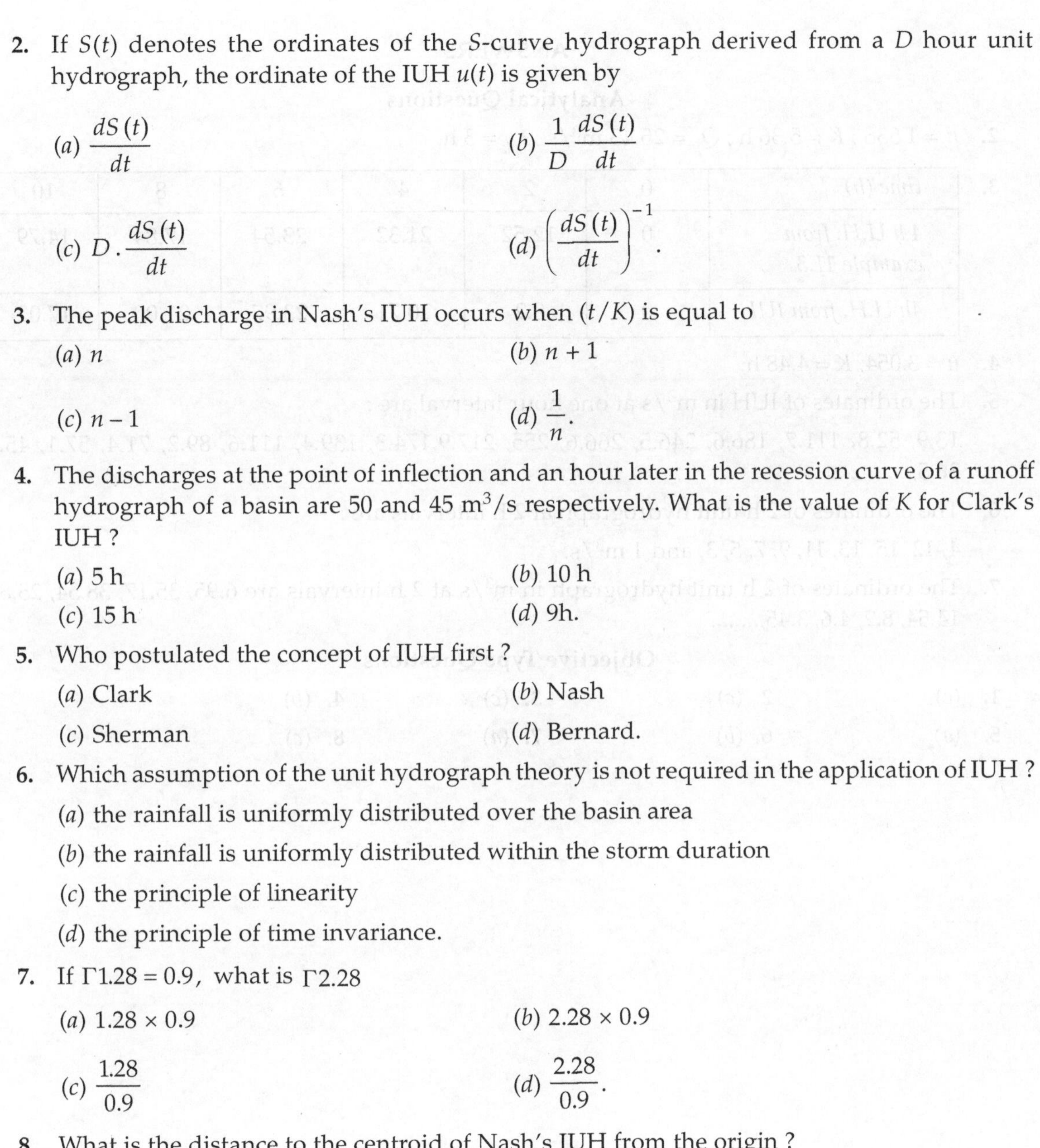

2. If $S(t)$ denotes the ordinates of the S-curve hydrograph derived from a D hour unit hydrograph, the ordinate of the IUH $u(t)$ is given by

   (a) $\dfrac{dS(t)}{dt}$ (b) $\dfrac{1}{D}\dfrac{dS(t)}{dt}$

   (c) $D.\dfrac{dS(t)}{dt}$ (d) $\left(\dfrac{dS(t)}{dt}\right)^{-1}$.

3. The peak discharge in Nash's IUH occurs when (t/K) is equal to

   (a) n (b) $n+1$

   (c) $n-1$ (d) $\dfrac{1}{n}$.

4. The discharges at the point of inflection and an hour later in the recession curve of a runoff hydrograph of a basin are 50 and 45 m^3/s respectively. What is the value of K for Clark's IUH ?

   (a) 5 h (b) 10 h

   (c) 15 h (d) 9h.

5. Who postulated the concept of IUH first ?

   (a) Clark (b) Nash

   (c) Sherman (d) Bernard.

6. Which assumption of the unit hydrograph theory is not required in the application of IUH ?

   (a) the rainfall is uniformly distributed over the basin area

   (b) the rainfall is uniformly distributed within the storm duration

   (c) the principle of linearity

   (d) the principle of time invariance.

7. If $\Gamma 1.28 = 0.9$, what is $\Gamma 2.28$

   (a) 1.28×0.9 (b) 2.28×0.9

   (c) $\dfrac{1.28}{0.9}$ (d) $\dfrac{2.28}{0.9}$.

8. What is the distance to the centroid of Nash's IUH from the origin ?

   (a) $\dfrac{n}{K}$ (b) $\dfrac{n}{K^2}$

   (c) nK (d) nK^2.

ANSWERS

Analytical Questions

2. $n = 1.536$; $K = 5.36$ h ; $Q_p = 25.43$ m^3/s ; $t_{pk} = 3$ h.

3.

time (h)	0	2	4	6	8	10
4 h U.H. from example 11.3	0	12.52	21.32	23.54	17.84	14.79
4h U.H. from IUH	0	8.2	20.71	23.97	21.00	17.02

4. $n = 3.054$, $K = 4.48$ h.

5. The ordinates of IUH in m^3/s at one hour interval are :

   13.9, 52.8, 111.7, 186.6, 246.5, 266.6, 255, 217.9,174.3, 139.4, 111.6, 89.2, 71.4, 57.1, 45.7, 36.5, ...

6. The ordinates of 2 h unit hydrograph at 2 h intervals are :

   4, 12, 15, 13, 11, 9, 7, 5, 3, and 1 m^3/s.

7. The ordinates of 2 h unit hydrograph in m^3/s at 2 h intervals are 6.95, 35.17, 38.54, 25.85, 14.54, 8.2, 4.6, 3.45,........

Objective Type Questions

1. (*c*) 2. (*c*) 3. (*c*) 4. (*b*)

5. (*a*) 6. (*b*) 7. (*a*) 8. (*c*).

Chapter 13 FLOOD ROUTING

13.1 INTRODUCTION

Flood routing may be defined as the procedure whereby the shape of a flood hydrograph at a particular location on the stream is determined from the known or assumed flood hydrograph at some other location upstream.

As the discharge in a stream increases stage also increases and with it the volume of water in temporary storage in the channel. When the flood wave recedes water is depleted from the channel storage and the stage in the stream falls. As a result, a flood wave moving down a channel lengthens its base time and consequently its peak gets reduced. The flood wave is then said to be attenuated. The effect of reservoir storage or lake storage is also similar. Flood routing is the technique in hydrology to compute the effect of storage on the shape and movement of the flood wave.

Flood routing is used in establishing the height of a flood peak at a downstream location in short term flood forecasting, estimating the protection that would result from construction of a reservoir, determining required levee heights for flood protection, determining the adequacy of spillways, predicting the behaviour of a river after a change in channel conditions and in the derivation of synthetic unit graphs. (Clark's method which was discussed in section 12.5)

Flood routing may be divided into two basic types namely (*i*) the *reservoir routing* and (*ii*) the *channel routing* or the *streamflow routing. The* reservoir routing analyses the effect of reservoir storage on the flood hydrograph, while the channel routing analyses the effect of storage of a specified channel reach on the flood hydrograph.

13.2 RESERVOIR ROUTING

Letting *I* and *Q* to denote the inflow into and outflow from a reservoir, and *S* the storage in the reservoir, the continuity equation in the differential form for the reservoir is given by

$$I - Q = \frac{dS}{dt} \qquad ...(13.1)$$

Alternatively, the same can be written as

$$\bar{I}\Delta t - \bar{Q}\Delta t = \Delta S \qquad ...(13.2)$$

where $\bar{I}$ is the average inflow rate in a small time interval Δt, $\bar{Q}$ is the average outflow rate in the same time interval and ΔS is the corresponding change in the storage of the reservoir during the same time interval. If suffixes 1 and 2 are used to denote a given quantity at the beginning and the end of the time interval and if the inflow and outflow have straight line variation within the time interval, Eq. (13.2) can be written as

$$\left(\frac{I_1 + I_2}{2}\right)\Delta t - \left(\frac{Q_1 + Q_2}{2}\right)\Delta t = S_2 - S_1 \qquad ...(13.3)$$

Proper units must be chosen for storage to maintain compatibility in Eq. (13.3). For example, if I and Q are expressed in m^3/s and Δt in days, then storage must be expressed in cumec- days. In Eq. (13.3), the first term on left hand side represents the volume of water entering the storage, the second term represents the volume of water leaving the storage and term on the right hand side represents the change in storage, which can be either positive or negative depending on whether $\bar{I}$ is more than or less than $\bar{Q}$.

In Eq. (13.3), I_1 and I_2 are known from the given inflow hydrograph to be routed through the reservoir, Q_1 and S_1 are the initial outflow from the reservoir and the initial storage in the reservoir which are either known or assumed and Q_2 and S_2 are the two unknown quantities which must be determined. Thus to solve for Q_2 and S_2 one more relation is needed. In reservoirs the water surface is level (or horizontal) and it can be assumed that the storage in the reservoir is independent of inflow and the outflow is dependent only on storage. And it is this relation between storage and outflow which provides the additional relationship required for the solution of S_2 and Q_2. For this reason, the reservoir routing is also called *level pool routing*. All the reservoir routing methods will make use of this storage discharge relation in some convenient form or other. Before we discuss the routing methods, we describe how the storage discharge relation can be obtained from the topographical data of the reservoir site and the discharge characteristics of the reservoir outlets such as spillways and sluices.

Establishing Storage-Discharge Relationship

The area at the reservoir site is surveyed in detail and a contour map is prepared with a contour interval of 2 to 3 m depending on the size of the reservoir. From this map the areas enclosed by various contours are planimetered and a curve of elevation *versus* area is prepared. Such a curve is very useful in estimating the evaporation losses from the reservoir since the area enclosed by any contour will also represent the water spread area if the water is stored in the reservoir upto that contour elevation, though it is not directly useful in the routing procedure. Once the areas enclosed by contours are known the incremental volumes of water stored between any two successive contours can be determined using one of the following equations.

$$\Delta \forall = \frac{h}{2}(A_1 + A_2) \text{ (Trapezoidal formula)} \qquad ...(13.4)$$

$$\Delta \forall = \frac{h}{3}\left(A_1 + A_2 + \sqrt{A_1 A_2}\right) \text{ (Cone formula)} \qquad ...(13.5)$$

$$\Delta \forall = \frac{h}{6}\left(A_1 + 4A_m + A_2\right) \text{ (Prismoidal formula)} \qquad ...(13.6)$$

where A_1 and A_2 are the areas corresponding to two successive contour values and h is the difference between them (or the contour interval). A_m is the area enclosed by a contour line midway between the two adjacent contours. If these incremental volumes are accumulated upto any contour value, then that sum represents the storage volume of the reservoir up to the elevation of that contour. This information is used to prepare a curve of elevation *versus* storage such as shown in Fig. (13.1*a*). The outflow from the reservoir corresponding to any elevation can be determined using the discharge equations of the openings. For example,

$$Q_{sp} = C\,L\,(H - H_{sp})^{3/2} \qquad ...(13.7)$$

$$Q_{sl} = n \times c_d \times A\sqrt{2g(H - H_{sl})} \qquad ...(13.8)$$

where
Q_{sp} = the discharge from the spillway
C = coefficient of the spillway
L = effective length of the spillway
H_{sp} = sill level of the spillway
H = water level in the reservoir
Q_{sl} = the discharge from the sluices
n = the number of sluices
A = area of each sluice
H_{sl} = sill level of sluices
C_d = coefficient of discharge of sluices

If the reservoir is provided with both spillway and sluices, $(Q_{sp} + Q_{sl})$ will give the discharge corresponding to the elevation H. When the outflow is plotted against elevation, a curve as shown in Fig. (13.1*b*) is obtained which is known as the elevation-discharge curve. The curve of Fig. (13.1*a*) and Fig. (13.1*b*) are then utilised to prepare a curve of storage *versus* discharge as shown in Fig. (13.1*c*). It is this curve which is used in reservoir routing.

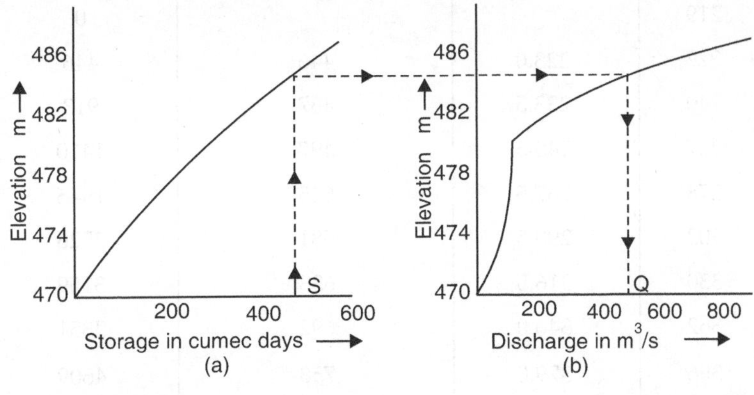

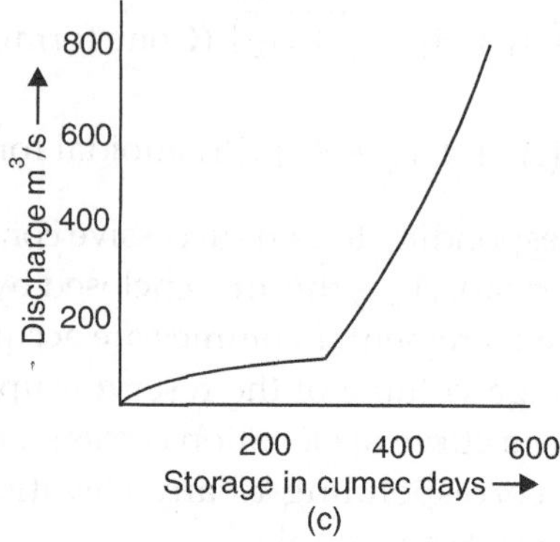

FIGURE 13.1 Storage-discharge relation for a reservior.

Example 13.1. *The topographic surveys at a proposed reservoir site yielded the following data :*

Contour elevation, m :	*470*	*472*	*474*	*478*	*480*	*482*	*484*	*486*
Contour area hectares :	*219*	*227*	*257*	*278*	*303*	*330*	*362*	*396*

There are 2 circular sluices with a diameter of 2.5 m and with their centres at an elevation of 470 m. A spillway with an effective crest length of 20 m is also provided with its sill at 480 m. The C_d for sluices may be taken as 0.8 and for spillway C = 2.25. Prepare the Storage-discharge curve for the reservoir.

Solution. The calculations to obtain the storage *versus* elevation curve are given in Table 13.1. The contour interval is uniform and equal to 2 m. Trapezoidal formula has been used to compute the incremental volumes. The storage in hectare-metres is converted into cumec days using the conversion that 1 cumec- day = 8.64 ha-m. This is because storage expressed in cumec-days will be usually convenient in the routing problems. The elevation-storage curve prepared from this table is shown in Fig. 13.1 (*a*).

Table 13.1 Calculations for Storage-Elevation Curve

Elevation in m	*Area A in hectares*	$\frac{(A_1 + A_2)}{2}$	*Incremental storage* $\Delta \forall = h\left(\frac{A_1 + A_2}{2}\right)$	*Cumulative storage Ha-m*	*Storage cumec-days*
470	219			0	0
472	227	223.0	446	446	51.62
474	240	233.5	467	913	105.67
476	257	248.5	497	1410	163.19
478	278	267.5	535	1945	225.12
480	303	290.5	581	2526	292.36
482	330	316.5	633	3159	365.63
484	362	346.0	692	3851	445.72
486	396	379.0	758	4609	533.45

The discharge from the sluices is given by

$$Q_{sl} = n \times c_d \times A \times \sqrt{2g(H - H_{sl})}$$

$$= 2 \times 0.8 \times \frac{\pi}{4} \times (2.5)^2 \times \sqrt{19.62} \times \sqrt{(H - H_{sl})}$$

$$= 34.79 \sqrt{(H - 470)}$$

The discharge from the spillway is given by

$$Q_{sp} = C \times L \times (H - H_{sp})^{3/2}$$

$$= 2.25 \times 20 \times (H - 480)^{3/2}$$

$$= 45\ (H - 480)^{3/2}$$

$$Q = Q_{sl} + Q_{sp}$$

The computations to obtain the discharge from the reservoir for different elevations are given in Table 13.2 and the elevation-discharge curve prepared from this table is shown in Fig. 13.1(*b*).

Table 13.2. Calculations for Elevation-discharge Curve

Elevation *H* *m*	$(H - H_{sl})$ *m*	Q_{sl} m^3/s	$(H - H_{sp})$ *m*	Q_{sp} m^3/s	$Q = Q_{sl} + Q_{sp}$ m^3/s
470	0	0	—	—	0
472	2	49.2	—	—	49.2
474	4	69.6	—	—	69.6
476	6	85.2	—	—	85.2
478	8	98.4	—	—	98.4
480	10	110.0	0	0	110.0
482	12	120.5	2	127.3	247.8
484	14	130.2	4	360.0	490.2
486	16	139.2	6	661.4	800.6

From Tables 13.1 and 13.2, the plotting positions for the points of storage-discharge curve are obtained as given in Table 13.3.

Table 13.3. Data for Storage-discharge Curve

Elevation *m*	*Storage* *cumec-day*	*Discharge* m^3/s
470	0	0
472	51.62	49.2
474	105.67	69.6
476	163.19	85.2
478	225.12	98.4

480	292.36	110.0
482	365.63	247.8
484	445.72	490.2
486	533.45	800.6

The curve prepared by plotting the storages in Col. (2) of Table 13.3 against the discharges in Col. (3) is shown in Fig 13.1 (*c*). It may be pointed out here that there may be storage below the contour level of 470 m in the reservoir. But it is essentially a dead storage and is irrelevant as for as the routing is concerned since it cannot affect the flood hydrograph.

I.S.D. Method. The Inflow-storage-Discharge method was first developed by L.G. Puls of the U.S. Army Corps of Engineers. According to this method, Eq. (13.3) is rearranged and written as

$$\frac{I_1 + I_2}{2} \cdot \Delta t + \left(S_1 - \frac{Q_1}{2}\Delta t\right) = \left(S_2 + \frac{Q_2}{2}\Delta t\right) \quad ...(13.9)$$

In the above equation all the quantities on left hand side are known and hence the quantity $\left(S_2 + \frac{Q_2}{2}\Delta t\right)$ is determined. From this quantity the value of Q_2 can be found out from the storage-discharge relationship as explained below.

From the available storage discharge curve, the curves of $\left(S - \frac{Q}{2}\Delta t\right)$ *versus* Q and $\left(S + \frac{Q}{2}\Delta t\right)$ *versus* Q are developed. Such curves, as shown in Fig. 13.2, may be called the *routing curves*. From the known value of Q_1, the value of $\left(S_1 - \frac{Q_1}{2}\Delta t\right)$ is read from Q *Vs* $\left(S - \frac{Q}{2}\Delta t\right)$ curve. This value is then added to $\left(\frac{I_1 + I_2}{2}\right) \cdot \Delta t$ to give $\left(S_2 + \frac{Q_2}{2}\Delta t\right)$. Entering the graph with this value of $\left(S_2 + \frac{Q_2}{2}\Delta t\right)$, the value of Q_2 is read out from Q *Vs* $\left(S + \frac{Q}{2}\Delta t\right)$ curve. The value of Q_2 thus determined becomes Q_1 for the next time interval. The procedure is repeated for all the subsequent routing periods till the complete outflow hydrograph is obtained. In preparing the routing curves care should be taken to see that $\left(S + \frac{Q}{2}\Delta t\right)$ and $\left(S - \frac{Q}{2}\Delta t\right)$ will have consistent units. For example, if Q is in m^3/s and Δt is in days S should be in cumec-days. Actually the routing can be performed with only one routing curve, that is Q *Vs* $\left(S + \frac{Q}{2}\Delta t\right)$. This is because $\left(S - \frac{Q}{2}\Delta t\right)$ for any Q can be obtained by first reading $\left(S + \frac{Q}{2}\Delta t\right)$ from the Q *Vs* $\left(S + \frac{Q}{2}\Delta t\right)$ curve and then substracting $Q \cdot \Delta t$ from this value.

If the inflow and outflow hydrographs are now plotted on the same time scale, it will be observed that the peakflow of outflow hydrograph is less than the peak-flow of the inflow hydrograph. In other words, the peak-flow is reduced. Similarly, the time to peak in the outflow hydrograph is more than the time to peak in the inflow hydrograph. These are the effects of reservoir storage on the movement of flood wave through the reservoir. The reduction in peak is known as the *attenuation* and the difference in times to peak is known as the *reservoir lag*, which are schematically represented in Fig. 13.3. However, it must be noted here that the attenuation and the reservoir lag are dependent on the initial outflow from the reservoir which existed at the time of arrival of flood. Example 13.2 illustrates the I.S.D method of reservoir routing.

It may also be noted from Fig 13.3 that the peak flow in the outflow hydrograph occurs at the intersection point of the inflow and outflow hydrographs. The reason for this is easy to guess. In the early stages (*i.e.*, from 6th hour onwards in Fig. 13.3) the inflow is higher than the outflow indicating the accumulation of storage which continues till the two curves meet. After the intersection point, the outflow curve is above the inflow curve which indicates the withdrawal from the storage. Therefore the storage would be maximum at the intersection point. If the outflow from the reservoir is unregulated , there will be one to one correspondence between outflow and storage and hence the discharge is maximum when the storage in the reservoir is maximum. Hence, in the case of unregulated reservoir the peak flow in the outflow hydrograph always occurs at the intersection point. In other words, the peak flow of outflow hydrograph will always lie on the falling limb of the inflow hydrograph. However, in the case of channel routing (as will be seen later in the next section) this need not be true since the storage in a channel reach is a function of both inflow and outflow.

Example 13.2. *Route the following flood hydrograph through the reservoir of Example 13.1 and determine the attenuation and the reservoir lag.*

Time h	0	6	12	18	24	30	36	42	48	54	60	66	72
Flow m³/s	50	180	270	360	410	370	300	230	155	90	60	35	20

Also find out what is the maximum elevation reached by the water surface in the reservoir. Assume that the outflow from the reservoir just before the flood arrived was 200 m³/s.

Solution. The ordinates of storage-discharge curve of Example 13.1 are taken from Table 13.3 and the calculations needed for the preparation of the routing curves are given in Table 13.4. Since the inflow ordinates are given at 6 h intervals, the routing time interval is also taken to be 6 h or 0.25 days. Therefore Δt = 0.25 days.

Table 13.4. Calculations for the Routing Curves of I.S.D. Method

S *cumec-days*	Q *m³/s*	$\frac{Q}{2}.\Delta t$ (= 0.125 Q) *cumec-days*	$\left(S+\frac{Q}{2}\Delta t\right)$ *cumec-days*	$\left(S-\frac{Q}{2}\Delta t\right)$ *cumec-days*
0	0	0	0	0
51.62	49.2	6.15	57.77	45.47
105.67	69.6	8.70	114.37	96.97
163.19	85.2	10.65	173.84	152.54
225.12	98.4	12.30	237.42	212.82
292.36	110.0	13.75	306.11	278.61
365.63	247.8	30.98	396.61	334.65
445.72	490.2	61.28	507.00	384.44
533.45	800.6	100.08	633.53	433.37

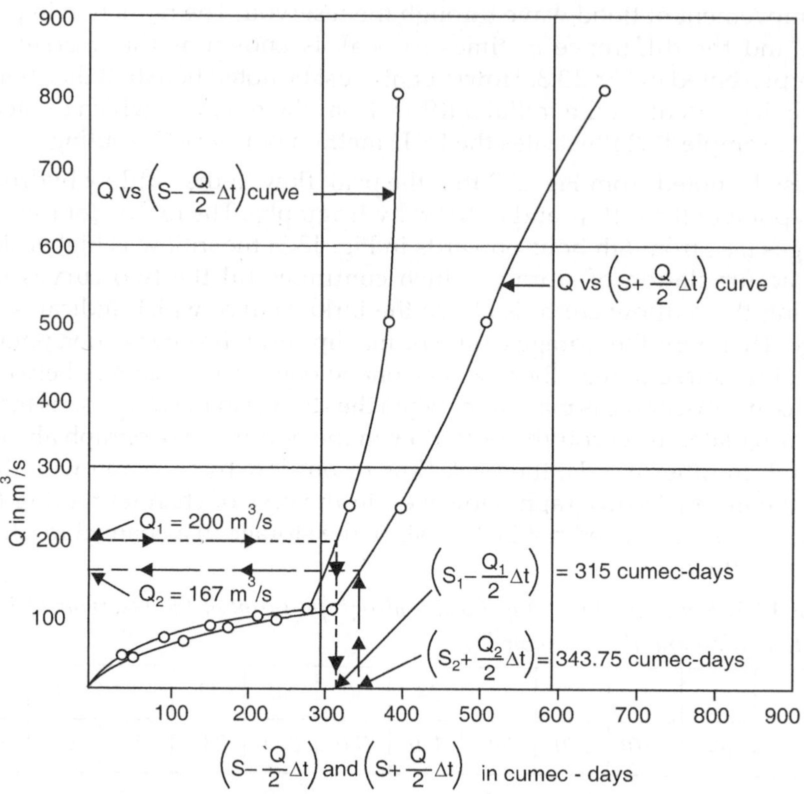

FIGURE 13.2 Routing curves of I.S.D. method.

The routing curves prepared using the data in Cols. (2), (4), and (5) of Table 13.4 are shown in Fig. 13.2. The reader is advised to prepare these curves to as large a scale as possible so that the errors due to the graphical interpolations are kept at minimum.

(*i*) For the first time interval between 0 and 6 hours

$I_1 = 50\ \text{m}^3/\text{s}, \qquad I_2 = 180\ \text{m}^3/\text{s}$

$Q_1 = 200\ \text{m}^3/\text{s}, \qquad \Delta t = 0.25\ \text{day}.$

For $Q_1 = 200\ \text{m}^3/\text{s}$, from Q *Vs* $\left(S - \frac{Q}{2}\Delta t\right)$ curve of Fig. 13.2, we obtain that $\left(S_1 - \frac{Q_1}{2}\Delta t\right)$ = 315 cumec days.

$$\frac{I_1 + I_2}{2} . \Delta t = \frac{50 + 180}{2} \times 0.25 = 28.75 \text{ cumec-days}$$

$$S_2 + \frac{Q_2}{2} . \Delta t = \frac{I_1 + I_2}{2} . \Delta t + \left(S_1 - \frac{Q_1}{2}\Delta t\right)$$

$$= 28.75 + 315 = 343.75 \text{ cumec-days.}$$

For $\left(S_2 + \frac{Q_2}{2}\Delta t\right) = 343.75$, from Q *Vs* $\left(S + \frac{Q}{2}\Delta t\right)$ curve of Fig. 13.2, we obtain $Q_2 = 167\ \text{m}^3/\text{s}$. This becomes Q_1, for the second time interval.

(*ii*) For the second time interval between 6 and 12 hours,

$$I_1 = 180 \text{ m}^3/\text{s}, \qquad I_2 = 270 \text{ m}^3/\text{s}$$
$$Q_1 = 167 \text{ m}^3/\text{s}, \qquad \Delta t = 0.25 \text{ day.}$$

For $Q_1 = 167$ m^3/s, from Q *Vs* $\left(S - \frac{Q}{2}\Delta t\right)$ curve of Fig. 13.2., we obtain that $\left(S_1 - \frac{Q_1}{2}\Delta t\right)$ = 302 cumec-days.

$$\frac{I_1 + I_2}{2}.\Delta t = \frac{180 + 270}{2} \times 0.25 = 56.25 \text{ cumec-days}$$

$$S_2 + \frac{Q_2}{2}.\Delta t = \frac{I_1 + I_2}{2}.\Delta t + \left(S_1 - \frac{Q_1}{2}\Delta t\right)$$

$$= 56.25 + 302 = 358.25 \text{ m}^3/\text{s}.$$

For $\left(S_2 + \frac{Q_2}{2}\Delta t\right) = 358.25$ m^3/s, from Q *Vs* $\left(S_2 + \frac{Q}{2}\Delta t\right)$ curve of Fig. 13.2, we obtain that Q_2 = 189 m^3/s. This becomes Q_1 for the third time interval and so on.

This procedure is repeated for the remaining time intervals of the inflow hydrograph in a tabular form as shown in Table 13.5. After all the ordinates of the outflow hydrograph are obtained, the elevation of water surface in the reservoir corresponding to every discharge are read from the Elevation-discharge curve of Fig.13.1 (*b*).

Table 13.5. Routing Through a Reservoir By I.S.D. Method

Time	*Inflow*	$\frac{I_1 + I_2}{2}.\Delta t$	$S_1 - \frac{Q_1}{2}\Delta t$	$S_2 + \frac{Q_2}{2}.\Delta t$	*Outflow*	*Elevation*
t *hours*	*I* *m³/s*	*cumec-days*	*cumec-days*	*cumec-days*	*Q* *m³/s*	*H* *metres*
(1)	(2)	(3)	(4)	(5)	(6)	(7)
0	**50**				**200**	481.30
6	**180**	28.75	315	343.75	167	480.83
12	**270**	56.25	302	358.25	189	481.15
18	**360**	78.75	311	389.75	237	481.84
24	**410**	96.25	330	426.25	313	482.54
30	**370**	97.50	348	445.50	355	482.88
36	**300**	83.75	356	439.75	342	482.78
42	**230**	66.25	354	420.25	300	482.43
48	**155**	48.13	345	393.13	242	481.92
54	**90**	30.63	332	362.63	196	481.25
60	**60**	18.75	314	332.75	150	480.58
66	**35**	11.88	295	306.88	110	480.00
72	**20**	6.88	280	286.88	106	479.28

The figures in bold face are the given data.

Columns (1) and (2) = ordinates of the given inflow hydrograph.

Column (3) $= \frac{I_1 + I_2}{2} \cdot \Delta t = \frac{I_1 + I_2}{8}$, since $\Delta t = \frac{1}{4}$ day

Column (4) $= \left(S_1 - \frac{Q_1}{2}\Delta t\right)$ obtained from Q Vs $\left(S - \frac{Q}{2}\Delta t\right)$ curve of Fig. 13.2 against the value of outflow in Col. (6) of the previous row.

Column (5) $= \left(S_2 + \frac{Q_2}{2}\Delta t\right)$ = Col. (3) + Col. (4).

Column (6) = Q obtained from Q Vs $\left(S + \frac{Q}{2}\Delta t\right)$ curve of Fig. 13.2 against the value of $\left(S_2 + \frac{Q_2}{2}\Delta t\right)$ in Col. (5).

Column (7) = Elevation of water surface in the reservoir obtained from Elevations-discharge curve of Fig. 13.1 (*b*) against the value of Q in Col. (6).

The inflow and outflow hydrographs and also the graph of water surface elevation against time are shown in Fig. 13.3, from which the following results are obtained.

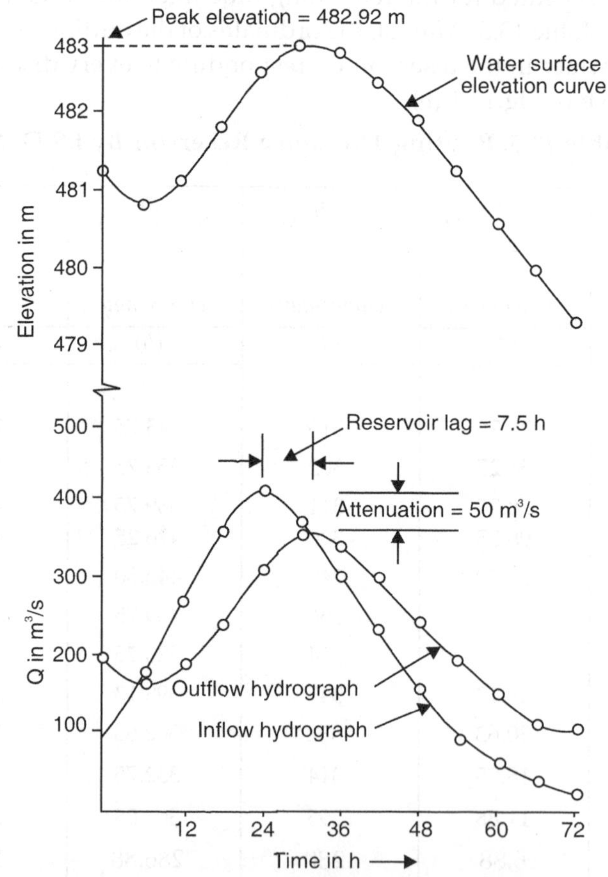

FIGURE 13.3 Inflow and routed hydrographs of example 13.2.

Attenuation = 50 m^3/s

Reservoir lag = 7.5 h

Max. water surface elevation = 482.92 m.

As pointed out earlier, the attenuation and reservoir lag are very much dependent on the initial outflow. For example, if the routing is performed with an initial outflow of only 50 m^3/s instead of 200 m^3/s, it will be observed that the attenuation and lag are 177 m^3/s and 18 h respectively.

Modified Puls Method. This method is also known as the *storage indication method*. In this method the basic continuity equation of (13.3) is rearranged to take the form given by

$$\frac{2S_2}{\Delta t} + Q_2 = (I_1 + I_2) + \left(\frac{2S_1}{\Delta t} - Q_1\right) \qquad ...(13.10)$$

The left hand side of Eq. (13.10) is known as the *storage indication quantity*. After selecting the routing time interval Δt, the routing curve of Q *Vs* $\left(\frac{2S}{\Delta t} + Q\right)$ such as shown in Fig. (13.4) is prepared from the available storage-discharge curve of the reservoir. For smoothly rising hydrographs, a minimum value of $\Delta t = \frac{t_p}{5}$ is recommended in which t_p is the time to peak in the inflow hydrograph. In the case of computer aided calculations, $\Delta t = \frac{t_p}{10}$ to $\frac{t_p}{20}$ may be used. To start with, for the first time interval I_1, I_2 and Q_1 are known. From the routing curve, $\left(\frac{2S_1}{\Delta t} + Q_1\right)$ is read against the known Q_1, and from this quantity $2Q_1$ is subtracted to get $\left(\frac{2S_1}{\Delta t} - Q_1\right)$ which is then added to (I_1+I_2) to provide an estimate of $\left(\frac{2S_2}{\Delta t} + Q_2\right)$ Again with this known quantity, the value of Q_2 is obtained from the routing curve. The value of Q_2 thus determined becomes Q_1, and $\left[\left(\frac{2S_2}{\Delta t} + Q_2\right) - 2Q_2\right]$ becomes $\left(\frac{2S_1}{\Delta t} - Q_1\right)$ for the next time interval. The procedure is repeated for all the subsequent routing periods till the complete outflow hydrograph is obtained.

The routing curve of this method has the discharge units on both the axes unlike the routing curve of I.S.D. method which has the units of discharge on one axis and units of storage on the other. But for the minor graphical interpolation errors. Modified Puls and I.S.D. methods should give identical results since both the methods make use of the same storage-discharge relationship of the reservoir.

The following example demonstrates the procedure of the Modified Puls method of routing a flood hydrograph through a reservoir.

Example 13.3. *Solve example 13.2 using Modified Puls method.*

Solution. As in the previous example a routing time interval of 0.25 day is adopted in this example also. The calculations needed to prepare the routing curve of Q *Vs* $\left(\frac{2S}{\Delta t}+Q\right)$ are shown in Table 13.6 and the routing curve is shown in Fig. 13.4.

Table 13.6. Calculations for the Routing Curve of Modified Puls Method

Q m^3/s	S *cumec-days*	$\frac{2S}{\Delta t} = 8 \times S$ m^3/s	$\left(\frac{2S}{\Delta t}+Q\right)$ m^3/s
0	0	0	0
49.2	51.62	412.96	462.16
69.6	105.67	845.36	914.96
85.2	163.19	1305.52	1390.72
98.4	225.12	1800.96	1899.36
110.0	292.36	2338.88	2448.88
247.8	365.63	2925.04	3172.84
490.2	445.72	3565.76	4055.96
800.6	533.45	4267.6	5068.2

(*i*) For the first time interval between 0 and 6 hours

$I_1 = 50 \text{ m}^3/\text{s}$ $\quad$ $I_2 = 180 \text{ m}^3/\text{s}$

$Q_1 = 200 \text{ m}^3/\text{s}$ $\quad$ $\Delta t = 0.25$ days.

For $Q_1 = 200 \text{ m}^3/\text{s}$, from Q *Vs* $\left(\frac{2S}{\Delta t}+Q\right)$ curve of Fig. 13.4, we obtain

$$\frac{2S_1}{\Delta t}+Q_1 = 2920 \text{ and} \left(\frac{2S_1}{\Delta t}-Q_1\right) = \left(\frac{2S_1}{\Delta t}+Q_1\right) - 2Q_1$$

$$= 2920 - 2 \times 200 = 2520 \text{ m}^3/\text{s}$$

$$I_1 + I_2 = 50 + 180 = 230 \text{ m}^3/\text{s}$$

$$\frac{2S_2}{\Delta t}+Q_2 = I_1 + I_2 + \left(\frac{2S_1}{\Delta t}-Q_1\right)$$

$$= 230 + 2520 = 2750 \text{ m}^3/\text{s}$$

For $\left(\frac{2S_2}{\Delta t}+Q_2\right) = 2750$; from routing curve of Fig. 13.4,

$$Q_2 = 167 \text{ m}^3/\text{s}.$$

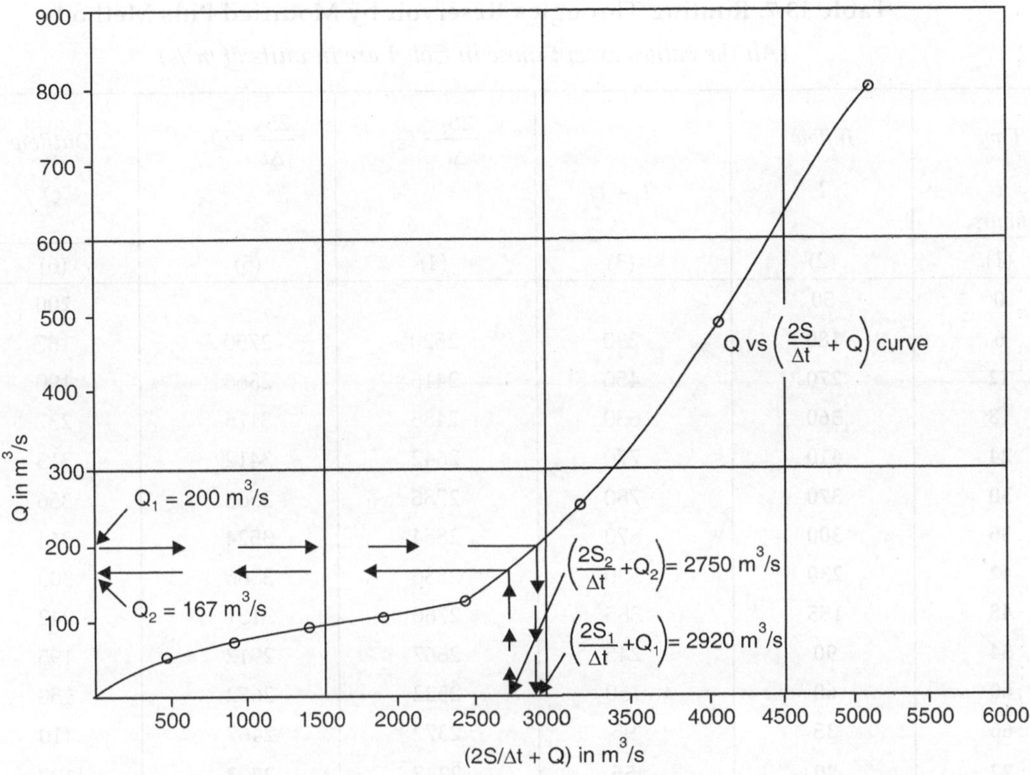

FIGURE 13.4 Routing curve for modified Puls method.

(*ii*) For the second time interval between 6 and 12 hours

$I_1 = 180$ m³/s $\qquad I_2 = 270$ m³/s

$Q_1 = 167$ m³/s $\qquad \Delta t = 0.25$ days.

From previous step $\frac{2S_1}{\Delta t} + Q_1 = 2750$;

$$\frac{2S_1}{\Delta t} - Q_1 = 2750 - 2 \times 167 = 2416 \text{ m}^3/\text{s}.$$

$$I_1 + I_2 = 180 + 270 = 450 \text{ m}^3/\text{s}$$

$$\frac{2S_2}{\Delta t} + Q_2 = 450 + 2416 = 2866 \text{ m}^3/\text{s}.$$

For $\left(\frac{2S_2}{\Delta t} + Q_2\right) = 2866$, from the routing curve of Fig. 13.4, we obtain that $Q_2 = 190$ m³/s. This becomes Q_1 for the third time interval and so on.

This procedure is repeated for the remaining intervals of the inflow hydrograph in a Tabular form as shown in Table 13.7.

Table 13.7. Routing Through a Reservoir by Modified Puls Method

(All the values except those in Col. 1 are in units of m³/s)

Time t hours	Inflow I	$I_1 + I_2$	$\frac{2S_1}{\Delta t} - Q_1$	$\frac{2S_2}{\Delta t} + Q_2$	Outflow Q
(1)	(2)	(3)	(4)	(5)	(6)
0	**50**				**200**
6	**180**	230	2520	2750	167
12	**270**	450	2416	2866	190
18	**360**	630	2486	3116	237
24	**410**	770	2642	3412	313
30	**370**	780	2786	3566	356
36	**300**	670	2854	3524	344
42	**230**	530	2836	3366	300
48	**155**	385	2766	3151	242
54	**90**	245	2667	2912	195
60	**60**	150	2522	2672	150
66	**35**	95	2372	2467	110
72	**20**	55	2247	2302	107

The figures in bold face are the given data.

Column (1) and (2) = ordinates of the given inflow hydrograph

Column (3) $= I_1 + I_2$

Column (4) $= \frac{2S}{\Delta t} - Q_1$ obtained as $\left[\left(\frac{2S_2}{\Delta t} + Q_2\right) - Q_2\right]$ of the previous row.

Column (5) $= \left(\frac{2S_2}{\Delta t} + Q_2\right) =$ Col. (3) + Col. (4)

Column (6) $= Q$ obtained from Q *Vs* $\left(\frac{2S}{\Delta t} + Q\right)$ curve of Fig. 13.4 against the value of $\left(\frac{2S_2}{\Delta t} + Q_2\right)$ in Col. (5).

The outflow hydrograph obtained in this example is almost identical to the one in the previous example. Hence the answers for attenuation, reservoir lag and the maximum water surface elevation remain the same.

Other Methods. Apart from the two semi-graphical methods described above, there are many other methods of routing through reservoirs. They include trial and error method, graphical method etc. For details of some of these methods the reader is advised to consult Handbook of Applied Hydrology by Ven Te Chow.

Reservoir Routing with Regulated Outflow. A reservoir with an uncontrolled outlet, which is exclusively meant for the purpose of flood control is called a *detention basin* or a *detention reservoir*. Such structures are usually small in size and very uncommon, at least in India. Most large reservoirs are generally provided with some type of outflow control in the form of gated spillways. The continuity equation for such reservoirs may be modified as

$$\frac{I_1 + I_2}{2}\Delta t - \frac{Q_1 + Q_2}{2}\Delta t - Q_R.\Delta t = S_2 - S_1 \qquad ...(13.11)$$

where Q is the uncontrolled outflow and Q_R is the regulated outflow. Equation (13.11) can be rearranged in the storage indication form as

$$(I_1 + I_2) - 2Q_R + \left(\frac{2S_1}{\Delta t} - Q_1\right) = \left(\frac{2S_2}{\Delta t} + Q_2\right) \qquad ...(13.12)$$

The solution of Eq. (13.12) is identical with Eq. (13.10) except for the inclusion of Q_R.

If the gates are set at fixed openings so that the discharge is a function of head, the solution requires a family of $\frac{2S}{\Delta t} + Q$ curves for various gate openings. The routing method is again the same as that of Eq. (13.10) except that we have to use the routing curve appropriate to the existing gate opening.

13.3 CHANNEL ROUTING

The channel routing, or *stream channel routing* as it is called often, uses mathematical relations to calculate outflow from a stream channel once inflow, lateral contributions and channel characteristics are known. The terms channel routing and flood routing are used interchangeably. This is because most of the channel routing applications are in flood flow analysis, flood control design and flood forecasting.

The specific length of a stream channel between the upstream section where the hydrograph is known and the downstream section where the hydrograph is to be determined is called a *channel reach*. The hydrograph at the upstream end of the reach is the inflow hydrograph and the hydrograph at the downstream end of the reach is the outflow hydrograph. Lateral contributions consist of tributary inflows joining the reach at different points, and/or the contributions due to interflow and groundwater flow which may be distributed over the part or full length of the reach. When there is no lateral inflow into the channel reach, the unsteady flow in the channel is described by the following two equations which are known as Saint-Venant equations.

$$\frac{\partial Q}{\partial x} + \frac{\partial A}{\partial t} = 0 \qquad ...(13.13)$$

$$\frac{\partial V}{\partial t} + V\frac{\partial V}{\partial x} + g\left(\frac{\partial y}{\partial x} - S_o + S_f\right) = 0 \qquad \text{...(13.14)}$$

where x = the distance measured along the flow direction from some reference point to the section under consideration

Q = the discharge at the section

A = area of the flow at the section

t = the time variable

V = the average velocity of flow for the section

y = the depth of flow at the section

g = acceleration due to gravity

S_o = channel bed slope, and

S_f = slope of the energy line.

Eq. (13.13) is the continuity equation which expresses the conservation of mass and Eq. (13.14) is the momentum equation which expresses the conservation of momentum. The solution of these two partial differential equations will provide the answer for the flood routing problem. Methods of flood routing based on these equations are known as the *hydraulic routing* methods. Though the hydraulic routing methods are theoretically more exact, in general they assume a uniform channel section and therefore they become cumbersome when applied to irregular natural channels. A discussion on hydraulic routing methods, such as method of characteristics etc., falls under the realm of books on Hydraulics and hence is not attempted here.

In contrast, the *hydrologic routing* method makes use of only the continuity equation in the form of Eq. (13.3). The Muskingum method is the most widely used hydrologic channel routing method which is now described.

Muskingum Method. As in the reservoir routing, in the channel routing also a second independent relationship is required to solve for Q_2 and S_2 in Eq. (13.3). This is provided by the Muskingum storage equation which assumes that the storage in the channel reach at any time is dependent both on inflow and outflow.

As shown in Fig. 13.5, the storage in the channel reach consists of two parts, namely, the *prism storage* and the *wedge storage*. The prism storage is formed by a volume of constant cross

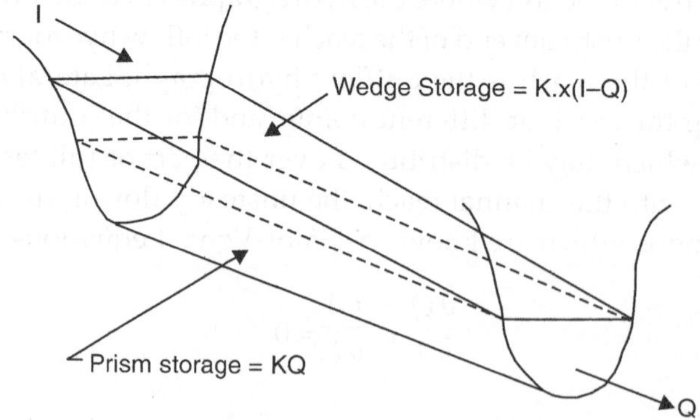

FIGURE 13.5 Prism and wedge storages in a channel reach.

section along the length of the prismatic channel. Assuming that the cross-sectional area of the flood flow is directly proportional to the discharge at the section, the volume of prism storage is equal to KQ, where K is a constant of proportionality, evidently possessing dimensions of time. The wedge storage may be taken as a fraction of the volume of the prism corresponding to $(I - Q)$. In the earlier stages when $I > Q$, the wedge storage is positive and when $I < Q$, during later stages a negative wedge is formed. The volume of wedge storage is then equal to $K.x.\ (I - Q)$, where x is a weighing factor having the range $0 \leq x \leq 0.5$.

The total storage is therefore equal to the sum of the two components,

$$S = KQ + Kx\ (I - Q) \qquad \text{...(13.15)}$$

which can be rearranged to give the storage function for the Muskingum method

$$S = K\ [xI + (1 - x)\ Q] \qquad \text{...(13.16)}$$

The value of x depends on the shape of the wedge. It ranges from 0 for reservoir type storage to 0.5 for a full wedge. When $x = 0$, there is no wedge and this represents the case of a level-pool reservoir. In natural streams x lies between 0 and 0.3 with a mean value of around 0.2. Great accuracy in determining x may not be necessary because the results of the method are relatively insensitive to the value of this parameter. The parameter K represents the *time of travel* of the flood wave through the channel reach. The value of K depends on the length of the reach and other roughness characteristics. The values of K and x of a reach are to be determined from a pair of observed inflow and outflow hydrographs. The methods of determining K and x are explained later. This storage equation was developed by G.I. Mc Carthy and others in connection with the studies on Muskingum River in U.S.A. and hence the name Muskingum method.

Using Eq. (13.16), the change in storage is now given by

$$S_2 - S_1 = K\ [x\ (I_2 - I_1) + (1 - x)(Q_2 - Q_1)] \qquad \text{...(13.17)}$$

Substituting Eq. (13.17) in Eq. (13.3) and rearranging the terms, we get

$$Q_2 = C_0\ I_2 + C_1\ I_1 + C_2\ Q_1 \qquad \text{...(13.18)}$$

where C_0, C_1 and C_2 are called the *routing coefficients* given by

$$C_0 = \frac{(\Delta t/K) - 2x}{2(1-x) + (\Delta t/K)} \qquad \text{...(13.19)}$$

$$C_1 = \frac{(\Delta t/K) + 2x}{2(1-x) + (\Delta t/K)} \qquad \text{...(13.20)}$$

$$C_2 = \frac{2(1-x) - (\Delta t/K)}{2(1-x) + (\Delta t/K)} \qquad \text{...(13.21)}$$

It may be easily verified that $C_0 + C_1 + C_2 = 1.0$, which fact can be used as an arithmetical check on the computed values of the routing coefficients. It should be noted that while computing C_0, C_1 and C_2 both K and Δt must be in the same units of time. Given an inflow hydrograph, the initial outflow, a chosen time interval Δt, and the routing parameters K and x, the routing coefficients can be calculated with Eq. (13.19) to (13.21) and the outflow hydrograph with Eq. (13.18).

Crest segment Routing. If we are not interested in routing the entire inflow hydrograph but we want to determine only the attenuation and lag time between the peaks, we can do so by routing only that portion of the inflow hydrograph which gives them. That is we directly route the portion of the inflow hydrograph starting from few time intervals prior to the peak to few time intervals after the peak.

Applying the Muskingum routing equation to $(n-1)$th time interval

We can write

$$Q_n = C_0 I_n + C_1 I_{n-1} + C_2 Q_{n-1} \quad ...(13.22)$$

We can express Q_{n-1} as

$$Q_{n-1} = C_0 I_{n-1} + C_1 I_{n-2} + C_2 Q_{n-2} \quad ....(13.23)$$

Substituting Eq. (13.23) in Eq.(13.22), we get

$$Q_n = C_0 I_n + (C_1 + C_2 C_0) I_{n-1} + C_1 C_2 I_{n-2} + C_2^2 Q_{n-2}$$

Let $(C_1 + C_2 C_0)$ be denoted by p. Then

$$Q_n = C_0 I_n + p I_{n-1} + C_1 C_2 I_{n-2} + C_2^2 Q_{n-2} \quad ...(13.24)$$

Again we can express Q_{n-2} as

$$Q_{n-2} = C_0 I_{n-2} + C_1 I_{n-3} + C_2 Q_{n-3} \quad ...(13.25)$$

Substituting Eq. (13.25) in Eq. (13.24), we get

$$Q_n = C_0 I_n + p I_{n-1} + C_1 C_2 I_{n-2} + C_2^2 (C_0 I_{n-2} + C_1 I_{n-3} + C_2 Q_{n-3})$$

$$= C_0 I_n + p I_{n-1} + I_{n-2} C_2 (C_1 + C_2 C_0) + C_2^2 C_1 I_{n-3} + C_2^3 Q_{n-3}$$

or

$$Q_n = C_0 I_n + p I_{n-1} + p C_2 I_{n-2} + C_2^2 C_1 I_{n-3} + C_2^3 Q_{n-3} \quad ...(13.26)$$

Similarly replacing Q_{n-3} by $(C_0 I_{n-3} + C_1 I_{n-4} + C_2 Q_{n-4})$ in Eq. (13.26) and simplifying

$$Q_n = C_0 I_n + p I_{n-1} + p C_2 I_{n-2} + p C_2^2 I_{n-3} + C_2^3 C_1 I_{n-4} + C_2^4 Q_{n-4}$$

If we continue by successively replacing Q_{n-j} by $(C_0 I_{n-j} + C_1 I_{n-j-1} + C_2 Q_{n-j-1})$ till, we get Q_1, we obtain

$$Q_n = C_0 I_n + \sum_{m=0}^{n-3} p C_2^m I_{n-1-m} + C_2^{n-2} C_1 I_1 + C_2^{n-1} Q_1 \quad ...(13.27)$$

Thus for example, for $n = 6$ we have

$$Q_6 = C_0 I_6 + \sum_{m=0}^{3} p C_2^m I_{5-m} + C_2^4 C_1 I_1 + C_2^5 Q_1$$

or

$$Q_6 = C_0 I_6 + p I_5 + p C_2^m I_4 + p C_2^2 I_3 + p C_2^3 I_2 + C_2^4 C_1 I_1 + C_2^5 Q_1$$

The routing constants C_0, C_1 and C_2 are all less than one and their higher powers will be very small. With little loss of accuracy we can stop the summation after 4 terms. Therefore we can write

$$Q_n = C_0 I_n + pI_{n-1} + pC_2 I_{n-2} + pC_2^2 I_{n-3} \qquad ...(13.28)$$

with $n \geq 5$. Thus we see that Eq. (13.28) uses the ordinates of the inflow hydrograph only and hence we can start routing from any time interval we choose and obtain the corresponding ordinates of the outflow hydrograph. Of course if we use more than 4 terms or all the terms the loss of accuracy will be less.

Example 13.4. *Using the crest segment routing method determine the attenuation in peak flow and lag time between the peaks of a hydrograph when it passes through a stream reach whose Muskingum parameters are given as K = 9 h and x = 0.125. The ordinates of the hydrograph in m³/s at 4 h interval are 50, 82, 140, 197, 233, 250, 230, 204, 180, 154, 127, 106, 89, 74, 65*

Solution. $\Delta t = 4$ h; $K = 9$ h; $x = 0.125$

$$\frac{\Delta t}{K} = \frac{4}{9} = 0.4444;\ 2(1-x) + \frac{\Delta t}{K} = 2(1 - 0.125) + 0.4444 = 2.1944$$

$$C_0 = \frac{\frac{\Delta t}{K} - 2x}{2(1-x) + \frac{\Delta t}{K}} = \frac{0.4444 - 0.25}{2.1944} = 0.0886$$

$$C_1 = \frac{\frac{\Delta t}{K} + 2x}{2(1-x) + \frac{\Delta t}{K}} = \frac{0.4444 + 0.25}{2.1944} = 0.3164$$

$$C_2 = \frac{2(1-x) - \Delta t/K}{2(1-x) + \Delta t/K} = \frac{2(1-0.125) - 0.4444}{2.1944} = \frac{1.3056}{2.1944}$$

$$= 0.595$$

Check: $C_0 + C_1 + C_2 = 0.0886 + 0.3164 + 0.595 = 1.$

$$Q_n = C_0 I_n + p I_{n-1} + pC_2 I_{n-2} + pC_2^2 I_{n-3} \qquad ...(13.28)$$

$$C_0 = 0.0886$$

$$p = C_1 + C_2 C_0 = 0.3164 + 0.595 \times 0.0886 = 0.3691$$

$$pC_2 = 0.3691 \times 0.595 = 0.2196$$

$$pC_2^2 = pC_2 \times C_2 = 0.2196 \times 0.595 = 0.1307$$

Now

$$Q_5 = C_0 I_5 + p I_4 + pC_2 I_3 + pC_2^2 I_2$$

$$= 0.0886 I_5 + 0.3691 I_4 + 0.2196 I_3 + 0.1307 I_2$$

$$= 0.0886 \times 233 + 0.3691 \times 197 + 0.2196 \times 140 + 0.1307 \times 82$$

$$= 134.82 \text{ m}^3/\text{s}$$

$$Q_6 = C_0 I_6 + pI_5 + pC_2 I_4 + pC_2^2 I_3$$
$$= 0.0886 \times 250 + 0.3691 \times 233 + 0.2196 \times 197 + 0.1307 \times 140$$
$$= 169.71 \text{ m}^3/\text{s}$$
$$Q_7 = C_0 I_7 + pI_6 + pC_2 I_5 + pC_2^2 I_4$$
$$= 0.0886 \times 230 + 0.3691 \times 250 + 0.2196 \times 233 + 0.1307 \times 197$$
$$= 189.57 \text{ m}^3/\text{s}$$
$$Q_8 = C_0 I_8 + pI_7 + pC_2 I_6 + pC_2^2 I_5$$
$$= 0.0886 \times 204 + 0.3691 \times 230 + 0.2196 \times 250 + 0.1307 \times 233$$
$$= 188.32 \text{ m}^3/\text{s}$$
$$Q_9 = C_0 I_9 + pI_8 + pC_2 I_7 + pC_2^2 I_6$$
$$= 0.0886 \times 180 + 0.3691 \times 204 + 0.2196 \times 230 + 0.1307 \times 250$$
$$= 174.43 \text{ m}^3/\text{s}$$

Similarly, we get

$$Q_{10} = 0.0886 \times 154 + 0.3691 \times 180 + 0.2196 \times 204 + 0.1307 \times 230$$
$$= 155 \text{ m}^3/\text{s}$$

We can stop the crest segment routing since the peak in the outflow hydrograph is already obtained. For the time intervals of interest, the inflow and outflow hydrographs are sketched in Fig. 13.6.

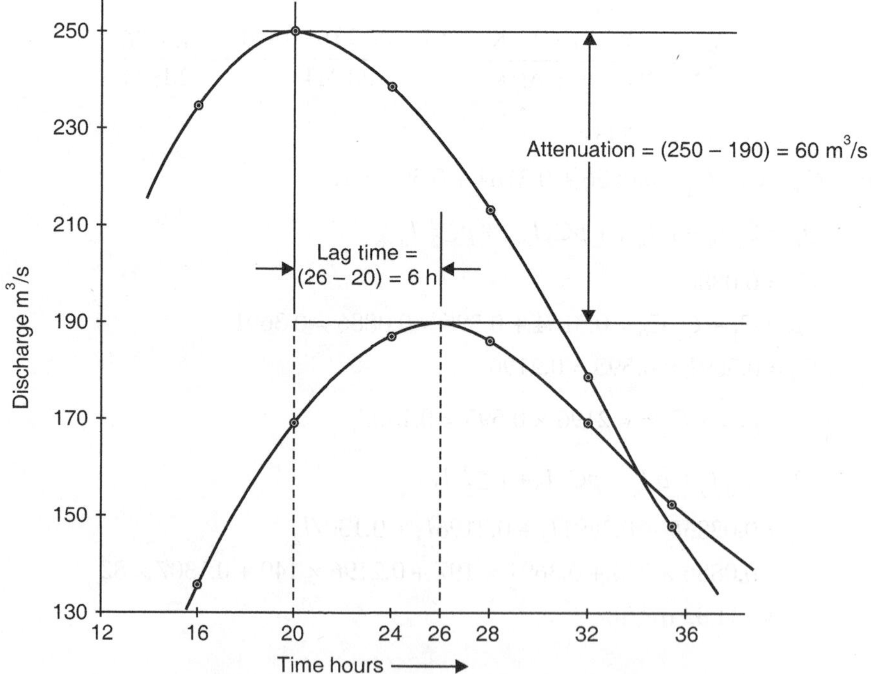

FIGURE 13.6 Crest segment routing of example 13.6.

Peak in the outflow hydrograph occurs between Q_7 and Q_8. That is at the middle of 7th time interval or at 24 + 2 = 26 h.

Peak in the inflow hydrograph occurs as I_6 or at 20 h.

∴ Lag time = 26 – 20 = 6 h

Similarly the peakflow in inflow hydrograph = 250 m³/s and the peakflow in outflow hydrograph = 190 m³/s

∴ Attenuation = 250 – 190 = 60 m³/s.

Selection of Routing Time Interval. The continuity equation in the form of Eq. (13.3) assumes that the average rate of flow during the interval Δt is given by $\left(\frac{I_1 + I_2}{2}\right)\Delta t$, which implies that the hydrograph is a straight line, or can be approximated as a straight line, in the interval Δt.

Therefore the controlling factor in selecting the routing period is that it must be sufficiently short to ensure that this assumption is not seriously violated. The time of travel for a reach may be taken as the time lag between the peak of inflow hydrograph and peak of the corresponding outflow hydrograph. The time of travel can be obtained from a pair of observed inflow and outflow hydrographs. The routing period should never be greater than the time of travel through the reach. Because if it is so, it would be possible for the flood peak to pass completely through the reach during a routing period. If the routing period is shorter than is really necessary the work is increased. Generally a routing period between $\frac{1}{2}$ and $\frac{1}{3}$ of the time of travel will work well.

Example 13.5. *Route the flood hydrograph given below through a channel reach and derive the outflow hydrograph. The values of K and x for the reach may be taken as 12 h and 0.278 respectively.*

Time-h	0	4	8	12	16	20	24	28	32
Flow-m³/s	42	68	116	164	194	200	192	170	150
Time-h	36	40	44	48	52	56			
Flow m³/s	128	106	88	74	62	54			

Solution. The routing coefficients are computed as given below.

$$K = 12 \text{ h}, \Delta t = 4 \text{ h}, x = 0.278$$

$$\frac{\Delta t}{K} = \frac{4}{12} = \frac{1}{3}.$$

$$C_0 = \frac{\left(\frac{\Delta t}{K}\right) - 2x}{2(1-x) + (\Delta t/K)}$$

$$= \frac{\frac{1}{3} - 2 \times 0.278}{2(1-0.278) + \frac{1}{3}} = \frac{0.3333 - 0.556}{1.444 + 0.3333} = -0.125$$

$$C_1 = \frac{\left(\frac{\Delta t}{K}\right) + 2x}{2(1-x) + (\Delta t/K)} = \frac{0.3333 + 0.556}{1.444 + 0.3333} = 0.50$$

$$C_2 = \frac{2(1-x) - (\Delta t/K)}{2(1-x) + (\Delta t/K)} = \frac{1.444 - 0.3333}{1.444 + 0.3333} = 0.625$$

Check. $C_0 + C_1 + C_2 = -0.125 + 0.5 + 0.625 = 1.0$

Now the routing equation is

$$Q_2 = C_0 I_2 + C_1 I_1 + C_2 Q_1$$

or

$$Q_2 = -0.125\, I_2 + 0.5\, I_1 + 0.625\, Q_1$$

To start with for the first time interval between 0 and 4 hours, the initial outflow Q_1 is assumed to be same as I_1. Thus we have

$$I_1 = 42 \text{ m}^3/\text{s}, I_2 = 68 \text{ m}^3/\text{s}, Q_1 = 42 \text{ m}^3/\text{s}$$

$\therefore$ $$Q_2 = -0.125 \times 68 + 0.5 \times 42 + 0.625 \times 42 = 38.75 \text{ m}^3/\text{s}.$$

This becomes Q_1 for the next time interval.

For the second time interval, we have

$$I_1 = 68 \text{ m}^3/\text{s}, I_2 = 116 \text{ m}^3/\text{s}, Q_1 = 38.75 \text{ m}^3/\text{s}$$

$\therefore$ $$Q_2 = -0.125 \times 116 + 0.5 \times 68 + 0.625 \times 38.75 = 43.72 \text{ m}^3/\text{s}.$$

Subsequent ordinates of the outflow hydrograph are obtained successively following the same procedure. All the computations are shown in Table 13.8.

Table 13.8. Channel Routing by Muskingum Method

(All the values except those in Col. 1 are in m³/s)

Time h	*Inflow* I	$-0.125\, I_2$	$0.5\, I_1$	$0.625\, Q_1$	*Outflow Q*
0	42				42*
4	68	– 8.5	21	26.25	38.75
8	116	– 14.5	34	24.22	43.72
12	164	– 20.5	58	27.33	64.83
16	194	– 24.25	82	40.52	98.27
20	200	– 25.0	97	62.42	133.42
24	192	– 24.0	100	83.39	159.39
28	170	– 21.25	96	99.62	174.37
32	150	– 18.75	85	108.98	175.23
36	128	– 16.0	75	109.52	168.52
40	106	– 13.25	64	105.33	156.08
44	88	– 11.00	53	97.55	139.55
48	74	– 9.25	44	87.22	121.97
52	62	– 7.75	37	76.23	105.48
56	54	– 6.75	31	65.93	90.18

* Assumed value.

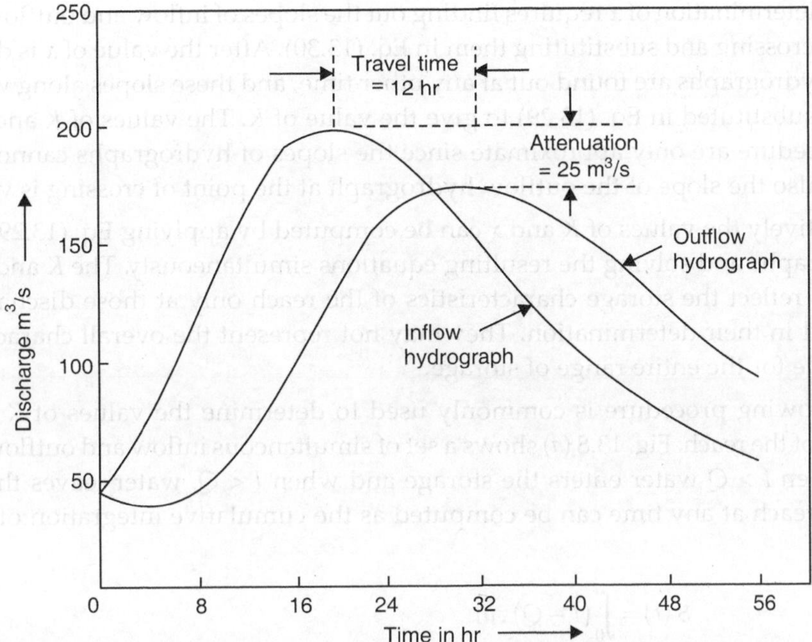

FIGURE 13.7 Inflow and outflow hydrographs of example 13.4

The given inflow hydrograph and the derived outflow hydrograph are plotted in Fig. 13.6. Also indicated on Fig. 13.7 are the time of travel and the attenuation.

Unlike in reservoir routing, the peak flow of the outflow hydrograph in channel routing does not occur at the time where inflow and outflow hydrographs intersect.

Determination of Muskingum Parameters K and x. When a pair of observed inflow and outflow hydrographs is available for a reach, the values of K and x of the reach can be determined using these graphs. When these two graphs are plotted it will be seen that both will cross on the recession side of the inflow hydrograph.

The routing equation is

$$I - Q = \frac{dS}{dt}$$

$$S = K\,[xI + (1 - x)Q]$$

$$\therefore \quad \frac{dS}{dt} = K\left[x.\frac{dI}{dt} + (1 - x)\frac{dQ}{dt}\right]$$

or

$$I - Q = K\left[x.\frac{dI}{dt} + (1 - x)\frac{dQ}{dt}\right] \quad \text{...(13.29)}$$

When the two hydrographs are crossing $I = Q$ and $I - Q = 0$.

Making use of this condition in Eq. (13.29), we get

$$x = \frac{\frac{dQ}{dt}}{\frac{dQ}{dt} - \frac{dI}{dt}} \quad \text{...(13.30)}$$

Hence determination of x requires finding out the slopes of inflow and outflow hydrographs at the point of crossing and substituting them in Eq. (13.30). After the value of x is determined, the slopes of the hydrographs are found out at any other time, and these slopes along with the known value of x are substituted in Eq. (13.29) to give the value of K. The values of K and x determined using this procedure are only approximate since the slopes of hydrographs cannot be found out precisely and also the slope of the outflow hydrograph at the point of crossing is very small.

Alternatively the values of K and x can be computed by applying Eq. (13.29) at two points on the hydrographs and solving the resulting equations simultaneously. The K and x values from this procedure reflect the storage characteristics of the reach only at those discharge conditions which are used in their determination. They may not represent the overall characteristics of the reach applicable for the entire range of storage.

The following procedure is commonly used to determine the values of K and x that are representative of the reach. Fig. 13.8 (*a*) shows a set of simultaneous inflow and outflow hydrographs of a reach. When $I > Q$ water enters the storage and when $I < Q$, water leaves the storage. The storage in the reach at any time can be computed as the cumulative integration of the difference graph. That is

$$S(t) = \int_0^t (I - Q) \, . \, dt \quad \text{...(13.31)}$$

The storage is then plotted against time as shown in Fig. 13.8 (*b*).

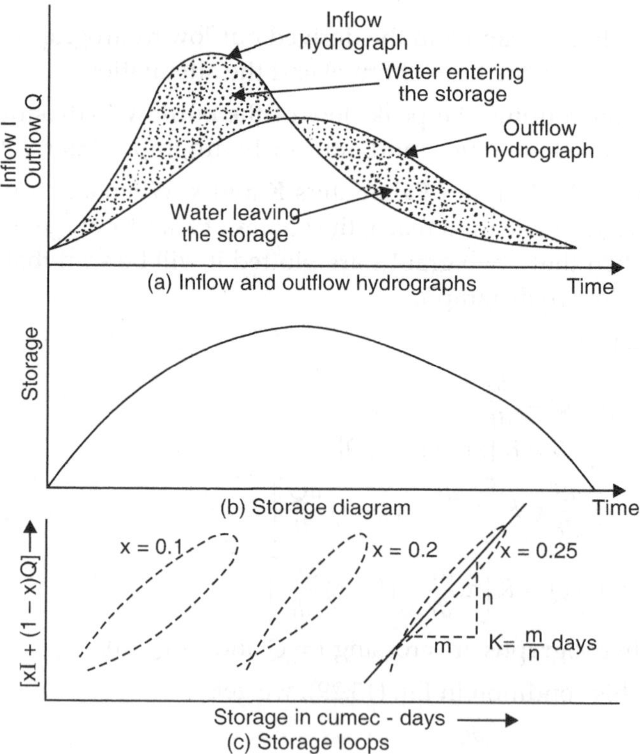

FIGURE 13.8 Determination of muskingum parameters K and x

Now assuming a trial value of x the value of the expression $[xI + (1 - x) Q]$ is evaluated at various points of time and these values are plotted against the corresponding storage read from Fig. 13.8 (*b*). Such a plot, as shown in Fig. 13.8 (*c*) usually results in a loop. Similar plots are made with different values of x till the plot is very nearly a straight line the slope of which gives an estimate of K and the best values of x is taken to be that which causes the plot to be a straight line. In this analysis if storage is expressed in cumec-days and discharge in m^3/s, K is obtained in days.

Example 13.6. *The inflow and outflow hydrograph ordinates of a particular channel reach are given in m^3/s at 6 h intervals in Cols, (2) and (3) of Table 13.9. Determine the Muskingum parameters K and x for this reach.*

Solution. The necessary computations to find storage S and the quantity $[xI + (1 - x) Q]$ at different times are also shown in Table 13.9 along with the given inflow and outflow hydrograph.

Table 13.9. Determination of Muskingum Parameters of Example 13.5

Time	*I*	*Q*	*(I – Q)*	$(I - Q)_{ave}$	ΔS	$S = \Sigma\Delta S$	$[xI + (1 - x) Q]$ *in* m^3/s *with*		
h	m^3/s	m^3/s	m^3/s	m^3/s	*cumec-days*	*cumec-days*	*x= 0.3*	*x = 0.2*	*x = 0.1*
(1)	(2)	(3)	(4)	(5)	(6)	(7)	(8)	(9)	(10)
0	35	35	0	—	—	—	—	—	—
6	135	57	78	39.0	9.75	9.75	80.40	72.60	64.80
12	440	200	240	159.0	39.75	49.50	272.00	248.00	224.00
18	670	455	215	227.5	56.88	106.38	519.50	498.00	476.50
24	685	612	73	144.0	36.00	142.38	633.90	626.60	619.30
30	535	617	– 82	– 4.5	– 1.13	141.25	592.40	600.60	608.80
36	385	510	– 125	– 103.5	– 25.88	115.37	472.50	485.00	497.50
42	235	375	– 140	– 132.5	– 33.13	82.24	333.00	347.00	361.00
48	140	246	– 106	– 123.0	– 30.75	51.49	214.20	224.80	235.40
54	95	158	– 63	– 84.5	– 21.13	30.36	139.10	145.40	151.70
60	60	102	– 42	– 52.5	– 13.13	17.23	89.40	93.60	97.80
66	42	68	– 26	– 34.0	– 8.50	8.73	60.20	62.80	65.40
72	35	56	– 21	– 23.5	– 5.88	2.85	49.70	51.80	53.90

Column (1), (2) and (3) = given inflow and outflow hydrographs

Column (4) = $(I - Q)$ = net inflow into the reach = Col. (2) – Col. (3)

Column (5) = $(I - Q)_{ave}$ = average net inflow which is taken as the average of the net inflows at the beginning and end of the time interval. For example, between 0 and 6 hours $(I - Q)_{ave}$

$$= \frac{0 + 78}{2} = 39 \text{ m}^3/\text{s}.$$

Column (6) = ΔS = incremental storage volume in the reach = $(I - Q)_{ave} \times \Delta t$.

$$\text{Since } \Delta t = 0.25 \text{ days, } \Delta S = \frac{(I - Q)_{ave}}{4} \text{ cumec-days.}$$

Column (7) = Storage in the reach = $\Sigma \Delta S$

For Example storage up to 18 h is the sum of all the incremental storages up to that time = 9.75 + 39.75 + 56.88 = 106.38 cumec-days.

Column (8) = $[xI + (1 - x)\,Q]$ with $x = 0.30$

For example when $t = 6$ h, $I = 135\ \text{m}^3/\text{s}$; $Q = 57\ \text{m}^3/\text{s}$.

$\therefore \quad [xI + (1 - x)Q] = 0.3 \times 135 + 0.7 \times 57 = 80.4\ \text{m}^3/\text{s}$.

Now a graph is plotted with $[xI + (1 - x)Q]$ of Col. (8) on x-axis and the storage S of Col. (7) on y-axis as shown in Fig. 13.9 (*a*). The points indicate the formation of a loop and so in the next trial $[xI + (1 - x)\,Q]$ is computed with $x = 0.20$. These values are given in Col. (9). The graph between values of Col. (7) and Col. (9) is shown in Fig. 13.9 (*b*). There is improvement but still the loop persists though it has become narrower. In the third trial $[xI + (1 - x)Q]$ *is* computed with $x = 0.1$, and entered in Col. (10). The plot between values of Col. (7) and Col. (10), as shown in Fig. 13.9 (*c*), is almost a straight line.

Therefore the correct value of x may be taken as 0.1 for this reach. The slope of the line in Fig. 13.9 (*c*) gives the value of K, which is obtained as

$$K = \frac{50\ \text{cumec-days}}{200\ \text{m}^3/\text{s}} = \frac{1}{4}\ \text{days}$$

$$= 6\ \text{h}.$$

This tallies with the peak to peak time of 6 h in the inflow and outflow hydrographs. Therefore, for this reach the value of $x = 0.1$ and $K = 6$ h may be adopted.

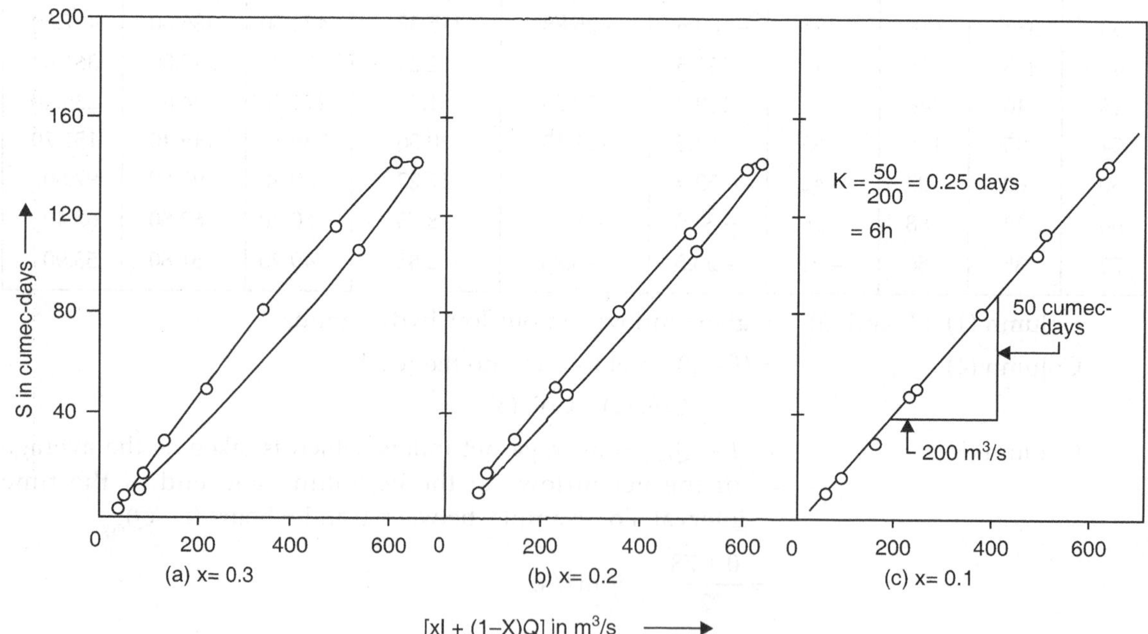

FIGURE 13.9 Determination of Muskingum parameters of example 13.6.

Other Methods of Channel Routing. There are other channel routing methods which include the electronic routing machines, the graphical methods etc. They are not discussed here. Interested readers can get the details of these methods in Handbook of Applied Hydrology by Ven Te Chow.

REVIEW EXERCISES

Descriptive Questions

1. Define flood routing. What are the uses of flood routing ?
2. What is the differential equation of storage ? What principle is it based on ?
3. Explain how the reservoir characteristics curves are prepared ?
4. Describe the various steps involved in the I.S.D. method of reservoir routing.
5. Describe the Modified Puls method of reservoir routing.
6. What is a routing curve ? What is the storage indications quantity ? What is the appropriate time interval to be adopted in reservoir routing ?
7. In reservoir routing with unregulated outflow, the peak outflow occurs when the inflow and outflow coincide. Explain why ?
8. Differentiate between
   (*a*) Hydraulic routing and hydrologic routing
   (*b*) Channel routing and reservoir routing
   (*c*) Prism storage and wedge storage.
9. Derive the Muskingum routing equation and the expressions for the routing coefficients C_0, C_1 and C_2.
10. Explain the method of determining the Muskingum parameters K and x of a reach from a pair of observed inflow and outflow hydrographs.
11. In the Muskingum storage equation what do the parameter K and x represent ?
12. Write down the two basic differential equations of hydraulic routing method and explain each term.
13. Explain the method of finding K and x parameter by drawing the tangents to the hydrograph at the point of crossing.
14. Describe at least one other method of channel routing not discussed in this book.
15. Discuss how the routing time interval Δt is chosen for Muskingum routing through a channel reach.

Analytical Questions

1. The inflow hydrograph into a linear reservoir is triangular in shape with time base of 20 h and a peak flow of 240 m^3/s occurring at 8 h. Assume that the storage constant of the reservoir is 2 h and the outflow from the reservoir at the time of arrival of the inflow is zero. Route the inflow hydrograph with $\Delta t = 2$ h and find the peak outflow.

2. The storage discharge characteristics of a reservoir with an uncontrolled spillway are as given below.

Storage above the spillway crest cumec-hours	0	278	556	833	1110	1389	1667
Outflow from the spillway m³/s	0	17	48	88	136	190	250

A flood with ordinates of 20, 50, 100, 130, 150, 140, 110, 90, 70, 50, 30, and 20 m^3/s at 1 hour interval arrives when the outflow from the reservoir is 10 m^3/s. Route the flood using I.S.D. method and find the peak outflow.

3. The storage and outflow characteristics of a reservoir provided with an uncontrolled spillway with its crest at 490 m are listed below.

Elevation *m*	*Storage* *Mm³*	*Discharge* *m³/s*
490	0	0
490.6	0.96	6.27
491.2	2.10	17.75
491.8	3.52	32.60
492.4	5.10	50.19
493.0	6.80	70.15
493.6	8.60	92.21
494.2	10.50	116.20
494.8	12.62	141.97
495.4	14.88	169.40

The following flood arrives into the reservoir when the outflow is zero.

Time h	0	6	12	18	24	30	36	42	48	54	60
Flow m³/s	0	255	184	136	99	74	54	40	28	17	8

Route the hydrograph using Modified Puls method and find the attenuation, reservoir lag and the maximum elevation reached by the water surface.

4. Determine the actual free board for the reservoir with the following data.

Dam crest elevation = 370 m

Spillway crest elevation = 365 m

Spillway rating is $Q = 1.7\, L\, H^{3/2}$

Effective length of spillway = 18 m.

Elevation-Storage Data:

Elevation m	365	366	367	368	369	370
Storage Mm³	3.00	3.05	3.15	3.35	3.75	4.25

Design flood hydrograph;

Time h	0	1	2	3	4	5	6	7	8	9	10	11
Flow m³/s	0	10	15	30	55	85	105	125	150	135	110	95

Time h	12	13	14	15	16	17	18	19	20	21
Flow m³/s	72	55	38	29	14	9	7	2	1	0

Assume the initial reservoir pool level at spillway crest.

5. Solve Exercise 18 assuming the initial outflow from the reservoir as 50 m^3/s.
6. A small reservoir has a plan area of 400 hectares at spillway crest level and the banks are essentially vertical for several meters above the crest level. The spillway has a length of 15 m with a co-efficient of 3.0. Taking the inflow hydrograph of Exercise 15 as the inflow hydrograph into the reservoir compute the maximum discharge from the reservoir and the maximum water level to be expected if the water surface in the reservoir is at spillway crest level initially.
7. Show that the Muskingum routing equation can be written in the form

$$Q_2 = Q_1 + C_1 (I_1 - Q_1) + C_2 (I_2 - I_1)$$

where $C_1 = \dfrac{\Delta t}{K(1-x) + 0.5\,\Delta t}$

$$C_2 = \frac{0.5\,\Delta t - Kx}{K(1-x) + 0.5\,\Delta t}.$$

8. Derive the Muskingum routing equation assuming that the storage in the reach is given by

$$S = aI + bQ.$$

9. The observed inflow and outflow hydrographs of a reach are as given under.

Date	*Time h*	*Inflow m³/s*	*Outflow m³/s*
27-10-84	16	16.8	16.8
	24	56.8	23.7
28-10-84	8	154.5	59.8
	16	261.4	130.5
	24	328.2	214.5
29-10-84	8	336.4	283.6
	16	305.9	307.3
	24	207.3	300.0
30-10-84	8	145.5	250.1
	16	111.4	179.1
	24	87.3	139.5
31-10-84	8	65.5	106.8
	16	53.6	81.8
	24	43.2	64.5

(*a*) Determine the value of x using Eq. (13.30) and hence the value of K.

(*b*) Evaluate the constants K and x using the graphical method of Fig. 13.8.

(*c*) Compare the values obtained in (*b*) with those in (*a*).

10. Using the inflow hydrograph of the above problem as the inflow to a reach with $K = 24$ h and $x = 0.20$, find the peak flow and time of peak in the outflow hydrograph.

11. Assuming that a storm over the catchment shown in the figure below produces simultaneously at A and B the runoff hydrograph listed as the inflow hydrograph of Exercise 24, determine the combined maximum runoff at C. The values of K and x for the reach AC may be taken as 24 hours and 0.20 respectively. Assume no lateral flow into the stream between A and C.

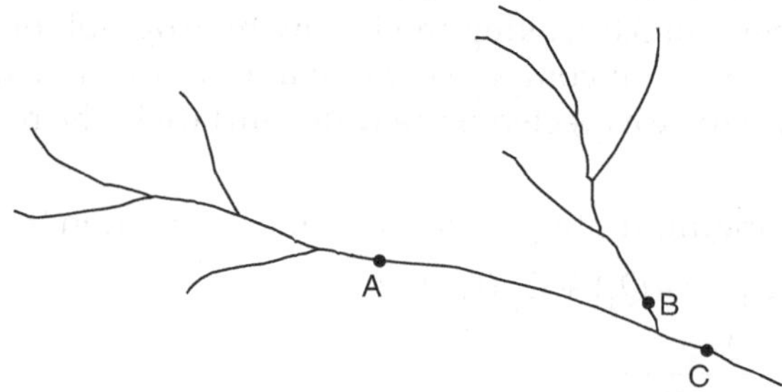

12. Compute the Muskingum routing coefficient with $K = \Delta t$ and $x = 0.5$ and comment on the results.

13. Given the following inflow hydrograph to a certain stream channel reach calculate the outflow by the Muskingum method.

Time h	0	1	2	3	4	5	6	7	8	9	10	11	12
Inflow m³/s	10	20	40	80	120	150	120	60	50	40	30	20	10

Assume initial outflow as 10 m³/s and $K = 4$ h and $x = 0.25$.

14. Show that the Muskingum parameters x and K can be expressed in terms of the routing constants as

$$x = \frac{C_1 - C_0}{2(1 - C_0)} \text{ and } K = \frac{\Delta t\,(1 - C_0)}{C_1 + C_2}$$

15. Show that the nth ordinate Q_n in the outflow hydrograph can be directly computed, in the case of Muskingum routing, from the equation

$$Q_n = C_0 I_n + \sum_{m=0}^{n-3} p C_2^m I_{n-1-m} + C_2^{n-2} C_1 I_1 + C_2^{n-1} Q_1$$

where C_0, C_1 and C_2 are the usual Muskingum constants and $p = (C_1 + C_2\,C_0)$

16. Compute the Muskingum routing constants C_0, C_1 and C_2 assuming that the channel storage is given by $S = KQ$. Can you justify this assumption.

17. The following are the ordinates in m^3/s of an inflow hydrograph into a stream reach at 24 h intervals.

120, 180, 260, 414, 480, 560, 604, 580, 460, 404, 320, 290, 260, 240, 224, 210.

Route the flood through the reach if $K = 2$ days and $x = 0.2$ and find out attenuation in peak and the time lag in peak flows.

18. For the hydrograph of analytical question 24 apply crest segment routing and obtain the attenuation in peak and the lag time between peaks. Assume $K = 24$ h and $x = 0.20$.

19. The inflow hydrograph into a stream reach and the corresponding outflow hydrograph are as given below. Determine the values of K and x for the reach.

Time h	*Inflow I m^3/s*	*Outflow Q m^3/s*
6	28	28
12	113	37
18	270	42
24	388	88
30	352	170
36	288	224
42	232	249
48	184	246
54	155	232
60	133	212
66	116	191
72	102	174
78	88	164
84	76	144
90	68	127
96	59	110

20. In the analytical question 20, now that you know the values of K and x, route the inflow hydrograph and obtain the outflow hydrograph. Compare the outflow hydrograph obtained from routing and the actual outflow hydrograph by plotting them on the same graph.

Objective Type Questions

1. The hydrologic routing methods are based on
   (*a*) continuity equation only
   (*b*) continuity equation and energy equation
   (*c*) continuity, momentum and energy equations
   (*d*) momentum equation only.

2. The peak of the outflow hydrograph coincides with the intersection of inflow and outflow hydrographs in the case of
   (*a*) reservoir routing with controlled outflow
   (*b*) reservoir routing with uncontrolled outflow
   (*c*) channel routing
   (*d*) any flood routing.
3. Which of the following is a storage indication quantity ?
   (*a*) $\left(\frac{I_1 + I_2}{2}\right) \Delta t$ (*b*) $\left(\frac{I_1 - Q_1}{2}\right) \Delta t$
   (*c*) $I_1 + I_2 + \left(\frac{2S_1}{\Delta t} - Q_1\right)$ (*d*) $\left(\frac{I_1 + I_2}{2}\right) \Delta t + (2\, S_1 - Q_1 \Delta t)$.
4. Modified Puls method of reservoir routing is also known as
   (*a*) Inflow storage discharge method (*b*) Muskingum method
   (*c*) Storage indication method (*d*) none of the above.
5. Hydraulic routing methods make use of
   (*a*) energy equation only (*b*) continuity equation only
   (*c*) momentum equation only (*d*) both continuity and momentum equations.
6. The Saint Venant equations represent
   (*a*) continuity and momentum equations (*b*) continuity equation in two different forms
   (*c*) momentum equation in two different forms
   (*d*) momentum and energy equations.
7. Which of the following is the Muskingum storage equation ?
   (*a*) $S = K\,[xQ + (1 - x)I\,]$ (*b*) $S = K\,[xQ - (1 - x)I\,]$
   (*c*) $S = K\,[xI + (1 - x)Q\,]$ (*d*) $S = K\,[xI - (1 - x)Q\,]$.
8. In the channel routing by the Muskingum method, the value of the routing coefficients C_0 and C_1 are estimated as – 0.2 and 0.5 respectively. The value of the third coefficient C_2 would be
   (*a*) 0.2 (*b*) – 0.5
   (*c*) 0.3 (*d*) 0.7.
9. The condition satisfied by the three routing coefficients of the Muskingum method is
   (*a*) $C_0 + C_1 + C_2 = 0$ (*b*) $C_0 + C_1 + C_2 = 1$
   (*c*) $C_0 C_1 C_2 = 1$ (*d*) $\frac{C_0}{C_1} = \frac{C_1}{C_2}$.
10. For natural stream channels the value of the Muskingum parameters x will generally be
   (*a*) between 0 and 0.3 (*b*) between 0.3 and 0.6
   (*c*) more than 0.5 (*d*) more than 1.0.

ANSWERS

Analytical Questions

1. The ordinates of the outflow hydrograph in m^3/s at 2h intervals are :
0, 20, 66.67, 122.22, 180.74, 206.92, 188.97, 156.32, 118.78, 79.59, 39.86, 13.29, 4.43, 1.48, 0.49 ; Peak outflow = 206.92 m^3/s.
2. The ordinates of the outflow hydrograph in m^3/s at 1 h interval are :
10, 11.5, 15.2, 24.4, 36.6, 48, 58.4, 64, 66.2, 65.3, 62, 56.7 etc. Peak outflow = 66.2 m^3/s.
3. The ordinates of the outflow hydrograph in m^3/s at 6 h intervals are :
0, 21.85, 64.52, 86.71, 93.85, 93.20, 85.6, 76.57, 66.7, 56.5, 46.9, etc.
Peak flow = 93.85 m^3/s ; Attenuation = 161.15 m^3/s ;
Reservoir lag = 18 h ; Max. elevation = 493.64 m.
4. The ordinates of the outflow hydrograph in m^3/s at 1 h interval are :
0, 5.24, 12.85, 22.97, 42.90, 70, 93.26, 110.43, 131.8, 140.24, 126.24, 107.5, 88.5, etc.
Max. water surface elevation = 367.74 m
Actual free board = 2.26 m.
5. The ordinates of the outflow hydrograph in m^3/s at 6 h interval are :
50, 67.45, 102.77, 116.89, 117, 109.3, 98.58, 86.67, 74.35, 62.4, 51.23 etc.
Peak flow = 117 m^3/s ; Attenuation = 138 m^3/s
Reservoir lag = 18 h ; Max. elevation = 494.22 m.
6. The ordinates of the outflow hydrograph in m^3/s at 6 hour interval are :
0, 22.7, 84.5, 114, 115, 104, 89, 73.5, 60, 48, 37 etc.
Peak flow =118 m^3/s ; Attenuation = 137 m^3/s
Reservoir lag = 15 h ; Max elevation = 1.9 m above spillway crest.

8. $Q_2 = C_0I_2 + C_1I_1 + C_2Q_1$

where $C_0 = \dfrac{0.5\,\Delta t - a}{0.5\,\Delta t + b}$

$C_1 = \dfrac{0.5\,\Delta t + a}{0.5\,\Delta t + b}$

$C_2 = \dfrac{b - 0.5\,\Delta t}{0.5\,\Delta t + b}$

9. (*a*) $x = 0.20$; $K = 16.5$ h (with slopes determined at $t = 24$ h of 29-10-84)
(*b*) $x = 0.20$; $K = 12$ h.
10. The ordinates of the routed hydrograph at 8 h intervals are :
16.8, 15.42, 26.31, 66.82, 131.61, 199.11, 247.51, 271.05, 251.2, 215.94, 180.72, 149.26, 120.79, 97.98 m^3/s, etc.
Peak flow of 271.05 m^3/s occurs at 24 h on 29-10-84.

11. 553.41 m^3/s occurs at 16 h on 29-10-84.
12. $C_0 = C_2 = 0$ and $C_1 = 1$, Therefore $Q_2 = I_1$. The inflow hydrograph is bodily shifted by one time period.
13. The ordinates of the outflow hydrograph in m^3/s at one hour interval are :
    10, 8.57, 8.98, 12.12, 25.81, 48.44, 81.75, 84.11, 78.65, 71.9,
    64.21, 55.86, 47.04, etc.
16. $C_0 = C_1 = \Delta t/(2K + \Delta t)$; $C_2 = \dfrac{2K - \Delta t}{2K + \Delta t}$
17. The ordinates of the outflow hydrograph at 24 h interval are :
    120, 122.9, 153.9, 211.7, 311.2, 395.4, 475.9, 535.8, 551.1,
    505.1, 452.9, 388.2, 340, 300.9, 271.2, 248.1
    Attenuation = 52.9 m^3/s ; Lag time = 2 days.
18. Attenuation = 106 m^3/s ; Lag time = 2 days.
19. $K = 28$ h ; $x = 0.4$.

Objective Type Questions

1. (*a*) 2. (*b*) 3. (*c*) 4. (*c*) 5. (*d*)
6. (*a*) 7. (*c*) 8. (*d*) 9. (*b*) 10. (*a*)

Chapter 14 DESIGN FLOOD

14.1 INTRODUCTION

Any flow which is relatively high and which overtops the natural or artificial banks in any reach of a river may be called a flood. Floods are produced when the capacity of the river channel is inadequate to carry off the abnormal quantity of water arising from heavy rainfall causing the river to overflow its banks and inundate the surrounding low-lying country. When the banks are overtopped, water spreads over the flood plain and comes into conflict with the activities of man. Since the flood plain is a desirable location for man and his activities, it is important that floods be controlled so that the damage does not exceed an acceptable limit. Reservoirs, levees, channel improvements, etc. are some of the flood control measures. For an economic and efficient design of these measures floods are to be estimated with reasonable accuracy. Design of culverts, road and rail bridges, drainage works and irrigation diversion works, also need a reliable estimate of the flood at the site concerned. The ideal solution would be achieved in all the above cases if the hydrologist could supply the hydraulic engineer with an estimate of maximum flood which would occur at the site under consideration during a specified period of time in future. Unfortunately, the hydrologist cannot make such a certain estimate. Because the runoff is essentially a random process, it can be estimated only with a specified probability. For what flood must the structures be actually designed is decided by the designing agency taking safety, economy and other factors into consideration. For example, a highway culvert which may be replaced within 10 to 20 years whose failure might not lead to any serious dislocation in the traffic would be designed to withstand relatively moderate floods. On the other hand, a large dam impounding large volume of water must be designed to withstand extremely large flood which may be greater than the flood that would be reasonably expected during its useful life. Here the design is dictated not by the value of the structure or the inconvenience brought about by it destruction, but by the fact that the people below must be protected against the possibility that failure of the dam during a major flood would claim many lives and much valuable property. Between these two extremes there is a wide variety of situations of different degrees of complexity.

The *maximum probable flood*, abbreviated as MPF, has been defined as the flood that may be expected from the most severe combination of critical meteorologic and hydrologic conditions that are reasonably possible in the region. This is very large flood indeed and is very rarely used in design except for reservoir spillways where failure would lead to great damage and loss of life. The determination of the MPF involves a detailed study of storm patterns, transposition of storms to a position that will give maximum runoff and computation of maximum flood by the unit hydrograph method.

A *design flood* is the flood discharge adopted for the design of a structure after careful consideration of economic and hydrologic factors. As the magnitude of the design flood increases, the capital cost of the structure also increases but the probability of annual damage will decrease. The most economical design flood can be found after studying the various alternatives. The design flood in many a case is less than the maximum probable flood because the MPF is not economically practicable. Several design floods may be adopted in a single structure, as for example the design flood for a spillway of a reservoir may be much greater than the design flood adopted for temporary coffer dams at the initial stages of the structure. Central water commission has given certain recommendations regarding the design flood to be adopted with due consideration to the size and importance of the project.

In general the methods used in the estimation of the design flood can be grouped as under.

(*i*) Increasing the observed maximum flood by a certain percentage

(*ii*) Envelope curves

(*iii*) Empirical flood formulae

(*iv*) Rational Method

(*v*) Unit hydrograph application

(*vi*) Frequency analysis (or Statistical methods).

For an economic design consistent with safety it is necessary to have the complete design flood hydrograph rather than merely the value of flood peak. Methods (*i*) to (*iv*) of the above list can give only the flood peak. Method (*v*) gives the complete flood hydrograph but it cannot be used for very large areas. In method (*vi*) the flood hydrograph may be constructed by making its peak discharge equal to the estimated design flood, and other features to resemble the observed hydrographs in the available record. The method to be adopted in any specific case is mostly decided by the nature of data that is available. A discussion of these methods will now follow.

14.2 INCREASING THE OBSERVED FLOOD

When the flood data is not available, or when the available data is short, the maximum flood that occurred during the past, and the number of years during which this flood was the highest are found out by local enquiries or from the observation of past records. This is then increased by a certain percentage and adopted as the design flood. The percentage increase naturally depends on the number of years over which the observed flood was maximum. If it is over a longer period the percentage increase may be less and for shorter periods the increase may have to be more. This method is adhoc in nature and has not much theoretical justification and should be resorted to under helpless situations.

14.3 ENVELOPE CURVES

When the observed peak floods in different catchments of a hydrometeorologically homogeneous region are plotted against their respective areas on a log-log sheet it will usually show that all the points lie below an enveloping curve which can be drawn by eye. After the envelope curve is drawn

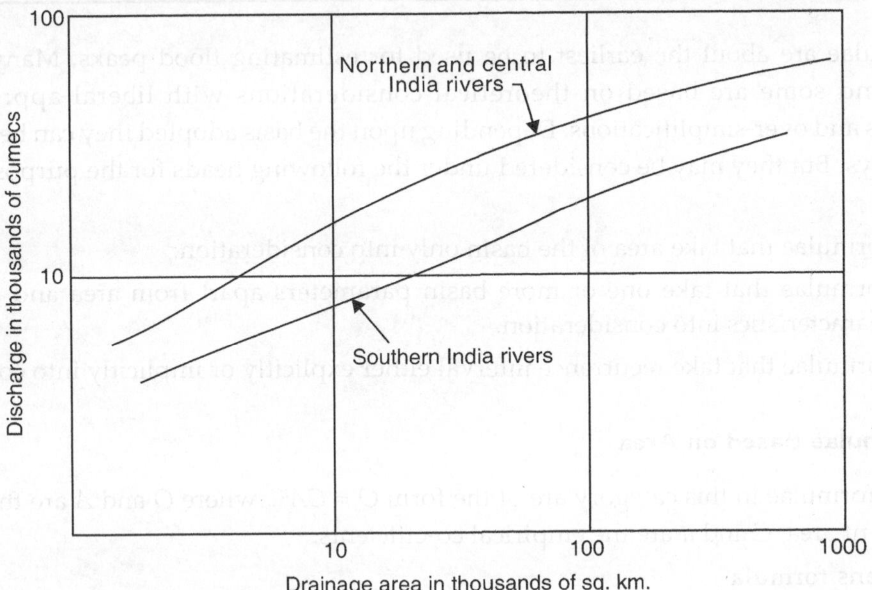

FIGURE 14.1 Envelope curves.

the maximum flood for the catchment under consideration can be read by entering the plot with the known area of the catchment. Fig. 14.1 shows the envelope curves for Indian Catchments, presented by Kanwarsain and Karpov (1967). Baird and Mcilluraith (1951) gave the following equation for the enveloping curve of maximum floods throughout the world.

$$Q = \frac{3010\,A}{(277 + A)^{0.78}} \qquad \text{...(14.1)}$$

where Q is in m^3/s and A is in km^2.

The catchments included in the preparation of envelope curve should be in the hydrometeorologically homogeneous region and the data used should represent similar type of floods, that is, rain floods or snowmelt floods. When only a part of the area happened to contribute to a maximum flood only that area must be used in the plot. The points through which the envelope curve passes are called control points. Care should be taken to see that these points do not represent abnormal conditions like a breach of dam.

An inherent disadvantage of the envelope curve lies in the fact that it is based on past recorded floods and there is always a possibility that still higher flood may occur in future. No return period can be assigned to the flood estimated from envelope curve. As more and more data accumulates the curves are to be redrawn. The results from envelope curve should not be relied upon wholly. Design flood values, if they are to be MPF, should be normally higher than those obtained from envelope curves.

14.4 EMPIRICAL FORMULAE

Flood formulae are about the earliest to be used for estimating flood peaks. Many are purely empirical and some are based on theoretical considerations with liberal approximations, assumptions and over-simplifications. Depending upon the basis adopted they can be classified in different ways. But they may be considered under the following heads for the purpose of present discussion.

(*a*) Formulae that take area of the basin only into consideration.

(*b*) Formulae that take one or more basin parameters apart from area and also rainfall characteristics into consideration.

(*c*) Formulae that take recurrence interval either explicitly or implicitly into consideration.

14.4.1 Formulae Based on Area

Most of the formulae in this category are of the form $Q = CA^n$, where Q and A are the floodpeak and catchment area, C and n are the empirical co-efficients.

Dickens formula

$$Q = CA^{3/4} \qquad \text{...(14.2)}$$

where A is in km^2 and C is a constant depending on other catchment parameters which may be taken as 11.4 for North Indian basins, 13.9 to 19.5 for Central Indian basins and 22.2 to 25.0 for the catchments lying in Western ghats. Q is in m^3/s.

Ryves formula

$$Q = CA^{2/3} \qquad \text{...(14.3)}$$

where Q is in m^3/s and A is in km^2. This formula is specially developed by Ryves for the catchment areas in Madras State and is widely used in South India. The value of the constant C may be adopted as follows, though some observed floods have given as high a value as 30 for C.

Areas within 80 km from the coast 6.75

Areas within 80 to 2400 km from the coast 8.45

Limited areas near hills 10.00

Inglis formula. This formula is applicable for catchments of Maharashtra

$$Q = \frac{124\,A}{\sqrt{A+10.4}} \qquad \text{...(14.4)}$$

Again Q is in m^3/s and A is in km^2.

Ali Nawaz Jung Bahadur formula

$$Q = CA^{\left(0.993 - \frac{1}{14}\log A\right)} \qquad \text{...(14.5)}$$

where Q is in m^3/s and A is in km^2. This formula has been developed by Ali with special application to catchments in old Hyderabad State. C takes a value of 48 to 60.

14.4.2 Formulae with Area and Other Factors

The formulae which come under this category are those developed by Craig, Lillie, Rhind, etc.

Craig's formula

$$Q = 10\,k\,BVI\,\ln\left(\frac{4.97\,L^2}{B}\right) \qquad \text{...(14.6)}$$

where Q = Maximum flood in m^3/s
L = The greatest length of catchment in km
B = The average width of catchment in km
V = The velocity of runoff in m/s
k = The co-efficient of runoff
I = The average intensity of rainfall in cm/h.

Lillie's formula. Lillie developed a formula by dividing the catchment area into series of triangles with the apex of each triangle at the basin outlet and expressing each triangle in terms of its apical angle θ and its vertical length L. His formula is

$$Q = 0.058\,V\,R\,k\,\Sigma\theta\,.\,L \qquad \text{...(14.7)}$$

where R = A rainfall coefficient = $2 + \dfrac{P}{38.1}$
Q = Max. flood discharge in m^3/s
V = Velocity of flow in m/s
P = Annual rainfall in cm
k = A co-efficient (0.8934 + log L)
L is in km.

Rhinds formula

$$Q = 0.098\,C\,S\,R\,(0.386\,A)^p \qquad \text{...(14.8)}$$

where Q = Max. flood in m^3/s

S = The average fall of the last 5 km of the river above the basin outlet in m/km

R = The greatest average annual rainfall recorded at some point in the area in cm

p = A variable index

L = The greatest length of the catchment in km

C = A co-efficient depending on the ratio R/L.

Chamiers formula

$$Q = 3.5\, IkA^{3/4} \qquad ...(14.9)$$

where Q is in m^3/s, I is the average rate of rainfall in cm/h for such duration as will allow of the flood water flowing to the outlet from the farthest point of catchment, k is runoff co-efficient and A is catchment area in sq. km.

Iszkowski's formula

$$Q = 0.03171\, CMHA \qquad ...(14.10)$$

where Q = Maximum flood in m^3/s

H = Annual rainfall in metres

C = Co-efficient of runoff

A = Catchment area in km^2

M = A co-efficient relating maximum to average discharge. With a range of 0.17 to 0.8.

14.4.3 Formulae with Recurrence Interval

The formulae which give the flood discharge in terms of the recurrence interval are preferable as the flood event is actually random phenomenon.

Fullers formula

$$Q_{av} = CA^{0.8} \qquad ...(14.11)$$

$$Q = Q_{av} \times (1 + 0.8 \log T_r) \qquad ...(14.12)$$

$$Q_{max} = Q\,(1 + 2.66\, A^{-0.3}) \qquad ...(14.13)$$

where A is area in sq. km ; T_r is the return period in years ; Q_{av} is yearly average flood ; Q is maximum 24 h flood with return period of T_r years and Q_{max} is the maximum instantaneous discharge. The coefficient C varies from 0.026 to 2.77.

Hortons formula

$$Q = \frac{114\, T_r^{0.25}}{A} \qquad ...(14.14)$$

where Q is flood discharge in m^3/s per km^2 with a return period of T_r years A is the drainage area in sq. km.

Pettis formula

$$Q = C\left(P\overline{B}\right)^{1.25} \qquad ...(14.15)$$

where Q = The flood discharge with return period of 100 years in m^3/s

P = The one day rainfall with 100 years return period in cm

$\overline{B}$ = The average width of the basin in km

C = A co-efficient varying from about 1.51 in humid areas to 0.195 in desert areas.

The list of formulae presented here is by no means exhaustive. There are many more formulae reported from different parts of the world under each category. All these formulae should generally be used in those regions where they are developed. Even then they should be used with caution. As more and more data accumulates with time the values of constants developed originally with meager data should be re-examined before the formula is applied. When sufficient data is available it is better to use the unit hydrograph and frequency analysis methods and empirical formulae should not be used at all.

If the maximum flood discharge can be expressed as a function of all variables which influence it, it would be just an ideal solution. Unfortunately the factors affecting the flood discharge, as already described earlier, are too many and too complex that it is almost impossible to obtain such an ideal solution.

14.5 RATIONAL METHOD

In this method it is assumed that the maximum flood flow is produced by a certain rainfall intensity which lasts for a time equal to or greater than the period of *concentration time*. This concentration time is nothing but the time required for the surface runoff from the remotest part of the catchment area to reach the basin outlet. When a storm continues beyond concentration time every part of the catchment would be contributing to the runoff at outlet and therefore it represents condition of peak runoff. The runoff rate corresponding to this condition is given by

$$Q = C\,A\,I \qquad \text{...(14.16)}$$

where A is the area of the catchment, I is the intensity of rainfall and C is a runoff coefficient to account for the abstractions form the rainfall. In Eq. (14.16), if A and I are substituted in units of acres and inches/h, the runoff is obtained in ft^3/s without requiring any conversion factor. For this reason Eq. (14.16) is called the rational formula. However, in SI system of units when A is in km^2 and I is in cm/h, the flow rate in m^3/s is given by

$$Q = 2.778\,C\,A\,I \qquad \text{...(14.17)}$$

The maximum rainfall intensity, as discussed in the Intensity-Frequency-Duration analysis in section (5.11), depends on duration and frequency. The intensity of rainfall used in Eq. (14.17) should therefore be corresponding to a duration equal to concentration time and desired return period. This requires an estimate of concentration time which is usually provided by an empirical equation given by Kirpich (1940).

$$t_c = 0.0195\,L^{0.77}\,S^{-0.385} \qquad \text{...(14.18)}$$

where t_c is in minutes, L is the maximum length of travel of water along the water course in m and S is the slope expressed as the ratio of difference in elevation between the remotest point and the catchment outlet to the length L. There are other equations available to estimate concentration time.

Once the concentration time has been determined, the rainfall intensity can be determined using the intensity-duration-frequency relation applicable to the catchment area in question, chosing an appropriate rainfall frequency. For example, using Eq. (5.9) or Fig. 5.20. The frequency (return period) adopted may vary from 5 to 50 years depending on the type and importance of the structure. Once the rainfall intensity and catchment area have been obtained, a runoff coefficient applicable to the given design condition is selected and the peakflood estimated from Eq. (14.17). Some typical values of the runoff coefficient C are given in Table 14.1.

A given catchment may have distinctly different characteristics requiring the use of different runoff coefficient for different sub-areas within the catchment. In such cases the individual subcatchments are delineated and their runoff coefficients are identified. The runoff coefficients are weighed in proportion to the areas of their respective subcatchments and then the peak flood is estimated. That is

$$Q = 2.778\, I\, (\Sigma\, C_i\, A_i) \qquad ...(14.19)$$

in which C_i is the runoff coefficient of ith subcatchment and A_i is the drainage area of ith subcatchment.

The rational method is found to give good results for small areas upto about 50 km^2. It is generally used to estimate the peak flood in the design of urban drainage systems, storm sewers, air port drainage, and in the design of small culverts and bridges.

Table 14.1. Values of C for Use in Rational Method

Soil type	*Land use*		
	Cultivated	*Pasture*	*Forest*
With above average infiltration rates ; usually sandy or gravelly.	0.20	0.15	0.10
With average infiltration rates ; no clay pans ; loams and similar soils.	0.40	0.35	0.30
With below average infiltration rates ; heavy clay soils or soils with a clay pan near the surface ; shallow soils above in pervious rock.	0.50	0.45	0.40

Example 14.1. *An engineer is required to design a drainage system for an airport with an area of 2.5 km^2 for a 50 year return period. The 50 year rainfall intensity in that region is given by*

$$I = \frac{35}{(t + 10)^{0.38}}$$

where I is the intensity in cm/h and t is the duration in minutes. If the concentration time for the area is estimated as 50 minutes for what discharge must he design the system ?

Solution. Concentration time t_c = 50 min

$$\text{Intensity of rainfall for this duration} = I = \frac{35}{(50 + 10)^{0.38}}$$

$$= \frac{35}{60^{0.38}} = \frac{35}{4.74} = 7.384 \text{ cm/h}$$

Since the airport is fully paved, it may be considered impervious and the runoff coefficient C may be taken as unity. Therefore

$$Q = 2.778\ C\ A\ I$$
$$= 2.778 \times 1 \times 2.5 \times 7.384$$
$$= 51.28\ \text{m}^3/\text{s}.$$

∴ The engineer must design the drainage system for a discharge of 51.28 m³/s.

Example 14.2. *A small watershed consists of 1.5 km² of cultivated area (C = 0.20), 2.5 km² under forest (C = 0.10) and 1.0 km² under grass cover (C= 0.35). There is a fall of 22 m in a watercourse length of 1.8 km. The intensity-frequency-duration relation for the area may be taken as*

$$I = \frac{80\ T_r^{0.2}}{(t+13)^{0.46}}$$

where I is in cm/h, T_r is in years and t is in minutes. Estimate the peak rate of runoff for a 25 years frequency.

Solution. Length of water course = 1.8 km = 1800 m

Fall in the elevation = 22 m

$$\therefore \quad S = \frac{22}{1800}$$

$$t_c = 0.0195\ L^{0.77}\ S^{-0.385}$$

$$= 0.0195\ (1800)^{0.77} \left(\frac{22}{1800}\right)^{-0.385}$$

$$= 34 \text{ minutes}$$

$$I = \frac{80\ (25)^{0.2}}{(34+13)^{0.46}} = 25.9\ \text{cm/h}$$

$$A_1 = 1.5\ \text{km}^2 \qquad C_1 = 0.20$$
$$A_2 = 2.5\ \text{km}^2 \qquad C_2 = 0.10$$
$$A_3 = 1.0\ \text{km}^2 \qquad C_3 = 0.35$$

Substituting these values in Eq. (14.19)

$$Q = 2.778 \times 25.9\ (1.5 \times 0.2 + 2.5 \times 0.1 + 1.0 \times 0.35)$$
$$= 64.76\ \text{m}^3/\text{s}.$$

Example 14.3. *A culvert is proposed across a stream draining an area of 185 hectares. The catchment has a slope of 0.004 and the length of travel for water is 1150 m. Estimate the 25 year flood if the rainfall is given by*

$$I = \frac{1000\ T_r^{0.2}}{(t+20)^{0.7}}$$

where I is in mm/h, T_r is in years and t is in minutes. Assume a runoff coefficient of 0.35.

Solution.

$$L = 1150 \text{ m}$$

$$S = 0.004$$

$$t_c = 0.0195\,(1150)^{0.77}\,(0.004)^{-0.385}$$

$$= 37.2 \text{ minutes.}$$

$$I = \frac{1000\,(25)^{0.2}}{(37.2 + 20)^{0.7}} = 112.05 \text{ mm/h} = 11.205 \text{ cm/h}$$

$$A = 185 \text{ hectares} = 1.85 \text{ km}^2$$

$$C = 0.35$$

$$Q = 2.778 \times 0.35 \times 1.85 \times 11.205 = 20.12 \text{ m}^3\text{/s.}$$

14.6 UNIT HYDROGRAPH APPLICATION

The application of unit hydrograph to convert a given effective rainfall into corresponding runoff hydrograph has already been described in section (11.11) of Chapter 11 and demonstrated through example (11.8). So, when a unit hydrograph is available for the catchment under consideration, it can be applied to the *design storm* to yield the design flood hydrograph. The storm with a specified return period which is adopted for the estimation of design flood may be called a design storm. The design storm may be determined after a comprehensive study of major storms recorded in the region. The intensity-frequency-duration relations and the depth-area-duration relations which were discussed in sections (5.11) and (5.13) of Chapter 5, if available, will be highly useful in this regard. When the unit hydrograph is not available for the catchment, a synthetic unit hydrograph may be developed based on the data of neighbouring basins. Similarly when sufficient data is not available the design storm may be determined using storm transposition method, that is, the application of storm from one area to some other area within the same region of meteorological homogenity. For large and important projects the probable maximum precipitation is used as the design storm. When the probable maximum precipitation or PMP is used in the design flood estimation the resulting flood hydrograph gives the maximum probable flood. The procedure involved in the determination of PMP was discussed in section (5.16) of Chapter 5.

As already pointed out earlier, the unit hydrograph method cannot be applied for large areas, say more than 5000 km^2, with reliability. In such cases the basin is subdivided into a number of sub-basins corresponding to each tributary, and the design flood hydrograph for each sub-basin is derived using the respective unit hydrograph. These flood hydrographs of sub-basins are then routed through the channel reaches and superimposed with common time scale to obtain the flood hydrograph at the point of interest. If the area is very large it may not be correct to assume that the design storm produces rainfall over the entire basin simultaneously. Hence in such cases the design storm is assumed to occur only on a particular combination of the sub-basins which is considered reasonably probable and the resulting flood hydrograph is taken as the design flood hydrograph.

One of the important assumptions of unit hydrograph theory is the principle of linearity. This assumption is not strictly valid. If the unit hydrograph used in the design flood estimation is based on average storms only, the peak discharge in the design flood hydrograph is increased by certain percentage before it is actually implemented in the design to account for the effects of non-linearity in the basin. The percentage increase depends on the size of the storm used for derivation of unit hydrograph and the size of the design storm adopted for the design.

Whenever possible it is advisable to use the unit hydrograph method to obtain the design flood. It gives not only the flood peak but also the complete flood hydrograph which is essentially required in fixing the spillway capacity after incorporating the effect of storage of the reservoir on flood peak through flood routing.

When the area of the basin is very large it is better to use the frequency analysis methods described in the next sub-section.

14.7 FREQUENCY ANALYSIS

Frequency analysis makes use of the observed data in the past to predict the future flood events along with their probabilities or return periods. It is based on the assumption that combination of the numerous factors which produce floods are a matter of pure chance and therefore are subject to analysis according to the mathematical theory of probability. The procedure of frequency analysis was discussed in detail in section (3.5) of Chapter 3.

14.7.1 Selection of Data

If frequency analysis is to provide useful answers, it must be based on adequate and accurate data. The observed data is merely a sample of the total population of floods that have occurred and may be expected to occur in future. If the sample is too small the predictions about the future floods cannot be expected to be reliable. Studies on synthetic data have revealed that with a 20 year record the probability to overestimate the design flood is 80 per cent and that 45 per cent of the times the estimated value will exceed the true value by 30 per cent. Therefore records shorter than 20 years should not be used in the frequency analysis. The data used in the frequency analysis should not only be accurate in measurement but also be homogeneous. Dams, diversions, urbanisations and other land-use changes may introduce inconsistencies in the record. Records affected by such changes should not be used without adjusting them until the whole record is representing a common set of watershed conditions, usually either natural conditions or the current conditions.

For frequency analysis the flood data may consist of either *annual series* or the *partial duration series* (or simply the *partial series*). The choice depends on the purpose of study. If the extreme floods are of primary concern, wherein the flood magnitudes with an exceedence probability of 0.5 or less are estimated, the annual series is used. In the annual series the largest flood observed in each water year only is taken. It ignores the second and lower order events of each year which may be even greater than annual floods of other years. When estimates of very frequent events with return periods of less than 5 years are required, (for example in the design of coffer dams, urban drainage etc.) the partial series is preferable to annual series. In the partial series all flood events above a selected base value are included. The base is usually chosen so that not more than two or

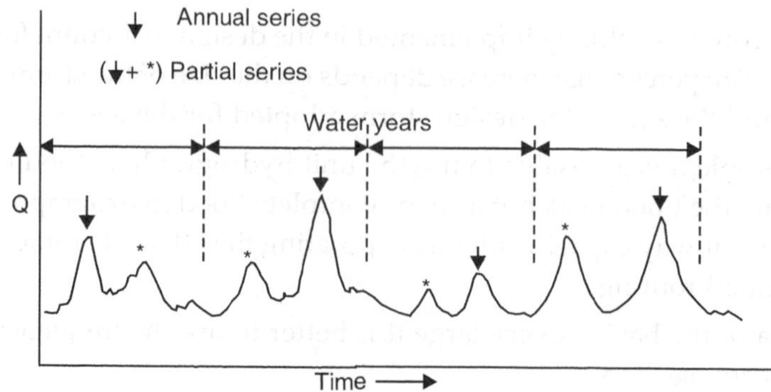

FIGURE 14.2 Composition of annual and partial series.

three events are included for each year. An objection to the partial series is that the events included may not be independent. However, when two consecutive events are not independent only the higher of the two peaks is considered. Fig. 14.2, schematically shows the composition of annual series and partial series.

14.7.2 Probability Plotting

After either the partial or annual series is compiled the items are arranged in descending order of magnitude and assigned an order or rank m starting with 1 for the highest observed flood. The *return period* or the *recurrence interval* is calculated for each event using the Weibul's formula

$$T_r = \frac{n+1}{m} \qquad \text{(From Eq. 3.32)}$$

where T_r is the return period in years, m is the order and n is the number of years in the record. This procedure is applicable to either annual or partial series. The words return period and recurrence interval are used interchangeably. They represent the average number of years within which a given event will be equalled or exceeded. Historical data on events occurring prior to the beginning of continuous record should be included in the analysis only if the evidence is sufficient to assign magnitude, order and return period. For example, even though the continuous record is available only for the period from 1955 to 1984, it may be an established fact that 25000 m^3/s of flood of 1929 was the greatest since 1880, and has not since been exceeded. Under these circumstances all required information, that is $m = 1$, $n = 105$, and $x = 25000$ m^3/s, is known and this flood can be plotted in accordance with Eq. (3.32). Since the return period and the exceedence probability are reciprocals of each other, we have

$$P = \frac{1}{T_r} \qquad \text{(From Eq. 3.30)}$$

Now if a graph is drawn between the flood magnitude and its return period on simple plane co-ordinates the resulting plot is called a *probability* or an *empirical distribution.* This graph may be extrapolated to get the design flood for any specific return period. Results obtained from such extrapolation are only approximate because the extrapolation is usually influenced by a few end points as shown in Fig. 14.3. Sometimes the graph may be drawn between the flood magnitude with linear scale on y-axis and the corresponding return period with logarithmic scale on x-axis. Such a plot as shown in Fig. 14.4 may yield a straight line fitted to all the points except the lowest values. Although extrapolation is simpler in this case, extrapolation is not necessarily more accurate than the previous unless return period follows logarithmic law. In general probability plots may be used to find the floods with return periods less than $n/5$ years. For long return periods it is better to fit a theoretical distribution to the data. The most commonly used distributions are Gumbel's distribution and log pearson type III distribution which will be discussed in the next sub-sections.

Example 14.4. *The observed annual flood peaks of a stream for a period of 40 years from 1941 to 1980 in m³/s are given below.*

395, 619, 766, 422, 282, 990, 705, 528, 520, 436,
697, 624, 496, 589, 598, 359, 686, 726, 527, 310,
408, 721, 814, 459, 440, 632, 343, 634, 464, 373,
289, 371, 522, 342, 446, 366, 699, 560, 450, 610

Construct the probability plot for the annual flood on ordinary graph and on semi-log graph. Determine the flood magnitude with a return period of 100 years in each case.

Solution. The given data is entered in Cols. (1) and (2) of Table 14.2. In Col.. (3) the floods are arranged in descending order and their ranks are shown in Col. (4). The return periods of floods in descending order are computed and entered in Col. (5).

Table 14.2. Frequency Analysis of Flood Data of Example 14.4

Year	*Flood peak* m^3/s	*Flood peaks in descending order* m^3/s	*Order* m	*Return period* $T_r = \frac{41}{m}$	*Reduced variate* $y = -\ln\ln\frac{T_r}{T_r - 1}$
(1)	(2)	(3)	(4)	(5)	(6)
1941	395	900	1	41	3.7013
1942	619	814	2	20.5	2.9955
1943	766	766	3	13.67	2.5775
1944	422	726	4	10.25	2.2764
1945	282	721	5	8.20	2.0398
1946	990	705	6	6.83	1.8432
1947	705	699	7	5.85	1.6742
1948	528	697	8	5.01	1.5022
1949	520	686	9	4.55	1.3936
1950	436	634	s10	4.10	1.2744
1951	697	632	11	3.73	1.1644
1952	624	624	12	3.4	1.0547
1953	496	619	13	3.15	0.9625
1954	589	610	14	2.93	0.8735
1955	598	598	15	2.73	0.7849
1956	359	589	16	2.56	0.7025

1957	686	560	17	2.41	0.6235
1958	726	528	18	2.28	0.5494
1959	527	527	19	2.16	0.4753
1960	310	522	20	2.05	0.4019
1961	408	520	21	1.95	0.3297
1962	721	496	22	1.86	0.2595
1963	814	464	23	1.78	0.1923
1964	459	459	24	1.71	0.1290
1965	440	450	25	1.64	0.0608
1966	632	444	26	1.58	0.0022
1967	343	440	27	1.52	– 0.0701
1968	634	436	28	1.46	– 0.1441
1969	464	422	29	1.41	– 0.2112
1970	373	408	30	1.37	– 0.2693
1971	289	395	31	1.32	– 0.3486
1972	371	373	32	1.28	– 0.4186
1973	522	371	33	1.24	– 0.4960
1974	342	366	34	1.21	– 0.5603
1975	446	359	35	1.17	– 0.6570
1976	366	343	36	1.14	– 0.7406
1977	699	342	37	1.11	– 0.8380
1978	560	310	38	1.08	– 0.9565
1979	450	289	39	1.05	– 1.1133
1980	610	282	40	1.03	– 1.2630

The probability plot of flood magnitude of Col. (3) against its return period of Col. (5) on an ordinary graph sheet is shown in Fig. 14.3. From this plot the flood discharge with a return period of 100 years is read as 1270 m^3/s.

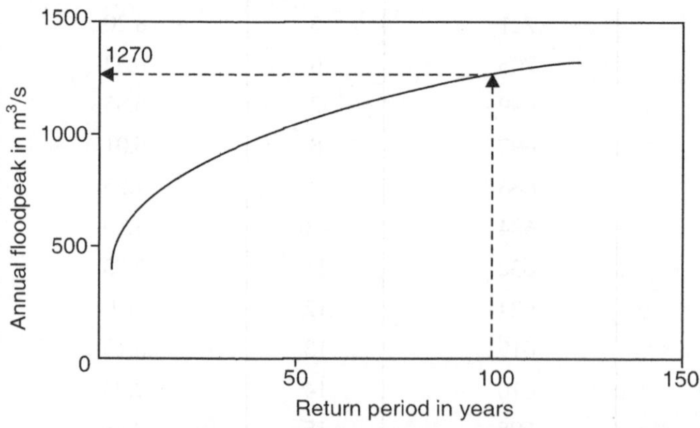

FIGURE 14.3 Ordinary probability plot.

The probability plot of flood discharge of Col. (3) on linear scale against its return period of Col. (5) on logarithmic scale is shown in Fig. 14.4. From this graph the flood discharge with a return period of 100 years is obtained as 1060 m^3/s.

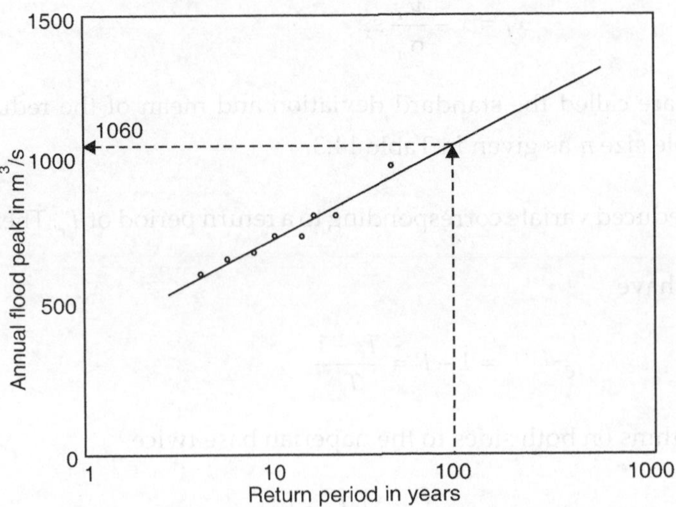

FIGURE 14.4 Logarithmic probability plot.

The computations given in Col. (6) are not meant for this example. They are required for example 14.5.

14.7.3 Gumbel's Distribution

This is one of the most widely employed distributions to describe the flood data. It has been explained in detail in section (3.3) of Chapter 3. As per this distribution the exceedence probability is given by

$$P = 1 - e^{-e^{-y}} = \frac{1}{T_r} \quad \text{...(14.20)}$$

where y is called the reduced variate given by

$$y = a\,(x - x_f) \quad \text{...(14.21)}$$

where a and x_f are the parameters of the distribution which can be obtained from sample statistics through the method of moments. When the sample size is large, say $n > 200$, they are given by

$$a = \frac{1.28255}{s_x} \quad \text{...(14.22)}$$

$$x_f = \bar{x} - 0.45005\, s_x \quad \text{...(14.23)}$$

where $\bar{x}$ and s_x are the mean and standard deviation of the sample computed from Eqs. (3.23) and (3.24), respectively. However, for small samples, the parameters a and x_f are to be estimated using the following equations given by Gumbel.

$$a = \frac{\sigma_n}{s_x} \quad \text{...(14.24)}$$

$$x_f = \bar{x} - \frac{\bar{y}_n}{\sigma_n} s_x \quad \text{...(14.25)}$$

in which σ_n and $\bar{y}_n$ are called the standard deviation and mean of the reduced extremes which depend on the sample size n as given in Table 14.3.

Let y_T be the reduced variate corresponding to a return period of T_r. Then, noting that $P = \frac{1}{T_r}$, from Eq. (14.20), we have

$$e^{-e^{-y_T}} = 1 - P = \frac{T_r - 1}{T_r}$$

Taking logarithms on both sides to the naperian base twice

$$y_T = -\ln\left[-\ln\left(\frac{T_r - 1}{T_r}\right)\right]$$

or

$$y_T = -\ln\left[\ln\left(\frac{T_r}{T_r - 1}\right)\right] \quad \text{...(14.26)}$$

Alternatively, converting the logarithms to the base 10, we can write

$$y_T = -\left[0.834 + 2.303 \log\log\left(\frac{T_r}{T_r - 1}\right)\right] \quad \text{...(14.26 a)}$$

Let x_T denote the magnitude of the flood with a return period of T_r years. Then from Eq. (14.21), we have

$$y_T = a\,(x_T - x_f) \quad \text{...(14.27)}$$

Substituting Eqs. (14.24) and (14.25) in Eq. (14.27) and simplifying, we obtain

$$x_T = \bar{x} + \left(\frac{y_T - \bar{y}_n}{\sigma_n}\right) . s_x \quad \text{...(14.28)}$$

or

$$x_T = \bar{x} + K_T . s_x \quad \text{...(14.29)}$$

where K_T is known as the *frequency factor* given by

$$K_T = \left(\frac{y_T - \bar{y}_n}{\sigma_n}\right) \quad \text{...(14.30)}$$

It has been shown by Ven Te Chow that most frequency distribution functions applicable in hydrologic studies can be expressed by Eq. (14.29). Obviously the frequency factor K_T depends on the type of distribution and the return period.

The procedure to estimate the design flood for any return period using Gumbel's distribution is as given below.

Step 1. From the given data on flood peaks for n years, the mean $\overline{x}$ and standard deviation s_x are computed from Eqs. (3.23) and (3.24) respectively.

Step 2. For the known sample size n, the values of $\overline{y}_n$ and σ_n are obtained from Table 14.3.

Step 3. For the given return period T_r, the reduced variate y_T is found from Eq. (14.26).

Step 4. With the values of $\overline{y}_n$ and σ_n obtained in step 2, and y_T obtained in step 3, the frequency factor K_T is calculated using Eq. (14.30).

Step 5. With values of $\overline{x}$ and s_x obtained in step 1 and K_T obtained in step 4, the magnitude of the flood x_T is computed using Eq. (14.29).

Table 14.3 Mean $\overline{y}_n$ and Standard Deviation S_n of Reduced Extremes of Gumbel's Distribution

n	$\overline{y}_n$	σ_n	n	$\overline{y}_n$	σ_n
8	0.4843	0.9043	49	.5481	1.1590
9	0.4902	0.9288	50	.54854	1.16066
10	0.4952	0.9497	51	.5489	1.1623
11	0.4996	0.9676	52	.5493	1.1638
12	0.5035	0.9833	53	.5497	1.1653
13	0.5070	0.9972	54	.5501	1.1667
14	0.5100	1.0095	55	.5504	1.1681
15	0.5128	1.02057	56	.5508	1.1696
16	0.5157	1.0316	57	.5511	1.1708
17	0.5181	1.0411	58	.5515	1.1721
18	0.5202	1.0493	59	.5518	1.1734
19	0.5220	1.0566	60	.55208	1.17467
20	0.52355	1.06283	62	.5527	1.1770
21	0.5252	1.0696	64	.5533	1.1793
22	0.5268	1.0754	66	.5538	1.1814
23	0.5283	1.0811	68	.5543	1.1834
24	0.5296	1.0864	70	.55477	1.18536
25	0.53086	1.09145	72	.5552	1.1873
26	0.5320	1.0961	74	.5557	1.1890
27	0.5332	1.1004	76	.5561	1.1906
28	0.5343	1.1047	78	.5565	1.1923
29	0.5353	1.1086	80	.55688	1.19382

30	0.53622	1.11238	82	.5572	1.1953
31	0.5371	1.1159	84	.5576	1.1967
32	0.5380	1.1193	86	.5580	1.1980
33	0.5388	1.1226	88	.5583	1.1994
34	0.5396	1.1255	90	.5586	1.20073
35	0.54034	1.12847	92	.5589	1.2020
36	0.5410	1.1313	94	.5592	1.2032
37	0.5418	1.1339	96	.5595	1.2044
38	0.5424	1.1363	98	.5598	1.2055
39	0.5430	1.1388	100	.56002	1.20649
40	0.54362	1.14132	150	.56461	1.22534
41	0.5442	1.1436	200	.56715	1.23598
42	0.5448	1.1458	250	.56878	1.24292
43	0.5453	1.1480	300	.56993	1.24786
44	0.5458	1.1499	400	.57144	1.25450
45	0.5463	1.15185	500	.57240	1.25880
46	0.5468	1.1538	750	.57377	1.26506
47	0.5473	1.1557	1000	.57450	1.26851
48	0.5477	1.1574	∞	.57722	1.28255

A simple graphical test to confirm whether the observed flood data follows Gumbel's distribution or not is as follows : After arranging the observed data in descending order of magnitude and assigning the return period for each flood, the reduced variate corresponding to each return period is computed from Eq. (14.26). Now a plot is made between the flood magnitude and its reduced variate on an ordinary graph paper. If an eye fit to this plot suggests a straight line, then it can be concluded that the Gumbel's distribution is a good fit for the flood data since y and x are linearly related through Eq. (14.21).

Extreme Value Probability Paper. A probability paper which plots the Gumbel's extreme value distribution as a straight line can be easily constructed as follows. If any random variable x is following Gumbel's extreme value distribution, then as per Eq. (14.21) the relation between the reduced variate y and the variable x is linear. Therefore a plot between y and x on an ordinary graph will be a straight line. Now the scale of the axis of the reduced variate y is transformed into the corresponding return period through the relation $y = -\ln \ln \left(\frac{T}{T-1}\right)$. Thus for example, the points corresponding to $y = 0.37, 1.5, 2.25, 3.2, 3.9$ and 4.6 on the axis of reduced variate are marked as 2, 5, 10, 25, 50, and 100 years respectively. The other axis representing the flood magnitude, however, will have ordinary scale only. This probability paper which is shown in Fig. 14.5 is also known as the *Powell's probability paper*. Also shown in Fig. 14.5 are the points obtained by plotting the floodpeak of Col. (3) against the return period of Col. (5) of Table 14.2. These points indicate a linear trend and hence a straight line is fitted by eye to these points. This justifies the use of Gumbel's distribution in Example 14.5 that follows.

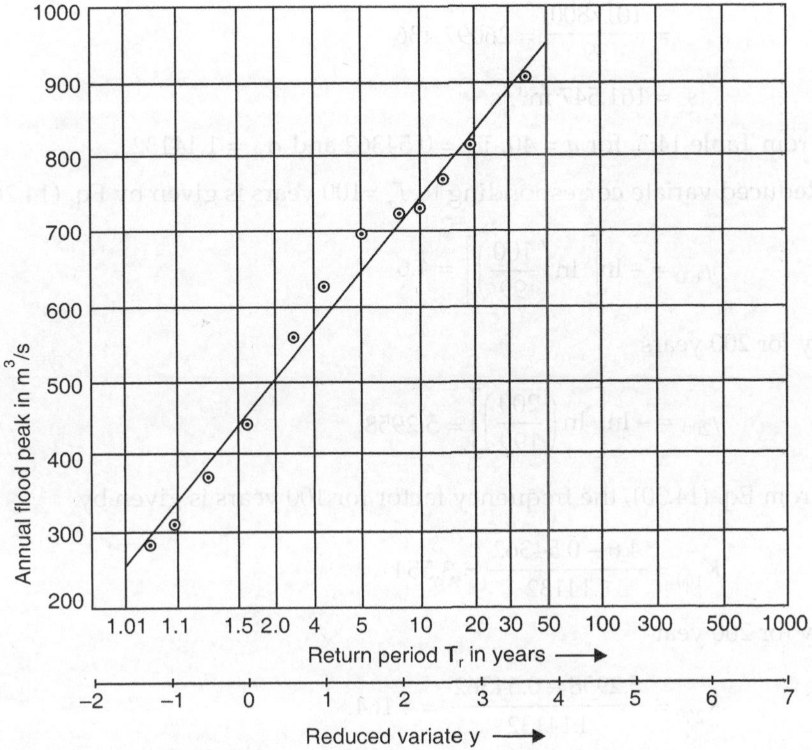

FIGURE 14.5 Extreme value probability paper.

From Table 14.3, for a large sample (as $n \to \infty$) we note that $\overline{y}_n = 0.5772$ and $\sigma_n = 1.28255$. Substituting these values in Eq. (14.30), the frequency factor K_T is given by $(y_T - 0.5772)/1.28255$. Therefore from Eq. (14.29) we get that when $x_T = \overline{x}$, the reduced variate $y_T = 0.5772$ which gives a return period of 2.33 years. In other words, as per Gumbel's distribution, the mean annual flood will have a return period of 2.33 years. After plotting the flood magnitude against return period on extreme value probability paper and fitting a straight line, the mean annual flood may be estimated from this line as the value corresponding to return period of 2.33 years.

Example 14.5. *Assuming that Gumbel's distribution fits the data of Example 14.4 estimate the 100 year and 200 year flood.*

Solution. In the last column of tabular statement of Example 14.4 the reduced variate for all the return periods are given. These are obtained from Eq. (14.26). The reader may verify that a plot between the flood magnitude [ordered of Col. (3)] and its corresponding reduced variate of Col. (6) can be fitted by an approximate straight line indicating that Gumbel's distribution is a reasonable fit for the data under consideration. The other computations are as given below.

Step 1.

$$\overline{x} = \frac{395 + 619 + 766 + \ldots\ldots\ldots + 610}{40} = 530 \text{ m}^3/\text{s}$$

$$s_x^2 = \frac{(395 - 530)^2 + (619 - 530)^2 + \ldots\ldots\ldots\ldots + (610 - 530)^2}{39}$$

$$= \frac{1017800}{39} = 26097.436$$

$$s_x = 161.547 \text{ m}^3/\text{s}.$$

Step 2. From Table 14.3, for $n = 40$, $\bar{y}_n = 0.54362$ and $\sigma_n = 1.14132$.

Step 3. Reduced variate corresponding to T_r =100 years is given by Eq. (14.26) as

$$y_{100} = -\ln\left[\ln\left(\frac{100}{99}\right)\right] = 4.6$$

Similarly for 200 years

$$y_{200} = -\ln\left[\ln\left(\frac{200}{199}\right)\right] = 5.2958.$$

Step 4. From Eq. (14.30), the frequency factor for 100 years is given by

$$K_{100} = \frac{4.6 - 0.54362}{1.14132} = 3.554$$

Similarly for 200 year

$$K_{200} = \frac{5.2958 - 0.54362}{1.14132} = 4.164.$$

Step 5. From Eq. (14.29), the 100 year flood is computed as

$$x_{100} = 530 + 3.554 \times 161.547 = 1104.14 \text{ m}^3/\text{s}.$$

Similarly the 200 year flood is given by

$$x_{200} = 530 + 4.164 \times 161.547 = 1202.68 \text{ m}^3/\text{s}.$$

Example 14.6. *From the analysis of available data on annual flood peaks of a small stream for a period of 35 years, the 50 year and 100 year flood have been estimated to be 660 m³/s and 740 m³/s using Gumbel's method. Estimate the 200 year flood for the stream.*

Solution. From Table 14.3, for a record length of $n = 35$ years

$$\bar{y}_n = 0.54034 \text{ and } \sigma_n = 1.12847$$

Using Gumbel's method

$$y_{50} = -\ln\ln\left(\frac{50}{49}\right) = 3.90194$$

$$K_{50} = \frac{3.90194 - 0.54034}{1.12847} = 2.9789$$

Similarly

$$y_{100} = -\ln\ln\left(\frac{100}{99}\right) = 4.60015$$

$$K_{100} = \frac{4.60015 - 0.54034}{1.12847} = 3.59762$$

$$x_{50} = \bar{x} + K_{50}\, s_x$$

or $$660 = \bar{x} + 2.9789\, s_x \quad ...(a)$$

$$x_{100} = \bar{x} + K_{100}\, s_x$$

or $$740 = \bar{x} + 3.59762\, s_x \quad ...(b)$$

Solving the two simultaneous equations (*a*) and (*b*) for $\bar{x}$ and s_x, we obtain $\bar{x}$ = 274.83 m^3/s and s_x = 129.3 m^3/s.

Now $$y_{200} = -\ln\ln\left(\frac{200}{199}\right) = 5.29581$$

$$K_{200} = \frac{5.29581 - 0.54034}{1.12847} = 4.2141$$

$$x_{200} = \bar{x} + K_{200}\, s_x$$

$$= 274.83 + 4.2141 \times 129.3 = 819.71 \text{ m}^3/\text{s}.$$

∴ The 200 year flood for the stream would be 820 m^3/s.

Table 14.4. Frequency Factor K for Use with Log Pearson Type III Distribution

Co-efficient of skew g	*Recurrence interval in years*							
	2	*5*	*10*	*25*	*50*	*100*	*200*	*1000*
	Per cent chance of exceedence							
	50	*20*	*10*	*4*	*2*	*1*	*0.5*	*0.1*
3.0	– 0.396	0.420	1.180	2.278	3.152	4.501	4.970	7.250
2.5	– 0.360	0.518	1.250	2.262	3.048	3.845	4.652	6.600
2.2	– 0.330	0.574	1.284	2.240	2.970	3.705	4.444	6.200
2.0	– 0.307	0.609	1.302	2.219	2.912	3.605	4.298	5.910
1.8	– 0.282	0.643	1.318	2.193	2.848	3.499	4.147	5.660
1.6	– 0.254	0.675	1.329	2.163	2.780	3.388	3.990	5.390
1.4	– 0.225	0.705	1.337	2.218	2.706	3.271	3.828	5.110
1.2	– 0.195	0.732	1.340	2.087	2.626	3.149	3.661	4.820
1.0	– 0.164	0.758	1.340	2.043	2.542	3.022	3,489	4.540
0.9	– 0..148	0.769	1.339	2.018	2.498	2.957	3.401	4.395
0.8	– 0.132	0.780	1.336	1.998	2.453	2.891	3.312	4.250
0.7	– 0.116	0.790	1.333	1.967	2.407	2.824	3.223	4.105
0.6	– 0.099	0.800	1.328	1.939	2.359	2.755	3.132	3.960
0.5	– 0.083	0.808	1.323	1.910	2.311	2.686	3.041	3.815
0.4	– 0.066	0.816	1.317	1.880	2.261	2.615	2.949	3.670

0.3	– 0.050	0.824	1.309	1.849	2.211	2.544	2.856	3.525
0.2	– 0.033	0.830	1.301	1.818	2.159	2.472	2.763	3.380
0.1	– 0.017	0.836	1.292	1.785	2.107	2.400	2.670	3.235
0	0	0.842	1.282	1.751	2.054	2.326	2.576	3.090
– 0.1	0.017	0.836	1.270	1.716	2.000	2.252	2.482	2.950
– 0.2	0.033	0.850	1.258	1.680	1.945	2.178	2.288	2.810
– 0.3	0.050	0.853	1.245	1.643	1.890	2.104	2.394	2.675
– 0.4	0.066	0.855	1.231	1.606	1.834	2.029	2.201	2.540
– 0.5	0.083	0.856	1.216	1.567	1.777	1.955	2.108	2.400
– 0.6	0.099	0.857	1.200	1.528	1.720	1.880	2.016	2.275
– 0.7	0.116	0.857	1.183	1.488	1.663	1.806	1.926	2.150
– 0.8	0.132	0.856	1.166	1.448	1.606	1.733	1.837	2.035
– 0.9	0.148	0.854	1.147	1.407	1.549	1.660	1.749	1.910
– 1.0	0.164	0.852	1.128	1.366	1.492	1.588	1.664	1.800
– 1.2	0.195	0.844	1.086	1.282	1.379	1.449	1.501	1.625
– 1.4	0.225	0.832	1.041	1.198	1.270	1.318	1.351	1.465
– 1.6	0.254	0.817	0.994	1.116	1.166	1.197	1.216	1.280
– 1.8	0.282	0.799	0.945	1.035	1.069	1.087	1.097	1.130
– 2.0	0.307	0.777	0.895	0.959	1.980	0.990	0.995	1.000
– 2.2	0.330	0.752	0.844	0.888	0.900	0.905	0.907	0.910
– 2.5	0.360	0.711	0.771	0.793	0.798	0.799	0.800	0.802
– 3.0	0.396	0.636	0.660	0.666	0.666	0.667	0.667	0.668

14.7.4 Log-Pearson Type III Distribution

Although this distribution has little theoretical basis, it is being widely used because it is fitting well to the flood peaks in many cases. It has been adopted as a standard distribution to fit the flood peak by several federal agencies in USA. The procedure to be adopted in using this distribution to arrive at the flood discharge of any given return period is as follows.

Step 1. The given flood discharge data series is converted into their logarithms.

$$y_i = \log (x_i) \qquad ...(14.31)$$

Step 2. The mean, standard deviation, and the skewness co-efficient of the y series are estimated using the following equations.

$$\bar{y} = \frac{1}{n}\sum y_i \qquad ...(14.32)$$

$$s_y = \left(\frac{\Sigma (y_i - \bar{y})^2}{(n-1)}\right)^{1/2} \qquad ...(14.33)$$

$$g = \frac{n\Sigma (y_i - \overline{y})^3}{(n-1)(n-2)\, s_y^{\,3}} \qquad ...(14.34)$$

In the above equations all the summations run from 1 to n.

Step 3. For the given return period T_r and the estimated skewness co-efficient g the value of K_T is selected from Table 14.4. This table has been prepared based on Log-Pearson type-III distribution.

Step 4. The logarithm of the design flood is computed from

$$y_T = \overline{y} + K_T \,.\, s_y \qquad ...(14.35)$$

Step 5. The design flood itself is now given by

$$x_T = \text{antilog}\,(y_T) = 10^{y_T} \qquad ...(14.36)$$

Example 14.7. *Determine 100 year and 200 year flood using log-Pearson type III distribution for the data of example (14.4).*

Solution. *Step 1.* The given flood discharges are converted into logarithms through the transformation $y_i = \log x_i$. The y series is

2.5966, 2.7917, ...

....

.... 2.6532, 2.7853.

Step 2. The statistics of the y-series are

$$\overline{y} = 2.70507$$

$$s_y = 0.1327$$

$$g = +\,0.023$$

Step 3. For $T_r = 100$ years and $g = +\,0.023$, from Table 14.4

$K_{100} = 2.343$ and from the same Table $K_{200} = 2.5976$.

Step 4. From Eq. (14.35), the logarithm of 100 year flood is given by

$$y_{100} = 2.70507 + 2.343 \times 0.1327 = 3.016$$

Similarly

$$y_{200} = 2.70507 + 2.5976 \times 0.1327 = 3.0498.$$

Step 5. From Eq. (14.36), the 100 year flood is given by

$$x_{100} = (10)^{3.016} = 1037.53 \text{ m}^3/\text{s}$$

Similarly $x_{200} = (10)^{3.0498} = 1121.5 \text{ m}^3/\text{s}$.

14.7.5 Reliability of Estimates of Distribution Characteristics

Though we take the sample characteristics such as $\overline{x}$ and s_x as the estimates of the corresponding characteristics of the probability distribution function μ and σ, they cannot be equal to the characteristics of the distribution since we are not using the entire population in computing them.

This is because while the sample size n is a finite number the size of the population approaches ∞. Obviously the reliability of characteristics computed from the sample will increase as the sample size becomes larger.

Consider a long record of hydrological variable such as the annual rainfall at a place. Suppose this record is 500 years long and suppose this record is divided into 25 samples each 20 year long. The mean of each sample is computed and denoted as $\bar{x}_1, \bar{x}_2, \ldots\ldots, \bar{x}_i, \ldots\ldots\ldots, \bar{x}_{25}$. These 25 values of $\bar{x}$ can be used to identify the distribution of the mean $\bar{x}$. In other words $\bar{x}$ is treated as a random variable which has its own distribution.

Central Limit Theorem. The central limit theorem can be stated as follows. If S_n is the sum of n independently and identically distributed random variables X_i, each having the mean μ and variance σ^2, then in the limit as n approaches infinity, the distribution of S_n approaches normal distribution with mean $n\mu$ and variance $n\sigma^2$. The distribution of the random variable X_i can be any distribution. If we define S_n as the average of the n variables, that is $S_n = (X_1 + X_2 + \ldots\ldots + X_i + \ldots\ldots + X_n)/n$, then from central limit theorem S_n (which is nothing but $\bar{x}$) will be normally distributed with mean μ and variance (σ^2/n).

Standard Error. As seen from the central limit theorem the sample mean $\bar{x}$ will be normally distributed with mean $\bar{x}$ and standard deviation $\left(s_x/\sqrt{n}\right)$. The standard deviation of mean $\bar{x}$ computed from a sample size n is called the standard error of the mean and will be equal to $\left(s_x/\sqrt{n}\right)$. Similarly the standard error of the standard deviation is given by $\left(s_x/\sqrt{2n}\right)$. As the sample size n increases the standard error decreases.

Confidence Limits. The value of a random variable for any specified exceedence probability estimated from the distribution fitted to the observed data will be in error due to the limited sample size. Therefore the limits on either side of the estimated value are needed to define an interval within which the true value is contained with a specified probability. The intervals so defined are called confidence limits, confidence intervals, confidence bands, error limits or control curves. They indicate the limits above and below the estimated value within which the true value will lie with a specified confidence level.

Suppose we want to define the confidence interval for the flood estimates for a T year return period with a confidence level of α (in per cent). The limits of the confidence band can be calculated as given here. The flood x_T corresponding to the desired return period T is computed from Eq. (14.29). Then the lower and upper limits of the confidence interval for a chosen confidence level α are given by

$$x_{TL} = x_T - G(\alpha)\, S_e \qquad \ldots(14.37)$$

$$x_{TU} = x_T + G(\alpha)\, S_e \qquad \ldots(14.38)$$

Table 14.5 Values of G (α) from the Standard Normal Distribution Table

α *in per cent*	50	68	80	90	95	99
G (α)	0.674	1.00	1.282	1.645	1.96	2.58

where G (α) is a function of α and can be determined from the table of standard normal distribution. For example if α = 50% we have to select the normal variates from the normal distribution table such that the area of the normal curve between the variates is 50%. It can be easily found that between – 0.674 and 0.674 the area of the curve is 50%. Table (14.5) gives the values of G (α) for some commonly used values of α. S_e is the probable error given by

$$S_e = b\frac{s_x}{\sqrt{n}} \quad ...(14.39)$$

where $$b = \left(1 + 1.3\,K_T + 1.1\,K_T^2\right)^{0.5} \quad ...(14.40)$$

Example 14.8. *Assuming Gumbel's distribution is a good fit for the data of Example (14.4) construct flood v_s Return period curve by computing the floods for return periods of 10, 50, 100, 200, 500 years. Also show on the same figure the confidence band for 50% and 90% probabilities.*

Solution. From the Examples (14.4) and (14.5) the following information is imported

$n = 40$ years, $\bar{x} = 530\ \text{m}^3/\text{s}$; $s_x = 161.547\ \text{m}^3/\text{s}$

$\bar{y}_n = 0.54362$; $\sigma_n = 1.14132$

The computations are shown in Table (14.6). How the values in various columns are obtained is detailed below for one return period T of 100 years. Which appears in the middle row.

Col. (1) T is taken as 100 years.

Col. (2) $y_T = -\ln\ln\left(\frac{T}{T-1}\right) = -\ln\ln\left(\frac{100}{99}\right) = 4.6$

Col. (3) $K_T = \frac{y_T - \bar{y}_n}{\sigma_n} = (4.6 - 0.54362)/1.14132 = 3.554$

Col. (4) $x_T = \bar{x} + K_T\, s_x = 530 + 3.554 \times 161.547 = 1104.1\ \text{m}^3/\text{s}$

Col. (5) $S_e = b\,.\,s_x/\sqrt{n}$; $b = \left(1 + 1.3\,K_T + 1.1\,K_T^2\right)^{1/2}$

$$S_e = \left(1 + 1.3 \times 3.554 + 1.1 \times 3.554^2\right)^{1/2} \times 161.547/\sqrt{40} = 113\ \text{m}^3/\text{s}$$

Col. (6) G (50%). $S_e = 0.674 \times 113 = 76$

Col. (7) $x_{TL} = x_T - G(\alpha)\,S_e = 1104.1 - 76 = 1028.1\ \text{m}^3/\text{s}$
Col. (8) $x_{TU} = x_T + G(\alpha)\,S_e = 1104.1 + 76 = 1180.1\ \text{m}^3/\text{s}$ } 50% confidence limits

Col. (9) G (90%). $S_e = 1.645 \times 113 = 186$

Col. (10) $x_{TL} = x_T - G(\alpha)\,S_e = 1104.1 - 186 = 918.1\ \text{m}^3/\text{s}$
Col. (11) $x_{TU} = x_T + G(\alpha)\,S_e = 1104.1 + 186 = 1290.1\ \text{m}^3/\text{s}$ } 90% confidence limits

Table 14.5(*a*) Calculation of 50% and 90% Confidence limits of Example 14.8

T	y_T	K_T	x_T	S_e	α = 50% G (α) = 0.674			α = 90% G(α) = 1.645		
years					0.674 S_e	x_{TL}	x_{TU}	1.645 S_e	x_{TL}	x_{TU}
1	2	3	4	5	6	7	8	9	10	11
10	2.25	1.495	771.5	59	40	731.5	811.5	97	674.5	868.5
50	3.9	2.941	1005.1	97	65	940.1	1070.1	160	845.1	1165.1
100	4.6	3.554	1104.1	113	76	1028.1	1180.1	186	918.1	1290.1
200	5.3	4.164	1202.7	129	87	1115.7	1289.7	212	990.7	1414.7
500	6.2	4.956	1330.6	150	101	1229.6	1431.6	247	1083.6	1577.6

A plot between return period *T* of Col. (1) and flood magnitude x_T of Col. (4) on Powells probability paper will give the required flood frequency curve. This is a straight line and labelled as (3) in Fig. (14.6). Plot between *T* of Col. (1) and x_{TL} of Col. (7) and x_{TU} of Col. (8) which are labelled as (1) in Fig. 14.6 will define the confidence band for 50% confidence level. Similarly curves obtained by plotting x_{TL} of Col. (10) and also x_{TU} of Col. (11) against return period *T* which are labelled as curves (2) in Fig. 14.6 will represent the confidence band for 90% probability level.

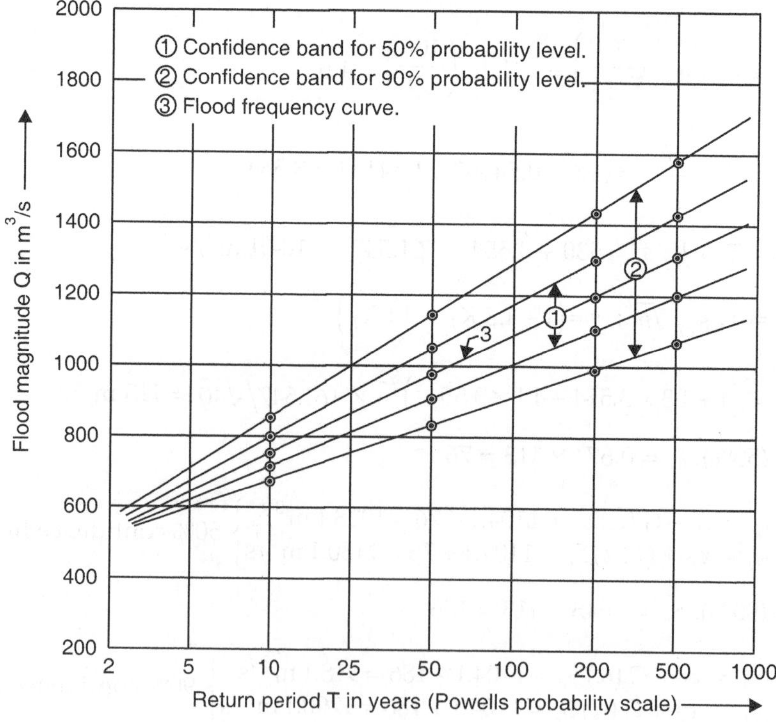

FIGURE 14.6 Flood frequency and confidence band curves of example 14.8

14.7.6 Selection of Design Return Period

The return period established by frequency analysis described so far indicates only the average interval between the occurrence of floods equal to or greater than a given magnitude. There is no implication for a T_r year flood to occur at the end of every T_r year period. In the terminology of probability theory, $\frac{1}{T_r}$ indicates the probability with which the T_r year flood may be equalled or exceeded in any one year. Therefore if it is desired to select a design flood which is not likely to occur during the life period of the structure, it becomes necessary to use a return period greater than the estimated useful life.

Suppose that a structure is designed for T_r year flood and its useful life period is N years. Then the probability that the design flood is equalled or exceeded (and hence the probability that the structure may fail) in any year is $\frac{1}{T_r}$. Alternatively the probability that the structure does not fail in any year is $\left(1-\frac{1}{T_r}\right)$. Assuming that the annual flood peaks are independent events, the probability that the structure does not fail in the next N years is $\left(1-\frac{1}{T_r}\right)^N$. Hence the probability that the structure may fail in any one of the next N years, which is nothing but the risk in the design, is given by

$$R = 1-\left(1-\frac{1}{T_r}\right)^N \qquad \text{...(14.41)}$$

The above equation may be used to evaluate the risk involved in adopting T_r year flood for a project with useful life of N years. Alternatively it can also be used to determine the required return period of the design flood for a given risk R and given life period N. For example if a structure with a life period of 50 years is designed also for a 50 year flood, the risk of failure is 0.636. In other words the chances of its failure in the next 50 years is 63.6% and not 2% as usually thought of. If we want to reduce the risk to 0.10, the structure has to be designed for 475 year flood.

Example 14.9. *A coffer dam is designed for a 25 year flood and constructed. If it takes 5 years to complete the construction of main dam, what is the risk that the coffer dam may fail before the end of the construction period ? What return period in the design of coffer dam would have reduced the risk to 10% ?*

Solution. From Eq. (14.41), the risk of failure is given by

$$R = 1-\left(1-\frac{1}{T}\right)^N$$

Here T = 25 years and N = 5 years. Therefore,

$$R = 1-\left(1-\frac{1}{25}\right)^5 = 0.1846 = 18.46\%$$

If the risk is to be reduced to 10%, we have

$$0.1 = 1 - \left(1 - \frac{1}{T}\right)^5, \text{ which gives}$$

$$T = 47.96 \text{ years say } 50 \text{ years.}$$

Example 14.10. *The analysis of a 30 year flood data at a point on a river yielded $\bar{x}$ = 1200 m³/s and s_x = 650 m³/s. For what discharge would you design the structure at this point to provide 95% assurance that the structure would not fail in the next 50 years ?Use Gumbel's method.*

Solution. Assurance = 95%

Therefore risk $R = 100 - 95 = 5\%$

or $R = 0.05$

The life period $N = 50$ years.

∴ From Eq. (14.41),

$$0.05 = 1 - \left(1 - \frac{1}{T_r}\right)^{50}$$

$$\left(1 - \frac{1}{T_r}\right)^{50} = 1 - 0.05 = 0.95$$

$$1 - \frac{1}{T_r} = (0.95)^{1/50} = 0.9989746$$

$$\frac{1}{T_r} = 1 - 0.9989746 = 0.0010253$$

$$T_r = \frac{1}{0.0010253} = 975.3 \text{ years.}$$

Now Gumbel's method is used to estimate the flood with a return period of 975.3 years. From Table 14.3 for $n = 30$, we obtain $\bar{y}_n = 0.53622$ and $\sigma_n = 1.11238$. From Eq. (14.26),

$$y_T = -\ln\ln\left(\frac{975.3}{974.3}\right) = 6.88223$$

From Eq. (14.30),

$$K_T = \frac{6.88223 - 0.53622}{1.11238} = 5.705$$

From Eq. (14.29), the design flood is given by

$$x_T = 1200 + 5.705 \times 650 = 4908.25 \text{ m}^3/\text{s}.$$

∴ The structure has to be designed for a discharge of 4910 m³/s.

14.8. REGIONAL FLOOD FREQUENCY ANALYSIS

The method of frequency analysis described in section 14.7 is of no value if the catchment under investigation is not having any observed record. Regional flood frequency analysis is the commonly used method to overcome this problem associated with ungauged catchments. The application of this method consists of developing two curves using the flood data of the gauged catchments in the region. The first curve shows the mean annual peak flood *versus* the catchment area. The second curve shows *peak flow ratio* versus the return period. The peak flow ratio is the ratio of peak flow for a given return period to the mean annual flood. Once these two curves are developed for the region, a flood frequency curve for any other ungauged catchment in the same region can be constructed. The procedure to develop the two curves consists of the following steps :

1. The data on annual flood peaks for several gauged stations, usually about 10 to 15, are assembled and a time base which is common to all stations is selected.
2. For each station, the observed flood peaks are arranged in descending order, ranks are assigned and return periods are computed using Weibul's formula.
3. For each station, annual peak flood is plotted against its return period on an extreme value probability paper and a straight line is fitted by visual inspection.
4. For each station, the mean annual flood, that is, the flood with a return period of 2.33 years is determined.
5. Now the first curve is prepared by plotting the mean annual flood of each site against its catchment area (see Fig. 14.7). Generally an equation of the form $\overline{Q} = CA^n$ may be fitted to this curve. Hence, when plotted on a log—log sheet, it assumes a straight line shape.

   Sometimes the mean annual flood $\overline{Q}$ may be correlated with other parameters, apart from the catchment area, such as basin slope, mean elevation, per cent of area covered by lakes, and mean annual precipitation over the catchment.
6. Several return periods such as 1.25, 2, 5, 10, 25, 50, etc., are chosen and for each return period and for each station the annual flood peak Q are obtained from the frequency curves prepared in step 3 and these are converted into peak flow ratios by dividing them by the respective mean annual flood obtained in step 4. Typical results expected in this step are as shown in Table 14.6.

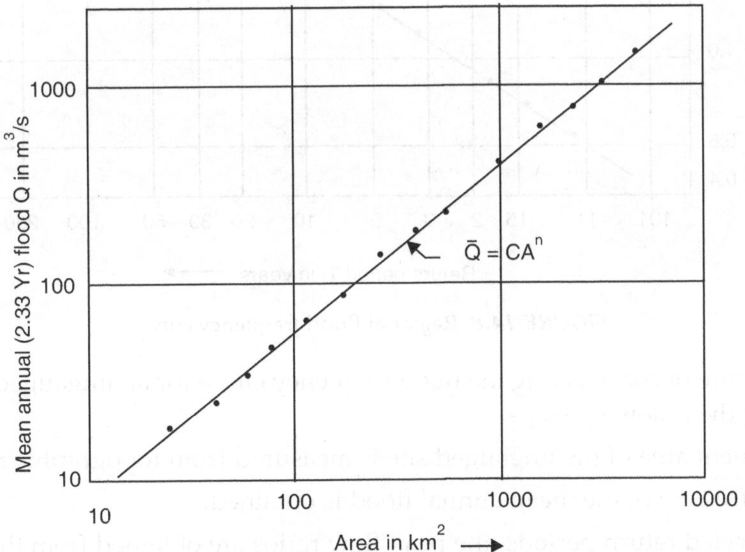

FIGURE 14.7 Mean annual flood versus catchment area.

Table 14.6. Peak Flowratios for Regional Frequency Analysis

Station	*$Q/\overline{Q}$ for a return period equal to*					
No.	*1.25*	*2*	*5*	*10*	*25*	*50*
1	0.49	0.90	1.45	1.82	2.28	2.62
2	0.51	0.92	1.44	1.79	2.23	2.56
3	0.55	0.92	1.40	1.73	2.14	2.44
4	0.45	0.90	1.50	1.88	2.38	2.74
5	0.50	0.91	1.46	1.84	2.32	2.68
Median	0.50	0.91	1.45	1.82	2.28	2.62

7. For each return period, the median value of the peak flow ratios for all stations is found out. These median peak flow ratios are shown in the last row of Table 14.6.
8. Now the second curve, known as the *regional flood frequency curve* is prepared by plotting the median peak flow ratio against the return period on the extreme value probability paper and drawing a best fit straight line (see Fig. 14.8).

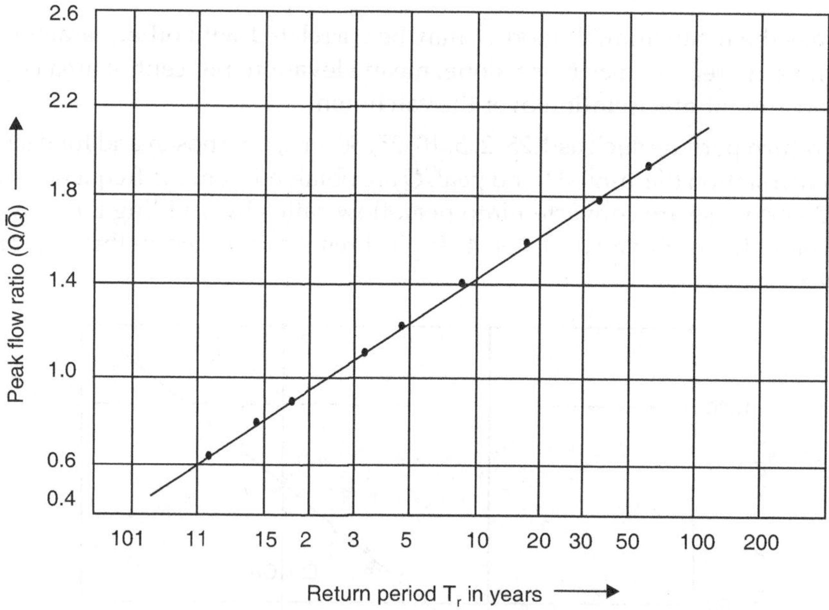

FIGURE 14.8 Regional flood frequency curve.

The procedure of constructing the flood frequency curve for an ungauged catchment in the region consists of the following steps.

1. The catchment area of the ungauged site is measured from topographic maps.
2. Using the first curve the mean annual flood is obtained.
3. For the selected return periods, the peak flow ratios are obtained from the second curve.

4. The flood peaks are determined by multiplying the peak flow ratios obtained in step 3 by the mean annual flood obtained in step 2.
5. The flood frequency curve for the ungauged catchment is then prepared by plotting the flood peaks against their return periods on extreme value probability paper.

The streams selected for regional flood frequency analysis should be as nearly similar in hydrologic characteristics as possible. They should have similar vegetal cover, land use, topographic conditions and geologic characteristics, and the range of the catchment area should not be excessive. They should also have similar rainfall and evapotranspiration regimes.

The regional flood frequency analysis assumes that for a given return period the peak flow ratio is same for all stations. This is true only when the coefficient of variation of flood peak is same for all the sites. Dividing Eq. (14.29) throughout by $\overline{x}$, we get

$$\frac{x_T}{\overline{x}} = 1 + K_T \cdot \frac{s_x}{\overline{x}} \qquad \text{...(14.42)}$$

For a given return period K_T is same for all the sites. Therefore the peak flow ratio $(x_T/\overline{x})$ will be same only when $\frac{s_x}{\overline{x}}$, which is nothing but the coefficient of variation, is same for all the sites. This is the weakness of the regional flood frequency approach.

REVIEW EXERCISES

Descriptive Questions

1. Distinguish between
   (*a*) Maximum probable flood and Design flood
   (*b*) Annual series and Partial series
   (*c*) Return period and Exceedence probability.
2. What are the methods of estimating design flood ? What are their limitations ?
3. What are envelope curves ? How are they prepared ?
4. Give three empirical formulae that are commonly used to estimate the design flood. Also specify the regions in India where they are applied and range of co-efficient values in the equations.
5. Explain the method of design flood estimation using the rational method.
6. What is concentration time ? How is it estimated ?
7. Explain the method of design flood estimation using the unit hydrograph method. How is it modified when the area is large ?
8. Describe the method of estimating a T_r- year flood using Gumbel's distribution.
9. Describe the method of estimating a T_r-year flood using Log-Pearson type III distribution.
10. What do you understand by the frequency factor ? How is it determined for Gumbel's distribution ?

11. How do you evaluate the risk in the design of a project which is designed for T_r-year flood and whose life period is N-years ?
12. How do you construct the extreme value probability paper ?
13. Describe the method of regional flood frequency analysis for constructing flood frequency curve of an ungauged catchment.
14. What is a design storm ? What is PMP ?
15. How is the rational formula modified when the catchment area is made up of different sub-areas with different runoff coefficients ?

Analytical Questions

1. The maximum rainfall intensity of 50 year return period over a region is given by

$$I = \frac{52}{(t+8)^{0.40}}$$

where I is cm/hr and t is duration in minutes. For a watershed of 50 km^2 in this region with a concentration time of 2 hours, estimate the flood peak using rational formula. Assume the runoff co-efficient C= 0.55.

2. The intensity-frequency-duration relationship for the catchment area is given by

$$I = \frac{800\, T_r^{0.2}}{(t+15)^{0.7}}$$

in which I = rainfall intensity in mm/h, T_r = return period in years, and t = rainfall duration in minutes. Estimate the 50 year flood using rational method for a catchment with an area of 750 hectares, slope of 0.004. The length of travel of water along the water course may be taken as 950 m. Take C = 0.35.

3. A catchment is divided into five sub-areas as given below :

Sub-area (km^2)	1.5	3.0	2.0	1.0	2.5
Runoff coefficient C	0.7	0.6	0.4	0.55	0.9

Calculate the 25 year flood using the rational method. Assume a concentration time of 35 minutes and use the following intensity-frequency duration function.

$$I = \frac{75.0\, T_r^{0.22}}{(t+12)^{0.85}}$$

where I is in cm/h and t is in minutes.

4. The observed annual flood peaks of a stream for the period 1939 to 1973 in m^3/s are as given below.

588, 432, 648, 396, 420, 500, 336, 900, 300, 600, 504, 396, 518, 420, 528
384, 698, 610, 408, 372, 480, 672, 720, 912, 624, 336, 324, 360, 696, 456,
636, 684, 756, 507, 312.

(*a*) Construct the probability plotting of this data on an ordinary graph paper and hence estimate 50 year and 100 year flood.

(*b*) Construct probability plotting on a semilog paper with return period on logarithmic axis and then estimate 50 year and 100 year flood.

(*c*) Estimate the 50 year and 100 year floods using Gumbel's extreme value distribution.

(*d*) Estimate the 50 year and 100 year floods using Log-Pearson type III distribution.

5. From the historical data of annual flood peaks of a catchment the mean and standard deviation are estimated as 20000 m^3/s and 10,000 m^3/s. An existing structure on this catchment has been designed for 40000 m^3/s. What could be its return period. Assume Gumbel's extreme value distribution with $\sigma_n = 1.06$ and $\bar{y}_n = 0.52$.

6. The data of observed flood peaks of a river for a period of 30 years is found to plot as a straight line on semi-log paper with return period plotted on the logarithmic scale. The largest and smallest floods in the record are 1170 m^3/s and 195 m^3/s respectively. If 1350 m^3/s is selected as a design flood, what is the probability of its being exceeded during the next 20 years.

7. The annual flood peak of a stream is estimated to have 50 year and 100 year floods of 2400 m^3/s and 2730 m^3/s respectively. What is 200 year flood for the same stream.

8. From a partial series of flood peaks on a stream it is found that in a period of 23 year a discharge of 490 m^3/s or more occurred 36 times and a discharge of 660 m^3/s or more occurred 8 times. What are the return periods of these floods ? What are the corresponding exceedence probabilities.

9. The 4*h* unit hydrograph of a basin may be approximated as a triangle with a base period of 48 h and a peak discharge of 900 m^3/s at the end of 12th hour. The design storm is having a total duration of 12 h with rainfall intensities of 2.5 cm/h, 4 cm/h and 3 cm/h in three successive 4 h periods. Determine the design flood discharge and its time of occurrence. Assume that the basin is completely saturated before the occurrence of design storm.

10. Given the following statistics of annual flood peaks of a river, compute the 100-year flood using (*a*) normal (*b*) Gumbel and (*c*) Log-Pearson type III distributions.

n = 45 years ; mean = 2700 m^3/s ; standard deviation = 1300 m^3/s ;

mean of logarithms = 3.1 ; standard deviation of logarithms = 0.4, skewness coefficient of logarithms = – 0.35.

11. The construction of a main dam takes 4 years for completion. For what return period would you design a coffer dam to provide 95% assurance that the coffer dam will not fail before the construction of main dam ? What would be the return period if the assurance is increased to 99% ?

12. What is the probability for 200 year flood to be exceeded in any of the next 20 years ?

13. From enquiry with local people it is found that a flood of magnitude 2550 m^3/s exceeded 5 times in 49 years and that with a magnitude of 10000 m^3/s exceeded twice. Determine the return periods and exceedence probabilities for these floods. Assuming that the two points corresponding to these two floods exactly lie on the flood frequency curve plotted on extreme value probability paper, estimate the 100 year and 200 year floods.

14. Based on analysis of 30 years data on annual floods, which gave a mean of 1200 m^3/s and standard deviation of 575 m^3/s, a structure with a useful life period of 50 years was designed for a flood of 3000 m^3/s. What is the risk involved in the design ? Later it was found that there was an error in the computation of statistics and the correct mean and correct standard deviation were estimated as 1075 m^3/s and 500 m^3/s respectively. Determine the corresponding reduction in the risk. Use Gumbel's method.

15. In a regional flood frequency analysis the equation fitted to the first curve is given by $\overline{Q} = 2.75\, A^{0.625}$, where the mean annual flood $\overline{Q}$ is in m^3/s and A is the catchment area in km^2. The coordinates of two points on the second curve are given below.

$Q/\overline{Q}$	1.85	2.66
Return period	10	50

Determine the 100 year flood for an ungauged catchment in the same region with a catchment area of 2500 km^2.

Objective Type Questions

1. A flood is given a rank m in an observed record of n years. Then according to Weibul's formula, its return period is given by

(*a*) $T_r = \dfrac{n}{m}$ (*b*) $T_r = \dfrac{n}{m+1}$

(*c*) $T_r = \dfrac{n}{m-1}$ (*d*) $T_r = \dfrac{n+1}{m}$.

2. According to Hazen's formula, the return period is given by

(*a*) $T_r = \dfrac{n}{m}$ (*b*) $T_r = \dfrac{2n}{2m-1}$

(*c*) $T_r = \dfrac{2n+1}{m}$ (*d*) $T_r = \dfrac{n}{2m+1}$.

3. According to California formula, the return period is given by

(*a*) $T_r = \dfrac{n}{m}$ (*b*) $T_r = \dfrac{2n}{2m-1}$

(*c*) $T_r = \dfrac{2n+1}{m}$ (*d*) $T_r = \dfrac{n}{2m+1}$.

4. The most commonly used formula for computing return period is

(*a*) California formula (*b*) Hazen's formula

(*c*) Weibul's formula (*d*) Beard's formula.

5. The most commonly used probability distribution to fit the flood data is

(a) Normal distribution (b) Gambel's extreme value distribution

(c) Log-Normal distribution (d) Gamma distribution.

6. A flood with a return period of 100 years is the flood which occurs

(a) every 100th year

(b) the maximum observed flood in the past 100 years

(c) once in every 100 years on the average

(d) only after 100 years in the immediate future.

7. The probability that a T_r year flood occurs in any year is

(a) $\frac{1}{T_r}$ (b) $\left(\frac{1}{T_r}\right)^2$

(c) $\log\left(\frac{1}{T_r}\right)$ (d) e^{-T_r}.

8. In Gumbel's distribution, the relation between the reduced variate y and the return period T_r is given by

(a) $e^{-e^{-y}} = 1 - \frac{1}{T_r}$ (b) $e^{-e^{-y}} = \frac{1}{T_r}$

(c) $e^{-e^{-y}} = 1 + \frac{1}{T_r}$ (d) $e^{-e^{-y}} = \frac{1}{T_r} - 1$.

9. A structure is designed for T_r year flood. It has an estimated useful life period of N years. Then the probability that it will not fail during its life period is given by

(a) $\left(1 - \frac{1}{T_r}\right)$ (b) $\left(1 - \frac{1}{T_r}\right)^N$

(c) $1 - \left(1 - \frac{1}{T_r}\right)^N$ (d) $\left(1 - \frac{1}{N}\right)^{T_r}$.

10. A structure with a useful life period of N years is designed for a T_r year flood. Then the risk in the design is given by

(a) $\frac{1}{T_r}$ (b) $\left(\frac{1}{T_r}\right)^N$

(c) $1 - \left(1 - \frac{1}{T_r}\right)^N$ (d) $1 - \left(\frac{1}{T_r}\right)^N$.

11. If A is the catchment area and C is a coefficient, the Ryve's formula for flood peak is given by

(*a*) $Q = C\ A^{1/3}$ (*b*) $Q = C\ A^{2/3}$

(*c*) $Q = C\ A^{1/4}$ (*d*) $Q = C\ A^{3/4}$.

12. The Dicken's formula for flood peak is given by

(*a*) $Q = C\ A^{1/3}$ (*b*) $Q = C\ A^{2/3}$

(*c*) $Q = C\ A^{1/4}$ (*d*) $Q = C\ A^{3/4}$.

13. If the annual flood peak follows the Gumbel's distribution, then the return period of the mean annual flood is equal to (where n is the length of record)

(*a*) 2.33 years

(*b*) $\frac{n}{2}$ years

(*c*) rank of the median of the observed floods

(*d*) n years.

14 The reduced variate y_T of Gumbel's distribution corresponding to a return period of 100 years is equal to

(*a*) 3.9 (*b*) 4.6

(*c*) 5.0 (*d*) 5.3.

15. If x_T and K_T are the flood and frequency factor corresponding to a return period of T_r years and $\overline{x}$ and s_x are the mean and standard deviation of the floods, then

(*a*) $x_T = \frac{\overline{x}}{K_T . s_x}$ (*b*) $x_T = \overline{x} - K_T\ s_x$

(*c*) $x_T = \overline{x} + K_T . s_x$ (*d*) $x_T = \frac{K_T . s_x}{\overline{x}}$

16. A concrete pavement of an area 50 ha has a concentration time of 20 minutes for which the max. intensity of rainfall is estimated to be 180 mm/h. What is the peak discharge from the area using rational method ?

(*a*) 5 m^3/s (*b*) 25 m^3/s

(*c*) 50 m^3/s (*d*) 100 m^3/s.

17. The data required to estimate the T_r year flood using Log-Pearson type III distribution is

(*a*) mean, standard deviation of the floods and length of record

(*b*) mean, standard deviation and skew coefficient of the floods

(*c*) mean, standard deviation and skew coefficient of the logarithms of the floods

(*d*) mean and standard deviation of the logarithms of the floods.

18. From a straight line fitted to the plot between the return period on logarithmic scale and flood peak on ordinary scale, the 10 year and 100 year floods are obtained as 550 m^3/s and 1100 m^3/s respectively. What would be the magnitude of 1000 year flood.

(*a*) 2200 m^3/s (*b*) 1650 m^3/s

(*c*) 11000 m^3/s (*d*) 4400 m^3/s.

19. Which of the following is the correct relation between the reduced variate y_T and the corresponding return period T_r.

(*a*) $e^{-e^{-y_T}} = 1 - \frac{1}{T_r}$ (*b*) $y_T = -\ln\ln\left(\frac{T_r}{T_r - 1}\right)$

(*c*) $y_T = -\left[0.834 + 2.303 \log\log\left(\frac{T_r}{T_r - 1}\right)\right]$

(*d*) all the above.

20. A catchment is made of 60% area with runoff coefficient 0.40 and remaining 40% area with runoff coefficient 0.60. What is the weighted runoff coefficient to be used is Rational formula ?

(*a*) 0.6 (*b*) 0.24

(*c*) 0.5 (*d*) 0.48.

ANSWERS

Analytical Questions

1. 570.36 m^3/s 2. 86.06 m^3/s 3. 103.42 m^3/s
4. (*a*) 1050 m^3/s and 1310 m^3/s (*b*) 990 m^3/s and 1050 m^3/s
(*c*) 1013 m^3/s and 1114 m^3/s (*d*) 960 m^3/s and 1050 m^3/s.
5. 14.52 years 6. 58 years, 0.293 7. 3059 m^3/s
8. $\frac{2}{3}$ years and 3 years ; 0.125 in any month and 0.33 in any year
9. 30600 m^3/s at the end of 20th hour
10. (*a*) 5723.8 m^3/s (*b*) 6807.3 m^3/s (*c*) 8445.0 m^3/s
11. 78.5 years ; 398.5 years
12. 0.0954 or 9.54%
13. 10 years and 0.1 ; 25 years and 0.04
100 year flood = 20980 m^3/s
200 year flood = 26436 m^3/s
14. 0.2608 or 26.08% 15. 1097.74 m^3/s

Objective Type Questions

1. (*d*)	2. (*b*)	3. (*a*)	4. (*c*)	5. (*b*)
6. (*c*)	7. (*a*)	8. (*a*)	9. (*b*)	10. (*c*)
11. (*b*)	12. (*d*)	13. (*a*)	14. (*b*)	15. (*c*)
16. (*b*)	17. (*c*)	18. (*b*)	19. (*d*)	20. (*d*).

Chapter 15 SEDIMENTATION

15.1 INTRODUCTION

When soil is moved from its current location *erosion* is said to have taken place. The soil particles which are detached and moved are referred to as *sediments*. The study of erosion and transport of sediments is an important subject in engineering hydrology and it encompasses the related fields such as *fluvial geomorphology* and river engineering. The process of erosion and sedimentation has important economic and environmental consequences. Estimating rates of erosion, transport of sediment and its subsequent deposition on the floodplains or in reservoirs, lakes and estuaries is one of the tasks of the hydrologists. This chapter briefly deals with the methods of estimating erosion and sediment deposition.

15.2 THE PROCESS OF EROSION

Erosion has been defined as a process of detachment of soil from its present location and subsequent transportation to other locations by the erosive agents. For erosion by water the agents are rainfall and runoff. The ease with which a soil erodes or the natural susceptibility of a soil to erosion is called soil erodibility. Some soils erode more easily than others. The difference is due to soil properties.

The presence of sediments in streams and rivers has its origin in soil erosion. The erosion is a result of series of complex and interrelated natural proceses, which would loosen and move away the soil and rock materials under the action of water, wind and other geological factors. The long term effect of erosion is the removal of soil and rock particles from exposed surfaces (*i.e.*, denudation of land surface), their transport to lower elevations, and eventual deposition.

According to the timespan involved, erosion can be classified as *normal* or *geologic erosion* and *accelerated* or *human induced erosion*. Normal erosion has been occurring at variable rates since the first solid materials formed on the surface of the earth. Normal erosion is extremely slow in most places and is largely a function of climate, parent rocks, precipitation, topography, and vegetative cover. Accelerated erosion occurs at a much faster rate than normal, usually through reduction of vegetative cover. Deforestation, cultivation, forest fires, and systematic destruction of natural vegetation result in accelerated erosion.

According to its source, erosion can be classified as (1) *sheet erosion*, (2) *rill erosion*, (3) *gully erosion* and (4) *channel erosion*.

Sheet erosion is the wearing away of a thin layer on the land surface, especially between rills, primarily by overland flow. Rill is nothing but a very small stream. The detachment of the soil particles takes place by the impact of raindrops. The kinetic energy of the drops can splash soil particles into the air. On level ground the particles are redistributed more or less uniformly in all directions but on slope grounds more particles will fall on downslope. If the overlandflow is taking place some of the falling soil particles of the splash will be entrained in the flowing water. Though overland flow is predominantly laminar and hence cannot cause detachment of soil particles from the soil mass, it can move loose particles already existing on the soil surface. Thus the splash and overland flow processes are responsible for sheet erosion. If d and v are the diameter and velocity of the rain drop respectively, its kinetic energy is proportional to d^3v^2. The erosive power of large drops found in intense storms may be few thousand times that of smaller drops. Also, the overland flow is more likely to occur during intense storms. This conforms with the observation that it is only intense storms which account for most of the sheet erosion. It is difficult to detect sheet erosion except when the soil surface is lowered below old soil marks on fence posts or when tree roots are exposed. Sheet erosion is the same as interrill erosion.

Rill erosion is the removal of soil by small concentration of flowing water. It may cause intensive soil movement from a limited part of the land surface. Rills carry the runoff from the interrill areas as well as the rain falling directly on them. Rills may develop due to topographic variations, tillage operations or irregularities on the surface. They are easily smoothed by agricultural practices.

Gully erosion is the removal of soil from the rivulets which are formed due to sufficient accumulation of overland flow on the slope grounds. If the turbulence in the flow through rivulets is strong enough it may dislodge particles from bed and banks leading to gully erosion. Gully erosion results in massive removal of soil by large concentrations of runoff and may yield large volumes of sediment.

Channel erosion refers to erosion occurring in stream channels in the form of stream bank erosion or streambed degradation and valley trenching. In many watersheds channel erosion is more significant than sheet erosion.

Gross erosion is the summation of erosion from all sources within the catchment area. It includes interrill erosion or sheet erosion, rill erosion from tilled cropland, meadows, pastures, construction sites, industrial areas, abondoned areas, mined areas and mine wastes, gully erosion, channel erosion, road and roadside erosion and floodplain scour.

15.3 FACTORS AFFECTING EROSION

The most important factors which influence the erosion from any area are rainfall regime, vegetal cover, soil type, land slope and land use.

As already mentioned, erosion is more in intense storms. This is because both raindrop size and velocity will be more in intense storms.

Vegetation absorbs the energy of the raindrop falling on it, and it reduces the size and velocity of the raindrops reaching the ground surface. It provides mechanical protection to soil against gully erosion. A good vegetal cover generally improves infiltration capacity through the addition of organic matter to the soil. Higher infiltration capacity results in less overland flow. Because of all these reasons, the erosion from a soil surface covered by good vegetation is very much less compared to barren lands.

A cohesive soil will resist splash erosion more than loose soils. Generally, splash erosion increases with an increasing fraction of sand in the soil because of the loss of cohesion.

The splash erosion and overland flow velocities are greater on steep slopes. Therefore erosion rates are greater on steeper slopes than on flat slopes.

Land use is also an important factor that controls rate of erosion. Poor cropping practices and removal of vegetation by fire will accelerate erosion. Proper soil conservation practices such as terracing and benching on steep hill slopes, contour bunding and check dams in the uplands may greatly reduce the erosion from catchments.

Taking into account the effect of soil erodibility, vegetal cover, land slope, slope length and rainfall intensity on soil erosion, Musgrave (1947) developed an empirical equation for the soil loss due to sheet and rill erosion which is expressed as

$$A = C\,R\,S^{1.35}\,L^{0.35}\,P_{30}^{1.75} \qquad ...(15.1)$$

where A is the long term average soil loss due to sheet and rill erosion in inches per year, C is the inherent erodibility of the soil in inches per year (soil erodibility factor), R is a relative soil-cover factor (crop-management factor), S is the degree of the slope in percent, L is the length of the slope in feet, and P_{30} is the 2-year 30 minutes rainfall amount in inches. The value of A can be converted into tons per acre by multiplying by an assumed volume weight of tons per acre-inch.

All the factors discussed above are considered in the universal soil loss equation developed by the Agricultural Research Service of the United States Department of Agriculture (1961), given by

$$A = RKLSCP \qquad ...(15.2)$$

in which A = average annual soil loss due to sheet and rill erosion in tons per hectare per year, R = rainfall factor, K = soil erodibility factor, L = slope length factor, S = slope gradient factor, C = crop management factor, and P = erosion-control-practice factor.

15.4 SEDIMENT TRANSPORT

The sediment generated in the uplands of the catchment reaches the streams and rivers wherein sediment transport takes place with flowing water as the carrier. Sediment transport refers to the mechanisms by which sediment is moved downstream by flowing water either in suspension (known as *suspended load*), or by rolling and sliding along the river bottom (known as *bed load*), or by bouncing along the bed (known as *saltation*). Saltation is a transitional stage between bed load and suspended load and it is usually considered as a part of bed load. *Wash load* refers to sediments comprising very minute particles of colloidal sizes which always remain in suspension and hence are seldom found on the bed of channels. It is turbulence in water which is responsible for suspended load, while the shear stresses developed by flowing water on the channel bed account for the bed load. The sediment which moves as bed load at one section may move as suspended load at another section. The transport of sediment continues in the downstream direction until the flow is no longer able to carry the sediment and at that time sediment deposition occurs. Generally, this happens at the entrance to reservoirs where the flow is decelerated due to the impounding of water by the dam structure. Deposition is also likely to occur naturally, for instance, downstream of sudden decrease in energy slope or in situations where the capacity of flow to carry sediment is substantially diminished. In the absence of these natural or human made features, sediment transport by the flow may continue unabated until it reaches the ocean, at which time the flow looses its kinetic energy and sediment deposition there results in growth of deltas.

Water flowing over a stream bed has a velocity gradient in the vertical direction. Due to this velocity gradient water exerts a shear stress on the particles lying on the stream bed. This shear stress is known as the *boundary shear stress*, or *bottom shear stress* or *tractive stress* and is denoted by τ_0. For wide channels it can be approximated by the following equation

$$\tau_0 = \gamma\, d\, S_0 \qquad \text{...(15.3)}$$

in which γ = specific weight of water, d = depth of flow, and S_0 = equilibrium energy slope of the channel flow. When the bottom shear stress exceeds a threshold value, the particles on the bed begin to move. This threshold value is known as the *critical bottom shear stress* or *critical tractive stress* and is denoted by τ_c. *Duboys formula* is the earliest equation which has been in use for many years to predict the bed load. It is given by

$$q_s = C_s . \frac{\tau_0}{\gamma}(\tau_0 - \tau_c) \qquad \text{...(15.4)}$$

in which q_s = bed-material transport rate per unit channel width in kilograms per second per meter, C_s = an empirical coefficient depending on the size and shape of the sediment particles, and τ_0 and τ_c are the prevailing bottom shear stress and critical tractive stress respectively.

Material in suspension will have varying *concentration* over the depth of flow. For a given volume of water-sediment mixture, the suspended sediment concentration is the ratio of the weight of dry sediment to the weight of water-sediment mixture expressed in parts per million. Thus, from a sample taken at any given depth, the suspended-sediment concentration C_s at that depth in ppm may be obtained as

$$C_s = \frac{\text{weight of dry sediment in the sample}}{\text{weight of the sediment laden water sample}} \times 10^6 \quad ...(15.5)$$

The suspended-sediment concentration varies with the flow depth, usually being higher near the stream bed and lower near the free surface. Once the variation of sediment concentration is known, it can be integrated over the depth of flow to obtain the total suspended load. The evaluation of suspended load by this approach is not discussed here. Interested readers may find the details in the book "Hydraulics of Sediment Transport", by **Graf** (1971).

Sediment Measurement. Sediment concentration in the flowing water is measured by collecting sediment ladden water in a sampler at any desired depth of flow in the stream and then finding the weight of sediment present in the water. Let w_1 be the weight of empty sampler and w_2 be the weight of sampler when it is filled with sediment ladden water sample. Then $(w_2 - w_1)$ gives the weight of sediment ladden water. The collected sample is filtered and the sediment is dried. Let w_3 be the weight of the dried sediment. Then the ratio $\frac{w_3}{(w_2 - w_1)}$ gives the sediment concentration as a dimensionless ratio and when multiplied by 10^6 gives the sediment concentration in ppm (parts per million parts) or milligram/litre.

A good sampler must cause least disturbance to streamflow, avoid errors from short-period fluctuations in sediment concentration and give results which can be related to velocity measurement. The sampler shown in Fig. 15.1, which was developed by Iowa Hydraulic Laboratory satisfies the above requirements.

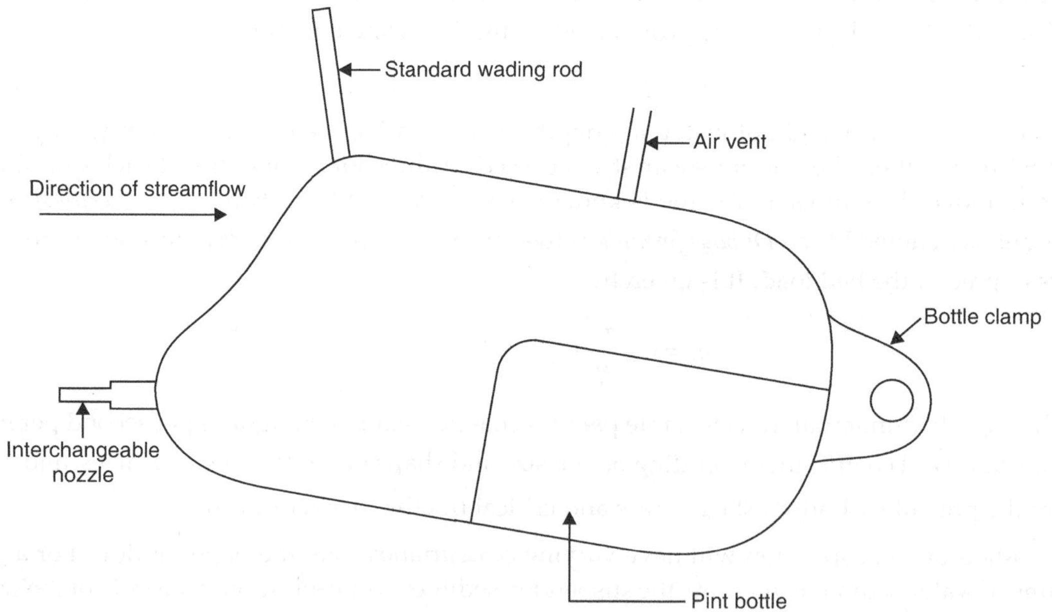

FIGURE 15.1 Suspended-sediment load sampler.

The sampler consists of a streamlined shield which encloses a glass bottle as a sample container. Nozzle tips of various sizes may be used to control the rate at which the bottle fills. A vent is provided in the shield which permits the escape of air as water enters the bottle and controls the inlet velocity so that it is approximately equal to the local stream velocity. The sample is firmly fixed to the bottom of a standard wading rod. This arrangement is very convenient for use in small streams. The large sized samplers which are supported by a cable are preferred in the case of large streams. Such large samplers are fitted with tail vanes to keep the sampler facing the direction of flow.

To collect the sample at any vertical the sampler is lowered through the stream at constant vertical speed until the bottom is reached and it is then raised to the surface at constant speed. The result is an integrated sample with the relative quantity collected at any depth in proportion to the velocity at that depth. The duration of the traverse is determined by the time required to nearly fill the sample bottle and can be computed from the filling rate curves for the particular nozzle when the stream velocity is known. A number of traverses are made at intervals across the section to determine the total suspended-sediment load for the section. Because of the shape of the sampler about 10 cm of depth near the bed is not sampled. This may lead to considerable error in shallow streams.

Depth integration type sampling is not possible because of large depths in the stream or high velocity. In such circumstances point samplers are used at various depths. Point samplers are also used when it is required to study the distribution of sediment in streams.

Samplers used to sample bed load must rest on the stream bed and trap the moving bed load without disturbing the flow. Most portable samplers consist of boxes or bags of wiremesh with supporting frame and tail vane to keep entrance pointed into the current. Permeable plastic bags are used when the bed material consists of fine particles which would pass through wire mesh. The problem is, any device used to collect the sample is certain to cause disturbance to the flow.

Grated openings in the stream bed may be used as permanent bed load traps on small streams. The bed load flowing past the trap will fall into it. The trapped material is later excavated and measured.

Sometimes turbulance producing weirs are designed and put across the stream which will throw the bed load into suspension locally so that it can be sampled with suspended-sediment load samplers. Comparison of the samples thus obtained with those from a section upstream of the weir where only suspended load is measured will indicate the quantity of bed load. This method is suitable only when the bed material is relatively fine.

The average annual sediment production from a catchment is dependent on many factors such as climate, soil type, land use, topography, and the presence of reservoirs. Complete data to undertake thorough analysis is difficult to obtain. However, the mean annual suspended load Q_s in tons is usually expressed by the following simple empirical equation

$$Q_s = a\,Q^n \qquad \text{...(15.6)}$$

in which Q is the mean annual discharge in the stream in m^3/s and a and n are the regression constants to be obtained from the observed sediment rates. From the relations of this nature errors as large as 50% may be expected.

15.5 SEDIMENT YIELD

The amount of sediment eroded and removed from the sources is known as the *gross sediment production*. This is usually measured in terms of weight of sediment per unit drainage area per unit time — for example in tons / km^2 / year. Whereas *sediment yield* refers to the actual delivery of eroded soil particles to a given downstream point. Since the eroded particles may be deposited before they reach the downstream point of interest, the sediment yield will be less than gross sediment production. The ratio of sediment yield to gross sediment production is called the *sediment delivery ratio*. In engineering applications, the quantity of sediment eroded at the sources is not as important as the quantity of sediment delivered to a downstream point *i.e.*, the sediment yield. The sediment yield may be calculated by multiplying the gross sediment production by sediment delivery ratio which varies in the range 0 to 0.1. Sediment yield is expressed in terms of weight per unit time past a certain point, for instance, tons per year at the catchment outlet.

15.6 RESERVOIR SEDIMENTATION

When the sediment laden water of the streams enters the reservoir, it is temporarily detained there and its velocity and turbulence are greatly reduced. Therefore its ability to entrain sediment is substantially impaired and the sediment deposition takes place in the reservoirs. Sediment deposition occurs in the vicinity of reservoirs typically as shown in Fig 15.2. First, deposition of the

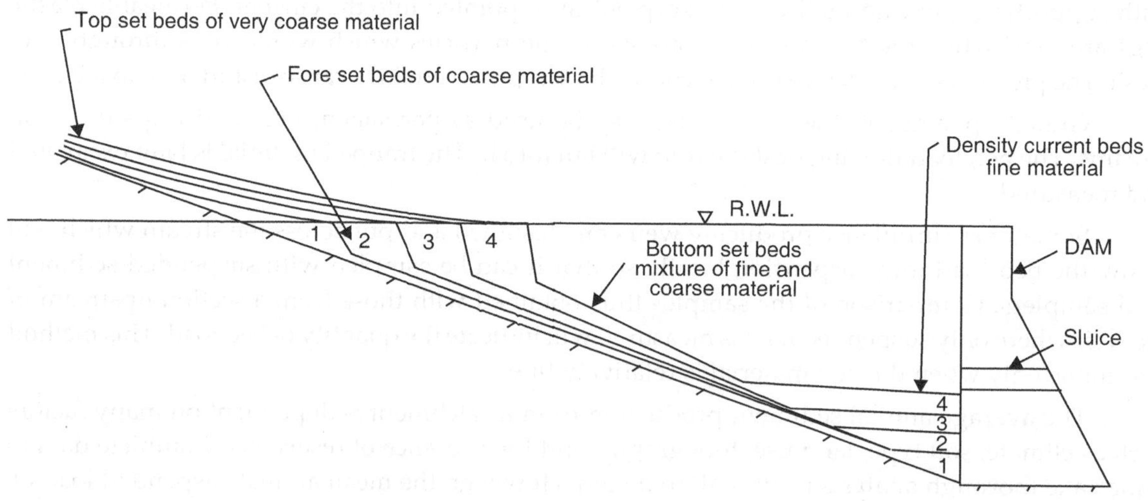

FIGURE 15.2 Sediment deposition in reservoirs.

coarse sediments takes place in a delta near the entrance to the reservoir. In course of time this delta grows in the direction of dam until it eventually fills the entire reservoir volume. Fine sediments which remain in suspension can be carried through to the dam and eventually settle in the deepest part of the reservoir. If the inflowing sediment laden water is denser than the surface water in the reservoir, it moves slowly below the surface in the form of a *density current* or *turbidity current* and carry its load to the dam even though the sediment may not be visible at the reservoir surface. The fine sediments which are deposited near the dam and the density currents carrying sediments may be flushed out by opening the scour sluices of the dam.

The sediment deposition in reservoirs is of considerable economic and practical interest since it reduces the storage capacity of the reservoirs. The *design life* of a reservoir is the period required for the reservoir to fulfil its intended purpose. Due to reservoir sedimentation, provision must be made to guarantee the full design reservoir storage capacity for the planned design life. This necessitates either (1) Cleaning out reservoir sediments at predetermined intervals of time during the life of reservoir or (2) providing a reservoir storage capacity large enough to store all the accumulated sediment deposits without encroachment on the useful storage volume. In either case calculation of sediment deposition rates become part of the design of reservoir storage projects.

The *trap efficiency* of the reservoir refers to the ability of the reservoir to entrap sediment being transported by the flowing water. It is defined as the ratio of sediment trapped to the incoming sediment, in percentage. Trap efficiency is a function of (1) the ratio of reservoir capacity to mean annual runoff volume and (2) the sediment characteristics. Trap efficiency must increase with the residence time of the sediment laden water in the reservoir. The ratio of reservoir capacity to inflow indirectly provides an index of this residence time. Since most sediment enters a reservoir during periods of high inflow, much of it will be discharged over the spillway if the capacity-inflow ratio is small. Therefore trap efficiency must be less for reservoirs with small capacity-inflow ratio. On the other hand, if the capacity-inflow ratio is large, there is less spillage and sediment laden water will be detained in the reservoir for extended periods which results in higher trap efficiency. These

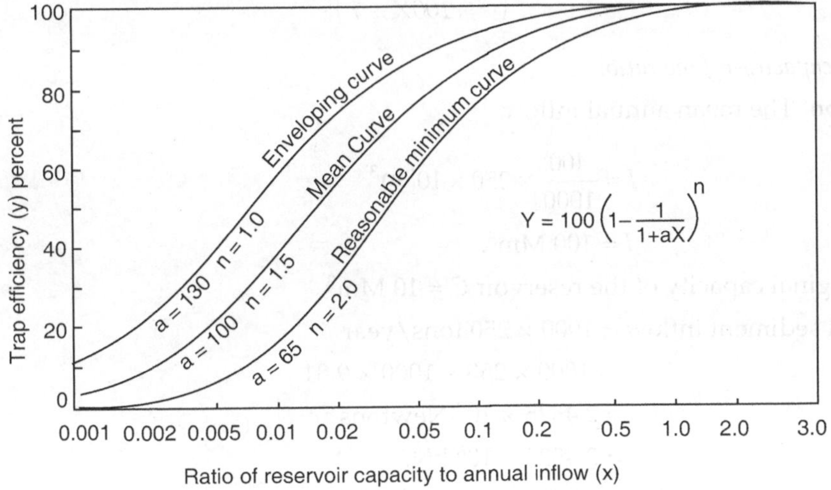

FIGURE 15.3 Trap efficiency versus capacity-inflow ratio.

variations are reflected in trap efficiency *versus* capacity-inflow ratio curves of Fig. 15.3 (Burne 1953). As the reservoir is filled with sediment, its storage capacity (and hence its capacity-inflow ratio) reduces and therefore the trap efficiency will decrease with time. For this reason, while determining the sediment trapped in the reservoir and reduction in the reservoir capacity it is necessary to make the computations for several time intervals with appropriate adjustment of the trap efficiency as demonstrated in Example 15.1.

The volume occupied by the sediment in the reservoir will depend on the specific weight of the deposited material. The specific weight varies with the type of sediment and the age of deposits. Older sediments have more time to consolidate under a superimposed load from the more recent deposits and therefore their specific weight will be greater. According to Lane and Koelzer (1943), the specific weight of the deposit γ_t after t years is given by

$$\gamma_t = \gamma_1 + K \log t \qquad ...(15.7)$$

where γ_1 is the specific weight of the initial deposit and K is the consolidation coefficient. If the deposit consists of mixture of materials, a weighted average specific weight should be calculated. That is

$$\gamma_t = (\gamma_{1,1} + K_1 \log t)x_1 + (\gamma_{1,2} + K_2 \log t)x_2 + (\gamma_{1,3} + K_3 \log t)x_3 .... \qquad ...(15.8)$$

where suffixes 1, 2 and 3 are used for different materials like sand, silt and clay and x_1, x_2 and x_3 are the fractional parts of these materials in total sediment.

Example 15.1. *A reservoir has a capacity of 10 Mm³ and is fed by a catchment of area 250 km². Mean annual runoff at the site is 400 mm. The annual sediment yield and the specific weight of the sediment deposits are estimated to be 1000 metric tons per km² and 12 kN/m³ respectively. Calculate the time it will take for the reservoir to fill up with sediments. The trap efficiency Y may be approximated by*

$$Y = 100\left(1 - \frac{1}{100X + 1}\right)^{1.5}$$

where X is the capacity-inflow ratio.

Solution. The mean annual inflow

$$I = \frac{400}{1000} \times 250 \times 10^6 \text{m}^3$$

or $$I = 100 \text{ Mm}^3.$$

The original capacity of the reservoir C = 10 Mm³.

Annual sediment inflow = 1000 × 250 tons/year

$$= 1000 \times 250 \times 1000 \times 9.81$$

$$= 2.4525 \times 10^9 \text{ Newtons}$$

$$= 2.4525 \times 10^6 \text{ kN.}$$

Volume of annual sediment inflow

$$= \frac{2.4525 \times 10^6}{12}$$

$$= 0.2044 \times 10^6 \text{ m}^3$$

$$= 0.2044 \text{ Mm}^3.$$

Volume of sediment trapped in the reservoir every year

$$= 0.2044 \times \frac{Y}{100}$$

$$= 0.002044\, Y \text{ Mm}^3.$$

No. of years required to fill up a volume of ΔV Mm³ is now given by

$$\Delta t = \frac{\Delta V}{0.002044\, Y} = \frac{489.24\, \Delta V}{Y}.$$

Choosing a uniform interval $\Delta V = 2$ Mm³

$$\Delta t = \frac{489.24 \times 2}{Y} = \frac{978.48}{Y} \text{ years.}$$

The remaining calculations are shown in Table 15.1.

Table 15.1. Calculation of Reservoir Sedimentation of Example 15.1

Reservoir Capacity Mm³	*Average capacity Mm³*	X = C/I *ratio*	*Trap efficiency* $Y = 100\left(1 - \frac{1}{100X+1}\right)^{1.5}$ *per cent*	$\Delta t = \frac{978.48}{Y}$ *years*
10				
8	9	0.09	85.38	11.46
6	7	0.07	81.85	11.95
4	5	0.05	76.07	12.86
2	3	0.03	64.95	15.07
0	1	0.01	35.35	27.68
				Σ= 79.02

∴ The time required for the reservoir to fill up with sediments

= 79 years

Had we used only one interval of 10 Mm³, then average capacity

= 5 Mm³

$$\frac{C}{I} = 0.05,\ Y = 76.07\% \text{ and } t = \frac{489.24\, V}{Y}$$

or

$$t = \frac{489.24 \times 10}{76.07} = 64.31 \text{ years.}$$

REVIEW EXERCISES

Descriptive Questions

1. Describe the difference between normal and accelerated erosion.
2. Differentiate between sheet, rill, gully and channel erosions.
3. Discuss briefly the factors affecting erosion.

4. Distinguish between bed load, suspended load and wash load. What is saltation ?
5. What is bottom shear stress ? How is it computed in wide channels ? What is critical tractive stress ?
6. What do you understand by density currents ?
7. What is design life of storage reservoir project ? How do you account for the sediment deposition in the reservoir while fixing its storage capacity ?
8. Define trap efficiency of reservoir. Describe how the time required for the reservoir to fill up with sediments is calculated.
9. Describe with neat sketches how the suspended-sediment load is measured.
10. How is the bed load in the streams measured ?

Analytical Questions

1. A reservoir has a capacity of 6 Mm^3 and a drainage area of 250 km^2. The average annual runoff is 400 mm and the sediment yield is 1250 tons per km^2. The sediment has an average in-place specific weight of 1500 kg/m^3. Find the time required to reduce the reservoir capacity to 2 M m^3. Adopt a uniform volume increment of 1 Mm^3. The trap efficiency of the reservoir may be assumed as given below. What is the percentage error in your answer if a single increment is used in the calculations ?

$\frac{C}{I}$	0.055	0.045	0.035	0.025
V	78.5	75.5	71.0	64.0

2. A reservoir has a capacity of 50 Mm^3 and average annual inflow of 75 Mm^3. Area of the catchment is 1000 km^2. Sediment yield and the in-place specific weight of the sediments are estimated at 550 tons per km^2 and 1375 kg/m^3 respectively. The trap efficiency may be approximated by $Y = 100\left(1 - \frac{1}{65X + 1}\right)^2$, where Y is the trap efficiency in percent and X is the capacity-inflow ratio. Using 3 increments, calculate how long will it take for the reservoir to become 60% filled with sediment?

Objective Type Questions

1. If d and v are the diameter and velocity of a raindrop respectively, then its erosive power is proportional to
   (a) d^2v^3 (b) d^3v^2
   (c) d^2v^2 (d) d^3v^3
2. In a wide channel, if d and S_0 denote the depth and energy slope of the flow, and γ is the specific weight of water, the bottom shear stress is given by
   (a) $\tau_0 = \gamma d\sqrt{S_0}$ (b) $\tau_0 = \gamma d/S_0$
   (c) $\tau_0 = \gamma\, dS_0$ (d) $\tau_0 = \gamma\sqrt{dS_0}$.

3. Pick up the wrong statement from the following :
   (*a*) Transport of sediment particles by bouncing along the bed is known as saltation.
   (*b*) The trap efficiency of the reservoir decreases with increase in the capacity-inflow ratio.
   (*c*) The sediment which moves as bed load at one section may move as suspended load at another section.
   (*d*) Deposition of coarse sediment takes place at the entrance to the reservoir.
4. If the initial specific weight of sediments is γ_1, the specific weight after *t* years is γ_t, and *K* is the consolidation coefficient, then,
   (*a*) $\gamma_t = \gamma_1 + K \log t$ (*b*) $\gamma_t = \gamma_1 + K.t$
   (*c*) $\gamma_t = \gamma_1 + K/t$ (*d*) $\gamma_t = \gamma_1 + K\sqrt{t}$.
5. Removal of soil from rivulets by concentrated overland flow is known as
   (*a*) sheet erosion (*b*) channel erosion
   (*c*) rill erosion (*d*) gully erosion.

ANSWERS

Analytical Questions

1. 26.73 years, 4.86% 2. 80.34 years.

Objective Type Questions

1. (*b*) 2. (*c*) 3. (*b*) 4. (*a*) 5. (*d*)

BIBLIOGRAPHY

Bouwer, H. (1978)	:	*Groundwater Hydrology,* McGraw-Hill Book Company, Tokyo.
Cassidy, W.C. (1977)	:	*Hydrology for Engineers and Planners,* Iowa State University Press, Iowa.
Chow, V.T. (1964)	:	*Handbook of Applied Hydrology,* McGraw-Hill Book Company, New York.
Chow, V.T. (1973)	:	*Open Channel Hydraulics,* McGraw-Hill Book Company
Chow, V.T. Maidment, D.R. and L.W. Mays (1988)	:	*Applied Hydrology,* McGraw-Hill Book Company, New York.
C.W.P.C. (1973)	:	*Estimation of Design Flood Peak,* Design Office Report No. 1/1973, New Delhi.
Eagleson, P.S. (1970)	:	*Dynamic Hydrology,* McGraw-Hill Book Company, New York.
Graf, W.H. (1971)	:	*Hydraulics of Sediment Transport,* McGraw-Hill Book Company, New York.
Gray, D.M. (1973)	:	*Principles of Hydrology,* McGraw-Hill Book Company, New York.
Jaya Rami Reddy, P. (1997)	:	*Stochastic Hydrology,* Laxmi Publications, New Delhi.
Jhonstone, D. and W.P. Cross (1949)	:	*Elements of Aplied Hydrology,* Ronald Press Company, New York.

Kazmann, R.G. (1972) : *Modern Hydrology 2nd edition,* Harper and Row Publishers, New York.

Kulandaiswamy, V.C. (1971) : *Lecture notes on Hydrology,* ISTE Summer School, Guindy.

Linsely, R.K.; Kohler, M.A. and J.L.H. Paulhus (1975) : *Applied Hydrology,* McGraw-Hill Book Company, New York.

Linsely, R.K.; Kobler, M.A. and J.L.H. Paulhus (1988) : *Hydrology for Engineers,* McGraw-Hill Book Company, New York.

Linsely, R.K.; J.B. Franzini (1972) : *Water Resources Engineering 2nd edition,* McGraw-Hill Book Company, New York.

Michael, A.M. (1978) : *Irrigation : Theory and Practice,* Vikas Publishing House, New Delhi.

Murthy, V.V.N. (1985) : *Land and Water Management Engineering,* Kalyani Publishers, New Delhi.

Mutreja, K.N., (1986) : *Applied Hydrology,* Tata McGraw-Hill, New Delhi.

Nemec, J.; (1973) : *Engineering Hydrology,* Tata-McGraw-Hill Publishing Company, New Delhi.

N.A.S. (1974) : *More Water for Arid Lands,* A report by National Academy of Sciences Washington.

Patra K.C., (2001) : *Hydrology and Water Resources Engineering,* Narosa Publishing House, New Delhi.

Ponce, V.M. (1989) : *Hydrology - Principles and Practices,* Prentice-Hall, Englewood Cliffs, New Jersy.

Raghunath, H.M. (1985) : *Hydrology*, Wiley Eastern, New Delhi.

Raudkivi, A.D. (1979) : *Hydrology,* Pergamon Press, Oxford.

Rao, K.L., (1979) : *Indias Water Wealth,* Orient Longman, New Delhi, India

Rodda, Downing and Lam (1976) : *Systematic Hydrology,* Newnes—Butterworths, Boston.

Schulz, E.F. (1973) : *Problems in Hydrology,* Water Resources Publications, Fort Collins, Colorado.

Singh, V.P., (1994) : *Elementary Hydrology,* Prentice Hall of India Pvt. Ltd., New Delhi

Smoot, G.F. ; and C.E. Novak (1969)	:	*Measurement of discharge by Moving boat, method U.S.G.S.* Technical Water Resources, Investigations Book 3.
Subramanya, K. (1984)	:	*Engineering Hydrology,* Tata-McGraw-Hill Publishing Company, New Delhi.
Todd, D.K. (1980)	:	*Groundwater Hydrology 2nd edition,* John Wiley & Sons, New York.
Varshney R.S. (1979)	:	*Engineering Hydrology,* Nem Chand & Bros., Roorkee.
Viessman, Harbaugh, and Knapp (1972)	: :	*Introduction to Hydrology,* Intext Educaltional Publishers, New York.
Ward R.C. (1975)	:	*Principles of Hydrology, 2nd edition,* McGraw-Hill Book Company, New York.
Weisner, C.J. (1983)	:	*Hydrometerology,* Chapman & Hall Ltd., London.
Wilson E.M. (1983)	:	*Engineering Hydrology,* Macmillan ELBS, London.
Wisler, C.D.; and E.F. Brater (1959)	:	*Hydrology* John Wiley & Sons, New York.

INDEX

D

E

F

G

P

R

S

T

U

V

W